Sustainable Agriculture Reviews

Volume 7

Series Editor
Eric Lichtfouse

For further volumes:
http://www.springer.com/series/8380

*Other Books by Dr. Eric Lichtfouse**

Sustainable Agriculture
Volume 1, 2009

Organic Farming, Pest Control and Remediation of Soil Pollutants
Sustainable Agriculture Reviews. Volume 1, 2009

Climate Change, Intercropping, Pest Control and Beneficial Microorganisms
Sustainable Agriculture Reviews. Volume 2, 2010

Sociology, Organic Farming, Climate Change and Soil Science
Sustainable Agriculture Reviews. Volume 3, 2010

Genetic Engineering, Biofertilisation, Soil Quality and Organic farming
Sustainable Agriculture Reviews. Volume 4, 2010

Biodiversity, Biofuels, Agroforestry and Conservation Agriculture
Sustainable Agriculture Reviews. Volume 5, 2010

Alternative Systems, Biotechnology, Drought Stress and Ecological Fertilisation
Sustainable Agriculture Reviews. Volume 6, 2010

Environmental Chemistry
Volume 1, 2005

Rédiger pour être publié! Conseils pratiques pour les scientifiques
2009

Forthcoming

Sustainable Agriculture
Volume 2

Environmental Chemistry for a Sustainable World
Part 1. Nanotechnology and Health Risk
Part 2. Remediation of Air and Water Pollution

---

*Editor-in-Chief, Agronomy for Sustainable Development. Editor-in-Chief, founder, Environmental Chemistry Letters. Editor, founder, Sustainable Agriculture Reviews. Lecturer, scientific writing. Publications: http://hal.archives-ouvertes.fr
e-mail: Eric.Lichtfouse@dijon.inra.fr

Eric Lichtfouse
Editor

Genetics, Biofuels and Local Farming System

Springer

Editor
Eric Lichtfouse
INRA
UMR Microbiologie du Sol et de l'Environnement
17, rue Sully
21000 Dijon
France
eric.lichtfouse@dijon.inra.fr

ISSN 2210-4410 e-ISSN 2210-4429
ISBN 978-94-007-1520-2 e-ISBN 978-94-007-1521-9
DOI 10.1007/978-94-007-1521-9
Springer Dordrecht Heidelberg London New York

Library of Congress Control Number: 2011931892

Cover design: SPi Publisher Services

Printed on acid-free paper

Springer is part of Springer Science+Business Media (www.springer.com)

Contents

Contributors

Claudio Altomare Institute of Sciences of Food Production, National Research Council, Via Giovanni Amendola 122/O, 70126 Bari, Italy, claudio.altomare@ispa.cnr.it

Rudy M. Baum Chemical & Engineering News, 115516th St., NW, Washington, DC 20036, USA, R_Baum@acs.org

Laurent Bedoussac INRA Toulouse, UMR 1248 AGIR Equipe VASCO, Castanet-Tolosan, France, laurent.bedoussac@toulouse.inra.fr

Nirupam Roy Choudhury Plant Molecular Biology Group, International Centre for Genetic Engineering and Biotechnology, Aruna Asaf Ali Marg, New Delhi 110 067, India, nirupam@icgeb.res.in

Branko Ćupina Faculty of Agriculture, University of Novi Sad, Trg Dositeja Obradovića 8, Novi Sad 21000, Serbia, cupinab@polj.uns.ac.rs

Joële Fustec UP-SP Laboratoire d'Écophysiologie Végétale et Agroécologie, Ecole Supérieure d'Agriculture – PRES Université Nantes Angers Le Mans, Angers, France, j.fustec@groupe-esa.com

Shaily Goyal Laboratory of Bio-Molecular Technology, Department of Botany, M.L. Sukhadia University, Udaipur 313001, India

Qazi Mohd. Rizwanul Haq Department of Biosciences, Jamia Millia Islamia (A Central University), New Delhi 110 025, India, qmrhaque@gmail.com

Eric̃ Justes INRA Toulouse, UMR 1248 AGIR Equipe VASCO, Castanet-Tolosan, France, Eric.Justes@toulouse.inra.fr

Kiriaki L. Kalburtji Faculty of Agriculture, Laboratory of Ecology and Environmental Protection, Aristotle University of Thessaloniki, Thessaloniki 541 24, Greece

David M. Kamau Department of Chemistry, Tea Research Foundation of Kenya, P.O. Box 820, Kericho 20200, Kenya

Samson M. Kamunya Department of Chemistry, Tea Research Foundation of Kenya, P.O. Box 820, Kericho 20200, Kenya

Nicholas E. Korres Department of Civil and Environmental Engineering, University College Cork, Lee Road, Cork, Ireland

Environmental Research Institute, University College Cork, Lee Road, Cork, Ireland, nkorres@yahoo.co.uk

Sofia K. Kostopoulou Faculty of Agriculture, Laboratory of Soil Science, Aristotle University of Thessaloniki, Thessaloniki 541 24, Greece

Đorđe Krstić Faculty of Agriculture, University of Novi Sad, Trg Dositeja Obradovića 8, Novi Sad 21000, Serbia, djordjek@polj.uns.ac.rs

Bowa O. Kwach Department of Chemistry, Maseno University, P.O. Box 333, Maseno 40105, Kenya

Andreas P. Mamolos Faculty of Agriculture, Laboratory of Ecology and Environmental Protection, Aristotle University of Thessaloniki, Thessaloniki 541 24, Greece, mamolos@agro.auth.gr

M. Martinov Department of Environmental Engineering, University of Novi Sad, Trg Dositeja Obradovića 6, Novi Sad 21000, Serbia

Aleksandar Mikić Institute of Field and Vegetable Crops, M. Gorkog 30, 21000 Novi Sad, Serbia, aleksandar.mikich@gmail.com

Manoranjan Mishra Department of Geography and NRM, Sikkim University India, Gangtok, Sikkim, India, geo.manu05@gmail.com

Solomon W. Msomba Crop Improvement Programme, Tea Research Institute of Tanzania, P.O. Box 2177, Dar Es Salaam, Tanzania

Jerry D. Murphy Department of Civil and Environmental Engineering, University College Cork, Lee Road, Cork, Ireland

Environmental Research Institute, University College Cork, Lee Road, Cork, Ireland, jerry.murphy@ucc.ie

Afsar Raza Naqvi Department of Biosciences, Jamia Millia Islamia (A Central University), New Delhi 110 025, India

Plant Molecular Biology Group, International Centre for Genetic Engineering and Biotechnology, Aruna Asaf Ali Marg, New Delhi 110 067, India, afsarrazaa@yahoo.co.in

Anna E. Nikolaidou Faculty of Agriculture, Laboratory of Ecology and Environmental Protection, Aristotle University of Thessaloniki, Thessaloniki 541 24, Greece

A.S. Nizami Department of Civil and Environmental Engineering, University College Cork, Lee Road, Cork, Ireland

Environmental Research Institute, University College Cork, Lee Road, Cork, Ireland

Amos W. Okal Department of Chemistry, Maseno University, P.O. Box 333, Maseno 40105, Kenya

P. Okinda Owuor Department of Chemistry, Maseno University, P.O. Box 333, Maseno 40105, Kenya, owuoropo@africaonline.co.ke

B.K. Padhi School of Environmental Sciences, Jawaharlal Nehru University, New Delhi 110067, India, bkpadhi@gmail.com

K. Padmavathy Department of Ecology and Environmental sciences, Pondicherry University, Puducherry 605014, India, ecopadma@gmail.com

Gyan Ranjan Panda Research Officer in Centre for Budget and Governance Accountability, New Delhi, India, panda.gyana77@gmail.com

Athina K. Pavlatou-Ve Faculty of Agriculture, Laboratory of Soil Science, Aristotle University of Thessaloniki, Thessaloniki 541 24, Greece

Borivoj Pejić Faculty of Agriculture, University of Novi Sad, Trg Dositeja Obradovića 8, Novi Sad 21000, Serbia, pejic@polj.uns.ac.rs

G. Poyyamoli Department of Ecology and Environmental sciences, Pondicherry University, Puducherry 605014, India, gpoyya9@gmail.com

K.G. Ramawat Laboratory of Bio-Molecular Technology, Department of Botany, M.L. Sukhadia University, Udaipur 313001, India, kg_ramawat@yahoo.com

Vito I. Sardo Department of Agricultural Engineering, University of Catania, via Santa Sofia, 100, Catania 95123, Italy, sardov@unict.it

Varsha Sharma Laboratory of Bio-Molecular Technology, Department of Botany, M.L. Sukhadia University, Udaipur 313001, India

Evangelia Sinapidou Department of Agricultural Development, Democritus University of Thrace, Pantazidou 193, Orestiada 68200, Greece, sineva01@hotmail.com

A. Singh Quantitative Sustainability Assessment, Department of Management Engineering, Technical University of Denmark, Lyngby, Denmark, apsinghenv@gmail.com

B.M. Smyth Department of Civil and Environmental Engineering, University College Cork, Lee Road, Cork, Ireland

Environmental Research Institute, University College Cork, Lee Road, Cork, Ireland

Frederick L. Stoddard Department of Agricultural Sciences, University of Helsinki, Helsinki, Finland, frederick.stoddard@helsinki.fi

T. Thamsiriroj Department of Civil and Environmental Engineering, University College Cork, Lee Road, Cork, Ireland

Environmental Research Institute, University College Cork, Lee Road, Cork, Ireland

Ioannis S. Tokatlidis Department of Agricultural Development, Democritus University of Thrace, Pantazidou 193, Orestiada 68200, Greece, itokatl@agro.duth.gr; itokatl@hotmail.com

Ivanka Tringovska Maritsa Vegetable Crops Research Institute, 32 Brezovsko shosse Str, Plovdiv 4003, Bulgaria, dwdt@abv.bg

Marie A. Uwimana Office Des Cultures Industrielles du Rwanda The' (OCIR The'), P.O. Box 1344, Kigali, Rwanda

J.Y. Wu Research and Development Center for Grasses and Environment, Beijing Academy of Agriculture and Forestry Sciences, Beijing, PR 100097, China

Ariel D. Zajdband School of Agronomy, University of Buenos Aires/CONICET, Av. San Martín 4453 (1414DSE), Ciudad de Buenos Aires, Argentina, zajdband@agro.uba.ar

Addicted to Growth

Rudy M. Baum

Abstract The sub-prime mortgage debacle. The Great Recession. Derivatives and hedge funds. The effective bankruptcy of Greece and the subsequent collapse of the euro. China's imminent bubble. The catastrophic oil spill in the Gulf of Mexico. Weeds resistant to glyphosate. Bacteria resistant to antibiotics. Global climate change. The common factor? Humans want too much. We are addicted to growth. The problem with being addicted to growth is that we live on a finite planet. In the new book "Eaarth: Making a Life on a Tough New Planet" Bill McKibben argues that anthropogenic climate change is already well advanced. It is not a problem for future generations. It is a problem for *us*. McKibben insists that we should be able to create social structures and an economic system that does not depend on growth. We will need to get smaller and less centralized, to focus not on growth but on maintenance, on a controlled decline from the perilous heights to which we've climbed.

The sub-prime mortgage debacle. The Great Recession. Derivatives and hedge funds. The effective bankruptcy of Greece and the subsequent collapse of the euro. China's imminent bubble.

The catastrophic oil spill in the Gulf of Mexico. Weeds resistant to glyphosate. Bacteria resistant to antibiotics. Global climate change.

The common factor? Humans want too much. Too many humans are greedy to the point of madness, and neither the global economy nor the global environment can withstand the onslaught of our greed.

Reprinted with permission of Rudy M. Baum. This note was originally published in Chemical & Engineering News, June 28, 2010, Volume 88, Number 26, p. 3.

R.M. Baum (✉)
Chemical & Engineering News, 115516th Street, NW, Washington, DC 20036, USA
e-mail: R_Baum@acs.org

E. Lichtfouse (ed.), *Genetics, Biofuels and Local Farming System*, Sustainable Agriculture Reviews 7, DOI 10.1007/978-94-007-1521-9_1,

Our greed, however, isn't the root cause of the problems we face. Our greed is a symptom of a far more fundamental flaw in the way humans organize their societies and their economies: We are addicted to growth. That addiction to growth stokes the greed that drives the endless and often pointless consumption that we have defined as economic success.

The problem with being addicted to growth is that we live on a finite planet. No matter what growth's apologists claim about finding more resources or harnessing new technology, an addiction to growth, by definition, must at some point collide with reality.

Proponents of endless growth insist that humans have always in the past overcome perceived resource limitations. This is a silly argument. We have been burning fossil fuels, the resources that underpin modern civilization, for a mere two centuries, a period of time that hardly qualifies as "always."

In the new book "Eaarth: Making a Life on a Tough New Planet," Bill McKibben, an environmentalist and a scholar in residence at Middlebury College, argues that anthropogenic climate change/global warming is already well advanced (McKibben 2010). It is not a problem for future generations. It is a problem for *us*. McKibben makes a persuasive argument that humans must begin, right now, to adapt to a radically changed planet. Earth, the planet that humans evolved on and which gave birth to human civilization, no longer exists. In its place is a radically changed place, "with melting poles and dying forests and a heaving, corrosive sea, raked by winds, strafed by storms, scorched by heat. An inhospitable place."

Most important, though, McKibben writes, is that it is a planet that will no longer tolerate growth. "Of all the things I've told you about our new planet… the most terrifying and strangest change would be the end of growth. Growth is what we do. Who ever dreamed it might come to an end?" he writes.

The first half of "Eaarth" is devoted to making the case that humans have already irrevocably changed the planet and that life in the future will have to be different because of those changes. The second half of the book focuses on what humans might do to achieve a good, sustainable existence on this new planet, as opposed to facing catastrophic collapse.

> "The trouble with obsessing over collapse," he writes, "is that it keeps you from considering other possibilities. Either you've got your fingers stuck firmly in your ears, or you're down in the basement oiling your guns. There's no real room for creative thinking. To its theologians, collapse is as automatic and involuntary as growth has been to its acolytes."

McKibben insists that there is another possibility, that we should be able to create social structures and an economic system that does not depend on growth. Near the end of "Eaarth," McKibben writes, "My point throughout this book has been that we'll need to change to cope with the new Eaarth we've created. We'll need, chief among all things, to get smaller and less centralized, to focus not on growth but on maintenance, on a controlled decline from the perilous heights to which we've climbed."

"Eaarth" is a manifesto, one that delivers a message that many people won't want to hear and that many will dismiss out of hand. Growth is a religion, and I think

McKibben underestimates how fervently many humans cling to that religion. It is a religion, however, that flies in the face of physical reality, and as such, cannot be maintained.

Thanks for reading.

References

McKibben B (2010) Earth: making a life on a tough new planet. Times Books, New York, April 2010, 272 p

Grass Biomethane for Agriculture and Energy

Nicholas E. Korres, T. Thamsiriroj, B.M. Smyth, A.S. Nizami, A. Singh, and Jerry D. Murphy

Abstract Many factors enforce the intensification of grassland utilization which is associated with significant environmental impacts subjected to various legislative constraints. Nevertheless, the need for diversification in agricultural production and the sustainability in energy within the European Union have advanced the role of grassland as a renewable source of energy in grass biomethane production with various environmental and socio-economic benefits. Here it is underlined that the essential question whether the gaseous biofuel meets the EU sustainability criteria of 60% greenhouse gas emission savings by 2020 can be met since savings up to 89.4% under various scenarios can be achieved. Grass biomethane production is very promising compared to other liquid biofuels either when these are produced by indigenous or imported feedstocks. Grass biomethane, given the mature and well known technology in agronomy and anaerobic digestion sectors and the need for rural development and sustainable energy production, is an attractive solution that fulfils many legislative, agronomic and environmental requirements.

Keywords Grass • Grass silage • Pasture management • Anaerobic digestion • Biofuels • Biomethane • Legislation • Life cycle assessment

N.E. Korres (✉) · T. Thamsiriroj · B.M. Smyth · A.S. Nizami · J.D. Murphy (✉)
Department of Civil and Environmental Engineering, University College Cork, Lee Road, Cork, Ireland

Environmental Research Institute, University College Cork, Lee Road, Cork, Ireland
e-mail: nkorres@yahoo.co.uk; jerry.murphy@ucc.ie

A. Singh
Quantitative Sustainability Assessment, Department of Management Engineering, Technical University of Denmark, Lyngby, Denmark
e-mail: apsinghenv@gmail.com

E. Lichtfouse (ed.), *Genetics, Biofuels and Local Farming System*, Sustainable Agriculture Reviews 7, DOI 10.1007/978-94-007-1521-9_2,

1 Introduction

Many factors enforce the intensification of grassland utilization (Kemp and Michalk 2007), which is associated with significant environmental impacts (Del Prado et al. 2006) and subjected to various legislative constraints (Anonymous 1997; UNECE 1999; EC 2000). Nevertheless, the need for diversification and sustainability in both agricultural and energy sectors has advanced the role of grassland with various environmental (Tilman et al. 2006) and socio-economic benefits (Baier and Grass 2001). The potential for biomethane as a transport fuel from grass and grass silage has been shown to be very promising (Smyth et al. 2009; Korres et al. 2010). This can enhance the regional and rural development with the creation of a domestic industry and employment opportunities (Del Rio and Burguillo 2009) in which sustainability of eco-friendly biomass production and conversion technologies is warranted (Singh et al. 2010a). Thus the role of grassland can be expanded beyond its traditional utilization as a source of animal feed. Among the methods of using grassland biomass for producing energy, biogas production currently is the most common practice in Europe. Surveys of agricultural biogas plants in Germany and Austria show, that grass silage, the second most frequent biomass feedstock after maize silage, is used as a feedstock in around 50% of the biogas plants (Weiland 2006). The aim of supplying crop feedstock for biomethane production is to achieve the highest possible methane yields per unit area, which are dependent on the biomass yield and the feedstock-specific methane yield (Prochnow et al. 2009). The biomethane yield from various plants across Europe using mainly grass and maize silage along with manures as feedstocks ranges from 10 to $1,150\,\mathrm{m^3\,h^{-1}}$ (Dena et al. 2009).

The process of anaerobic digestion produces biogas, which typically consists of around 55% methane, 45% carbon dioxide and a number of minor constituents. Biomethane is biogas that has been upgraded to the same standard as natural gas and typically has a methane content of about 97% (Korres et al. 2010). In this chapter certain aspects of grass biomethane regarding its production along with certain agronomic and technological issues are covered. Why grass biomethane regarding current EU legislative issues merits consideration particularly under European agricultural and energy sectors? What is the energy balance and greenhouse emissions savings of grass biomethane in other words is grass biomethane an energy feasible and sustainable biofuel? What are the advantages of grass biomethane in comparison with other indigenous and imported feedstocks for the production of biofuels? What is the most appropriate technology for grass biomethane production? What are the agronomic/husbandry factors for pasture utilization for both purposes? Farmers, potential investors and policy makers through this chapter can shape a clear idea on the production of grass biomethane.

1.1 Setting the Boundaries – Grassland Classification

Grassland holds an important role in global agriculture since occupies 69% of the global agricultural area or 26% of total land area (Reynolds 2005) and spans a range of climatic conditions from arid to humid (Verchot et al. 2006). Grassland use is characterized by various modes and intensities. Grassland is predominantly used in animal husbandry, as a principal source of food for ruminants, as well as for its ecological functions, such as protection of soil from erosion, ground water formation, habitat function and formation of diverse cultural landscapes (Prochnow et al. 2009). Amongst its various characteristics i.e. long persistency of high dry matter yield; intercropping potential with legumes and subsequent reduction in fertilizer application rates; low lignin content compared with other second-generation lignocellulosic materials for the production of biofuels; the lower rates of pesticide application; the disappearance of the "cereal and oilseed rape premium" of the former Common Agricultural Policy (CAP) system; the protection of grassland area in the present CAP cross-compliance system; and the unstable status of fossil energy and input prices, can increase the attractiveness of perennial grasses and legumes for biomethane production compared with first generation biofuels (Peeters 2009). Grassland is an important carbon sink (Tilman et al. 2006) and source of energy production (Murphy and Power 2009; Prochnow et al. 2009).

Grasslands vary greatly in their degree and intensity of management from extensively managed rangelands and savannahs to intensively managed continuous pasture and hay land (Verchot et al. 2006). More specifically, in Europe there are various types of grasslands, ranging from almost desertic types in south-east Spain through steppic and mesic types to humid grasslands/meadows, which dominate in the north and north-west (Silva et al. 2008). European grassland based on pasture management has been classified as rough mountain hill grazing, permanent and rotational or temporary grassland (Brockman and Wilkins 2003). The former is uncultivated grassland found in large enclosures on hills or unenclosed uplands, moorlands, heaths and downlands. It is characterised by high biodiversity, low stocking rates and low dry matter production. Permanent grassland is that observed in fields of relatively small enclosures with higher stocking rates compared to rough grazing. The last type of grassland is found within an arable rotation characterised by low species abundance and richness, and high stocking rates and production.

Lockhart and Wiseman (1988) distinguished two types of grassland, namely uncultivated and cultivated. The former consists of rough mountain and lowland heath grassland, whereas the latter includes permanent grassland (over 5 years old) and leys or temporary grassland (less than 5 years old). This, based on perennial ryegrass percentage, can be further distinguished as first (>30%), second (20–29%), third (<20%) grade and poor grassland. This type of grassland, according to Fossitt (2000), can be classified as improved grassland that is highly modified, intensively managed with low biodiversity. It is used for heavy grazing and/or silage production and includes regularly reseeded monoculture grasslands dominated by perennial ryegrass that is planted as part of an arable rotation. In this chapter the case of intensively cultivated grassland is considered in which grass can be used for both animal and biomethane production. Rough grazing grassland is excluded as a potential source for biomethane production due to low dry matter production and the acidic or peaty soil, which is difficult to cultivate (Brockman and Wilkins 2003).

1.2 Agricultural Transition Era and Sustainable Development

In recent years agriculture has undergone many changes with a significant food production increase, structural upheavals in many agricultural systems and concurrent restructuring of many rural communities (Wilson 2007). Many authors have argued that conventional agriculture or "productivism agricultural era" and its exclusive purpose, the production of food and fibre is at its terminal stage and that a new agricultural regime or "post-productivism agricultural era" with a wider purpose, including the "production" of nature and environmental management (Braun and Castree 1998; Marsden 1999) is emerging. Holmes (2006) reported that the transition of agricultural structure from "productivism" to "post-productivism era" will be achieved through its multifunctional role.

Multifunctionality is interpreted in various ways by different people (Wilson 2007). Marsden and Sonnino (2008), described three types of multifunctional agriculture: The first type is restricted to pluriactivity within an agro-industrial model. The second type rises from a 'post-productivist' paradigm as discussed above. Finally, the third type, as a part of a sustainable rural development paradigm in which agricultural production is seen to be intimately combined with the socio-economic health of rural areas, is recognised as an economic sector that must be integrated into the wider economy. The authors of this chapter cannot find any reason why one type shouldn't be interrelated with another under the broader frame of an agro-ecosystem as an inseparable part of rural areas. Hence, the multifunctional agricultural character is concerned with agricultural diversification, where agriculture is integrated into areas other than traditional farming. In other words, agricultural production can be integrated within an agro-industrial model under the wider concept of pluriactivity. This will result in generation of advantages towards increase non-farm income from the emerging opportunities such as grass biomethane production.

Furthermore, sustainable development, in other words the fulfilment through the optimal use of any available source within a production system, in the agricultural sector, that facilitates continuing benefits from land, water and biological resources to satisfy the human populations' current needs while preserving and bettering the base of all natural resources for the future generations (Siardos 1994; Boyazoglu 1998) can be promoted through grass biomethane production. The need for integration of potential sources implies that any evaluation of agroecosystem sustainability must consider the dynamics of multiple components (Belcher et al. 2004). The challenge of agricultural development is to counteract the dependence on non-renewable resources and environmental services whose seemingly unrestrained availability is made apparent by inadequate market forces and economic policies that hinder sustainability (Pezzey 1992).

2 Biofuels Use in Europe

Increasing fossil fuel prices, energy security concerns and environmental consciousness, especially related to climate change stabilization, have motivated countries to explore alternative energy sources, such as biofuels, fuels derived from biomass, (Zarrilli and Burnett 2008). The use of biofuels as a means of "greening" the transport sector, which is a sectors with high pollution impact (Reay and Grace 2007), by using renewable energy resources is strongly supported by European policy (EEA 2004). EU Directive 2009/28/EC on renewable energy sets a mandatory target for each European Union (EU) Member State for 10% of transport energy to be met with renewable sources in 2020 (EC 2009). The main biofuels are biodiesel and bioethanol (USDA 2006). The EU is by far the world's biggest producer of biodiesel, with Germany, France and Italy producing over half of the EU's biodiesel (Demirbas 2009) (Table 1).

The main feedstock of biodiesel production from the biggest EU producing countries is rapeseed whereas in USA, Brazil and Argentina, the biggest biodiesel producers outside EU, is soybean (Table 1).

The use of biogas as a transport fuel, via anaerobic digestion of various feedstocks, after upgrading to biomethane, has recently started to gain attention in many European countries, such as in Sweden, Austria, France and Switzerland (Mathiasson 2008). Additionally, biogas production from biomass has been strongly promoted in many developing regions including Asia, Latin America and some regions of West Africa (Eisentraut 2010). Agricultural and environmental related legislative issues along with sustainability criteria concerning renewables, as discussed in the following sections, prove that grass biomethane is an alternative that satisfies European conditions and merits considerable attention.

Table 1 Biofuel production worldwide (Smyth et al. 2010)

Country	Ethanol (billion l)	% Total	Principal feedstock	Country	Biodiesel (billion l)	% Total	Principal feedstock
USA	34	50.7	Corn	Germany	2.2	18.3	Rapeseed
Brazil	27	40	Sugarcane	USA	2	16.7	Soybean
China	1.9	2.8	Corn, wheat	France	1.6	13.3	Rapeseed
France	1.2	1.8	Sugar beet, cereals	Brazil	1.2	10	Soybean
Canada	0.9	1.3	Corn, wheat	Argentina	1.2	10	Soybean
Total top 5	65	97		Total top 5	8.2	68.3	
Total EU	2.8	4.2	Sugar beet, wheat	Total EU	8	66.7	Rapeseed, sunflower
Total world	67	100	–	Total world	12	100	–

2.1 *Why Grass Biomethane – Policy Drivers*

The renewable energy sector is amongst the fastest growing in the EU, with an annual turnover of €15 billion, more than 200,000 employees and more than 4.5 million green-power consumers (EC 2006). Within the renewable energy sector, substantial economic and environmental results are achieved through a combination of targets, fiscal incentives and market mechanisms (EC 2006). The European Commission introduced the Renewable Directive (EC 2009) on the promotion of the use of energy from renewable sources, which includes, among many other provisions, sustainability criteria for biofuels and other bioliquids. Amongst the sustainability criteria are requirements for lower limits of greenhouse gas emissions from biofuels compared to the fossil fuel replaced, restrictions on land use change and environmental requirements for agriculture (Zarrilli and Burnett 2008). Additionally, in the agricultural sector, policy enforces environmentally friendly production systems which will affect agricultural *status quo* and production chains.

2.1.1 Agricultural Legislation

Agriculture releases significant amounts of greenhouse gases such as CO_2, CH_4, and N_2O, along with ammonia (NH_3) emissions to the atmosphere (Paustian et al. 2004). The main mechanisms of gaseous emissions from agriculture comprise microbial decay or burning of plant litter and soil organic matter (Janzen 2004), decomposition of organic materials in oxygen-deprived conditions, notably from fermentative digestion by ruminant livestock and stored manures (Mosier et al. 1998), and the microbial transformation of nitrogen in soils and manures which is often enhanced where available nitrogen exceeds plant requirements, especially under wet conditions (Oenema et al. 2005).

EU crop production patterns have traditionally been heavily influenced by the Common Agricultural Policy with its high support prices, planting restrictions, intervention buying, stock management and rigid border controls (Schnepf 2006). The adverse effects of agricultural intensification on the environment under previous EU agricultural support mechanisms have directed attention towards environmental issues, and have influenced the development of CAP and the promotion of sustainable agriculture (Clergue et al. 2005). As such, the 2003 Mid Review of the CAP directed that agricultural support payments be conditional upon compliance with environmental standards and "Good Farming Practice". Community initiatives, amongst others, aim to limit agricultural pollution and to promote the development of the production and use of biofuels (Osterburg et al. 2005). With the same objectives, CAP reforms also established a special aid for energy crops grown on non-set-aside land. Energy crops, these grown for the production of biofuels, including biogas, or for use as biomass in the production of electric and thermal energy, are eligible for a premium of €45 per hectare.

Attaching conditions to the receipt of agricultural subsidies is a policy tool known as cross compliance and aims to improve standards in modern farming practices

(Farmer and Swales 2004). The 2003 reform of the EU's Common Agricultural Policy made cross compliance mandatory for all Member States (EC 2003 and EC 2004). The cross compliance standards consist of two strands, the Good Agricultural and Environmental Condition, which sets out minimum requirements for soil conditions and land maintenance, and the Statutory Management Requirements, which relate to animal welfare, and environmental, public, and plant health (Mussner et al. 2006). Land suitability is often a key for evaluating the potential of biofuels (Ragaglini et al. 2010). Cross compliance (EC 2003) requires Member States to ensure that land declared as under permanent pasture in 2003 is maintained under permanent pasture. Additionally, Article 3 of Regulation 796/2004 states that Member States should ensure that the ratio of land under permanent pasture in relation to the total agricultural area of the Member State does not decrease by 10% or more of the 2003 reference level (EC 2004). EU Member states are therefore under obligation not to allow any significant reduction in the total area of permanent pasture. Other measures to encourage environmentally friendly agricultural practices, such as the Rural Environment Schemes (REPS), were introduced under Council regulation EEC/2078/1992 (EC 1992). These, according to the report prepared by Directorate General for Agriculture and Rural Development, have to be applied beyond usual Good Farming Practice, which is defined as the level of environmental care that a reasonable farmer is expected to practice (EC 2005a). Under the Sixth Environment Action Programme (EC 2001), various Directives, such as the Water Framework Directive (EC 2000) and the Nitrates Directive (EEC 1991), introduced a series of measures that aim to reduce greenhouse gas emissions at farm level via lower stocking rates and could free up grassland for other purposes. Proposals for protecting waters through reducing nutrient losses include reducing stocking rates, harvesting grasslands for silage/hay instead of cattle grazing, and reducing the length of the grazing season. The adoption of batch storage for slurry has also been suggested (Chardon and Schoumans 2008), which could make it amenable to anaerobic digestion and biofuel production. Biofuels are also influenced by the Biodiversity Action Plan (Caslin 2009) which aims to improve or maintain biodiversity and prevent further biodiversity loss due to agricultural activities. Priorities include restricting intensive farming and establishing sustainable resource management.

Finally, Regulation EC/1698/2005 (EC 2005b) interlinks sustainable development and the diversification of the rural economy through, among other measures, the introduction of new technologies and innovation in non-food sectors such as renewable energy. This is of great importance, considering the efforts and policies that have been put in action to dismantle protectionist agricultural subsidies in order to combat rural poverty in Southern Europe, especially the livelihoods of marginal farmers, and policies to protect the environmental integrity of the countryside in the developed Northern European countries (Potter and Tilzey 2007).

Environmental legislative issues, changes in market dynamics imposed by the decoupling of the majority of direct aid (EC 2003), discussions about trade liberalization, e.g. Doha Round trade negotiations in the World Trade Organization, will most probably influence the dynamics of agri-food supply chains (McCorriston and Sheldon 2007). In addition, the need for a multifunctional agriculture for optimal use of any available resources and the achievement of maximum potential (Boyazoglu 1998) necessitate the rapprochement of grassland utilization.

In the face of climate change and growing demands for agricultural productivity, future pressures on grassland ecosystems will intensify (Watkinson and Ormerod 2001). Moreover, most grassland in the EU is devoted to meat (cattle and sheep) production where profitability is low and farmers often rely on EU single farm payments to survive, thus grassland farming can face considerable challenges in implementing new environmental measures without financial supports (Boyle 2008). As suggested in this chapter, diversification of grassland, except for livestock production, could include its utilization as a source of biomethane production for transport purposes. Major reform of CAP is expected in 2013 and this may have implications for energy crops and biofuels.

2.1.2 Energy Legislation – Biofuels as a Means for Bypassing Fossilization

Various policy goals, which focus on boosting the decarbonisation of transport fuels, diversifying fuel supply sources and developing long-term replacements for fossil oil, have motivated the European Union to promote the production and use of biofuels (Schnepf 2006). With increasing use of biomass for energy, questions arise about the energy efficiency of the renewable energy source and its validity as a means of reducing greenhouse gas emissions and dependence on fossil fuels (Haas et al. 2001; Gerin et al. 2008; Cherubini et al. 2009).

The policies effectively impose constraints on many conventional energy crop biofuels and reinforce the merits of using biomethane as a transport fuel. Renewable Directive 2009/28/EC (EC 2009) states that biofuels must meet the following criteria in order to be considered sustainable and to be counted for the purposes of meeting EU biofuels targets:

- Greenhouse gas emission savings from the use of biofuels and bioliquids shall be at least 35%, and by 2017 savings of at least 50% compared with fossil fuel replaced must be achieved. From 1 January 2018, greenhouse gas emissions savings should be at least 60% for biofuels and bioliquids produced in installations in which production started on or after 1 January 2017.
- Biofuels from peatlands and land with high biodiversity value or high carbon stock may not be used (e.g. permanent grassland).
- Impact of biofuel policy on social sustainability, food prices and other development issues is to be assessed, in particular for people living in developing countries.

2.2 *Life Cycle Assessment (LCA)*

To understand and manage the energy efficiency of renewable energy sources and related greenhouse gas emissions, the whole system should be considered

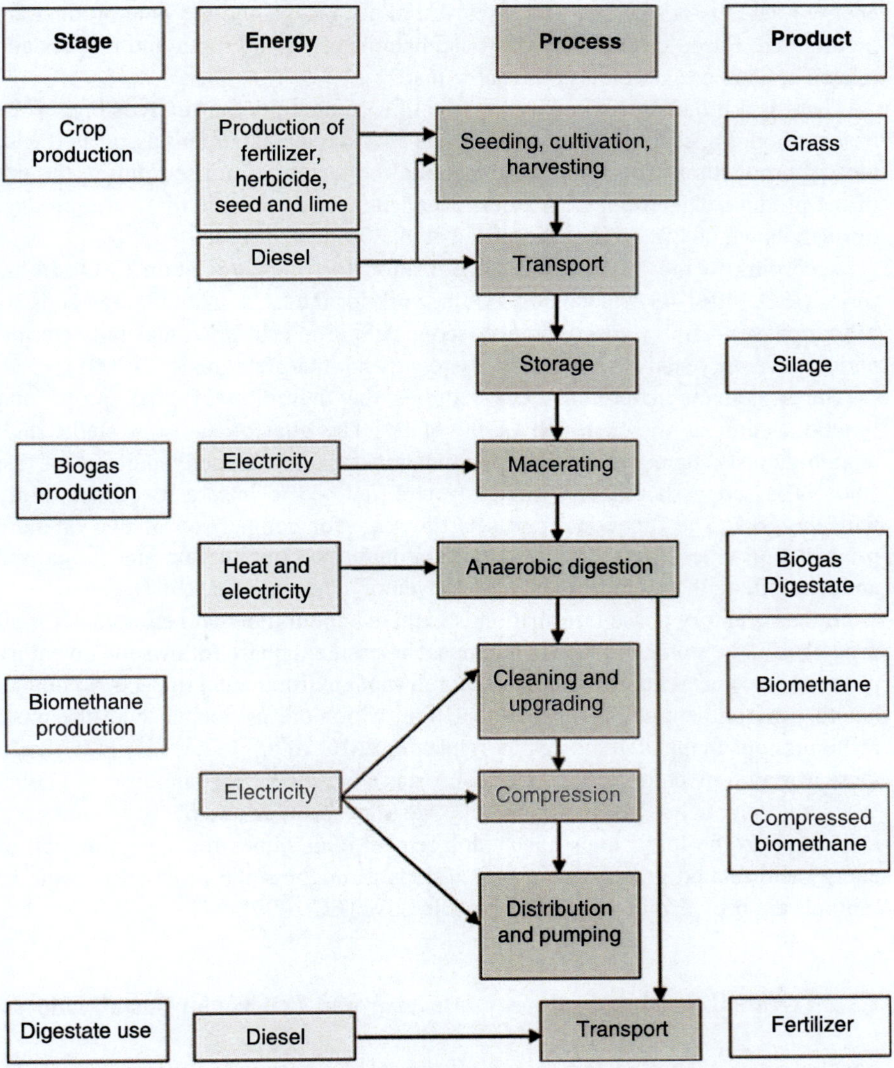

Fig. 1 Flow chart of grass biomethane production system (Smyth et al. 2009)

(Phetteplace et al. 2001). The production of energy crops for anaerobic digestion should result in a net production of renewable energy and a net reduction of greenhouse gas emissions (Gerin et al. 2008). The energy balance and related greenhouse gas emissions of the grass silage biomethane system therefore need to be quantified (Fig. 1).

In the figure above (Fig. 1) the whole cycle of grass biomethane production is represented in a clear and distinguishable way. Each production stage and energy related issues for the processes within each stage and the final product are highlighted. Based on this flow chart estimations of energy input and outputs and related greenhouse gas emissions can be made.

There is a broad agreement in the scientific community that LCA is one of the best methodologies for the evaluation of the environmental burdens associated with biofuel production (Consoli et al. 1993) and the resources utilised during the life of the product. Therefore, LCA offers a holistic and systematic view of a product through its whole life cycle (Payraudeau et al. 2007).

According to the International Organization for Standardization (ISO) 14000 series (ISO 14041-43) the technical framework for the LCA methodology consists of four phases, namely goal definition, scope definition and functional unit determination; inventory analysis; impact assessment; and interpretation (ISO 2006).

The systematic nature of LCA requires the definition of goal, scope and functional unit as the first step of the study. The goal of an LCA study shall unambiguously state the intended application to the intended audience of the study. The scope should be sufficiently well defined to ensure its compatibility with the goal. The functional unit sets the scale for comparison of two or more products, provides a reference to which the input and output data are normalised and harmonises the establishment of the inventory (Jensen et al. 1997).

In the inventory phase information is gathered about input and emissions for all processes in the studied system. The impact assessment phase follows the inventory phase and performs an assessment of all relevant environmental impacts on human health, environment and resources depletion, which are associated with the input and emissions mapped in the inventory phase (SAIC 2006). In the case of biofuels, there is requirement to express greenhouse gas emission savings in terms of Global Warming Potential (GWP) of carbon dioxide, nitrous oxide and methane emissions in relation to the fossil fuels under replacement. Guidelines for the estimation of energy and related greenhouse gas emissions throughout the production cycle of biofuels are provided in the Renewable Directive (EC 2009).

2.2.1 Grass Biomethane – Energy Efficiency and Criteria for Sustainability

For the determination of the energy efficiency of a renewable energy source, all energy inputs and outputs through the whole production cycle of the product need to be taken into consideration (Salter and Banks 2009). Oilseed rape, for example, covers about 80% of the set-aside land devoted to non-food energy crops for biodiesel production in the EU (Table 1), (Bauen 2005). Smyth et al. (2009) reported that the gross and net energy of rapeseed biodiesel and wheat bioethanol are much less than those for grass biomethane, palm oil biodiesel and sugar cane bioethanol (palm oil and sugar cane are non-indigenous European crops) (Fig. 2). Importing feedstock from tropical countries, such as Malaysia or Indonesia in the case of biodiesel, may not be an option. The increasing demand for palm oil production from these countries, which account for about 80% (FAOSTAT, undated) of global production, is contributing to deforestation at an annual rate of

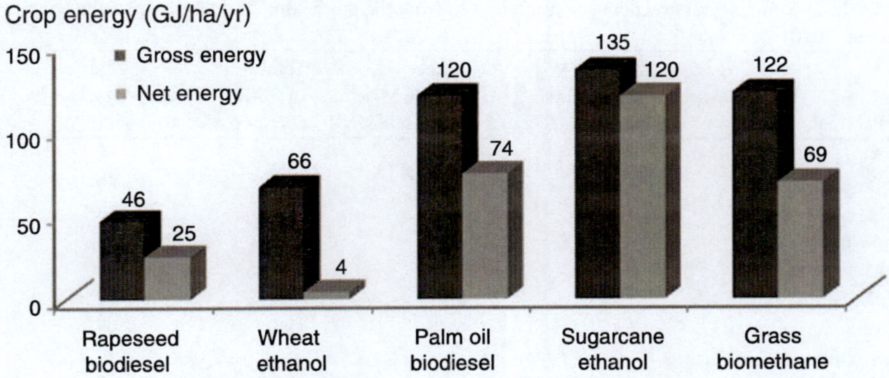

Fig. 2 Comparison of gross and net energy per hectare of selected energy crop biofuel systems (Smyth et al. 2009). Note: Net energy is defined by gross energy yield (biomass dry matter yield × specific methane yield) minus inputs (direct and indirect energy) for cultivation of feedstock and biomethane production

1.5% (Fargione et al. 2008). With land use change, there are no net GHG savings (Reinhard and Zah 2009). Consequently, palm oil biodiesel, for example, is not considered a biofuel according to Directive 2009/28/EC, which requires greenhouse gas savings of 60% by 2020. Additionally, increases in palm oil production cause habitat loss and drainage of peatlands (Wakker 2004; Greenpeace 2007), whereas land tenure conflict, particularly in Indonesia (Colchester et al. 2006), raise further concerns regarding the sustainable supply of biodiesel. Although, in theory, palm oil for biodiesel could be taken from existing supplies and new capacity achieved by development on previously cultivated land and/or via yield improvements, in practice, any incentivisation of palm oil-derived biodiesel for the EU market is likely to provide an indirect stimulus for land clearance (Thornley et al. 2009). Similar concerns regarding, amongst others, deforestation, decarbonisation and degradation of soils have arisen for sugar cane ethanol (Goldemberg et al. 2008). In comparison, grass biomethane, an indigenous European crop that performs much better than rapeseed biodiesel and wheat ethanol, and that has a similar energy balance to tropical based biofuels, seems an attractive alternative for the EU (Fig. 2).

Tilman et al. (2006) stated that biofuels derived from low-input native grassland perennials can provide more usable energy, greater greenhouse gas reductions and less agrochemical pollution per hectare than arable crops, such as corn grain ethanol or soybean biodiesel.

Finally, to promote non-food feedstock the Renewable Directive (EC 2009) considers the contribution made by biofuels produced from wastes, residues and lignocellulosic material to be twice that made by other biofuels for the purposes of demonstrating compliance with the 10% target. Many authors considered grass as a lignocellulosic feedstock for biomethane production (Peeters 2009; Eisentraut 2010; Singh et al. 2010b).

Table 2 Summary of greenhouse gas emissions from the production of grass biomethane (Korres et al. 2010)

Parameters	Direct[a] (g CO_2e MJ^{-1} energy replaced)	Indirect[b] (g CO_2e MJ^{-1} energy replaced)	Total (g CO_2e MJ^{-1} energy replaced)	Total (kg CO_2e ha^{-1} $year^{-1}$)
Agriculture				
Crop production	2.67	6.34	9.01	893
Herbicide volatilization	0.05	–	0.05	5.44
Lime dissolution	5.55	–	5.55	550
N_2O emissions	5.18	0.11	5.29	525
Total agricultural emissions	13.45	6.45	19.90	1,973
Transportation[c]	–	0.89	0.89	88
Biomethane production process				
Anaerobic digestion plant	18.25	7.24	25.49	2,524
Upgrading		12.64	12.64	1,251
Total processing emissions	18.25	19.88	38.13	3,775
Biogas losses	10.82		10.82	1,071
Total	45.75	27.11	69.74	6,904

The analysis was based on a grass biomethane system using 7,500 t $year^{-1}$ of grass silage (137.5 ha), with a net biomethane yield of 99 GJha^{-1} $year^{-1}$

[a]Direct greenhouse gas emissions are defined as the emissions from agronomic operations, e.g. ploughing, harrowing, rolling, application of agrochemicals, harvesting etc., and from energy consumed in anaerobic digestion plant e.g. heating of digesters

[b]Indirect greenhouse gas emissions are defined as the emissions from the production of pasture inputs, e.g. fertilizers, herbicides etc., and from the maceration, mixing and water pumping activities in anaerobic digestion plant

[c]Transportation includes emissions from lime and silage transportation to the field

2.2.2 Grass Biomethane – Greenhouse Gas Emissions

A greenhouse gas analysis conducted by Korres et al. (2010) determined the greenhouse gas emissions from grass biomethane, produced by anaerobic digestion and used as a transport fuel in place of diesel, as 69.74 $gCO_{2e}MJ^{-1}$ energy replaced or 6,904 $kgCO_{2e}ha^{-1}$ $year^{-1}$ (Table 2).

The largest contributors were emissions from crop production and from the anaerobic digestion process. Indirect emissions from the production of nitrogen and potassium fertilizers were the major contributors to agricultural emissions and, in the biomethane production process, the largest source of emissions was from digester heating. When compared with emissions from fossil diesel grass biomethane production under the base case scenario, which includes the production of grass silage and transportation of feedstock to anaerobic digestion plant and digestate back to field, greenhouse gas emissions savings were estimated to 21.5% (Fig. 3).

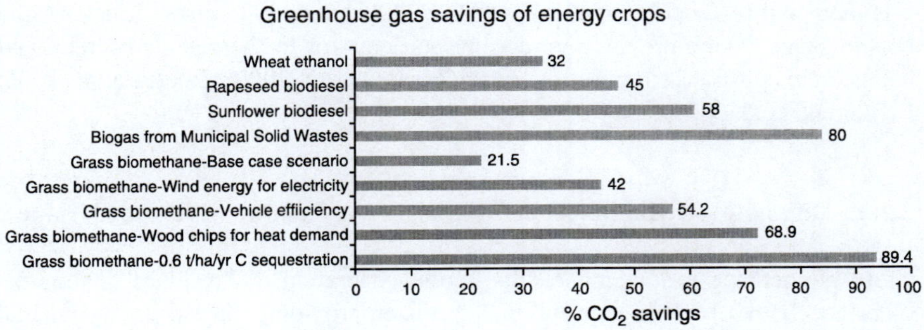

Fig. 3 Cumulative greenhouse gas emissions savings of grass biomethane over fossil diesel (Korres et al. 2010) (Note: Default values for wheat ethanol, rapeseed biodiesel, sunflower biodiesel and biogas from municipal wastes were adopted from 2009/28/EC Renewable directive (EC 2009))

Nevertheless, cumulative greenhouse gas emissions savings under various scenarios, i.e. incorporating electricity from wind, improved vehicle efficiency, energy from wood chips for anaerobic digestion heat demands and carbon sequestration of $0.6\,t\,C\,ha^{-1}\ year^{-1}$ (a minimum value for most European permanent crops and grasslands according to Freibauer et al. (2004) and Jones and Donnelly (2004)), resulted in greenhouse gas emissions savings of up to 89.4%. This easily meets the 60% greenhouse gas savings required in 2018 by EU Directive 2009/28/EC.

In contrast, liquid biofuels produced by indigenous European feedstock, such as wheat ethanol, rapeseed biodiesel and sunflower biodiesel, do not meet the requirement for 60% greenhouse gas savings (Fig. 3). Oilseed rape is a crop with high nitrogen and pesticide demand, which heavily influences the greenhouse gas balance for biodiesel production (Thornley et al. 2009). Additionally, the overall carbon balance for rapeseed biodiesel production using existing technology is poor and any preference for better performing feedstocks in terms of carbon reduction potential will constrain the contribution of rape seed to bioenergy targets (Thornley et al. 2009). Production of wheat bioethanol is accompanied by low greenhouse gas savings, mainly due to nitrous oxide emissions during cultivation, the fuel used in the ethanol production plant, the fate of by-products and the low biofuel yields of only 3,220 L/ha (Smith et al. 2005; Borjesson 2009). Studies on sunflower biodiesel have shown that areas, such as in the case of Tuscany in Italy, classified as suitable for biodiesel production could only provide efficient conversion rates to meet the current requirement for 35% greenhouse gas savings on 30% of arable land (Ragaglini et al. 2010). Other studies on sunflower biodiesel, although reporting promising environmental benefits, haven't highlighted in a clear way its potential as a sustainable renewable energy source (Sanz-Requena et al. 2010; Tsoutsos et al. 2010). Manure, on the other hand, is an easily available resource on farms and biogas from manure results in high greenhouse gas savings. However, the limited production rate, low biogas yields due to the high water content and high investment costs make the economical feasibility of manure biogas difficult (Gerin et al. 2008). Nevertheless, co-digestion of various substrates, including manure, often results in a higher methane yield (Jagadabhi et al. 2008), due to synergistic effects of the co-substrates, which provide the missing nutrients for methanogenic bacteria and balance the substrate composition (Mata-Alvarez et al. 2000; Umetsu et al. 2006; Lehtomaki et al. 2007).

2.2.3 Digestate – A Hidden Anaerobic Digestion Residue with Added Value

Grass biomethane production results in the generation of “residues” known as digestate. It can replace conventional fertilizer providing further environmental benefits to the biofuel process chain (Cherubini et al. 2009), because the nutrient cycle is almost closed (Seppala et al. 2009). Nitrogen use efficiency within intensive farming systems has been shown to be very low. For example, on intensive dairy farms in the Netherlands only 16% of the N inputs (feed, fertilizer etc.) are captured as outputs (meat, milk etc.) (Aarts et al. 2000). Improved N-use efficiency within farm systems is achieved by maximising utilisation of the N circulating within the system, e.g. utilising the N in organic manures, the N supplied by soil, the N supplied by biological fixation through clover and the use of digestate as fertilizer. This results in increased production efficiency and profitability due to lower costs of production while at the same time reduces losses to the environment.

Gerin et al. (2008) examined the production of maize and grass silage as a feedstock for biogas and the use of digestate as substitute for mineral fertilizer for both crops. Furthermore, Matsunaka et al. (2006) reported positive effects on dry matter increases of *Phleum pratense* (timothy grass) with the application of digestate from various organic materials. They also reported that digestate positively affects N uptake by grass, particularly when applied in spring. Additionally, Salter and Banks (2009) stated that digestate can be separated into liquid and fibre components. A proportion of liquid can be recirculated back to the anaerobic digestion process to increase its efficiency, and the remaining quantity can be processed into liquid biofertilizer and/or can be used as a multi-purpose press juice (Berglund and Borjesson 2006). The solid digestate can be processed into fibres, which can either be applied to land as a soil conditioner or processed into high value insulation boards (Grass 2004; Salter and Banks 2009). Under this scenario, a new concept of green biorefinery (Fig. 4) can be developed (Kamm et al. 1998; Narodoslawsky 1999; Kamm and Kamm 2004), which will further enhance the role of grassland as a major agricultural, industrial and economic resource. According to Grass (2004), a green biorefinery will also increase greenhouse gas emissions savings.

3 Anaerobic Digestion – Grass and Grass Silage as a Feedstock

Anaerobic digestion is the conversion of organic matter to a biogas under anaerobic conditions. Four successive biological processes are involved in the anaerobic degradation of organic matter: hydrolysis, acidogenesis, acetogenesis and

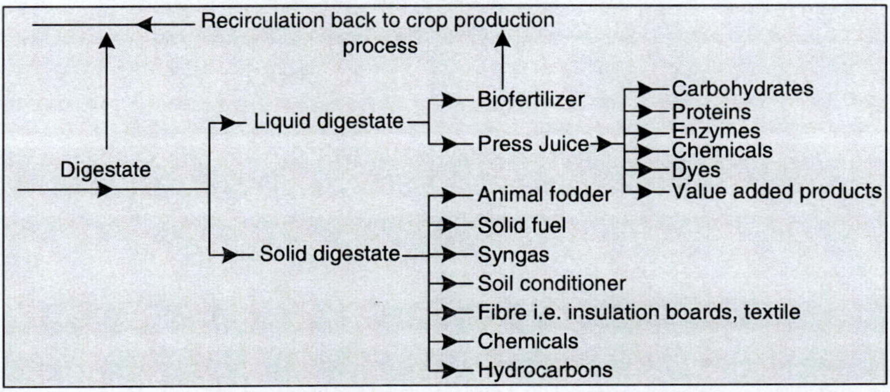

Fig. 4 Value-added products of digestate (Based on Kamm and Kamm 2004; Salter and Banks 2009)

methanogenesis. Complex polymers are converted into monomers by extra-cellular enzymes during hydrolysis while these monomers are transformed into volatile fatty acids (acetic, propionic and butyric acids) and hydrogen (H_2) during acidogenesis. Acetate, carbon dioxide (CO_2) and H_2 are produced from volatile fatty acids during acetogenesis and finally converted into methane (CH_4) during methanogenesis (Bernet and Beline 2009). The biogas produced during anaerobic digestion, mainly composed of CH_4 (55–80%) and CO_2 (20–45%) can be used as an energy source, generally as heat and/or electricity, or as a biofuel (Murphy and Power 2009). This process has been widely applied for years to the treatment of organic waste, including manure (Chynoweth et al. 1999; Burton and Turner 2003) and other farm waste, wastewater, industrial organic waste, municipal solid waste, agricultural residues, crops and crop residues (Vandevivere 1999). Nevertheless, as stated by Tabajdi (2008), both the production of biogas and the number of biogas installations are unevenly distributed in Europe, demonstrating untapped potential. Grass and grass silage have recently received considerable attention in the EU (Braun and Steffen 1997; Mahnert et al. 2005), as the crop can be utilised as a beneficial feedstock for the production of biomethane due to high yield, its perennial nature (a low energy input crop), the high volatile content and the associated relatively high biomethane yield (Hall 1997; Murphy and Power 2009; Seppala et al. 2009).

Many factors are involved in biomethane production chain namely agricultural and operational procedures along with pre- and post-treatments of the raw feedstock and digestate respectively (Fig. 5). Relevant factors to the scope of this chapter are the agronomic/husbandry factors which determine the quality of grass and grass silage for higher biomethane production.

The reader can be advised in more detail about operational measures, pre- and post treatments in Nizami et al. (2009).

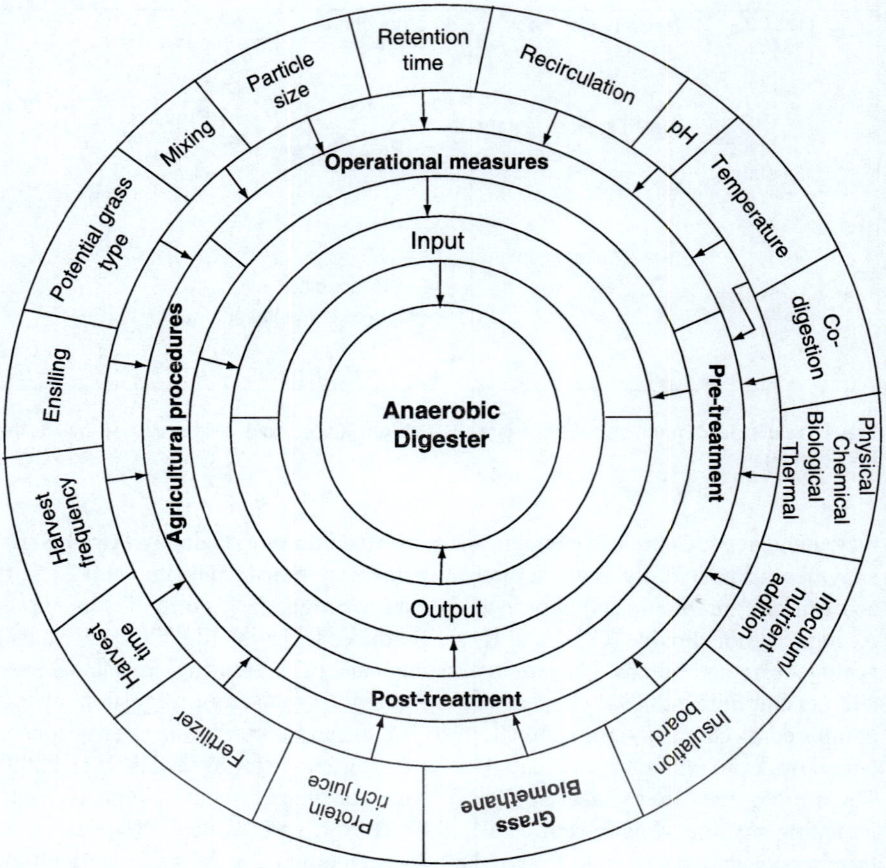

Fig. 5 Circular diagram of the main factors affecting biomethane yield of an anaerobic digester (Nizami et al. 2009)

3.1 Anaerobic Digester Designs

Anaerobic digester design and configuration are important components of anaerobic digestion technology, as they promote the efficient conversion of organics to gaseous products (Demirbas and Ozturk 2005). A range of digester types and configurations exists based on various process parameters. The parameters that classify anaerobic digesters are the moisture content of the feedstock (wet digestion vs. dry); the number of phases/stages of digestion activity (single or two/multi-stage); operating temperature (thermophillic or mesophilic); method of feeding the substrate (batch or continuous) and retention time (Vandevivere et al. 2003; Karagiannidis and Perkoulidis 2009; Nizami and Murphy 2010), (Fig. 6).

Various researchers have reviewed and compared different digesters suitable for anaerobic digestion of solid wastes (De Baere and Mattheeuws 2008; Vandevivere

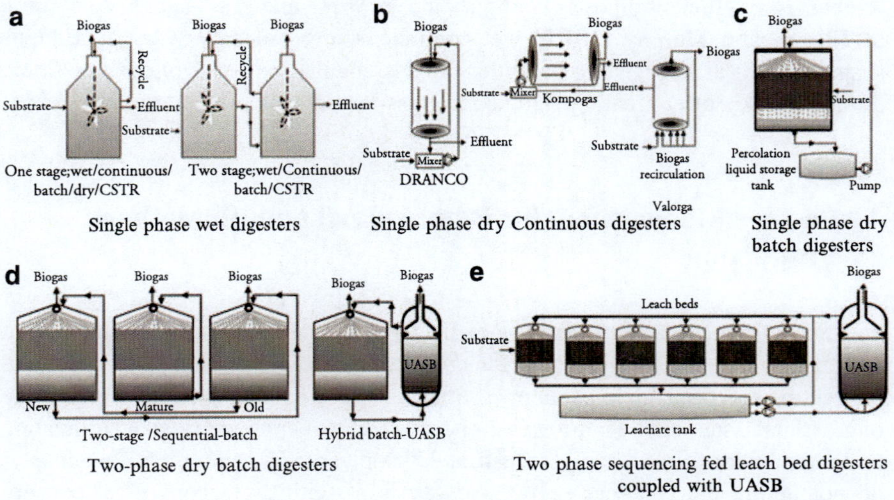

Fig. 6 Various types of anaerobic digesters (**a**) single phase wet digesters (**b**) single phase dry continuous digesters (**c**) single phase dry batch digester (**d**) two-phase dry batch digesters (**e**) two phase sequencing fed leach bed digesters coupled with UASB (Vandevivere et al. 2003; Nizami and Murphy 2010). Note: In single phase digesters, all phases of anaerobic digestion (hydrolysis, acidogenesis, acetogenesis and methanogenesis) occur within the same compartment in the digester. In two phase digesters, hydrolysis, acidogenesis and acetogenesis are separated from methanogenesis. In batch digesters, the feedstock is inserted once into the digester for a certain period of time to complete the digestion activity. In continuous digesters, the feedstock is constantly or regularly fed, either mechanically or by force of the new feed. In dry digesters, high solid feedstock with dry matter ranging from 20 to 50% is used as substrate. Wet digesters operate for substrates with total solids content less than 13%. *CSTR* Continuously Stirred Tank Reactor, *UASB* Upflow Anaerobic Sludge Blanket

et al. 2003). As Murphy and Power (2009) reported, digester design is based on the nature of the feedstock regarding its volatile solid content and is an important factor for anaerobic digestion efficiency. Digesters, for example, optimized for the organic fraction of municipal solid wastes (OFMSW) may not be ideal for grass silage because the volatile solids content of grass silage is significantly higher than that of OFMSW, i.e. up to 92% compared to values as low as 60%. Thus the digestate from grass may be quite liquid in nature (solids content of less than 5%) as opposed to digestate from OFMSW which may have a solids content of over 20%. This will lead to significant effects on materials handling. For example, vertical garage door batch digesters may be suitable for OFMSW, but not for grass.

Two-phase digester configuration supports high growth rates of hydrolytic and methanogenic bacteria (Gunaseelan and Nallathambi 1997; Verrier et al. 1987; Mata-Alvarez 1987); hence this design/configuration is most appropriate for high solid feedstocks, such as grass silage. Furthermore, work conducted by Lehtomaki (2006) and Yu (2002) suggests that incorporation of a high rate reactor, such as a Upflow Anaerobic Sludge Blanket, with a Continuously Stirred Tank

Reactor is an efficient digester configuration for grass and grass silage. According to Nizami and Murphy (2010), wet continuous, dry batch and dry continuous systems, such as sequencing batch leach-beds coupled with a Upflow Anaerobic Sludge Blanket offer greater potential for grass biomethane production.

4 Pasture Management for Both Animal and Biomethane Production

To resolve many varying dilemmas in the agricultural sector, and to attain what appears to be an increasingly difficult balance between sustainable grassland production and rural socio-economic status, there is a need both to raise the profile of grassland issues and to improve our understanding of applied grassland utilization under the wider frame of agro-industrial development. The authors of this chapter do not ignore the fact that energy security is a significant geopolitical concern for the EU but, equally, and perhaps of greater importance, is the issue of food security. It is therefore necessary to consider the possible impacts of biofuels, and more specifically grass biomethane, on food production and food prices before the examination on the suitability of main pasture practices for both systems.

4.1 Food versus Biofuels

In many developing countries, increased demand for food products has resulted in price pressures in markets. This pressure can be either direct, through growing demand and changes in consumption patterns as incomes rise, or indirect, as alternative uses of food crops, such as for biofuels, have led to higher domestic prices (OECD 2008a,b). There is no consensus on how large the impact of biofuels production is on food price increases. According to a study published by the World Bank, between 70 and 75% of the increase in the price of food commodities from January 2002 to June 2008 is attributed to biofuels and the related consequences of low grain stocks, large land use shifts, speculative activity and export bans (Mitchell 2008). A US Council of Economic Advisors study found that the contribution of biofuels to the price increase of agricultural commodities was 3% (CEA 2008) whereas a report from the Organisation for Economic Co-operation and Development (OECD) concludes that biofuels are responsible for 15% of the food price increases (OECD 2008a,b). Nevertheless, other factors can influence food price increases, including poor harvests due to extreme weather events; increased demand for meat and milk products, pushing up the demand for animal fodder; high energy prices raising the price of producing, processing and transporting food (Glauber 2008); and international and national agricultural policies. It is worth noting that since biofuels have dampened the increase in oil prices, they have limited one of the factors leading to an increase in food prices. Additionally, hyper-food inflation is unlikely because, amongst other reasons which are beyond the scope of this chapter, the globalization of the food economy and the greater diversification of crop and food production will act as a barrier to food price inflation (Alexander and Hurt 2007).

Ruminant livestock farming, the dominant grass-based farming system, is paramount to the development of depressed areas where grassland occupies a significant proportion of the agricultural area (Veysset et al. 2010). Recent reform of the CAP has removed the link between financial support and the obligation to retain specific animal numbers, and will probably enhance the declining trend in livestock populations (Smyth et al. 2010). Additionally, animal breeding and structural adaptations in agriculture (Rosch et al. 2009), increased cost of land, feed and energy, alongside restricted government support, will most probably result in reductions in livestock population (Flach 2009). According to Food and Agricultural Organization (FAO 2006), the numbers of cattle and sheep in the EU-25 declined by 10.3% and 11.4% respectively between 1990 and 2003. Also, the area of grasslands in the EU declined by 12.8% from 1990 to 2003 and only few Member States managed to prevent this trend (FAO 2006).

Rosch et al. (2009) reported that in some regions of Germany almost one-quarter of the grassland is not used in animal husbandry. Additionally, more traditional farmland areas, where socio-economic conditions for extensive agriculture are generally unfavourable, tend to be abandoned, particularly in central and Eastern Europe, where political and economic changes have negatively affected farming conditions (Silva et al. 2008). For example, a survey of Estonia in 2000 found that some 56% of the permanent grassland was abandoned (Silva et al. 2008).

Against this background, and along with the increased demand of biomass feedstock for energy purposes and the political and financial support for renewable energy, "surplus" grassland biomass could be used as an additional energy resource (Rosch et al. 2009). This will prevent abandonment of grassland and will enhance EU policy for sustainable energy and agricultural development.

4.2 *Evaluation of Grass for Both Animal and Biomethane Production*

Grassland use is characterized by various modes and intensities but its role as a principal source of food for ruminants dominates. Grassland husbandry has evolved as modern scientific approaches to farming have been developed. Agronomists and progressive farmers require the cultivation of highly productive grass species for highly productive pastures (Connolly 2001), which under appropriate pasture management allows full utilisation of the grassland (Walker 1995). Full utilisation of grass and grass silage can be achieved by using the crop both for animal feed and as a feedstock for biomethane production.

Forage quality is a function of nutrient value, amount of forage intake (consumption of forage by the animal), digestibility (the fraction of the dry matter that remains in the body on passage through the gut tract and is positively related with the dry matter intake by the animals) and partitioning of metabolised products within animals that is usually determined by animal performance when forages are fed to livestock (Wheeler and Corbett 1989; Buxton 1996; Brown 1999). Particularly in ruminants, where rumen fermentation modifies the actual diet received by the animal, forage and fibre chemical, physical and nutritional nature hold an important role (Van Soest et al. 1991). Many indices have been used for quality evaluation of grass and grass silage, including the factors affecting digestibility and voluntary intake such as crude or true protein (Keady et al. 2000; Krizsan and Randby 2007), water soluble carbohydrates concentration (O'Kiely et al. 2002) (carbohydrates classified based on cold water solubility of non-structural carbohydrates, namely monosaccharides, oligosaccharides and some polysaccharides so as to distinguish them from the starches) and fibre content, including neutral and acid detergent fibres (NDF and ADF respectively) (De Boever et al. 1993; Bach-Knudsen 1997; Keady et al. 2000; Nordheim-Viken and Volden 2009). Most of these indices reflect a particular agronomic factor such as growth stage at harvest (Peyraud et al. 1997; Van Dorland et al. 2006), nitrogen application (Peyraud et al. 1997; Keady et al. 2000), ensiling process (Nsereko et al. 1998; Dawson et al. 1999; Charmley 2001; Van Dorland et al. 2006) or pasture composition (Stypinski 1993; Soegaard 1993; Lee et al. 2009).

The main polymers found in the plant cell walls of lignocellulosic materials like grasses are cellulose, hemicellulose and lignin. The strong inter-linkages between these polymers, and non-covalent and covalent cross linkages between them, provide the plant a stable shape and structure (Perez et al. 2002). They are relatively resistant to hydrolysis and their degradation largely depends on microbial activity (Orr and Kirk 2003). The crude (structural carbohydrates including lignin and pectin) and neutral detergent fibre (the insoluble fibre mainly cellulose and hemicellulose) are the most used indices for determining the feed value for the ruminant (Van Soest et al. 1991; Bach-Knudsen 1997). Additionally, many authors have used acid detergent fibre for cellulose and lignin estimations (De Boever et al. 1993; Keady et al. 2000). According to Bach-Knudsen (1997), calculated and analysed values regarding the fibre components of fibre rich materials can deviate significantly.

Plant cell walls, composed mostly of structural carbohydrates and lignin, account for 40–80% of the organic matter in forage crops and, depending on their concentration, can limit feed intake and digestibility of forages (Buxton 1996). Neutral detergent fibre is resistant to mammalian enzyme degradation (Van Soest et al. 1991).

The crystalline structure of cellulose acts a barrier for microbial and enzymatic degradation in anaerobic digestion (Lehtomaki 2006). On the contrary, hemicellulose is easily hydrolysed by hemicellulase enzymes (Clavero and Razz 2002). Lignin is the most recalcitrant part of the structural carbohydrates because of its non-water soluble nature and resistance to microbial action and oxidative forces

(Lewis and Davin 1998; Hendriks and Zeeman 2009). Therefore, the biodegradability of grass and grass silage in an anaerobic digester is limited by the higher concentration of cellulose and lignin (acid detergent fibre values) (Lehtomaki and Bjornsson 2006). Inefficient biodegradation results in reduced solubilisation of grass silage, which limits the conversion of volatile solids to chemical oxygen demand and consequently biomethane production (Nizami and Murphy 2010). The components of grass in the cell cytoplasm are proteins and nitrogenous compounds, lipids and non-structural carbohydrates (McDonald et al. 1991). The protein, lipid and extracted fractions of carbohydrates, often known as water soluble carbohydrates, are the soluble parts of grass silage that are sources of energy in both the rumen and for anaerobic digestion microorganisms (Hendriks and Zeeman 2009). High specific methane yields can be achieved when crop substances are characterised by low lignin concentrations and high concentrations of easily degradable components in other words non-structural carbohydrates and soluble cell components (Amon et al. 2007a; Schittenhelm 2008). The direct relationship between pasture management practices and those attributes that influence the production of high value grass and grass silage suitable for animal and biomethane production necessitates the examination of these agronomic practices. This is an important step for sustainable development in both agricultural and energy sectors.

4.2.1 Grass Species and Biomethane Production

The advantages of perennial grasses over arable crops as a feedstock for biofuel production due to better energy balance and environmental benefits have resulted in a diversion of interest from arable crops to perennial grasses. Research on *Panicum vigratum* (switchgrass) (McLaughlin and Kszos 2005), *Miscanthus* × *Giganteous* (miscanthus) (Clifton-Brown et al. 2004), *Phalaris arundinacea* (reed canary grass) and *Phleum pratence* (timothy) (Lewandowski et al. 2003), *Andropogon gerardii* (big bluestem) (Weimer and Springer 2007), *Lolium perenne* (perennial ryegrass) (Smyth et al. 2009; Korres et al. 2010) as energy crops has been accelerated. Nevertheless, the selection of grass species and seed mixtures is determined by the purpose for which the sward is to be used, in this case for both animal and biomethane production, along with the prevailing environmental conditions (Feehan 2003). Furthermore, the physiology of grasses, considering for example their photosynthetic (PS) pathway, i.e. C-3 (cool season or temperate species) vs. C-4 (warm season or tropical species), imposes environmental specificity and hence differences in the adaptability, productivity (Niu et al. 2006) and qualitative aspects, such as water soluble carbohydrates and possible biomethane yield (Table 3), of the grasses.

There are distinctive differences between C-3 and C-4 species that affect their productivity, for example the former fix CO_2 in lower temperatures and they respond to N fertilizer early in the spring, whereas in high temperatures their growth rates are reduced. In contrast, C-4 species require less N to achieve the same light-saturated assimilation rate, leading to higher photosynthetic N use efficiency, are more efficient at fixing carbon dioxide in warm environments and are more tolerant of

Table 3 Potential perennial grasses as energy crops in Europe and methane yields per hectare and per ton of dry matter[a] (DM), Photosynthesis (PS)

Common name	Latin name	PS pathway	Methane ($m^3 ha^{-1}$)	Yield ($t\,DM\,ha^{-1}$)	Methane ($m^3\,t^{-1}$ DM)
Ryegrass	*Lolium perenne*	C-3	2,500–6,150	9–15[b]	278–410
Miscanthus	*Miscanthus* × *Giganteous*	C-4	1,432–5,450	5–44	124–286
Switchgrass	*Panicum vigratum*	C-4	900–7820[c]	5–23	180–340
Reed canary grass	*Phalaris arundinacea*	C-3	1,700–4,730	7–13	243–364
Timothy	*Phleum pratence*	C-3	1,362–5,800	9–18	151–322
Meadow foxtail	*Alopecurus pratensis*	C-3	1,463	6–13	112–243
Big bluestem	*Andropogon gerardii*	C-4	–	8–15	–
Cocksfoot	*Dactylis glomerata*	C-3	1,480–3,800	8–10	185–380
Tall fescue	*Festuca arundinacea*	C-3	1,462–2,000	8–14	183–143
Napier grass	*Pennisetum purpureum*	C-4	0.19–0.34[d]	27	–
Sudan grass	*Sorghum* × *drummondii*	C-4	2,130–6,060	10–20	213–303
Cypergrass	*Cyperus longus*	C-4	–	4–19	–

Data were adopted from Lewandowski et al. (2003); Prochnow et al. (2009); Braun et al. (undated); Seppala et al. (2009)

[a]Methane yields per ton of dry matter were estimated based on the methane yield ha^{-1} and potential dry matter yields ha^{-1}

[b]Yields of early, intermediate and late perennial ryegrass were reported equal to 16.7, 15.3 and $15\,t\,DM\,ha^{-1}\,year^{-1}$ respectively (Lockhart and Wiseman 1988)

[c]Based on 0.18–0.34 $m^3\,CH_4\,kg^{-1}$ dry matter (Chynoweth et al. 2001; Sampson 2006)

[d]$L\,CH_4\,g^{-1}$ of Volatile Solids (Wilkie 2008)

water stress conditions (Winslow et al. 2003; Lunt et al. 2007; Nippert et al. 2007). Additionally, as mentioned by White (1973), the enzymes which convert CO_2 into organic compounds in C-3 and C-4 (ribulose-1,5-diphosphate carboxylase and phosphoenolpyruvate carboxylase for C-3 and C-4 species respectively) are affected by temperature. Hence, temperate grasses require lower optimum growth temperatures in comparison to tropical species, a fact that can influence the distribution of grasses based on annual temperature fluctuations, their nutritive value and yield, along with biomethane potential.

Despite the positive greenhouse gas balance of miscanthus (Styles and Jones 2007; Lewandowski et al. 1995), impacts on biodiversity have raised a few concerns (Semere and Slater 2007a and b) about its sustainability as a potential energy crop in European temperate climates.

The energy ratios for switchgrass production are generally very positive (Thornley et al. 2009) and there are particular biodiversity benefits for pheasants,

Table 4 Dry Matter Intake (DMI), Dry Matter Digestibility (DMD) and Acid Detergent Fibres (ADF) concentration of C-3 and C-4 grasses, Dry matter (DM)

Grass type	DMI (g day^{-1} $kg^{0.75}$)[a]	DMD (%)	ADF (% of DM)
C-4 (sheep)	56	62	–
C-3 (sheep)	71	71	–
C-4 (sheep)	65.7	54.5	42.5
C-3 (sheep)	66.2	65.5	35.8
C-4 (cattle)	89.8	60	42.7
C-3 (cattle)	89.5	67	38.3

Adopted from Brown (1999)
[a]DM intake is linked to energy requirements that are proportional to 0.75 power of body weight (Allison 1985)

quail and rabbits. Agro-chemical inputs are low and nitrogen leaching rates and soil erosion rates are both low compared to arable crops (Thornley et al. 2009). The unfamiliarity of EU farmers with these crops and the lack of research data on the crops under European conditions are the main issues causing concern on their suitability.

Reed canary grass is widely distributed in temperate regions of Europe (Thornley et al. 2009) but the high nitrogen requirements (Geber 2002), concerns about invasiveness, the relatively low yields (Riche 2005) that results in a poor energy balance and the susceptibility to pests and diseases affect its sustainability as a potential energy crop (Thornley et al. 2009).

However, productivity of animals consuming mostly forage is directly related to the quality of the forage and the amount consumed (Penning et al. 1995; Buxton 1996; Fontaneli et al. 2001). C-4 grass species are lower in dry matter digestibility and usually in dry matter intake than C-3 species, mainly because of their higher fibre concentrations (Minson 1981; Reid et al. 1988) (Table 4). Additionally, the neutral detergent fibres of forages grown under high temperatures is usually less digestible than that of forages grown under lower temperatures because of increased lignification (Buxton and Fales 1994).

The lower digestibility of warm grasses due to their higher fibre content is indicative of possible lower biomethane yields, since digestibility of dry matter may be equated to the potential digestibility of the silage in the cattle paunch (Robson et al. 1989).

In temperate grassland regions, perennial ryegrass (*Lolium perenne*) is preferred for digestion because of its high digestibility (Robson et al. 1989), water soluble and non-structural carbohydrates content (Smith et al. 2002; Buxton and O'Kiely 2005) (Fig. 7) and reduced concentration of crude fibre (Nizami et al. 2009).

As Mahnert et al. (2005) reported perennial ryegrass produced the highest biogas yield (0.83–0.86 m^3 kg^{-1} VS added) in comparison to other grasses (both fresh and ensiled); for example, cocksfoot resulted a biogas yield of 0.65–0.72 m^3 kg^{-1} VS added. Tetraploid ryegrass varieties are recommended due to high sugars levels (Dieterich 2008). Tetraploid varieties remain an important component of grass seed mixtures because of their higher water soluble carbohydrate content, their increased

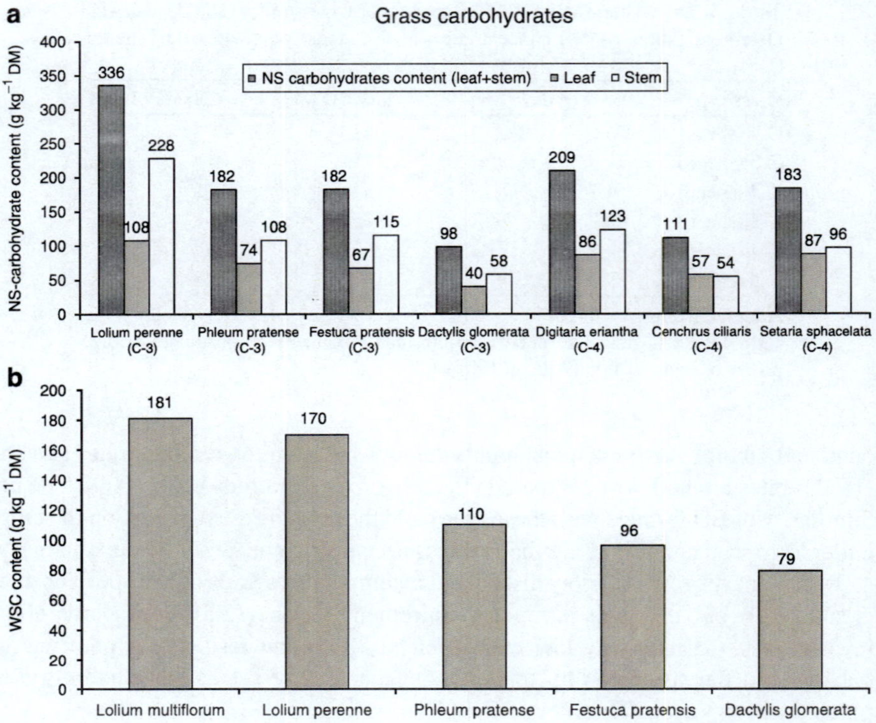

Fig. 7 *Non-structural* (NS) content of temperate (C-3) and tropical (C-4) grasses (**a**) and water soluble carbohydrate (WSC) content of temperate grasses (**b**) (Based on Buxton and O'Kiely 2005)

palatability (leading to higher intake by livestock) and their tolerance to drought. However, they tend to have lower tiller densities, resulting in more open swards and lower dry matter, compared with diploids (Anonymous 2008). Seppala et al. (2009) reported that cocksfoot, tall fescue and timothy are better than reed canary grass for biogas production in boreal conditions because they offer higher specific methane yields, higher dry matter yields per hectare and better regrowth ability. As reported by Mahnert et al. (2005), the supply of a high quality feedstock in combination with high yielding ryegrass cultivars for silage production is an essential prerequisite to obtain optimal gas yields.

Selection of the appropriate grass species should consider not only biomethane potential but also suitability for animal production systems. In Western Europe, perennial ryegrass is the most widely used grass species for grazing cattle, because of its high productivity, palatability and nutritive value (Taweel et al. 2005). Additionally, O'Kiely et al. (2005) stated that the main benefits of perennial ryegrass swards are that they produce high yields in response to fertilizer application, have high digestibility when harvested at the appropriate growth stage, are relatively easy to preserve as silage due to their superior content of sugar and persist as permanent swards where favourable management practices prevail.

4.2.2 Grass and Clover Mixed Pastures

As stated by Kelm et al. (2004), an integrated approach to crop production should consider both productivity and environmental trade-offs as for example energy efficiency, fertilization and strategies for reducing N_2O emissions through different cropping strategies (pure grass swards vs. clover/grass swards, fertilization strategies). In Europe, during the last decade, interest for feeding temperate forage legumes, both in fresh and conserved form, has grown particularly in extensive grass-based farming systems (Parente and Frame 1993). This renewed attention was due to changes in agricultural policy, which supports extensive and sustainable farming systems (EC 2003). Forage legumes are considered environmental friendly alternatives with regard to artificial nitrogen fertilizer (Evers et al. 1993) and also have benefits in their nutritional composition, both as protein (N) and energy sources for ruminants in low-concentrate input systems (Van Dorland et al. 2006). Woodmansee (1978) reported the fixation of 25 kg Nha^{-1} by the incorporation of *Trifolium* spp. (clover) in annual grassland whereas Brockman and Wilkins (2003) and Moller et al. (2002) reported a range between 3 and 150 kg N ha^{-1} depending on the percentage of clover into the sward. This is very important considering, as mentioned in the previous sections, the contribution of nitrogen fertilization in nitrous oxide emissions.

Additionally, Stypinski (1993), in a 5 year field experiment using three grass species (cocksfoot, timothy and ryegrass) in a mixture with and without clover, found that white clover (*Trifolium repens*) improved the pasture value with regard to protein content, energy content, the concentration of macro and micro-nutrients, as well as digestibility. However, the fibre content of the feed with clover was lower than that without clover (Fig. 8).

Mixed pastures can influence the level of intake and hence digestibility. Comparisons of grasses with legumes, such as white clover, are often associated with higher level of intake (Ribeiro et al. 2003). Baumont (1996), Rutter et al. (2004), and Assoumaya et al. (2007) have explained that forage legumes are reduced more quickly into small particles than grasses and that less time is needed to take and masticate a similar bite for clover than for grass. As suggested by Rutter et al. (2004), further research is required regarding this.

Nevertheless, digestibility of pure grass compared to grass mixed with white clover varies because it depends on the environmental conditions, the grass species involved, the defoliation management and other factors (Soegaard 1993).

As reported by Cavallero et al. (1993), grass species as in the case of cocksfoot vs. perennial ryegrass along with other factors can affect the establishment of a white clover population in the sward. These authors reported that, after 3 years of experimentation, a steady equilibrium was achieved for cocksfoot-clover but not for perennial ryegrass-clover mixture, under both continuous and rotational grazing.

Grass varieties with high tillering capacity, which results in dense pastures, should be avoided in grass/clover pastures (Sheehy and Culleton 2002). Breeding clover cultivars to withstand grass competition enables the establishment of grass/clover under a wide range of grass varietal selection. Tetraploids are

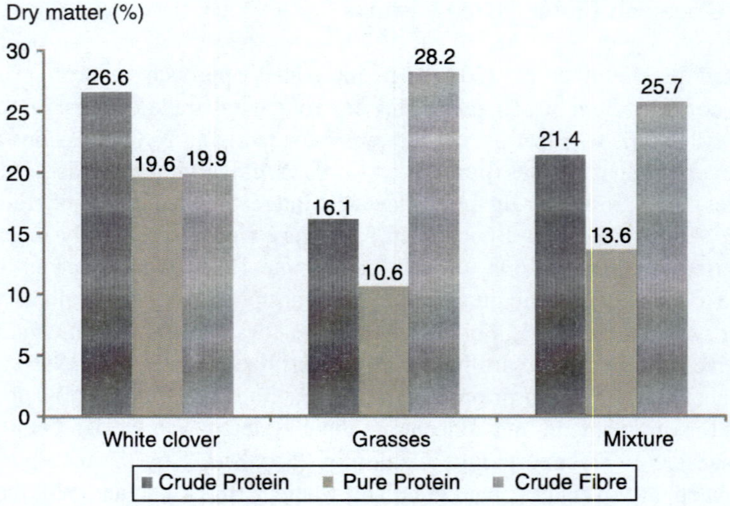

Fig. 8 Effects of white clover on qualitative characteristics of animal feed (Based on Stypinski 1993)

particularly suitable because of their open sward, hence permitting the development of the clover. As an example, a 50:50 mixture of a medium diploid and a tetraploid variety of ryegrass would be suitable for most situations (Sheehy and Culleton 2002). Frankow-Lindberg et al. (1996) reported that the marginal cost of grass/clover is half that of a grass/fertilised N production system, one of the main reasons for which is the reduced N fertilisation rate in the former.

Grass species may vary in terms of their chemical composition hence methane yields from grassland could possibly depend on the mixture of species within the vegetation (Prochnow et al. 2009). Mixtures of grasses result in increased methane yields compared to a single grass type such as in *Cynodon* spp. (Bermuda grass) (Gunaseelan and Nallathambi 1997). Additionally, Plochl and Heiermann (2006) reported methane production from forage and paddock mixtures of 297–370 and 246 $m^3 t^{-1}$ organic dry matter (ODM), respectively. The efficiency of anaerobic digestion can be considerably improved when using mixed feedstock, such as grass with legumes, because the neutral detergent fibre concentration of grasses is usually greater than that of legumes. This is caused mostly by differences in neutral detergent fibre concentration between grass and legumes leaves (Buxton 1996). Hence, increasing the proportion of legumes, particularly clover, and consequently the leaf to stem ratio of forage, results in lower cell wall concentration, or in other words reduced indigestible material and increased feedstock digestibility. This improves the efficiency of lignocellulosic decomposers and possibly increases biomethane production (Table 5).

Silage from mixed grass pastures with clover tends to produce higher methane yields than silage from mixed pastures without clover; however, the anaerobic digestion conditions are also significant factors for biomethane production.

Table 5 Effects of pasture type on methane production (Prochnow et al. 2005)

Substrate	Biogas yield (L/kg VS)	Methane yield (L/kg VS)	Conditions
Intensive grassland (monoculture fresh, silage)	700–720	–	Batch/35°C/25 d
Extensive grassland (fresh and silage)	540–580	–	
Extensive grassland (fresh and hay)	500–600	–	Semi-continuous, 35°C, 18–36 d, co-digestion
Extensive grassland (silage)	500–550	–	Continuous, 35°C, 20 d, co-digestion
Mixed pasture grassland (fresh and silage)	650–860	310–360	Batch, 35°C, 28 d, mono-digestion
Mixed pasture grassland (silage)	560–610	300–320	Semi-continuous, 35°C, 28 d, mono-digestion
Grasses and clover (silage)	532, 474, 427[a]	370, 326, 297[a]	Batch, 37–39°C, 58 d, mono-digestion
Intensive grassland (monoculture, silage)	–	390	Semi-continuous, 37°C, 25–60 d, co-digestion
Extensive grassland (silage)	–	220	Semi-continuous, 37°C, 25–60 d, co-digestion

[a]Harvesting mid-May (before anthesis); end of May (anthesis); mid-June (after anthesis) respectively. Note: VS = Volatile Solids

4.2.3 Fertilization Management

Fertilization of grassland, particularly nitrogen fertilization, for higher yields is probably one of the most important husbandry factors, because nitrogen facilitates many functions in plants, including photosynthesis, enzyme synthesis, and nucleic acid, protein and cell walls formation (Addiscot 2005). Inorganic nitrogen fertiliser is applied to grassland to ensure that economically viable yields are available for harvesting at a time when the feed value of the grass is adequate (O'Kiely et al. 2002). N fertiliser has been increasingly seen as management tool (Vellinga et al. 2004) for herbage yield and quality increases (Eckersten et al. 2007) and for efficient grazing and cutting planning. The increased growth rate following N application reduces the growth time required between grazing and silage cuts and results in more cuts per year (Van Burg et al. 1981). The amount of N losses through ammonia volatilisation, nitrate leaching, nitrous oxide emission and denitrification can be reduced by appropriate timing of fertilization and cutting (Jarvis 1996; Whitehead 2000).

The effects of N fertilizer on the chemical composition of grass have been studied extensively. Nordheim-Viken and Volden (2009), investigating the effects

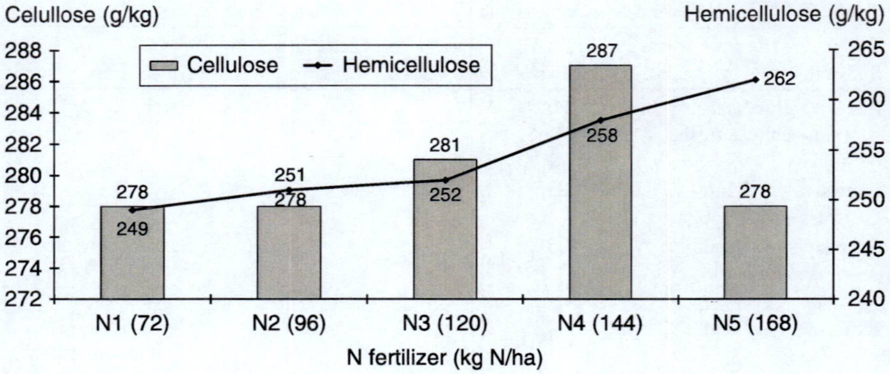

Fig. 9 Effects of N fertilization on cellulose and hemicellulose content in perennial ryegrass (Based on Keady et al. 2000)

of nitrogen fertilization rate during 3 years experimentation on timothy, observed that increases in fertilization rate tend to increase neutral detergent fibres content. Furthermore, Keady et al. (2000) reported increases of cellulose and hemicellulose content in perennial ryegrass (Fig. 9), and therefore increases in its digestibility, when N fertilizer application rate was increased.

In general, the results of nitrogen fertilization on fibre digestibility are moderate. As such, Peyraud et al. (1997) found that unfertilised perennial ryegrass had decreased fibre digestibility when provided as a feed for dairy cattle, but had moderate increases in neutral detergent fibres, acid detergent fibres and acid detergent lignins with nitrogen application.

Several authors have reported lower water soluble carbohydrates in grasses with increased N fertilization (Van Soest et al. 1978; Buxton and Fales 1994; Tremblay et al. 2005-cited by Nordheim-Viken and Volden 2009). O'Kiely et al. (2002) stated that lower water soluble carbohydrates content and increased buffering capacity of perennial ryegrass, due to increased N fertilization rate, negatively affects the ensilability of grass.

Dry matter yield and crude protein content is positively affected by increased N application rate in timothy (Belanger and McQueen 1999; Nordheim-Viken and Volden 2009), perennial ryegrass (Keady et al. 2000; O'Kiely et al. 2002), and cocksfoot (Mills et al. 2009).

There has been limited research into the effects of nitrogen management, specifically with regard to application rate (kg Nha^{-1}) and fertilizer form (organic, inorganic), on biomethane production from grass and grass silage. For the purpose of this chapter the potential grass biomethane production was estimated, based on the work conducted by Seppala et al. (2009) and Kaiser and Gronauer (2007), as between 0.325 and 0.339 m^3 CH_4 kg^{-1} volatile solids. This is the average biomethane production for a wide range of fertilizer forms (mineral, organic and combination) and grass species (cocksfoot, reed canary grass, timothy and tall fescue).

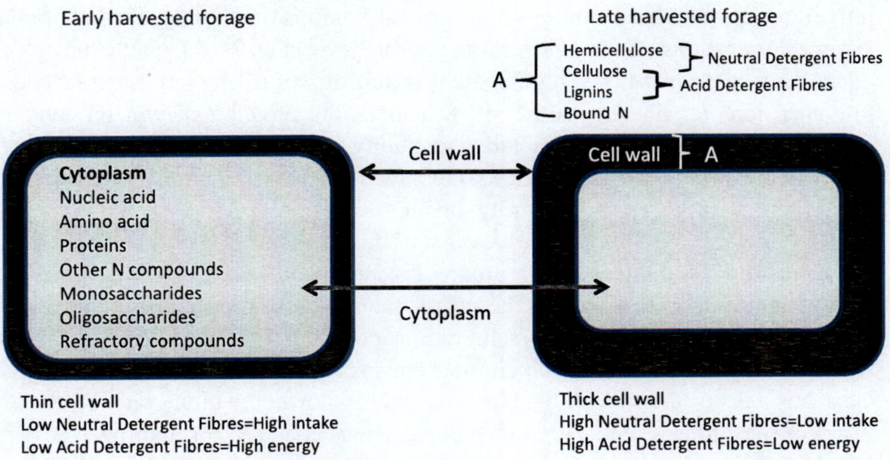

Fig. 10 Grass cell anatomical and biochemical characteristics that affect ruminant nutrition, according to harvesting date (Note: The relative changes in structural and chemical composition between early and late harvested forage depict generalised differences which (**a**) developed as plant matures, (**b**) exist between leaf and stem tissues and (**c**) exist between C-3 (cool season) and C-4 (warm season) plants. Refractory compounds: a diverse group of primarily secondary plant compounds, in addition to lignin, which affects the digestibility and/or nutritive value of plant tissue. Included are tannins, flavones, essential oils, steroids, saponins, waxes and alkaloids. These types of compounds have been variously isolated from the neutral detergent solubles and neutral detergent fibres fractions (Huston and Pinchak 2007))

4.2.4 Harvesting Date

The structural characteristics of forage are described in various ways. Botanists and agronomists approach plant cellular structure from the standpoint of biosynthesis whereas animal nutritionists emphasize the attributes of cells and tissues that enhance bio-degradation (Van Soest 1982) and liberation of nutrients. Most probably the same approach would be taken by an engineer, who investigates methods for higher biomethane production from biomass and more specifically from grass and grass silage. Harvesting date is of prime importance for grass species digestibility, intake, and for grass biomethane production.

As the grass matures, the proportion of the cell wall components (cellulose, hemicellulose and lignin) increases (De Boever et al. 1993), while the proportion of cell contents decreases (Bruinenberg et al. 2002) (Fig. 10). The leaves of grass are more digestible than the stems (Gilliland 1997; Bruinenberg et al. 2002) due to their higher protein content (Nissinen 2004-cited in Seppala et al. 2009). In the generative growth stage, the total solids (TS) yield increases to over 200 kg TS/ha per day; however, digestibility (Ito et al. 1997) and the amount of raw protein decrease by 0.5–1.0 percentage units per day (Nissinen 2004-cited in Seppala et al. 2009). Forage quality is reduced with maturity due to a decrease in leaf-to-stem ratio and an increase in fibre components (Ugherughe 1986). The development stage is an important factor in determining the chemical composition and quality of forage legumes; for example, in red clover forage (Vasiljevic et al. 2005), young red clover plants have large leaf mass, high content of moisture, protein and minerals and a low fibre content, whereas as the plant matures, the protein and mineral content, and consequently also the intake and digestibility, decline.

Nordheim-Viken and Volden (2009) recorded decreases in crude protein content in timothy with increased maturity under various environments. Additionally, De Boever et al. (1993) reported that most of the chemical, physical and biological characteristics of the grass were more favourable in early cut silage than in intermediate or late cut silage. As the sward matures, there are a higher proportion of stems due to decreases in leaf-to-stem ratio. According to Benvenutti et al. (2006), stems can have a barrier effect on bite size and instantaneous intake rate. The higher the stems density, the smaller the bite area and the slower the biting rate. This leads to a decrease of the instantaneous intake rate. Boval et al. (2007) confirms that stem length and stem proportion in the sward have a negative impact on biting rate.

Harvesting date is an important factor for feedstock specific methane yield (Prochnow et al. 2009). It has been reported by several authors that the specific methane of the feedstock decreases with advancing stage of maturity since increases in non-degradable fibre content limit biomethane production potential (Shiralipour and Smith 1984; Weiland 2001). Additionally, the decrease crude protein and crude lipid content with maturity (Holmes 1980) negatively affects methane percentages in biogas, as these components contribute to methane production (Weiland 2001).

Seppala et al. (2009), in an investigation of the methane potential of four grass species at various locations, fertilizer regimes and harvesting intervals, reported that the methane yield from the 1st harvest of all grasses was higher than for following harvests.

Nizami et al. (2009) reported that grass for silage is usually harvested at a less mature stage of growth (leafy and non-lignified) since the aim is to obtain a crop with a relatively high content of fermentable substrate and a low content of fibre. As Woolford (1984) stated, the crop at early growth stages usually exhibits a high leaf-to-stem ratio. This is supported in the case of *Pennisetum purpureum* (Napier grass) as reported by Gunaseelan and Nallathambi (1997). Amon et al. (2007a, b) on a multifaceted crop rotation aiming to increase the yield of methane per hectare reported that the first cut at vegetative stage was selected as the optimum option for harvesting. Kaparaju et al. (2002) found that *Trifolium* spp. (clover) produced 50% more methane per tonne of volatile solids at vegetative stage than at flowering stage. Nevertheless results recorded by Pouech et al. (1998) show that methane yield per tonne of volatile solids was 32% lower at vegetative stage than at flowering stage when similar experiments were conducted. The content of total and volatile solids in grass depends on several factors, such as location and origin, seasonal variations, cultivation practices, type of soil, pre-treatment of the biomass, and the nutrient composition of the grass (Bauer et al. 2007). These factors could affect the yield of grass biomethane in anaerobic digestion (Nizami and Murphy 2010).

The diurnal and seasonal variation of carbohydrates in grasses influences the harvesting date of the sward since studies on various grass species have shown that the concentration of total non-structural carbohydrates was lowest at 6 am and increased linearly to a high at 6 pm (White 1973). Additionally, as the same author stated, the seasonal variation of carbohydrates differs among grass species; in some species the level is lowest when the second or third leaf emerges, but in other species, the reserve level is lowest after seed ripening. Taking into consideration the importance of non-structural carbohydrates in animal and biofuel production systems, harvesting should be commenced at a time when the level of carbohydrates is highest.

4.2.5 Grass Silage

High forage systems, as opposed to concentrate feeding systems, have been shown to have beneficial effects in terms of meat quality, stability and sensory characteristics (Lee et al. 2009). Grass, and in particular grass silage, form the basal diet for the vast majority of ruminants in many parts of the world during the winter feeding period (Charmley 2001). The transformation of harvested herbage to silage is therefore considered as part of grassland management and, as mentioned in previous sections, pasture management that affects quantitative and qualitative characteristics of grass production would affect the same characteristics of grass silage. Furthermore, the use of various crop silages, such as fodder and sugar beets, grain crops and grass silages has been studied for a number of biogas processes (Amon et al. 2007a; Lehtomäki et al. 2008). In order to ensure constant quality and supply of substrate to an anaerobic digestion facility, the ensiling of grass as silage is preferable to the utilization of fresh grass (Nizami et al. 2009). Ensiled grass has lower organic matter losses and more independent of weather conditions than hay (dried grass) (Egg et al. 1993).

The basic principles for silage production are based on the reduction of fresh grass moisture content (dehydration) and the prevention of bacteria and fungi growth (which can deteriorate the final product), usually with the use of inhibitors or an acid medium (O'Kiely et al. 2002). Normally during ensiling, the fodder undergoes acid fermentation in which bacteria produce lactic, acetic and butyric acids from sugars present in the raw material. The net result is a reduction in pH which prevents the growth of spoilage microorganisms, the majority of which are intolerant of acid conditions (Woolford 1984). Plant aerobic respiration and enzymatic fermentation are the major processes determining silage quality and ensiling efficiency (Murdoch 1980), and possibly also anaerobic digestion. According to the same author the action of enzymes on the carbohydrate content of the herbage result in the production of heat, water and carbon dioxide, accompanied with a restriction of lactic acid formation. Additionally, increased heat results in increased silage temperature and a reduction in protein digestibility. Coinciding with and following plant respiration, bacterial fermentation occurs, and a major aim in silage making is to control this bacterial action. Silage which has undergone an undesirable fermentation is characterised by a relatively high butyric acid content and an extensive degree of proteolysis, resulting in reduced digestibility and low intake characteristics (Murdoch 1980). Silage fermentation also depletes soluble carbohydrate concentration (Chamberlain et al. 1985; Khalili and Huhtanen 1991; Charmley 2001). Moreover, there is a need to restrict clostridia growth, which consumes lactic acid (Murdoch 1980), and causes deterioration in silage quality (Woolford 1984). Ammonia may act as a simple index of silage fermentation quality (Charmley 2001), since is predominantly a product of clostridia fermentation of amino acids and its excess indicates a low quality product. Fermentation of non-structural carbohydrate in silage has a direct effect on the pattern of volatile fatty acids production in the rumen since the concentration of soluble sugars in silages that have undergone extensive homolactic fermentation is almost negligible (Charmley 2001). This affects both animal (Cushnahan and Mayne 1995; Keady and Murphy 1996) and biomethane production (Madhukara et al. 1993). Keady et al. (2002) reported that improvements in silage fermentation could be indicated by decreases in pH and ammonia-N due to various factors, such as the use of additives. Silage additives may influence methane yields indirectly via silage quality or directly by providing additional feed for lactic acid bacteria with sugars, by improving degradability of organic matter with enzymes or by inhibiting or promoting micro-organisms with acids (Kung et al. 2003). Nevertheless, effects of additives on grass silage methane potential are moderate. Prochnow et al. (2009) reported various silage additive treatments for different feedstocks, and their use resulted in increased biomethane production in some cases and had no effect in others. Additionally, other studies have shown contrasting results (Madhukara et al. 1997; Rani and Nand 2004; Neureiter et al. 2005).

5 Conclusion

Most grassland in the EU is devoted to meat (cattle and sheep) production where profitability is low and farmers often rely on EU single farm payments to survive; thus grassland farming can face considerable challenges in implementing new environmental measures without financial supports (Boyle 2008). Additionally, the response to climate change must not restrict the opportunities for the agricultural sector to create employment, develop export markets and expand into emerging markets (Walshe 2009). Effective solutions to climate change may require integrated multi-sectoral approaches based on new sources of energy and industrial feedstock, new technologies for emissions capture and reduction as well as new mechanisms for balancing an immediate need for food and energy with the long-term goal of sustainability (Teagasc 2008). Current economic and environmental concerns require the control of inputs in intensive grassland systems in order to maximize efficiency and to reduce potential pollution sources. The authors of this chapter do not ignore the fact that energy security is a significant issue and geopolitical concern for the EU; the issue of food security is equally important. We propose the integration of grass and grass silage production with a biomethane production system because such a system: (1) will diversify agricultural production (2) will improve agricultural competitiveness (3) is an environmental friendly technology well established across Europe (4) enhances security and sustainability of energy supply and (5) offers great opportunities for rural development by creating a domestic industry with associated employment opportunities.

Acknowledgements The authors are deeply indebted to the Irish Environmental Protection Agency for funding this study. They also acknowledge the suggestions and recommendations of the three anonymous reviewers and these of the chief editor which have improved the quality of this chapter greatly.

References

Aarts HFM, Habekotte B, Van Keulen H (2000) Efficiency of nitrogen (N) management in the "De Marke" dairy farming system. Nutr Cycl Agroecosyst 56:231–240, ISSN: 0178–2762

Addiscot TM (2005) Nitrate, agriculture and the environment. CABI Publishing, Wallingford

Alexander C, Hurt C (2007) Biofuels and their impact on food prices. Purdue Extension, Expert Review, ID-346-W, Purdue University. http://www.ces.purdue.edu/extmedia/ID/ID-346-W.pdf. Accessed May 2010

Allison CD (1985) Factors affecting forage intake by range ruminants: a review. J Range Manage 38(4):305–311, http://www.jstor.org/stable/3899409

Amon T, Amon B, Kryvoruchko V, Machmuller A, Hopfner-Sixt K, Bodiroza V, Hrbek R, Friedel J, Potsch E, Wagentristl H, Schreiner M, Zollitsch W (2007a) Methane production through anaerobic digestion of various energy crops grown in sustainable crop rotations. Bioresour Technol 98:3204–3212. doi:10.1016/j.biortech.2006.07.007

Amon T, Amon B, Kryvoruchko V, Zollitsch W, Mayer K, Gruber L (2007b) Biogas production from maize and dairy cattle manure-influence of biomass composition on the methane yield. Agric Ecosyst Environ 118:173–182. doi:10.1016/j.agee.2006.05.007

Anonymous (1997) Kyoto protocol to the United Nations Framework Convention on climate change: UNEP/IUC, Geneva 1997

Anonymous (2008) Grass and clover. Recommended list varieties for Ireland, Irish Deptartment of Agriculture, Fisheries and Food, Dublin

Assoumaya C, Sauvant D, Archimede H (2007) Etude comparative de l'ingestion et de la digestion des fourrages tropicaux et temperes. INRA Prod Anim 20(5):383–392

Bach-Knudsen KE (1997) Carbohydrate and lignin contents of plant materials used in animal feeding. Anim Feed Sci Technol 67:319–338. doi:10.1016/S0377-8401(97)00009-6

Baier U, Grass S (2001) Bioraffination of grass, anaerobic digestion 2001. In: 9th World Congress for anaerobic conversion for sustainability, Antwerp, 2001

Bauen A (2005) Biomass in Europe. In: Silveira S (ed) Bioenergy-realizing the potential. Elsevier, Amsterdam, pp 19–30

Bauer A, Hrbek R, Amon B, Kryvoruchko V, Machmüller A, Hopfner-Sixt K, Bodiroza V, Wagentristl H, Pötsch E, Zollitsch W, Amon T (2007) Potential of biogas production in sustainable biorefinery concepts. In: Proceedings of the 5th research and development conference of central- and Eastern European Institutes of Agricultural Engineering-Part 2, vol 2, pp 20–31, ISBN: 966-8302-16-08

Baumont R (1996) Palatabilite et comportement alimentaire chez le ruminant. INRA Prod Anim 9(5):349–358

Belanger G, McQueen RE (1999) Leaf and stem nutritive value of timothy grown with varying N nutrition in spring and summer. Can J Plant Sci 79:223–229

Belcher KW, Boehm MM, Fulton ME (2004) Agroecosystem sustainability: a system simulation model approach. Agric Syst 79:225–241. doi:10.1016/S0308-521X(03)00072-6

Benvenutti MA, Gordon IJ, Poppi DP (2006) The effect of the density and physical properties of grass stems on the foraging behaviour and instantaneous intake rate by cattle grazing an artificial reproductive tropical sward. Grass Forage Sci 61:272–281, ISSN: 0142–5242

Berglund M, Borjesson P (2006) Assessment of energy performance in the life cycle of biogas production. Biomass Bioenergy 30:254–266. doi:10.1016/j.biombioe.2005.11.011

Bernet N, Beline F (2009) Challenges and innovations on biological treatment of livestock effluents. Bioresour Technol 100:5431–5436. doi:10.1016/j.biortech.2009.02.003

Borjesson P (2009) Good or bad bioethanol from a greenhouse gas perspective-what determines this? Appl Energy 86:589–594. doi:10.1016/j.apenergy.2008.11.025

Boval M, Fanchone A, Archimede H, Gibb MJ (2007) Effect of structure of a tropical pasture on ingestive behaviour, digestibility of diet and daily intake by grazing cattle. Grass Forage Sci 62:44–54. doi:10.1111/j.1365-2494.2007.00560.x

Boyazoglu J (1998) Livestock farming as a factor of environmental, social and economic stability with special reference to research. Livest Prod Sci 57:1–14. doi:10.1016/S0301-6226(98)00193-6

Boyle G (2008) Sustainable grassland systems in Europe and the EU Water Framework Directive. In: Proceedings of sustainable grassland systems in Europe and the EU Water Framework Directive. Teagasc, Johnstown Castle Research Centre, Wexford, 12th–14th Nov 2008, p 11

Braun R, Steffen R (1997) Anaerobic digestion of agroindustrial byproducts and wastes. In: Verstraete W (ed) Workshop on anaerobic conversions for environmental protection, sanitation and re-use of residues. Sustainable Rural Environment and Energy Network, REUR Technical Series (FAO), No 52, FAO/SREN. FAO, Rome, pp 27–41

Braun R, Weiland P, Wellinger A (Undated) Biogas from energy crop digestion. IEA Bioenergy, Task 37-Energy from biogas and Landfill Gas

Braun B, Castree N (eds) (1998) Remaking reality: nature at the millennium. Routledge, London

Brockman JS, Wilkins RJ (2003) Grassland. In: Soffe R (ed) Primrose McConnell's the agricultural notebook. Blackwell Science Ltd, Oxford, pp 131–176

Brown RH (1999) Agronomic implications of C4 photosynthetis. In: Sage RF, Monson RK (eds) C4 plant biology. Academic, San Diego, pp 473–509

Bruinenberg MH, Valk H, Korevaar H, Struik PC (2002) Factors affecting digestibility of temperate forages from seminatural grasslands: review. Grass Forage Science 57:292–301

Burton CH, Turner C (2003) Anaerobic treatment options for animal manures. In: Beck JAF, Martinez J, Martens W, Pahl O, Piccinini S, Svoboda I (eds) Manure management-treatment strategies for sustainable agriculture. Silsoe Research Institute, Silsoe, pp 273–320

Buxton DR (1996) Quality-related characteristics of forages as influenced by plant environment and agronomic factors. Anim Feed Sci Technol 59:37–49, ISSN: 0377–8401

Buxton DR, O'Kiely P (2005) Preharvest plant factors affecting ensiling. Silage science and technology. Agronomy monograph 42, pp 199–250. Pub., American Society of Agronomy, Inc., Crop Science Society of America, Inc., Soil Science Society of America, Inc., Madison, WI, USA. ISSN: 0065–4663

Buxton DR, Fales SL (1994) Plant environment and quality. In: Fahey GC Jr et al (eds) Forage quality, evaluation, and utilization. American Society of Agronomy, Madison, pp 155–199

Caslin B (2009) Overview of biofuels in Ireland. Bioenergy News. Renewable Energy Information Office, Sustainable Energy Ireland, Autumn 2009, pp 12–14

Cavallero A, Grignani C, Reyneri A (1993) Comparison between continuous and rotational grazing for two grass-white clover mixtures in North Italian Plain. In: White clover in Europe: State of the art. REUR Technical Series 29. FAO Corporate Document Repository, Rome

CEA (2008) Council of Economic Advisors 2008 Testimony of Edward P. Lazear, Chairman. CEA before the US Senate Foreign Relations Committee hearing on responding to the global food crisis, 14 May 2008

Chamberlain DG, Thomas PC, Wilson W, Newbold CJ, McDonald JC (1985) The effect of carbohydrate supplements on ruminal concentrations of ammonia in animals given diets of grass silage. J Agric Sci 104:331–340. doi:10.1017/S0021859600044002

Chardon WJ, Schoumans OF (2008) Overview of potential European measures for agriculture under WFD, collected within COST Action 869: mitigation options for nutrient reduction in surface water and groundwaters. In: Proceedings of Sustainable grassland systems in Europe and the EU Water Framework Directive. Teagasc, Johnstown Castle Research Centre, Wexford, 12–14 Nov

Charmley E (2001) Towards improved silage quality: a review. Can J Anim Sci 81:157–168, ISSN: 0008–3984

Cherubini F, Bird ND, Cowie A, Jungmeier G, Schlamadinger B, Woess-Gallasch S (2009) Energy-and greenhouse gas-based LCA of biofuel and bioenergy systems: key issues, ranges and recommendations. Resour Conserv Recycl 53:434–447. doi:10.1016/j.resconrec.2009.03.013

Chynoweth DP, Wilkie AC, Owens JM (1999) Anaerobic treatment of piggery slurry-review. Asian Aust J Anim Sci 12:607–628

Chynoweth DP, Owens JM, Legrand RL (2001) Renewable methane from anaerobic digestion of biomass. Renewable Energy 22:1–8. doi:10.1016/S0960-1481(00)00019-7

Clavero T, Razz R (2002) Effects of biological additives on silage composition of mott dwarf elephantgrass and animal performance. Rev Cient XII(4):313–316, FCV-LUZ

Clergue B, Amiaud B, Pervanchon F, Lasserre-Joulin F, Plantureux S (2005) Biodiversity: function and assessment in agricultural areas. A review. Agron Sustain Dev 25:1–15. doi:10.1007/978-90-481-2666-8_21

Clifton-Brown JC, Stampfl PF, Jones MB (2004) Miscanthus biomass production for energy in Europe and its potential contribution to decreasing fossil fuel carbon emission. Glob Change Biol 10:509–518. doi:10.1111/j.1529-8817.2003.00749.x

Colchester M, Jiwan N, Andiko, Sirait M, Yunan-Firdaus A, Surambo A, Pane H (2006) Promised land: palm oil and land acquisition in Indonesia: implications for local communities and Indigenous Peoples. Forest Peoples Programme/Perkumpulan Sawit Watch, London/Bogor

Connolly V (2001) Breeding improved varieties of perennial ryegrass. Teagasc, Dublin. www.teagasc.ie. Accessed April 2010

Consoli F, Allen D, Boustead I, Fava J, Franklin W, Jensen AA, de Oude N, Parrish R, Perriman R, Postlethwaite D, Quay B, Séguin J, Vigon B (eds) (1993) Guidelines for life-cycle assessment: a "Code of practice". Society of Environmental Toxicology and Chemistry (SETAC). SETAC Workshop, Sesimbra, 31 Mar–3 Apr 1993

Cushnahan A, Mayne CS (1995) Effects of ensilage of grass on performance and nutrient utilization by dairy cattle. 1. Food intake and milk production. Anim Sci 60:337–345

Dawson LER, Ferris CP, Steen RWJ, Gordon FJ, Kilpatrick DJ (1999) The effects of wilting grass before ensiling on silage intake. Grass Forage Sci 54:237–247. doi:10.1046/j.1365-2494.1999.00176.x

De Baere LD, Mattheeuws B (2008) State-of-the-art 2008 – anaerobic digestion of solid waste. Waste Manag World 9(4). http://www.waste-management-world.com. Accessed April 2010

De Boever JL, De Smet A, De Brabander DL, Boucque CV (1993) Evaluation of physical structure. 1. Grass silage. J Dairy Sci 76(1):140–153

Del Prado A, Brown L, Schulte E, Ryan M, Scholefield D (2006) Principles of development of a mass balance N cycle model for temperate grasslands: an Irish case study. Nutr Cycl Agroecosyst 74:115–131. doi:10.1007/s10705-005-5769-z

Del Rio P, Burguillo M (2009) An empirical analysis of the impact of renewable energy deployment on local sustainability. Renew Sustain Energy Rev 13:1314–1325. doi:10.1016/j.rser.2008.08.001

Demirbas A (2009) Political, economic and environmental impacts of biofuels: a review. Appl Energy 86:S108–S117. doi:10.1016/j.apenergy.2009.04.036

Demirbas A, Ozturk T (2005) Anaerobic digestion of agricultural solid residues. Int J Green Energy 1:483–494. doi:10.1081/GE-200038719

Dena HM, Kontges A, Rostek S (2009) Biogaspartner – a joint initiative. Biogas grid injection in Germany and Europe-Market, Technology and Players. German Energy Agency, HP Druck, Berlin

Dieterich B (2008) Energy crops for anaerobic digestion (AD) in Westray. Report written for Heat and Power Ltd., Westray, Orkney. Available from http://cngireland.com/bio-methane%20info/Burkart%20Grass%20Report%20-%20Final170308.pdf. Accessed April 2010

EC (1992) Council Regulation (EEC) No 2078/92 of 30 June 1992 on agricultural production methods compatible with the requirements of the protection of the environment and maintenance of the countryside. Official Journal of the European Union L215/85

EC (2000) Directive 2000/60/EC of the European Parliament and of the Council of 23 October 2000 establishing a framework for Community action in the field of water policy. Official Journal of the European Communities L327/1-72

EC (2001) Commission of the European Communities, COM (2001) 31 final. Communication from the commission to the council, the european parliament, the economic and social committee and the committee of the regions on the sixth environment action programme of the european community "Environment 2010: Our Future, Our Choice". Proposal for a Decision of the European Parliament and of the Council laying down the Community Environment Action Programme 2001–2010

EC (2003) Council Regulation EC/1782/2003 for establishing common rules for direct support schemes under the common agricultural policy and establishing certain support schemes for farmers. Official Journal of the European Union, L270/1-L270/69

EC (2004) Commission Regulation (EC) No 796/2004 of April 2004 laying down detailed rules for the implementation of cross-compliance, modulation and the integrated administration and control system provided for in of Council Regulation (EC) No 1782/2003 establishing common rules for direct support schemes under the common agricultural policy and establishing certain support schemes for farmers. Official Journal of the European Union L 141

EC (2005a) Agri-environmental measures. Overview of general principles, types of measures and application. European Commission Directorate General for Agriculture and Rural Development, Unit G-4, Evaluation of measures applied to agriculture, studies, March 2005

EC (2005b) Council Regulation EC/1698/2005 on support for rural development by the European Agricultural Fund for Rural Development. Official Journal of the European Union, L 277/1

EC (2006) Environment fact sheet: energy for sustainable development. European Commission 2006. http://ec.europe.eu/environment/pubs/factsheets.htm. Accessed May 2010

EC (2009) Directive 2009/28/EC of the European Parliament and of the Council of 23 April 2009 on the promotion of the use of energy from renewable sources and amending and subsequently repealing Directives 2001/77/EC and 2003/30/EC. Official Journal of the European Union, L140/16-62

Eckersten H, Torssell B, Kornher A, Bostrom U (2007) Modelling biomass, water and nitrogen in grass ley: estimation of N uptake parameters. Eur J Agron 27:89–101. doi::10.1016/j.eja.2007.02.003

EEA (2004) Transport biofuels: exploring links with the energy and agriculture sectors. European Energy Agency Briefing 04. Available at www.eea.eu.int. Accessed April 2010

EEC (1991) Council Directive 1991/676/EEC of 12 December 1991 concerning the protection of waters against pollution caused by nitrates from agricultural sources. http://ec.europa.eu/environment/water/water-nitrates. Accessed May 2010

Egg R, Coble C, Engler C, Lewis D (1993) Feedstock storage, handling and processing. Biomass Bioenergy 5:71–94. doi:10.1016/0961-9534(93)90009-S

Eisentraut A (2010) Sustainable production of second generation biofuels. Information paper. International Energy Agency

Evers GW, Grichar WJ, Pohler CL, Schubert AM (1993) Tolerance of three annual forage legumes to selected postemergence herbicides. Weed Technology 7:735–739

FAO (2006) Food and Agricultural Organization, Statistical yearbook. FAOSTAT

FAOSTAT (Undated) Food and agricultural commodities production. Food and Agriculture Organization of the United Nations. http://faostat.fao.org/site/339/default.aspx. Accessed April 2009

Fargione J, Hill J, Tilman D, Polasky S, Hawthorne P (2008) Land clearing and the biofuel carbon debt. Science 319:1235–1238. doi:10.1126/science.1152747

Farmer M, Swales V (2004) The development and implementation of Cross Compliance in the EU 15: an analysis. Institute for European Environmental Policy

Feehan J (2003) Farming in Ireland. History, heritage and environment. University College Dublin, Faculty of Agriculture, Dublin, Ireland

Flach B (2009) EU-27 livestock and products annual. GAIN report Number: NL9022. USDA Foreign Agricultural Service. Global Agricultural Information Network

Fontaneli RS, Sollenberger LE, Staples CR (2001) Yield, yield distribution and nutritive value of intensively managed warm-season annual grasses. Agron J 93:1257–1262

Fossitt JA (2000) A guide to habitats in Ireland, The Heritage Council. (www.heritagecouncil.ie). Accessed May 2010

Frankow-Lindberg BE, Danielson DA, Moor C (1996) The uptake of white clover technology in farming practice. In: White clover in Europe: state of the art. REUR Technical Series 29. FAO Corporate Document Repository, Rome

Freibauer A, Rounsevell MDA, Smith P, Verhagen J (2004) Carbon sequestration in the agricultural soils of Europe. Geoderma 122:1–23. doi:10.1016/j.geoderma.2004.01.021

Geber U (2002) Cutting frequency and stubble height of reed canary grass (Phalaris arundinacea L.): influence on quality and quantity of biomass for biogas production. Grass Forage Sci 57(4):389–394. doi:10.1046/j.1365-2494.2002.00329.x

Gerin PA, Vliegen F, Jossart JM (2008) Energy and CO2 balance of maize and grass as energy crops for anaerobic digestion. Bioresour Technol 99:2620–2627. doi:10.1016/j.biortech.2007.04.049

Gilliland TJ (1997) Changes induced by defoliation in the yield and digestibility of leaves and stem of perennial ryegrass (*Lolium perenne* L.) during reproductive development. Eur J Agron 6:257–264. doi:10.1016/S1161-0301(96)02053-9

Glauber J (2008) Statement to the U.S. Congress Joint Economic Committee. Increasing food prices. Hearing, Washington, DC

Goldemberg J, Teixeira CS, Guardabassi P (2008) The sustainability of ethanol production from sugarcane. Energy Policy 36:2086–2097. doi:10.1016/j.enpol.2008.02.028

Grass S (2004) Utilisation of grass for production of fibres, protein and energy. In: Biomass and agriculture: sustainability, markets and policies. Biomass Project Services, Switzerland, pp 169–177

Greenpeace (2007) How the palm oil industry is cooking the climate. Greenpeace, Amsterdam

Gunaseelan V, Nallathambi S (1997) Anaerobic digestion of biomass for methane production: a review. Biomass Bioenergy 13:83–114. doi:10.1016/S0961-9534(97)00020-2

Haas G, Wetterich F, Kopke U (2001) Comparing intensive, extensified and organic grassland farming in southern Germany by process life cycle assessment. Agric Ecosyst Environ 83:43–53. doi:10.1016/S0167-8809(00)00160-2

Hall DO (1997) Biomass energy in industrialised countries-a view of the future. For Ecol Manage 91:17–45. doi:10.1016/S0378-1127(96)03883-2

Hendriks ATWM, Zeeman G (2009) Pretreatments to enhance the digestibility of lignocellulosic biomass, review. Bioresour Technol 100:10–18. doi:10.1016/j.biortech.2008.05.027

Holmes W (1980) Grass: Its Production and Utilisation; Blackwell Scientific Publications: Oxford, UK

Holmes J (2006) Impulses towards a multifunctional transition in rural Australia: gaps in the research agenda. J Rural Stud 22:142–160. doi:10.1016/j.jrurstud.2005.08.006

Huston JE, Pinchak WE (2007) Range animal nutrition. In: Briske DD, Heitschmidt RK (eds) Grazing management: an ecological perspective. University of California, Davis

ISO (2006) International Standardization Organization (ISO), Environmental management–life cycle assessment-principles and framework, ISO 14040

Ito M, Shimazu Y, Yamaguchi H, Ito M, Toyoda T (1997) Seasonal trends of tiller emergence and senescence in several temperate herbage grasses grown under sward conditions. Grassl Sci 43(1):7–13, http://www.affrc.go.jp/en/

Jagadabhi PS, Lehtomäki A, Rintala J (2008) Co-digestion of grass silage and cow manure in a CSTR by re-circulation of alkali treated solids of the digestate. Environ Technol 29:1085–1093. doi:10.1080/09593330802180385

Janzen HH (2004) Carbon cycling in earth systems - a soil science perspective. Agric Ecosyst Environ 104:399–417. doi:10.1016/j.agee.2004.01.040

Jarvis SC (1996) Future trends in nitrogen research. Plant Soil 181:47–56. doi:10.1007/BF00011291

Jensen AA, Hoffman L, Moller BT, Schmidt A, Christiansen K, Elkington J, van Dijk F (1997) Life-cycle assessment (LCA): a guide to approaches, experiences and information sources, vol 6, Environmental issues series. European Environment Agency, Copenhagen. ISBN 92-9167-079-0

Jones MB, Donnelly A (2004) Carbon sequestration in temperate grassland ecosystems and the influence of management, climate and elevated CO2. New Phytol 164(3):423–439, http://www.jstor.org/stable/1514751

Kaiser F, Gronauer A (2007) Methanproduktivität nachwachsender Rohstoffe in Biogasanlagen (Methane potential of renewable resources in biogas plants). Bayerische Landesanstalt für Landwirtschaft. http://www.lfl.bayern.de/publikationen/daten/informationen/p_27455.pdf. Accessed April 2010

Kamm B, Kamm M (2004) Principles of biorefineries. Appl Microbiol Biotechnol 64:137–145. doi:10.1007/s00253-003-1537-7

Kamm B, Kamm M, Soyez K (1998) The green biorefinery, concept of technology. In: First international symposium on green biorefinery, Neuruppin, Society of Ecological Technology and System Analysis, Berlin

Kaparaju P, Luostarinen S, Kalmari E, Kalmari J, Rintala J (2002) Codigestion of energy crops and industrial confectionery by-products with cow manure: batch scale and farm-scale evaluation. Water Sci Technol 45:275–280, ISSN: 02731223

Karagiannidis A, Perkoulidis G (2009) A multi-criteria ranking of different technologies for the anaerobic digestion for energy recovery of the organic fraction of municipal solid wastes. Bioresour Technol 100:2355–2360. doi:10.1016/j.biortech.2008.11.033

Keady TWJ, Murphy JJ (1996) Effects of inoculant treatment on ryegrass silage fermentation, digestibility, rumen fermentation, intake and performance of lactating dairy cattle. Grass Forage Sci 51:232–241. doi:10.1111/j.1365-2494.1996.tb02058.x

Keady TWJ, Mayne CS, Fitzpatrick DA (2000) Prediction of silage feeding value from the analysis of the herbage at ensiling and effects of nitrogen fertilizer, date of harvest and additive treatment on grass silage composition. J Agric Sci 134:353–368. doi:10.1017/S0021859699007674

Keady TWJ, Kilpatrick CM, Cushnahan A, Murphy JA (2002) The cost of producing and feeding forages. Proceedings of the XIII International Silage Conference, Auchincruive, Scotland, pp 322–323

Kelm M, Wachendorf M, Trott H, Volkers K, Taude F (2004) Performance and environmental effects of forage production on sandy soils. III. Energy efficiency in forage production from grassland and maize for silage. Grass Forage Sci 59:69–79. doi:10.1111/j.1365-2494.2004.00406.x

Kemp DR, Michalk DL (2007) Towards sustainable grassland and livestock management. J Agric Sci 145:543–564. doi:10.1017/S0021859607007253

Khalili H, Huhtanen P (1991) Sucrose supplements in cattle given grass silage-based diet. 1. Digestion of organic matter and nitrogen. Anim Feed Sci Technol 55:247–261. doi:10.1016/0377-8401(91)90064-Y

Korres NE, Singh A, Nizami AS, Murphy JD (2010) Is grass biomethane a sustainable transport biofuel? Biofuels Bioprod Bioref 4(3):310–325. doi:10.1002/bbb.228

Krizsan SJ, Randby AT (2007) The effect of fermentation quality on the voluntary intake of grass silage by growing cattle fed silage and the sole feed. J Anim Sci 85:984–996. doi:10.2527/jas.2005-587

Kung L, Stokes MR, Lin CJ (2003) Silage additives. In: Al-Amoodi L (ed) Silage science and technology, vol 42, Agronomy. American Society of Agronomy, Inc./Crop Science Society of America, Inc./Soil Science Society of America, Inc., Madison, pp 31–93

Lee MRF, Evan PR, Nute GR, Richardson RI, Scollan ND (2009) A comparison between red clover silage and grass silage feeding on fatty acid composition, meat stability and sensory quality of the M. Longissimus muscle of dairy cull cows. Meat Sci 81:738–744. doi:10.1016/j.meatsci.2008.11.016

Lehtomaki A (2006) Biogas production from energy crops and crop residues. PhD dissertation, Jyväskylä studies in biological and environmental science. University of Jyväskylä, Jyväskylä

Lehtomaki A, Bjornsson L (2006) Two-stage anaerobic digestion of energy crops: methane production, nitrogen mineralization and heavy metal mobilisation. Environ Technol 27:209–218. doi:10.1080/09593332708618635

Lehtomaki A, Huttunen S, Rintala JA (2007) Laboratory investigations on co-digestion of energy crops and crop residues with cow manure for methane production; Effect of crop to manure ratio. Resour Conserv Recycl 51:591–609. doi:10.1016/j.resconrec.2006.11.004

Lehtomäki A, Viinikainen TA, Rintala JA (2008) Screening boreal energy crops and crop residues for methane biofuel production. Biomass Bioenergy 36:3267–3278. doi:10.1016/j.biombioe.2007.11.013

Lewandowski I, Kicherer A, Vonier P (1995) CO_2 balance for the cultivation and combustion of miscanthus. Biomass Bioenergy 8:81–90. doi:10.1016/0961-9534(95)00008-U

Lewandowski I, Scurlock JMO, Lindvall E, Christo M (2003) The development and current status of perennial rhizomatous grasses as energy crops in the US and Europe. Biomass Bioenergy 25:335–361. doi:10.1016/S0961-9534(03)00030-8

Lewis NG, Davin LB (1998) The biochemical control of monolignol coupling and structure during lignan and lignin biosynthesis. In: Lewis NG, Sarkanen S (eds) Lignin and lignan biosynthesis. American Chemical Society, Washington, DC, pp 334–361

Lockhart JAR, Wiseman AJL (1988) Introduction to crop husbandry, 6th edn. Pergamon Press, Oxford

Lunt DJ, Ross I, Hopley PJ, Valdes PJ (2007) Modelling Late Oligocene C4 grasses and climate. Palaeogeogr Palaeoclimatol Palaeoecol 251:239–253. doi:10.1016/j.palaeo.2007.04.004

Madhukara K, Nand K, Raju NR, Srilahta HR (1993) Ensilage of mango peel for methane generation. Process Biochem 28:119123

Madhukara K, Srilatha HR, Srinath K, Bharathi K, Nand K (1997) Production of methane from green pea shells in floating dome digesters. Process Biochem 32:509–513

Mahnert P, Heiermann M, Linke B (2005) Batch- and semi-continuous biogas production from different grass species. Agric Eng Int: CIGR Ejournal, V11, Manuscript EE 05 010

Marsden T (1999) Rural futures: the consumption countryside and its regulation. Sociologia Ruralis 39:501–520. doi:10.1111/1467-9523.00121

Marsden T, Sonnino R (2008) Rural development and the regional state: denying multifunctional agriculture in the UK. J Rural Stud 24(4):422–431. doi:10.1016/j.jrurstud.2008.04.001

Mata-Alvarez J (1987) A dynamic simulation of a two-phase anaerobic digestion system for solid wastes. Biotechnol Bioeng 30:844–851. doi:10.1002/bit.260300706

Mata-Alvarez J, Mace S, Llabres P (2000) Anaerobic digestion of organic solid wastes: an overview of research achievements and perspectives. Bioresour Technol 74:3–16. doi:10.1016/S0960-8524(00)00023-7

Mathiasson A (2008) Vehicle gas utilization in Sweden – today and tomorrow. In: 2nd Nordic biogas conference. The Swedish Gas Association, Malmö

Matsunaka T, Sawamoto T, Ishimura H, Takakura K, Takekawa A (2006) Efficient use of digested cattle slurry from biogas plant with respect to nitrogen recycling in grassland. Int Congr Ser 1293:242–252

McCorriston S, Sheldon IM (2007) Trade liberalization and net distribution in vertically related markets. In: Swinnels JFW (ed) Global supply chains, standards and the poor. CAB International, Wallingford, pp 59–74
McDonald P, Henderson N, Heron S (1991) The biochemistry of silage, 2nd edn. Chalcombe Publications, Marlow
McLaughlin SB, Kszos LA (2005) Development of switchgrass (Panicum virgatum) as a bioenergy feedstock in the United States. Biomass Bioenergy 28:515–535. doi:10.1016/j.biombioe.2004.05.006
Mills A, Moot DJ, Jamieson PD (2009) Quantifying the effect of nitrogen on productivity of cocksfoot (*Dactylis glomerata* L.) pastures. Eur J Agron 30:63–69. doi:10.1016/j.eja.2008.07.008
Minson DJ (1981) The effects of feeding protected and unprotected casein on the milk production of cows grazing ryegrass. J Agric Sci 96:239–241. doi:10.1017/S0021859600032056
Mitchell D (2008) A note on rising food prices, Policy Research Working Paper 4682. The World Bank Development Prospects Group, July 2008, World Bank, Washington, DC
Moller-Hansen E, Hogh-Jensen H, Djurhuus J (2002) Biological nitrogen fixation in a grazed perennial grass/clover ley and correlation with herbage and soil variables. Eur J Agron 16:309–320. doi:10.1016/S1161-0301(01)00135-6
Mosier AR, Duxbury JM, Freney JR, Heinemeyer O, Minami K, Johnson DE (1998) Mitigating agricultural emissions of methane. Clim Change 40:39–80. doi:10.1023/A:1005338731269
Murdoch JC (1980) The conservation of grass. In: Holmes W (ed) Grass: its production and utilisation. British Grassland Society by Blackwell Scientific Publications, Oxford
Murphy JD, Power NM (2009) An argument for using biomethane generated from grass as a biofuel in Ireland. Biomass Bioenergy 33:504–512. doi:10.1016/j.biombioe.2008.08.018
Mussner R, Leipprand A, Schlegel S (2006) Facilitating the CAP Reform: compliance and competitiveness of European Agriculture. Deliverable 5. Mandatory Standards in 7 EU countries and 3 non-EU countries. Germany Country Report, Project No. SSPE-CT-2005-006489
Narodoslawsky M (1999) Green biorefinery. In: Second international symposium on green biorefinery "SUSTAIN", Feldbach
Neureiter M, dos Santos JTP, Lopez CP, Pichler H, Kirchmayr R, Braun R (2005) Effect of silage preparation on methane yields from whole crop maize silages. In: Ahring BK, Hartmann H (eds) Proceedings of the 4th international symposium on anaerobic digestion of solid waste, vol 1. BioCentrum-DTU, Copenhagen, pp 109–115
Nippert JB, Fay PA, Knapp AK (2007) Photosynthetic traits in C3 and C4 grassland species in mesocosm and field environments. Environ Exp Bot 60:412–420. doi:10.1016/j.envexpbot.2006.12.012
Nissinen O (2004) Wild flower seed yields in northern Finland. In: Lüscher A, Suter D (eds) Land use systems in Grassland dominated regions, vol 9. AGFF Zuerich, pp 249–254
Niu S, Zhang Y, Yuan Z, Liu W, Huang J, Wan S (2006) Effects of interspecific competition and nitrogen seasonality on the photosynthetic characteristics of C3 and C4 grasses. Environ Exp Bot 57:270–277. doi:10.1016/j.envexpbot.2005.06.004
Nizami AS, Murphy JD (2010) What is the optimal digester configuration for producing grass biomethane? Renew Sustain Energy Rev. doi:10.1016/j.rser.2010.02.006
Nizami AS, Korres NE, Murphy JD (2009) Review of the integrated process for the production of grass biomethane. Environ Sci Technol 43:8496–8508. doi:10.1021/es901533j
Nordheim-Viken H, Volden H (2009) Effects of maturity stage, nitrogen fertilization and seasonal variation on ruminal degradation characteristics of neutral detergent fibre in timothy (*Phleum pratense*). Anim Feed Sci Technol 149:30–59. doi:10.1016/j.anifeedsci.2008.04.015
Nsereko VL, Rooke JA, Newbold CJ, Wallace RJ (1998) Influence of protease inhibitors on nitrogen distribution in ensiled perennial ryegrass and the utilisation of silage nitrogen for growth by rumen bacteria in vitro. Anim Feed Sci Technol 76:51–63. doi:10.1016/S0377-8401(98)00211-9
O'Kiely P, Moloney A, O'Riordan EG (2002) Reducing the cost of beef production by increasing silage intake. Grange Research Centre, Project No. 4622, Beef production series No. 51, December 2002, ARMIS No. 4622

O'Kiely P, Conaghan P, Howard H, Moloney A, Black A (2005) Grazing and ensiling of energy-rich grasses with elevated sugar contents for sustainable production of ruminant livestock (Acronym: SweetGrass). Beef production series No. 80. RMIS No. 5002, Grange Livestock Research Centre, Teagasc

OECD (2008a) OECD-FAO Agricultural Outlook 2007–2016

OECD (2008b) Organisation for Economic Cooperation and Development 2008 Report on Economic Assessment of Biofuel Support Policy. Trade and Agriculture Directorate, OECD, Paris

Oenema O, Wrage N, Velthof GL, van Groenigen JW, Dolfing J, Kuikman PJ (2005) Trends in global nitrous oxide emissions from animal production systems. Nutr Cycl Agroecosyst 72:51–65. doi:10.1007/s10705-004-7354-2

Orr RM, Kirk JA (2003) Animal physiology and nutrition. In: Soffe RJ (ed) Primrose McConnell's the agricultural notebook, 20th edn. Blackwell Publishing, Oxford

Osterburg B, Nitsch H, Kristensen L (2005) Environmental standards and their linkage to support instruments of the EU Common Agricultural Policy. Paper prepared for presentation at the 99th seminar of the EAAE, "The future of Rural Europe in the Global Agri-Food System", Copenhagen, 22–27 Aug 2005

Parente G, Frame J (1993) Alternative uses of white clover. In: White clover in Europe: state of the art. REUR Technical Series 29. FAO Corporate Document Repository, Rome

Paustian K, Babcock BA, Hatfield J, Lal R, McCarl BA, McLaughlin S, Mosier A, Rice C, Robertson GP, Rosenberg NJ, Rosenzweig C, Schlesinger WH, Zilberman D (2004) Agricultural mitigation of greenhouse gases: science and policy options. CAST (Council on Agricultural Science and Technology) Report, R141 2004

Payraudeau S, van der Werf HMG, Vertes F (2007) Analysis of the uncertainty associated with the estimation of nitrogen losses from farming systems. Agric Syst 94:416–430. doi:10.1016/j.agsy.2006.11.014

Peeters A (2009) Importance, evolution, environmental impact and future challenges of grasslands and grassland-based systems in Europe. Grassl Sci 55:113–125. doi:10.1111/j.1744-697X.2009.00154.x

Penning PD et al (1995) Intake and behaviour responses by sheep, in different physiological states, when grazing monocultures of grass or white clover. Appl Anim Behav Sci 451–2:63–78. doi:10.1016/0168-1591(95)00602-O

Perez J, Munoz-Dorado J, De-la-Rubia T, Martinez J (2002) Biodegradation and biological treatments of cellulose, hemicellulose and lignin: an overview. Int Microbiol 5:53–63. doi:10.1007/s10123-002-0062-3

Peyraud JL, Astigarraga L, Faverdin P (1997) Digestion of fresh perennial ryegrass fertilized at two level of nitrogen by lactating cows. Anim Feed Sci Technol 64:155–171. doi:10.1016/S0377-8401(96)01056-5

Pezzey J (1992) Sustainable development concepts: an economic analysis. World Bank Environmental Paper xiv, p 71

Phetteplace HW, Johnson DE, Seidi AF (2001) Greenhouse gas emissions from simulated beef and dairy livestock systems in the United States. Nutr Cycl Agroecosyst 60:99–102. doi:10.1023/A:1012657230589

Plochl M, Heiermann M (2006) Biogas farming in Central and Northern Europe: a strategy for developing countries? Invited overview. Agric Eng Int: CIGR Ejournal 3:8

Potter C, Tilzey M (2007) Agricultural multifunctionality, environmental sustainability and the WTO: resistance or accommodation to the neoliberal project for agriculture? Geoforum 38:1290–1303. doi:10.1016/j.geoforum.2007.05.001

Pouech P, Fruteau H, Bewa H (1998) Agricultural crops for biogas production on anaerobic digestion plants. In: Proceedings of the 10th European conference on biomass for energy and industry, Wurzburg, 8–11 June 1998, pp 163–165

Prochnow A, Heiermann M, Drenckhan A, Schelle H (2005) Seasonal pattern of biomethanisation of grass from landscape management. Agric Eng Int: CIGR Ejournal 2005, VII, Manuscript EE 05 011, ISSN: 1682–1130

Prochnow A, Heiermann M, Plochl M, Linke B, Idler C, Amon T, Hobbs PJ (2009) Bioenergy from permanent grassland-A review: 1. Biogas. Bioresour Technol 100(21):4931–4944. doi:10.1016/j.biortech.2009.05.070

Ragaglini G, Triana F, Villani R, Bonari E (2010) Can sunflower provide biofuel for inland demand? An integrated assessment of sustainability at regional scale. Energy. doi:10.1016/j.energy.2010.03.009

Rani DS, Nand K (2004) Ensilage of pineapple processing waste for methane generation. Waste Manage 24:523–528. doi:10.1016/j.wasman.2003.10.010

Reay DS, Grace J (2007) Carbon dioxide: importance, sources and sinks. In: Reay DS, Hewitt CN, Smith KA, Grace J (eds) Greenhouse gas sinks. CAB International, Wallingford, pp 1–10

Reid RL, Jung GA, Thayne WV (1988) Relationships between nutritive quality and fiber components of cool season and warm season forages: a retrospective study. J Anim Sci 66:1275–1291

Reinhard J, Zah R (2009) Global environmental consequences of increased biodiesel consumption in Switzerland: consequential life cycle assessment. J Cleaner Prod 17(Supplement 1):S46–S56. doi:10.1016/j.jclepro.2009.05.003

Reynolds SG (2005) Grasslands of the world. In: Suttie JM, Reynolds SG, Batello C (eds) Grasslands of the world, Plant production and protection series no. 34. Food and Agriculture Organization of the United Nations, Rome

Ribeiro HMN, Delagarde R, Peyraud JL (2003) Inclusion of white clover in strip-grazed ryegrass swards: herbage intake and milk yield of dairy cows at different ages of sward regrowth. Anim Sci 77:499–510. doi:10.1016/j.anifeedsci.2004.12.009

Riche AB (2005) A trial of the suitability of switchgrass and reed canary grass as biofuel crops under UK conditions. ETSU B/CR/ 00655/00/00. 5th Interim Report, www.bis.gov.uk/files/file15011.pdf. Accessed May 2009

Robson MJ, Parsons AJ, Williams TE (1989) Herbage production: grasses and legumes. In: Holmes W (ed) Grass: its production and utilization, 2nd edn. Blackwell, Oxford, pp 7–88

Rosch C, Skarka J, Raab K, And SV (2009) Energy production from grassland-assessing the sustainability of different process chains under German conditions. Biomass Bioenergy 33:689–700. doi:10.1016/j.biombioe.2008.10.008

Rutter SM, Orr RJ, Yarrow NH, Champion RA (2004) Dietary preference of dairy cows grazing ryegrass and white clover. J Dairy Sci 87:1317–1324

SAIC (2006) Life cycle assessment: principles and practice. Scientific Applications International Corporation (SAIC), Report No. EPA/600/R-06/060. National Risk Management Research Laboratory, Office of Research and Development, US Environmental Protection Agency, Cincinnati

Salter A, Banks CJ (2009) Establishing an energy balance for crop-based digestion. Water Sci Technol 59:1053–1060

Sampson R (2006) Journee CRAAQ Québec, 2006. Available at http://www.craaq.qc.ca. Accessed May 2010

Sanz Requena JF, Guimaraes AC, Quiros Alpera S, Relea Gangas E, Hernandez-Navarro S, Navas Gracia LM, Martin-Gil J, Fresneda Cuesta H (2010) Life Cycle Assessment (LCA) of the biofuel production process from sunflower oil, rapeseed oil and soybean oil. Fuel Process Technol 92(2):190–199. doi:0.1016/j.fuproc.2010.03.004

Schittenhelm S (2008) Chemical composition and methane yield of maize hybrids with contrasting maturity. Eur J Agron 29:72–79. doi:10.1016/j.eja.2008.04.001

Schnepf R (2006) European Union biofuels policy and agriculture: an overview. CRS Report for Congress, Order Code RS22404, 16 March 2006

Semere T, Slater FM (2007a) Ground flora, small mammal and bird species diversity in miscanthus (Miscanthus×giganteus) and reed canary-grass (Phalaris arundinacea) fields. Biomass Bioenergy 31:20–29. doi:10.1016/j.biombioe.2006.07.001

Semere T, Slater FM (2007b) Invertebrate populations in miscanthus (Miscanthus×giganteus) and reed canary-grass (Phalaris arundinacea) fields. Biomass Bioenergy 31:30–39. doi:10.1016/j.biombioe.2006.07.002

Seppala M, Paavola T, Lehtomaki A, Rintala J (2009) Biogas production from boreal herbaceous grasses-specific methane yield and methane yield per hectare. Bioresour Technol 100:2952–2958. doi:10.1016/j.biortech.2009.01.044

Sheehy J, Culleton N (2002) Establishing and managing clover pastures, Teagasc. http://www.teagasc.ie/publications. Accessed May 2009

Shiralipour A, Smith PH (1984) Conversion of biomass into methane gas. Biomass 6:85–92. doi:10.1016/0144-4565(84)90011-8

Siardos G (1994) Sustainable agriculture and agricultural development. Medit 5:13

Silva JP, Toland J, Jones W, Eldridge J, Thorpe E, O'hara E (2008) Life and Europe's grassland: restoring a forgotten habitat. Environmental Directorate-General, European Commission. doi:10.2779/23028

Singh A, Korres NE, Murphy JD (2010a) Grass Biomethane: A sustainable alternative industry for grassland. In Grassland in a changing world. European Grassland Federation 2010, Kiel Germany, Grassland Science in Europe, Vol. 15:139–148

Singh A, Pant D, Korres NE, Nizami AS, Prasad S, Murphy JD (2010b) Key issues in life cycle assessment of ethanol production from lignocellulosic biomass: challenges and perspectives. Bioresour Technol 101:5003–5012. doi:10.1016/j.biortech.2009.11.062

Smith KF, Culvenor RA, Humphreys MO, Simpson RJ (2002) Growth and carbon partitioning in perennial ryegrass (Lolium perenne) cultivars selected for high water-soluble carbohydrate concentrations. Journal Agricultural Science 138(4):375–385

Smith TC, Kindred DR, Brosnan JM, Weightman RM, Shepherd M, Sylvester-Bradley R (2005) Wheat as a feedstock for alcohol production. Research Review no. 61. Home-Grown Cereals Authority

Smyth B, Murphy JD, O'Brien C (2009) What is the energy balance of grass biomethane in Ireland and other temperate northern European climates? Renew Sustain Energy Rev 13(9):2349–2360. doi:10.1016/j.rser.2009.04.003

Smyth B, O'Gallachoir BP, Korres NE, Murphy JD (2010) Can we meet targets for biofuels and renewable energy in transport given the constraints imposed by policy in agriculture and energy. J Cleaner Prod 18(16–17):1671–1685

Soegaard K (1993) Nutritive value of white clover. In: Frame J (ed) White clover in Europe: state of the art. REUR Technical Series 29, pp 17–23

Styles D, Jones MB (2007) Energy crops in Ireland: quantifying the potential life- cycle greenhouse gas reductions of energy-crop electricity. Biomass Bioenergy 31:759–772. doi:10.1016/j.biombioe.2007.05.003

Stypinski P (1993) The effects of white clover on chemical composition and nutritive value of companion grasses in grass/clover mixtures. In: Frame J (ed) White clover in Europe: state of the art. REUR Technical Series 29, pp 81–83

Tabajdi CS (2008) Report on sustainable agriculture and biogas: a need for review of EU legislation (2007/2017 (INI)). Committee on Agriculture and Rural Development. European Parliament 2004–2009. Session Document A6-0034/2008

Taweel HZ, Tas BM, Williams BA, Elgersma A, Dijkstra J, Tammingra S (2005) Fibre degradation rate of perennial ryegrass varieties measured using three techniques: in situ nylon bag, in vivo rumen evacuation and in vitro gas production. In: Proceedings XX International grassland congress: offered papers, Dublin, June–July 2005, pp 228

Teagasc (2008) Towards 2030. Teagasc's role in transforming Ireland's agri-food sector and the wider bioeconomy. Foresight Report, Teagasc, May 2008. www.teagasc.ie. Accessed May 2010

Thornley P, Upham P, Tomei J (2009) Sustainability constraints on UK bioenergy development. Energy Policy 37:5623–5635. doi:10.1016/j.enpol.2009.08.028

Tilman D, Hill J, Lehman C (2006) Carbon-negative biofuels from low-input high diversity grassland biomass. Science 314:1598–1600. doi:10.1126/science.1133306

Tremblay GF, Bélanger G, Drapeau R (2005) Nitrogen fertilizer application and developmental stage affect silage quality of timothy (Phleum pratense L.) Grass Forage Sci 60:337–355. doi:10.1111/j.1365-2494.2005.00483.x

Tsoutsos T, Kouloumpis V, Zafiris T, Foteinis S (2010) Life Cycle Assessment for biodiesel production under Greek climate conditions. J Cleaner Prod 18:328–335. doi:10.1016/j.jclepro.2009.11.002

Ugherughe PO (1986) Relationship between digestibility of *Bromus inermis* plant parts. J Agron Crop Sci 157:136–143. doi:10.1111/j.1439-037X.1986.tb00060.x

Umetsu K, Yamazaki S, Kishimoto T, Takahashi J, Shibata Y, Zhang C, Misaki T, Hamamoto O, Ihara I, Komiyama M (2006) Anaerobic co-digestion of dairy manure and sugar beets. Int Congr Ser 1293:307–310

UNECE (1999) Protocol to the 1979 convention on long-range transboundary air pollution to abate acidification, eutrophication and ground-level ozone: United Nations Economic Commissions for Europe (UNECE), Geneva

USDA (2006) GAIN Report EU-25 Bio-fuels. Biofuels Annual 2006. USDA Foreign Agricultural Service

Van Burg PFJ, Prins WH, Den Boer DJ, Sluiman WJ (1981) Nitrogen and intensification of livestock farming in EEC countries. In: Proceedings of the Fertiliser Society, York, No 199, pp 78

Van Dorland HA, Wettstein R, Leuenberger H, Kreuzer M (2006) Comparison of fresh and ensiled white and red clover added to ryegrass on energy and protein utilization of lactating cows. Anim Sci 82:691–700. doi:10.1079/ASC200685

Van Soest PJ (1982) Nutritional ecology of the ruminant. O&B, Corvallis

Van Soest PJ, Mertens DR, Deinum B (1978) Preharvest factors influencing quality of conserved forages. *Journal of Animal Science* 47:712–720

Van Soest PJ, Robertson JB, Lewis BA (1991) Methods for dietary fiber, Neutral detergent fiber and nonstarch polysaccharides in relation to animal nutrition. Symposium: carbohydrate methodology, metabolism and nutritional implications in dairy cattle. J Dairy Sci 74:3583–3597, PMID: 1660498

Vandevivere P (1999) New and broad applications of anaerobic digestion. Crit Rev Environ Sci Technol 29:151–173

Vandevivere P, De Baere L, Verstraete W (2003) Types of anaerobic digester for solid wastes. In: Mata-Alvarez J (ed) Biomethanization of the organic fraction of municipal solid wastes. IWA Press, London, pp 112–140

Vasiljevic S, Katic S, Mihailovic V, Cupina B, Milic D, Mikic A, Karagic Dj, Pataki I (2005) Effect of cutting date on quality of red clover forage. In: Proceedings XX International grassland Congress: offered papers, Dublin, June–July 2005, pp 269

Vellinga TV, André G, Schils RLM, Oenema O (2004) Operational nitrogen fertiliser management in dairy farming systems: identification of criteria and derivation of fertiliser application rates. Grass Forage Sci 59:364–377, ISSN 0142–5242

Verchot L, Krug T, Lasco RD, Ogle S, Raison J (2006) Grassland. In: 2006 IPCC Guidelines for national greenhouse gas inventories, Bonn

Verrier D, Roy F, Albagnac G (1987) Two-phase methanization of solid vegetable waste. Biolog Waste 22:163–177. doi:10.1016/0269-7483(87)90022-X

Veysset P, Lherm M, Bebin D (2010) Energy consumption, greenhouse gas emissions and economic performance assessments in French Charolais suckler cattle farms: model-based analysis and forecasts. Agric Syst 103:41–50. doi:10.1016/j.agsy.2009.08.005

Wakker E (2004) Greasy palms: the social and ecological impacts of large scale oil palm plantation development in South East Asia. An AIDEnvironment report for Friends of the Earth London Friends of the Earth

Walker JW (1995) Viewpoint: grazing management and research now and in the next millennium. J Range Manage 48:350–357, http://www.jstor.org/stable/4002488

Walshe P (2009) Address by Padraig Walshe, Irish Farm Association President, to the Joint Oireachtas Committee on Climate Change and Energy Security. The Irish Farmer Association, 11 Nov 2009

Watkinson AR, Ormerod SJ (2001) Grasslands, grazing and biodiversity: editor's introduction. J Appl Ecol 38:233–237, http://www.jstor.org/stable/2655793

Weiland P (2001) Fundamentals of methane fermentation-biology and feedstocks. VDI-Berichte Nr. 1620, S, pp 19–32

Weiland P (2006) Biomass digestion in agriculture: a successful pathway for the energy production and waste treatment in Germany. Eng Life Sci 6:302–309. doi:10.1002/elsc.200620128

Weimer PJ, Springer TL (2007) Fermentability of Eastern gamagrass, big bluestem and sand bluestem grown across a wide variety of environments. Bioresour Technol 98:1615–1621. doi:10.1016/j.biortech.2006.06.003

Wheeler JL, Corbett JL (1989) Criteria for breeding forages of improved feeding value: results of a Delphi survey. Grass Forage Sci 44:77–83. doi:10.1111/j.1365-2494.1989.tb01912.x

White LM (1973) Carbohydrate reserves of grasses: A review. J Range Manag 26(1):13–18

Whitehead DC (2000) Nutrient elements in Grassland: soil-plant-animal relationships. CABI Publishing, Wallingford

Wilkie AC (2008) Biomethane from biomass, biowaste and biofuels. In: Wall JD, Harwood CS, Demain A (eds) Bioenergy. ASM Press, Washington, DC, pp 195–205

Wilson GA (2007) Multifunctional agriculture. A transition theory perspective. CAB International, Wallingford. ISBN 978 1 84593 256 5

Winslow JC, Hunt RE, Piper SC (2003) The influence of seasonal water availability on global C3 versus C4 grassland biomass and its implications for climate change research. Ecol Modell 163:153–173. doi:10.1016/S0304-3800(02)00415-5

Woodmansee RG (1978) Additions and losses of nitrogen in grassland ecosystems. Bioscience 28(7):448–453, http://www.jstor.org/stable/1307227 Accessed May 2010

Woolford MK (1984) The silage fermentation. In: Microbiological Series 14. Marcel Dekker Inc., New York, Basel

Yu HW, Samani Z, Hanson A, Smith G (2002) Energy recovery from grass using two-phase anaerobic digestion. Waste Manage 22(1):1–5. doi:10.1016/S0956-053X(00)00121-5

Zarrilli S, Burnett J (2008) Making certification work for sustainable development: the case of biofuels. In: United Nations Conference on Trade and Development. UNCTAD/DITC/TED/2008/1. United Nations, New York and Geveva

Genetic Mechanisms Enhancing Plant Biodiversity

Evangelia Sinapidou and Ioannis S. Tokatlidis

Abstract Biodiversity is essential for an evolving ecosystem and as a resource for further development of natural products by breeding. At present agriculture is under pressure by the demand for increased crop production and the public anticipation for sustainable cultivation practices. Undeniably, the prerequisite for adaptation of an organism to changing environmental conditions is genetic variability. The answer to the concern that the persistent accumulation of desirable alleles in a few cultivars could erode genetic variability and ultimately impede further improvement, comes from the fields of genetic and epigenetic studies that have revealed a range of mechanisms which result in remarkable variability even in narrow gene pools.

We review small and large scale mutations and transposable element activity that create genetic variability, as well as epigenetic mechanisms that could give rise to variation not necessarily depending on DNA sequence alterations. Major points include the naturally occurring mutation rate that might explain the difficulty in controlling weeds with single-target herbicides, in addition to the advances in plant breeding through intentional mutation. Moreover, allele expression biases are presented in polyploid species, as well as the implication of transposable element activity in intra-species variation.

Another major point refers to the reduced expression levels of a locus that correlates with DNA methylation, a process that has further been associated with phenomena such as paramutation, parental imprinting, and heterosis. Intriguingly, transposable element activity in cases like during environmental stress, has been implied to be controlled by DNA methylation and demethylation systems causing genome restructure, together with the fact that methylated nucleotides are themselves hot-spots for mutations.

E. Sinapidou (✉) · I.S. Tokatlidis
Department of Agricultural Development, Democritus University of Thrace, Pantazidou 193, Orestiada 68200, Greece
e-mail: sineva01@hotmail.com; itokatl@agro.duth.gr; itokatl@hotmail.com

E. Lichtfouse (ed.), *Genetics, Biofuels and Local Farming System*, Sustainable Agriculture Reviews 7, DOI 10.1007/978-94-007-1521-9_3,

Other major points involve histones, proteins responsible for DNA packaging and organization, that are involved in gene activation and silencing, for example during stress conditions or at different developmental stages. Lastly, some RNA molecules are implicated both in endogenous gene regulation and the control of invading genetic entities, which is particularly important when using biotechnological methods for the development of novel crops through the introduction of transgenes. Overall, epigenetic changes seem to happen more frequently and be reversible, whereas spontaneous DNA mutations are often random and more stable.

In conclusion, plant biodiversity can serve as a resource for sustainable agriculture. It is important for plant breeders to take advantage of the range of modern tools and knowledge of plant genomes, so that breeding is less a 'hit and miss' process, but a more precise strategy, where successful selection for crop improvement is increasingly supported by understanding the genetic variation underlying the phenotype.

Keywords DNA methylation • Epigenetic mechanisms • Genetic variation • Histone modifications • Mutation • Plant breeding • RNA molecules • Sustainable agriculture • Transposable elements

1 Introduction

Biodiversity, defined as the range and variability of animals, plants and microorganisms, is imperative for the ecosystem in terms of its structure and function. Plant diversity, although inadequately documented and poorly understood, is still the key for further development of natural products. At present, there are two main forces driving agricultural production. First, in view of the rising human population there is demand for increased crop production both for food and non-food products, such as fibre and bio-energy crops. Then, there are also expectations for cultivation practices to be sustainable and contribute to the resilience of ecosystems, as well as improve not only economic, but also social and environmental conditions.

Traditionally, increased crop production is depended mainly on fertilizes and pesticides. However, there is now a tendency to move from this chemical-based agriculture towards biodiversity as a global resource for optimizing agricultural production sustainably, and in an environmentally and socially responsible way. Crop genetic diversity could elevate production levels and nutritional variety in different agro-ecological conditions. Moreover, enhanced biodiversity in agrosystems can promote biodiversity in other parts of the environment that are related to the cropland, like water bodies and adjacent ecosystems including animal, plant and microbial biodiversity.

Plant biodiversity through evolution and crop improvement via breeding involves a dynamic interaction between natural genetic resources and management practices in an effort to ensure agricultural biodiversity in an ever-changing environment. It is a universal truth that for any living organism to survive it is imperative to be able to adapt to the environmental conditions it is exposed at. However, if natural adaptation itself is complex, i.e. the result of a massive amount of morphological and physiological traits affected by alleles at numerous loci, it is even more complex under breeding conditions. There, not only essential biological processes are to be performed efficiently, but more importantly production requirements and quality preferences should be satisfied.

In this sense, one could define plant breeding as the science to manage genetic variability in order to achieve certain goals. In other words, breeders first determine the goals of the breeding programme for a particular environment and then after exploring if adequate genetic variability is available, choose the correct breeding method to develop elite genotypes for the target environments. This practice may lead to the assumption that the continuous accumulation of desirable alleles in a few elite lines could in theory ultimately restrict the potential for further improvement. Such a scenario could be devastating in case conditions are different from the ones the elite genotypes were selected for.

Advances in the field of both genetic and epigenetic studies have been mounding data on a suite of mechanisms, often overlapping, which play a crucial role in causing alterations not just on a single-gene basis, but moreover on a whole genome scale, resulting in remarkable variability even in narrow gene pools. This review gives a concise picture of the genetic and epigenetic mechanisms leading to enhanced biodiversity in plant populations and discusses the various implications in crop improvement and sustainable agriculture through a number of examples.

2 Sources of Genetic Variation

Natural selection is responsible for the evolution of a species and its adaptation to novel environmental challenges. On the other hand, plant breeding is performed for improved production and quality as well as adaptation to environmental stress factors both abiotic, such as extreme temperatures, drought, salinity, and biotic, such as pests and diseases. For a breeder, the germplasm resources in a crop consist of wild relatives, weed races and landraces, of unimproved or purified cultivars used in the past or still in use locally and lastly of improved germplasm from breeding programmes frequently in commercial production. Advances in plant biotechnology have recently also rendered available genetic material far beyond the related genera of a crop.

The ultimate source of continuous genetic variation that is crucial both in natural adaptation and plant breeding stems from mutation, since selection itself either natural or artificial can only increase or decrease the frequency of alleles already existing in a population. The goal in the first part of this review is not to give an extensive presentation on the phenomenon, but rather to sketch a concise portrait of two major sources of genetic variation, namely mutations and transposable element activity, as well as their impact in creating variability in plant populations, which could be exploited so as to promote sustainable agriculture systems.

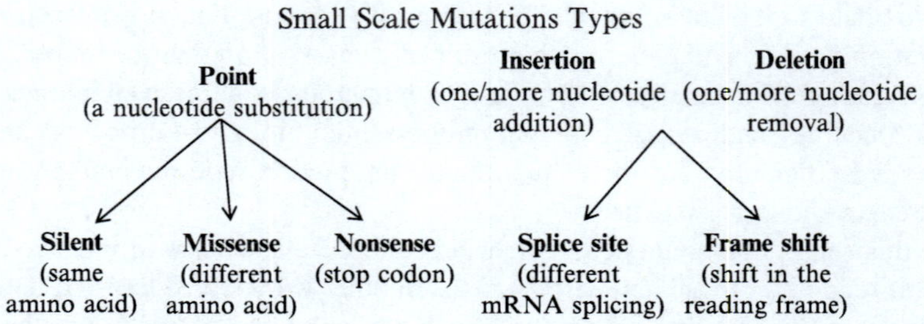

Fig. 1 According to DNA result, small scale mutations types include point mutations, insertions and deletions. These are further divided according to the result in the produced protein in silent, missense and nonsense mutations (point mutations), and in splice site and frame shift mutations (insertions and deletions)

2.1 Mutations Creating Genetic Variation

By effect on structure, mutations can be classified as small-scale mutations that affect one or a few nucleotides within a gene and large-scale mutations that change the number of chromosomes or the number of genes or the arrangement of genes in a chromosome.

The first class of small-scale mutations includes: (a) *Point mutations* often caused by chemicals or malfunction of the DNA replication resulting in the exchange of a single nucleotide for another (Freese 1959); (b) *Insertions*, that add one or more extra nucleotides into the DNA frequently caused by transposable elements so they can be reverted by the excision of the transposable element; and (c) *Deletions*, that remove one or more nucleotides from the DNA and can have the same effect as insertions, but are generally irreversible (Griffiths et al. 1993). A more detailed presentation of small scale mutations types appears in Fig. 1.

The second class of large scale mutations includes: (a) Mutations that cause changes in the structure of a chromosome; and (b) Mutations that cause changes in the number of chromosomes (Griffiths et al. 1993). Large scale mutations types are schematically presented in more detail in Fig. 2. Especially in the case of polyploidy, it could further be divided into *autopolyploidy*, when the multiplication of chromosome sets is within a species and *allopolyploidy,* when a hybrid is formed by the contribution of chromosome sets by two different species (Griffiths et al. 1993).

2.2 Variability Through Mutations and Its Implications in Sustainable Crop Management

Mutation is obviously a natural phenomenon; however, its frequency can be accelerated by human intervention. A recent study employing high-powered sequencing

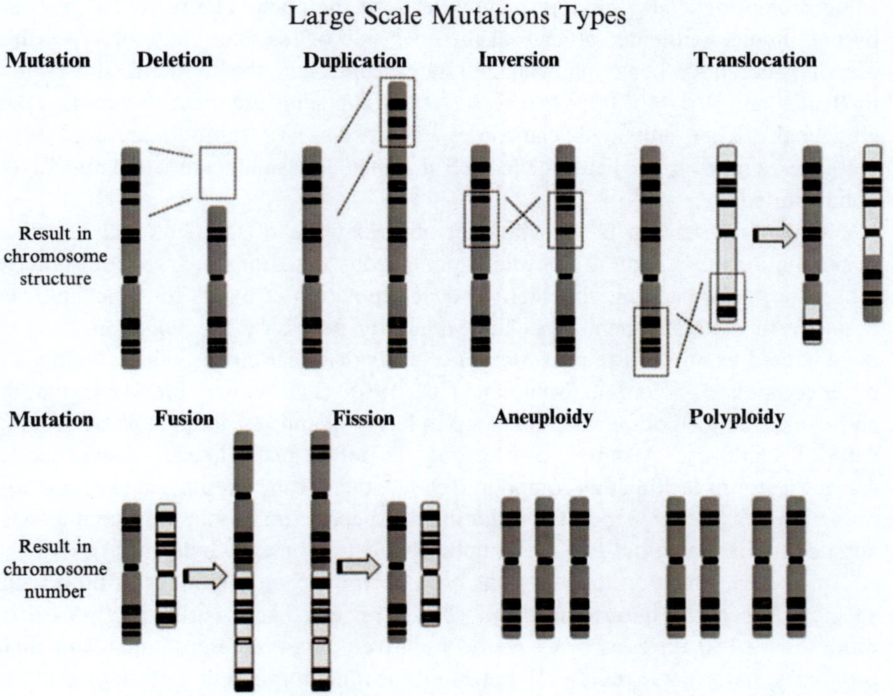

Fig. 2 Large scale mutations are divided in deletions, duplications, inversions and translocations that result in chromosome structure changes; and fusions, fissions, aneuploidy and polyploidy that result in chromosome number alterations

uncovered the rate of de novo mutations in the complete nuclear genomes of five *Arabidopsis thaliana* mutation accumulation lines that had been maintained by single-seed descend for 30 generations; that finding could help illuminate both the evolutionary history of plants and the selective breeding efforts in the future (Ossowski et al. 2010). In that study, the careful comparison of the full genome of each of the five *Arabidopsis* lines under examination showed that only about 20 base pairs had mutated in each line. Although this could sound as a modest mutation rate, it can have a big impact spread across thousands of seeds/individuals in one generation. Coupled with the fact that mutations occurred at about the same rate through the genome and not just in specific parts might help explain for example why efforts to keep some plants at bay with single-target herbicides are often only briefly successful.

Indeed, herbicides that target the enzyme acetolactate synthase (ALS) provide a characteristic example for the emergence of herbicide resistant plants caused by the selection pressure of the wide and sole application of such herbicides, which killed the susceptible plants letting the resistant ones reproduce without any competition. The mode of action of these herbicides is that they bind to the ALS enzyme, so it is no longer available for normal synthesis of certain amino acids resulting in some protein deficiency and ultimately death of the plant. The selection pressure by continuous herbicide use caused the evolution of resistant weed biotypes that carry an altered ALS enzyme, which no longer binds with the herbicide. In a review by Tranel and Wright (2002) five conserved amino acids are presented in the ALS enzyme that when substituted can confer resistance to ALS-inhibitor herbicides and there are 12 more amino acids in the ALS enzyme that result in a resistant phenotype if substituted.

Human intervention is not apparent only in ways of unintentional selection in populations that naturally develop spontaneous mutations, but geneticists have moreover performed experimentally induced mutation in plants for practical uses in order to generate themselves the variability necessary for selection. A well documented example is that of *Melilotus albus* or sweet clover, which contains a bitter compound –coumarin (Smith and Gorz 1965). This reduces blood coagulation and is used as medicine, when taken up in larger quantities, though, as by grazing cattle, it could cause continuous bleeding. Therefore plant breeders have tried to select for strains lacking this compound, while others made attempts to use mutation induction as a tool to get rid of it, imitating the success story of finding spontaneous, simple inherited alkaloid-free lupin mutants (Gustafsson and Gadd 1965).

Since then, mutation breeding has been performed on a number of crop plant species. Recently, Ahloowalia et al. (2004) reviewed the worldwide release of more than 2,250 varieties that have been derived either as direct mutants or their progenies. In the same review the impact of the mutation-breeding strategy on food production and quality enhancement during the past 70 years is discussed. In case studies presented, it is shown that the mutation-derived cultivars have traits, like modified oil and protein content, resistance to drought and diseases, even enhanced uptake of specific metals, which could have an impact on sustainable agriculture management. For example, short height genotypes in rice, wheat and corn resulted in higher yields through high planting density and relative increased doses of nitrogen application; early maturity of mutant cotton in Pakistan allowed a second crop in the same season, i.e. wheat, leading to higher yields; and changed oil profile of mutated sunflower allowed its more efficient use for food and industrial purposes and led to massive increase in the areas planted with this crop. It is notable that in China alone, from 1966 to 1990 about 80 mutant rice varieties were released that present novel characteristics of direct importance in rice production and of potential quality and nutritional value (Chen et al. 2006). Such varieties already used for commercial production include, leaf colour mutants that could facilitate the rapid identification of varietal purity of hybrid seeds and resistant starch mutants that are helpful in dietary prevention of diabetes.

Apart from small-scale or gene mutations, however, there are also large-scale or chromosome mutations that affect the structure or number of chromosomes. Especially the case of polyploidy is noticeably pervasive in plants with estimates suggesting that 50–80% of all angiosperms are polyploids (Otto and Whitton 2000) and perhaps all angiosperms had experienced polyploidy in their evolutionary history (Wendel 2000). On the contrary, polyploidy is rather infrequent among gymnosperms reaching close to 5% and can be as low as 1.5% in conifers (Khoshoo 1959).

Polyploidy is believed to have provided the fuel for an explosion in the evolution of new genes and subsequent speciation during the early evolution stages in animals and plants (Soltis and Soltis 1995). In fact, recent genomic studies suggested that a number of plants and animals are paleopolyploids, i.e. ancient polyploids, that later became diploids by sequence divergence between duplicated chromosomes (Wolfe 2001). Sequence-based approaches have demonstrated that each duplication event has been followed by a subsequent loss of much of the duplicated material, which could explain the deviation in colinearity among relatively closely related plants, like the cereals (Paterson et al. 2003). In wheat, loss of parental genes and fragments was demonstrated in synthetic allopolyploids, both immediately after polyploid formation in synthetic allotetraploids (Kashkush et al. 2002) and in later generations in synthetic allohexaploids (Ma et al. 2004). Equally intriguing is the differential retention of the duplicated genes. In Arabidopsis, it seems that some classes of genes have been preferentially retained and others lost (Blanc and Wolfe 2004). Looking at possible functions of retained duplicates, Thomas et al. (2006) suggested that the likelihood of duplicate retention correlated with the number of functional interactions among the gene products.

Furthermore, genome changes that accompany polyploid formation also have an impact on gene expression levels with the consensus being that gene expression in polyploids is usually non-additive and repression or silencing is often the case (Adams et al. 2004; Wang et al. 2006). Indeed, an experiment was designed using microarrays in order to identify thousands of genes that are differentially expressed between two tetraploid Arabidopsis entities, i.e. a synthetic allotetraploid *A. thaliana* line and a natural allotetraploid *A. arenosa*; later, a comparison of the average expression levels of these genes in both the tetraploid parents as well as two newly synthetic allotetraploids produced by them, showed that more than 94% of the repressed genes in the new allotetraploids matched the genes that were expressed at higher levels in only one of the parents (*A. thaliana*) and not the other (*A. arenosa*). Homeologous expression biases have also been measured in wheat (Mochida et al. 2003) and in cotton (Adams et al. 2003; Adams et al. 2004) and although the functional consequences of this phenomenon are not yet clear, its elucidation and possible control would be an appealing and most powerful tool in the hands of plant breeders. For example, an earlier detailed quantitative trait loci (QTL) mapping study in tetraploid cotton showed that most QTLs influencing fibre quality and yield were located on the D subgenome, which surprisingly derived from the ancestor that does not produce spinnable fibres, suggesting that the D genome loci have been "recruited" for the synthesis of fibre subsequent to polyploid formation (Jiang et al. 1998).

Polyploid crop plants mentioned by Stebbins (1950) include potato, coffee, banana, peanut, tobacco, wheat, oat, sugarcane, strawberry and plum. Plant breeders have long been interested in the artificial induction of polyploidy which was boosted by the discovery of the "colchcine method", so that polyploids could appear in clover, turnip, spinach, apple, grapes, sugarbeets, tea, watermelon, various forage grasses as well as ornamental plants (Elliot 1958). Yet it has been impossible to predict either which diploid genomes when merged in the same nucleus will coexist stably, or the phenotype of the resulting offspring. The genomic revolution and the divergent suite of modern tools available have expanded the knowledge of plant genomes and gave useful insights of the dynamics of polyploid plant genomes and the spectrum of phenomena that accompany polyploidy in model and crop plants. This could lead to a transformation from the 'hit and miss' process it is today to a more precise strategy where successful selection for crop improvement is increasingly supported by the understanding and unravelling of the complexities of genetic variation that underlies the phenotype.

2.3 Transposable Elements Creating Genetic Variation

Numerous data from a variety of organisms has been accumulating on the diversity and abundance of transposable elements (TEs) in natural populations, steadily building a picture of their undoubtedly significant role in genetic variation.

Transposable elements are classified in two major groups based on their mode of transposition. Class I of transposable elements, called retrotransposons, use reverse transcriptase to transpose via an RNA intermediate and include long terminal repeat retrotransposons (LTR) bearing direct repeats at the ends of the element, as well as long and short interspersed elements (LINES and SINES, respectively), that lack such repeats (reviewed by Slotkin and Martienssen 2007). Notably, retrotransposons are able to duplicate following each transposition and hence expand genomes. In angiosperms with large genomes more than 50% of the nuclear DNA has been shown to consist of LTR-retrotransposons and other repeats (Bennetzen et al. 2005).

On the other hand, Class II of transposable elements, called DNA transposons, transpose from DNA to DNA through the action of a protein termed transposase, which after recognizing the terminal inverted repeats (TIRs) flanking the transposable element, excises it out of the one position and integrates it into another. The transposition could be a "cut and paste" one if the opening created in the first position is closed without a copy of the transposable element left behind, or a "gap repair" when it is filled with a transposable element's copy (reviewed by Slotkin and Martienssen 2007). Recently, a different mode of duplication, via a rolling-circle mechanism similarly to some prokaryotic transposable elements, has been recorded in newly identified DNA transposons called helitrons, which do not contain terminal repeats and can be quite large. In addition, included in this class are also transposable element categories found in high copy numbers, like the miniature inverted-repeat transposable elements (MITEs) and the *Mutator*-like DNA elements (MULEs) (reviewed by Bennetzen 2005). Finally, depending on their self-sufficiency, both transposable element classes can be further divided into autonomous elements, if they themselves encode the products necessary for their transposition and non-autonomous ones, in the opposite case (reviewed by Slotkin and Martienssen 2007).

Active transposable elements are highly mutagenic causing changes (a) at gene level, mediated by insertions, excisions and/ or deletions either in the coding regions, the non-coding regions or the regulatory regions of genes; and (b) at genome level through alteration in genome size and reorganization in genome structure (reviewed by Kidwell and Lisch 2002). An interesting fact mentioned in Kidwell and Lisch's review (2002) is that, the frequency of mutations caused by transposable elements can exceed that of strong chemicals and radiation. It has been proposed that upon entering a genome, either through cross-fertilization or horizontal transfer, a transposable element usually duplicates freely until it produces itself a signal responsible for its epigenetic silencing (Slotkin et al. 2005). Similarly, release of constrains on transposable elements could equally lead to bursts in transposable element activity resulting in genome size increases. For example, 75% of the maize genome is said to consist of transposable elements (review by Kazazian 2004).

In terms of evolution, transposable elements could play an important role in generating the increased variation required for selection to lead towards individuals able to survive under stress. Indeed, transposable elements reactivation facilitated by the release of epigenetic silencing mechanisms (see Sect. 3) has been demonstrated in response to UV light, temperature, radiation, wounding, pathogen infection and polyploidization (reviewed by Capy et al. 2000).

Even more intriguing, though, has been the implication of transposable elements in intraspecies variation. Thus, intergenic regions between two different inbred strains of maize have been shown to be essentially unrelated due to alternate insertion of LTR-retrotransposons and helitrons (Wang and Dooner 2006). In fact, it has been suggested that as many as 10,000 genes or gene fragments may be unshared between maize inbred lines probably because of helitron activity (Morgante et al. 2005). In this case, it is notable that some lack of homology is the result of expressed gene fragments that have been captured within helitrons. Although there are puzzling differences between the species studied so far, i.e. maize, rice and Arabidopsis, this "gene" capture is not an uncommon phenomenon (Sweredoski et al. 2008); together with the observation that some helitrons in maize were found to produce transcripts which fused fragments from different genes, this hints at the possible role of helitrons in structural and even functional diversity in plant genomes through exon shuffling and ultimately emergence of new proteins (Morgante et al. 2005). In a similar manner, *Mutator*-like DNA elements (MULEs) have also been demonstrated to acquire gene fragments in maize and rice (reviewed by Bennetzen 2005). In rice, in particular, more than 3,000 MULEs containing genic sequences have been identified, of which 23% carried fragments from different genes and at least 5% were found to be expressed (Jiang et al. 2004).

The constant emergence of fascinating data about transposable elements has been revealing missing pieces from the mutations puzzle; however, more studies still remain crucial if geneticists are to further elucidate the role of the different transposable elements in plant evolution and develop means to utilize it towards more sustainable agro-systems.

2.4 Variability Through Transposable Elements and Its Implications in Sustainable Crop Management

Undoubtedly, the combination of classical genetic analyses with modern molecular approaches has shed considerable light on the way towards a better understanding of genetic as well as epigenetic changes which constitute contributing mechanisms to newly developed genetic diversity. This additional knowledge has changed the previous belief that genome is static; instead it is now well known that it is flexible, plastic and dynamic, modifying itself in response to environmental stress. It is thanks to such plasticity that gene pools, even theoretically small ones, have inherent mechanisms to provide a continuing source of novel genetic variability (Tsaftaris et al. 2008).

The essential contribution of these molecular functions toward a wider genetic diversity was clearly demonstrated in a number of studies concerning gene pools deemed to carry too narrow or even negligible genetic variability. In other words, significant genetic heterogeneity was found within three of the four sunflower (*Helianthus anuus*) inbred lines studied by Zhang et al. (1995). Also, Olufowote et al. (1997) investigating possible intra-cultivar variation in rice (*Oryza sativa* L.), found significant variation within 32 out of the 71 studied cultivars; landrace cultivars were more heterogeneous but variation was detected even within cultivars that had been purified by phenotypic evaluation. Moreover, Rasmusson and Phillips (1997) detected significant genetic variation within barley (*Hordeum vulgare* L.) cultivars, caused by various genetic and epigenetic changes. Similarly, Gethi et al. (2002) detected significant genetic heterogeneity within all of the six maize (*Zea mays* L.) inbred lines they studied. Finally, in flax (*Linum usitatissimum* L.) rapid modifications in the genome were observed during plant development under stress conditions, associated with changes in gene expression (Cullis 2005).

The sheer volume of accumulating data highlights the importance of genetic procedures, like mutations and transposable element activity, in providing new variation and expanding the existing biodiversity. Obviously, such endogenous genetic mechanisms constitute essential evolutionary processes in nature, which the on-going climate changes may intensify. Nevertheless, they can also have practical implications in agricultural systems. For instance, to date, sustainable agriculture relies on native or locally adapted cultivars, i.e. landraces, or on cultivars created for intensive agriculture. In the course of time, these valuable cultivars may lose their adaptation and even their identity and healthiness, due to accumulation of undesirable changes. However, plant breeding could, on the other hand, exploit this within landraces variation and have the opportunity to create new cultivars for the target environments in more sustainable agricultural systems. A good paradigm is the study by Tokatlidis et al. (2010) who employed a novel breeding method, the honeycomb breeding (Fasoula and Fasoula 2002), on two currently cultivated dry bean landraces. Selection applied either at a greenhouse to establish heat stress conditions or at the typical for the crop open-field conditions environment, led to single-plant progeny lines tolerant to heat stress and exhibiting up to 38% higher yield than the original material.

Even cultivars assumed fairly homogeneous may accumulate changes exploitable via breeding to either avoid degeneration or accomplish cultivar improvement, on condition that an appropriate breeding procedure is employed, i.e. one that accentuates the limited variation. Such a method is the honeycomb breeding (Fasoula and Fasoula 2002). Its major principle is application of selection at extremely low plant densities, based on the notion that absence of competition enhances phenotypic expression and hence the differentiation among genotypes is maximized. Indeed, exploitable through honeycomb selection, genetic intra-cultivar variation was detected in maize (Tokatlidis 2000), soft wheat (*Triticum aestivum* L.) (Fasoula 1990; Tokatlidis et al. 2004, 2006), cotton (*Gossypium hirsutum* L.) (Fasoulas 2000; Tokatlidis et al. 2008, 2011) and soybean [*Glycine max* (L) Merr.] (Fasoula and Boerma 2005, 2007).

The aforementioned studies illustrated the possibility to considerably upgrade the yielding capacity of cultivars, i.e. up to 20% in wheat (Fasoula 1990; Tokatlidis et al. 2006) and up to 12% in cotton (Fasoulas 2000; Tokatlidis et al. 2011). Particularly for sustainable agriculture, the following findings from honeycomb intra-cultivar selection are noteworthy. First, seed quality, i.e. protein and oil content in soybean cultivars (Fasoula and Boerma 2005, 2007), as well as fibre quality in cotton cultivars (Tokatlidis et al. 2008) were found improvable. Second, the potential to cope with drought was demonstrated in wheat and cotton cultivars, based on the physiological trait of 'carbon isotope discrimination' which reflects plant discrimination against $^{13}CO_2$ and assimilation of $^{12}CO_2$ during photosynthesis, and is connected with water use efficiency (Tokatlidis et al. 2004, 2008). Last but not least, selection within the susceptible cotton cultivar *Sindos 80* resulted in two lines tolerant to *Verticillium wilt* (Fasoulas 2000). These results also underlined the principal of the non-stop selection process, proposed by Fasoula and Fasoula (2002), which refers to the constant improvement of the crop yield and quality of released and adapted cultivars and can offer great advantages to sustainable agriculture. In maize, long-term selection studies for modified oil and protein showed that exploitable variation exists even after 90 generations of selection (Dudley and Lambert 2004).

3 Epigenetic Mechanisms as a Source of Variation

As it was mentioned in the beginning, genetic variation is the prerequisite for evolutionary progress, implying that genetic uniformity should in theory constrain severely the evolutionary potential of a given population. However, recent studies have demonstrated how novel patterns of gene expression may occur via a suite of epigenetic mechanisms even in the complete absence of sequence-based variation. In this sense, a definition proposed by Rapp and Wendel (2005) to describe the term 'epigenetics' encompasses the heritable alterations of phenotype, morphological or molecular that are not accompanied by changes in either the coding sequence of a gene or the upstream promoter region.

Although the concept of epigenesis is not something new – Tsaftaris et al. (2008) state it was initially proposed by Aristotle – only recently and aided by molecular techniques has the presentation of indisputable data towards this direction been achieved. Substantial and detailed reviews on this subject are now numerous, so this epigrammatic account may sound superficial to those already well-informed in the field of epigenetics. However, the purpose in the second part of this review is to draw a simple sketch of the developments in this field for the benefit of those who are less familiar with it, as well as the impact of such developments in sustainable crop management.

Three different ways have been reported through which epigenesis is expressed, namely DNA methylation, chromatin modification and RNA molecules.

3.1 DNA Methylation

The best-described epigenetic mechanism is DNA methylation, the chemical modification of cytosine (C) DNA bases with a methyl group (patterns of cytosine methylation is reviewed by Castiglione et al. 2002). In plants, DNA methylation would appear to be more complex than in other organisms. Thus, animals and fungi predominately methylate cytosine residues located in a CpG sequence context, with "p" referring to the phosphodiester bond between the cytosine and the guanine. In contrast, plants can encompass methylation at 'symmetric' sites, i.e. CpG and CpNpG where N is any nucleotide, hence methylatable cytosines appear in pairs on opposite strands, as well as 'asymmetric' sites (CpNpN), though this is less frequent (Grant-Downton and Dickinson 2005). More importantly, cytosine methylation is not static; instead tissue- and developmental stage-specific as well as stress-induced variation has been recorded (reviewed by Tsaftaris et al. 2008).

Generally, methylation of the 5′-carbon of the cytosine aromatic ring (Fig. 3) in a large number of sites at a locus, i.e. hypermethylation, correlates with a reduction in expression of the locus or its complete silencing. A most famous example is that of the *Lcycloidea* locus in *Linaria vulgaris*, where loss of expression results in 'peloria', i.e. a phenotype with more radially symmetrical flowers rather than strictly bilateral ones; naturally occurring variants with radial symmetry were found to have a methylated and transcriptionally silent *Lcyc* gene (Cubas et al. 1999).

At the chromosome level, genomes are organized into euchromatin, gene-rich as a rule, and heterochromatin, which is repeat-rich (Hall and Grewal 2003). The *Arabidopsis thaliana* five-pair chromosome genome contains considerable amounts of heterochromatin, which is repeat-rich and mostly located in the centromeric and pericentromeric regions (Fransz et al. 1996). Whole-genome, high resolution mapping of cytosine methylation in *A. thaliana* has revealed that such modifications co-locate with repeat sequences and centromeric regions, whereas fewer than 5% of expressed genes where shown to have methylated promoters. Intriguingly, it was also shown that about a third of genes that are both highly transcribed and constitutively expressed displayed methylation in their open reading frame,

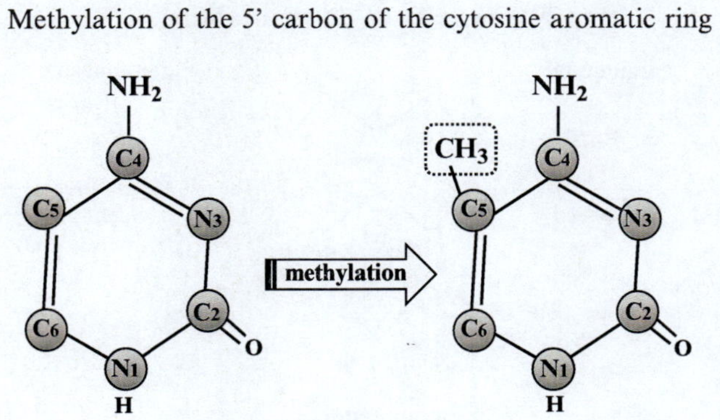

Fig. 3 Methylation of the 5' carbon of the cytosine aromatic ring in a large number of sites at a locus, i.e. hypermethylation, usually results in a reduction in expression of the locus

where by contrast, genes with methylated promoters had lower expression levels and frequently tissue-specific expression patterns (Zhang et al. 2006; Zilberman et al. 2007).

It is notable how epigenetically silent expression states can show remarkable stability through generations, supporting the concept of the epigenetic allele or "epiallele" that is responsible for a heritable difference in expression without harbouring a change in DNA sequence (Henderson and Jacobsen 2007). For example, hypermethylation at the *SUPERMAN* locus in *A. thaliana* that is involved in floral development is stable through many generations of inbreeding, though it can be unmethylated at a frequency of approximately 3% per generation (Jacobsen and Meyerowitz 1997).

Additionally, another notable characteristic of some epialleles is their ability to cause heritable change in the expression of other homologous sequences, i.e. alleles (Hall and Grewal 2003). One example is paramutation, an allele-dependent transfer of epigenetic information that results in the heritable alteration of the expression of one allele by another (Chandler and Stam 2004). The allele that induces the change is termed 'paramutagenic', whereas the other is referred to as 'paramutable' and later 'paramutated' when the phenomenon has taken place and is then symbolised with an apostrophe after the gene name (Hollick et al. 1997). One of the best-studied examples is the maize locus *b1* (Fig. 4), which encodes a transcription factor required for activation of the pigment anthocyanin pathway (Chandler and Stam 2004). The *B-I* paramutable allele of *b1* has an upstream tandem-repeat region and spontaneously gives rise to B′, i.e. silenced epialleles, at a low frequency (Patterson et al. 1993); they are more heavily methylated at cytosine bases in the repeat regions and are less frequently transcribed (Stam et al. 2002). When the B′ epiallele is brought together with a new copy of *B-I* allele by crossing, the *B-I* allele is paramutated to a silenced B′ state at a 100% frequency (Stam et al. 2002). It should be mentioned, however, not all paramutants are so efficient in establishing this new epigenetic state, nor are they so stable (review by Louwers et al. 2005).

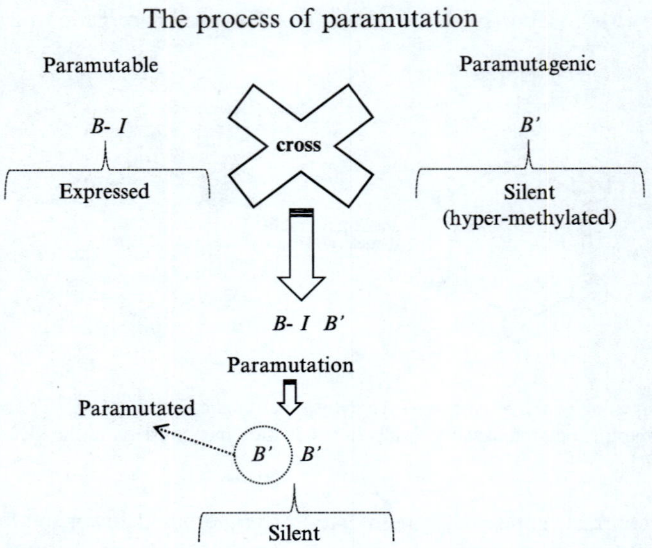

Fig. 4 In paramutation, the paramutagenic allele *B′* causes an alteration to the expression of the paramutable allele *B-I* when they are brought together by crossing. The previously expressed paramutable allele *B-I* in the presence of the silent, paramutagenic allele *B′* in the hybrid is paramutated and becomes silent, similarly to *B′*

DNA methylation has also been shown to be involved in parental imprinting, a process where only one of the parental alleles is active in an offspring (De Chiara et al. 1991). Since imprinted genes are active during gametogenesis and seed development, a short description of reproduction in plants is presented. In flowering plants, the male gametophyte, pollen, usually consists of a large vegetative cell that contains two sperm cells, each haploid (McCormick 2004). On the other hand, the female gametophyte usually consists of a seven-celled embryo sac, containing the two female gametes (Yadegari and Drews 2004), both of which, the egg and central cell, fuse with one sperm cell during the process of double fertilization (Weterings and Russell 2004). The now diploid fertilized egg cell produces the embryo, called also sporophyte, while the fertilized central cell leads to the formation of the endosperm that is thought to provide nutrition to the developing embryo (Haig and Westoby 1989). Due to the fact that the central cell contains two polar nuclei that fuse with the sperm nucleus, the endosperm is usually triploid (Baroux et al. 2002). In this sense, non imprinted genes could be expressed in the endosperm only in relative levels of 3x and 0x, i.e. expressed or not expressed, whereas imprinted genes could be expressed at a level of 2x or 1x depending on whether the maternal or paternal allele is active, respectively, and so imprinting could be a way to fine tune gene expression levels (Kermicle and Alleman 1990).

Through RT-PCR experiments Kinoshita et al. (1999) demonstrated that for the *MEA* locus from *A. thaliana*, which is transcribed in the early endosperm after fertilization only the maternally inherited alleles are active during seed development. Also reported to be parentally imprinted in *A. thaliana* is the gene *FWA*, whose 5′

region contains tandem repeats that are hypermethylated in all tissues except the ones where the maternal allele is present, namely the central cell and endosperm (Kinoshita et al. 2004). In the maternal gametophyte, the imprinting of both *MEA* and *FWA* genes is dependent on the activity of the *DEMETER* (*DME*) gene, which codes for a DNA glycosylate and has been suggested to selectively excise methylcytosine (Choi et al. 2002). In the past, DNA glycosylates have been reported to function in DNA repair and act as demethylases (Jost et al. 1995). Intriguingly, *MET1*, a methyltransferase in *A. thaliana* (Kankel et al. 2003) is acting antagonistically to *DME* in the control of maternal *MEA* expression, hinting that DME may activate *MEA* through demethylation (Xiao et al. 2003). This was further supported by the finding that the repressor of silencing *ROS1* gene, a factor involved in transcriptionally gene silencing, also participates in active DNA demethylation by a 5′-methylcytosine pathway (Morales-Ruiz et al. 2006).

Furthermore, even the phenomenon of heterosis, which describes the superior performance of heterozygous hybrid plants over their homozygous parental inbred lines, has been associated with epigenetic regulation and particularly DNA methylation levels. Having performed several studies in maize hybrids and their progenitors, Tsaftaris et al. (2008) pointed out that hybrids were generally less methylated than their parental inbreds and that was also true for the low-yielding inbreds. Recently selected inbreds, however, in particular those selected for high and stable yield under extremely low-density planting to preclude stress, i.e. in absence of competition, displayed low methylation levels. It should be noted that, minimal stress conditions through less dense planting have additionally resulted in activation by reduced methylation of the *Ac* element in maize (Tsaftaris and Kafka 1998).

Intriguingly, it has been shown that transposable elements (TEs) could also be managed by methylation and demethylation systems (Zhu et al. 2007). Since the first hint about their importance in shaping genomes by McClintock (1984), their modifying effects on the genome have been extensively discussed. The outcome of their activity in a structured genome could range from neutral to deleterious, so a mechanism that provides protection against high levels of transposition is crucial. A number of plant genomes tolerate such sequences, implying there are benefits in these cases. Indeed, integration of transposable elements within or close to coding genes has been shown to generate novel variation in plant genomes as previously discussed in Sect. 2.3.

Recently, Lippman et al. (2004) presented data pointing to the *FWA* locus imprinting being controlled by retrotransposon-derived tandem repeats at the 5′ region of the gene. It is tempting to assume that the dynamic control of transposable element activity is critical in facilitating a way for the genome to efficiently respond to developmental and environmental challenges through a restructure, which is supported by the increased mutation rate following the activation of transposable elements (Wessler 1996; Takeda et al. 1999). For example, cold stress was demonstrated to induce hypomethylation and transposition of a transposon (*Tam-3*) in *Antirrhinum majus* (Hashida et al. 2006). Also, methylation patterns have been associated with specialized phenomena such as the sex determination in plants (reviewed by Castiglione et al., 2002).

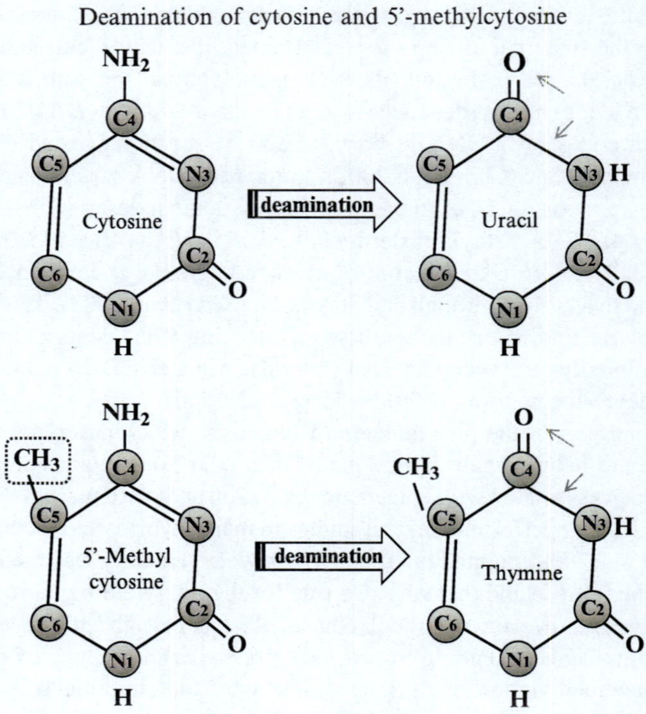

Fig. 5 When cytosine spontaneously deaminates in uracil, in DNA it is recognized as an erroneous base and is replaced by a new, correct cytosine. In contrast, the spontaneous deamination of 5′-methylcytosine to thymine is not easily detectable, so that incomplete and inefficient restoration of cytosines lead to more persistent mutations

To conclude, there seems to be a dynamic control of the genome depending on its methylation patterns, including heritable differences in expression without necessarily any actual changes in DNA sequences, as presented in the cases of imprinting and paramutation, which should be taken into consideration in the sustainable management of new cultivars, as well as when breeding for improved genotypes. However, in terms of evolution, methylation is an active player not only due to its local and seemingly immediate effects on gene expression, but moreover its more indirect consequences brought about by the release of transposable elements. Notably, cytosine methylation could itself be mutagenic, since evidence show that the mutational frequency of methylcytosine is substantially greater than unmodified cytosine (Rideout et al. 1990). In other words, methylated cytosine is a hot spot for mutations because they spontaneously deaminate to thymidines (Jones et al. 1992). As presented schematically in Fig. 5, unlike cytosine that spontaneously deaminates to uracil, which in DNA is recognized as an erroneous base and is replaced by a new, correct cytosine, the spontaneous deamination of 5-methylcytosine to thymine is not easily detectable, so that incomplete and inefficient restoration of cytosines lead to more persistent mutations (Shen et al. 1994).

3.2 Histone Modifications

Although the DNA sequence presents information in a linear fashion, deprived through experimental manipulation of any attached proteins, it seems that its association with such proteins could be imperative in unravelling the suite of mechanisms involved in how the sequence operates. In other words, there are certain proteins that enable the organization and packaging of DNA into functional chromosomes, which is a necessary step if the large genomes of eukaryotes are to fit inside the minute cell nuclei.

These proteins are called histones. They are basic, positively charged proteins, due to the abundance of positively charged arginine and lysine residues whose free amino group on their R group attracts protons (H^+), and thus ideal for binding tightly to the negative-charged phosphate groups of DNA, forming a DNA-protein complex referred to as chromatin (Van Holde 1988). There are five classes of histones presented schematically in Fig. 6, namely H1/H5, H2A, H2B, H3 and H4, that can be divided in two major groups, the core histones H2A, H2B, H3 and H4, and the linker histones H1 and H5 (Bhasin et al. 2006). Two of each of the core histones, in other words two H2A-H2B dimmers and a H3-H4 tetramer, assemble in an octameric nucleosome core particle, around which approximately 150 base pairs (bp) of DNA are wrapped in two turns (Luger et al. 1997). Then, the linker histone H1 binds the nucleosome and the entry and exit sites of the DNA, locking it in place, while an extra piece of DNA, termed linker DNA, joins adjacent nucleosomes together (Van Holde 1988). Repeated nucleosomes with intervening linker DNA

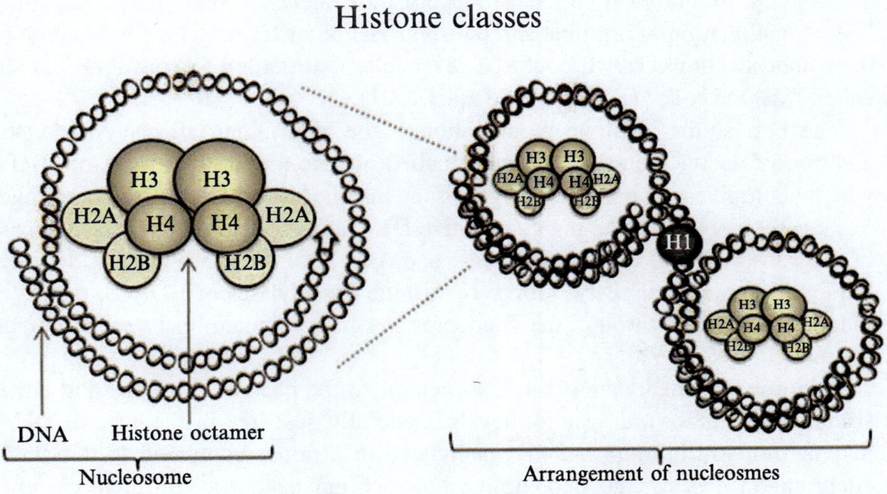

Fig. 6 Histones are divided into: (**a**) core histones H2A, H2B, H3 and H4 that form an octameric core particle, made by two H2A-H2B dimmers and a H3-H4 tetramer, and (**b**) linker histones H1/H5. DNA is wrapped around the histone octamer to form a nucleosome. Then, linker histone H1 binds the nucleosome and with an extra piece of DNA, called linker DNA, joins adjacent nucleosomes together

form a 10 nm-fibre that has a packing ratio of about five to ten (Felsenfeld and Groudine 2003). Still chromatin being too long to fit into the nucleus, it is further coiled in an even more compact fibre, the 30 nm-fibre (Woodcock 2005).

Histone proteins are highly conserved among eukaryotes, which highlights their importance in biological reactions in the cell nucleus, although some variant forms have been reported. These, however, usually carry out specific functions of the chromatin metabolism; for example H2A variant H2A.X mark DNA regions undergoing repair (Paull et al. 2000). Intriguingly, in tobacco, down-regulated by antisense linker histone H1, variants H1A and H1B showed phenotypic alterations like reduced plant size and aberrant floral development, but even more importantly a dramatic effect on male meiosis that greatly reduced pollen fertility (Prymakowska-Bosak et al. 1999). Moreover, a mutation in the core histone H2A in Arabidopsis *rat-5* gene increased the resistance of the genome to T-DNA integration, suggesting a role to genome protection from invasive, foreign DNA (Mysore et al. 2000).

Whilst the nucleosome is a very stable protein-DNA complex, it is not static; instead through a number of structural re-arrangements, nucleosomes can inhibit or enable transcription. In other words, modifications that may affect histone-DNA (Cosgrove et al. 2004) and histone-histone (Ye et al. 2005) interactions have been associated with either gene silencing or activation. The common nomenclature for histone modifications is: first, the name of the histone, e.g. H3; then, a single letter abbreviation for the amino acid and a number stating its position in the protein, e.g. K4 for the 4th Lysine from the N-terminal; and finally sometimes, the type of modification: me, for methylation, P, for phosphorylation, Ac, for acetylation and Ub, for ubiquitination. The core of histones H2A and H2B, as well as the long N-terminal "tails" of histones H3 and H4 that protrude from the nucleosome can be subjected to modifications, such as acetylation, methylation or ubiquitination of lysine, methylation of arginine and phosphorylation of serine. The combination of these modifications, which could be reversible, is proposed to constitute the so-called "histone code" (Jenuwein and Allis 2001).

The best studied histone modification is the N-terminal tail acetylation, the addition of acetyl groups by enzymes called histone acetyltransferases or HATs, which neutralizes the positive charges of the histone surface, reducing the strength of its association with the highly negative DNA and so enabling increased access of the transcriptional activity (Jenuwein and Allis 2001). On the contrary, deacetylation, the removal of acetyl groups by histone deacetylases or HDACs, results in reduced expression through the condensation of euchromatin to heterochromatin (Turner 2000) (Fig. 7).

Additionally, methylation can also neutralize the charge of lysines and either stimulate or hinder transcription levels. Generally, just like in the case of DNA, heterochromatin is more heavily methylated in histones compared to the active euchromatin region. One step further, heterochromatization of different chromosome regions depends not simply on the number of methylated histones, but more importantly on the number of methylated lysines in each histone, as well as on the degree of their methylation (mono-, di-, tri-) and that of the attached DNA (Dou et al. 2005; Tsaftaris et al. 2008). It has been demonstrated that di-methylation of core histone H3 on the lysine residues at positions 9 (H3K9) and 27 (H3K27)

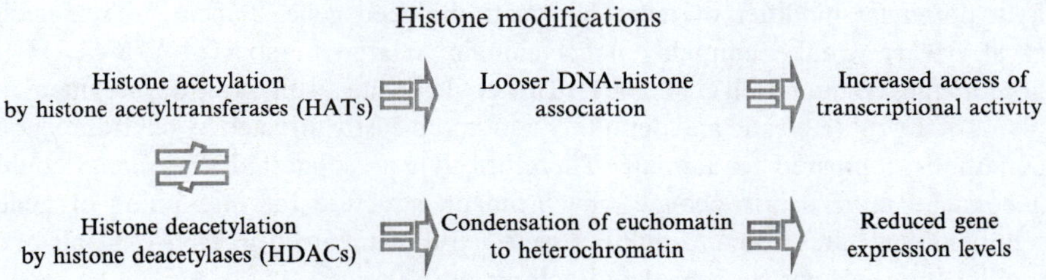

Fig. 7 Acetylation and deacetylation, i.e. the addition or removal of acetyl groups by enzymes, of histones can cause modifications that affect gene expression

represses transcription, which, however, is not true in the case of H3K4 where tri-methylation actually activates transcription (Zhang et al. 2007).

In a review of epigenetic regulation of stress responses in plants, Chinnusamy and Zhu (2009) present examples of gene expression repressed through histone deacetylases (HDACs), as well as examples of stress-responsive genes activation by the interaction of histone acetyltransferases (HATs) with transcription factors notably in response to biotic and abiotic stress stimuli. At the same time an example for the control of other physiological processes could be the vernalisation in plants, i.e. the competence to flower in the spring by exposure to low temperatures, that depends on repression of *FLC* genes after prolong exposure to cold. They code for a transcriptional regulator that represses a set of genes required for transition of apical meristem to a reproductive fate (Sheldon et al. 1999). Downregulation of the *FLC* locus remains epigenetically stable and has been correlated with increased core H3 histone di-methylation of lysine 9 and 27 (H3K9, H3K27) in domains of the locus (Bastow et al. 2004).

Although histone methylation plays an important role in condensing or relaxing chromatin structure, it has been suggested that a methyl group is relative small to neutralize itself the charge of lysine and arginine residues so much as to affect chromatin structure; instead it possibly creates binding sites for proteins that recognize the methylated residues (Tsaftaris et al. 2008). Indeed, there are enzymatic proteins that have binding domains, like the *bromo* domain for acetylation and the *chromo* domain for methylation, capable of recognizing modified histone residues and facilitating a number of histone alterations (Rusche et al. 2003).

Furthermore, protein complexes that have an active role in the regulation of developmental stages have been identified in plants and the Polycomb group (PcG) proteins present a well documented example (Alvarez-Venegas et al. 2003). They are best known for keeping genes silenced by a repressed chromatin state, which nevertheless is not static throughout development, but rather has a much more dynamic role since PcG target genes can become repressed or reactivated or exist in intermediate states (Schwartz and Pirrotta 2008).

Finally, there are proteins with a fundamental role in controlling heterochromatin formation not just in plants but also in animals and yeast and are termed SU(VAR)3–9-like proteins. The *Su(var)3–9* gene was identified in Drosophila as a dominant modifier of heterochromatin-induced gene silencing (Tschiersch et al. 1994). Unlike animals, plants contain a large number of SU(VAR)3–9 homologues (Baumbusch et al. 2001). This could be the result of the development in plants being more plastic and definitely a lot more easily affected by environmental conditions compared to animals. Therefore, it was suggested that plants could necessitate more subtle changes in chromatin structure for fine-tuning of gene regulation and thus contain multiple gene families involved in DNA and histone modification systems (reviewed by Reuter et al. 2005).

In conclusion, as far as sustainable management is concerned, it should be taken into account the fact that the combination of various histone modifications aptly termed "histone code" by Jenuwein and Allis (2001), constitutes a dynamic system that would be able to respond faster to external stimuli without the need to rely exclusively on DNA sequence alterations (Turner 2000).

3.3 RNA Molecules

Up until recently the role envisaged for RNA was almost exclusively as an intermediate between the DNA and protein formation. This rather simplistic notion, however, has steadily been changing since the identification of small non-coding RNA molecules that have been implicated in endogenous gene regulation as well as the control of invading genetic entities (reviewed by Bartel 2004).

There are two classes of small RNAs in plants; the micro RNAs (miRNAs) and the small interfering RNAs (siRNAs), that although sharing some similarities in their modes of action indicating toward analogous biochemical complexes mediating their function, differ in the process of their synthesis and more importantly their target sequences. In general terms, miRNAs are endogenously encoded and their targets are mRNAs from endogenous genes. These mRNAs are silenced by the miRNAs activity through changes in chromatin structure or the inhibition of translation or by altering mRNA stability. On the other hand, siRNAs are derived from double-stranded RNA either encoded in the genome or produced during viral infection. In turn, they silence such invading nucleic acids as transposons and viruses through RNA cleavage or changes in chromatin structure.

Initially, as presented schematically in Fig. 8, miRNAs come as part of longer transcripts termed primary miRNAs, i.e. pri-miRNAs (Lee et al. 2002), which in both plants and animals are transcribed by RNA polymerase II and include 5′ caps and 3′ poly(a) tails (Parizotto et al. 2004). The miRNA portion of the pri-miRNA transcript most likely forms a hairpin with signals for double stranded RNA (dsRNA)-specific nuclease cleavage, so that the dsRNA-specific ribonuclease –Dicer – can later digest the pri-miRNA to release hairpin, precursor miRNA, the so-called pre-miRNA (Meister and Tuschl 2004). Pre-miRNA is then processed to yield mature miRNA, a reaction that is also catalyzed by DICER in plants (Papp et al. 2003). Contrary to plants, where a single enzyme is involved in both processing steps in the nucleus, in animals the pri-miRNA is processed into a pre-miRNA

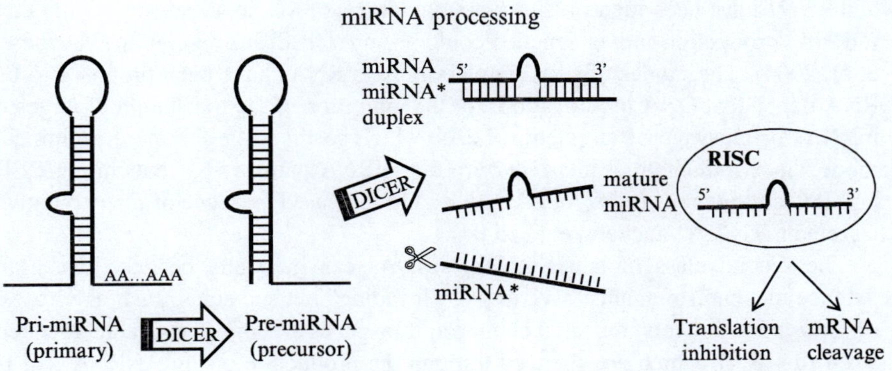

Fig. 8 Primary micro RNA (*pri-miRNA*) is digested by DICER (a dsRNA-specific ribonuclease) and precursor miRNA (*pre-miRNA*) is released. This then yields mature miRNA, in a process also catalysed by DICER in plants. Only one of the strands accumulates as mature miRNA, while the other (miRNA*) is degraded. Mature miRNAs incorporated in a complex called RNA-inducing silencing complex (RISC) and take part in mRNA silencing

stem-loop structure in the nucleus by the enzyme Dorsha (again an endoribonuclease III, like DICER), and then exported to the cytoplasm for processing into an miRNA duplex by DICER (Bartel 2004; Meister and Tuschl 2004). This could explain why the sizes of animal pre-miRNA hairpins are restricted to approximately 70 nucleotide-long structures, whereas Arabidopsis hairpins are variable and can be much larger (Reinhart et al. 2002). Also, unlike animals, where miRNAs are not usually completely complementary with their targets, plant miRNAs show more complete complementarity with their putative targets (Reinhart et al. 2002; Rhoades et al. 2002).

Then, after pre-miRNA has been processed by DICER (or DICER-LIKE, DCL1) only one of the strands of the product (the one with the lower stability at the 5′ end of the helix) accumulates as a mature miRNA that acts to regulate gene expression, while the other (symbolized miRNA*) seems to be rapidly degraded (Schwarz et al. 2003). The mature miRNAs are subsequently incorporated in a complex called RNA-induced silencing complex (RISC) and become active miRNA (Schwarz et al. 2003). A main component of this complex is a member of the ARGONAUTE protein family; there *ARGONAUTE1* (*AGO1*), for example, has been demonstrated to be imperative for the action of many miRNAs, since their accumulation is reduced in *ago1* mutants (Vaucheret et al. 2004). Intriguingly, at least ten ARGONAUTE-like proteins have been identified in the Arabidopsis genome suggesting the possibility that several different RNA-induced silencing complexes (RISCs) with diverse activities could be present in plants (review by Floyd and Bowman 2005).

Moreover, as a hint of potential crosstalk between miRNA and siRNA pathways, *AGO1* also acts in an siRNA-mediating pathway required for post transcriptional gene silencing (PTGS) and virus resistance (Morel et al. 2002). Notably, *DCL1* and *AGO1* have been identified as potential targets of the miRNAs, suggesting that miRNAs production and/ or function could be under feedback regulation (Vaucheret et al. 2004). The modes of gene regulation by miRNAs have been proposed to be RNA-dependent DNA modifications, or the inhibition of the translation of targeted mRNAs or cleavage of targeted mRNAs, which seems to be the more prominent mode, since mutants that interfere with the miRNA pathway by reducing DICER or RISC components (*ago1*, *dcl1*, etc.) lead to increased presence of many putative targeted mRNAs (Vaucheret et al. 2004).

The second class of small RNAs, siRNAs can generally be described as a defence mechanism against a variety of "invading" nucleic acids, such as viruses, transposable elements, repeated elements, transgenes and injected double stranded RNA (dsRNA), which are silenced through the production of long dsRNA that is later cleaved (Vance and Vaucheret 2001). A breakthrough in the quest to unravel the mechanisms involved in RNA silencing that is initiated in response to dsRNA came by the discovery by Fire et al. (1998) that dsRNA strongly induced RNA degradation in *Caenorhabditis elegans*, as well as the discovery by Hamilton and Baulcombe (1999) of small RNAs, that are specifically associated with RNA silencing in plants. Consistently, transcription of inverted-repeat structures and simultaneous expression of sense and antisense transgenes induce post transcriptional gene silencing (PTGS) in plants at a high frequency (Muskens et al. 2000).

Furthermore, RNA viruses that generate double stranded RNA (dsRNA) during their infection cycle are potent inducers of post transcriptional gene silencing (PTGS) in plants (Voinnet 2001). Double stranded RNA is cleaved by an RNase-III-like endonuclease, called Dicer (Bernstein et al. 2001) or Dicer-like (DCL), generating double-stranded siRNA of 21–25 nucleotides, with 2-nucleotide long 3′ overhangs and 5′ phosphate and a 3′ hydroxyl ends (Elbasir et al. 2001). This configuration seems to be necessary for the incorporation of each of the siRNA strands into the RNA-induced silencing complex (RISC), whereby it targets the complex to a complementary or partially complementary RNA sequence, which is later degraded (Martinez et al. 2002). It is, therefore, no surprise that plant viruses with RNA genomes that replicate via dsRNA intermediates have been demonstrated to encode proteins acting as suppressors of RNA silencing (reviewed by Roth et al. 2004).

In terms of crop improvement either following a traditional approach or by using biotechnological methods for the development of novel crops through the introduction of transgenes, stable and long-term activity of the new traits is an unquestionable requirement for agronomic production. Initial reports for reliable expression and transmission of transgenic traits (Budar et al. 1986) were answered shortly by observations for the unstable behaviour of transgenes (John and Amasino 1989). In addition, expression instability was not exclusive for transgenes but could affect homologous plant genes (Napoli et al. 1990; van der Krol et al. 1990), a phenomenon known as cosuppression. Both categories of gene silencing (GS), transcriptional (TGS) -linked to an altered epigenetic state due to signals in the promoter- and posttranscriptional (PTGS) - linked to an altered epigenetic state due to signals in the transcribed region- became interconnected with the discovery that RNA molecules are essential for the induction of both transgene methylation (Wassenegger et al. 1994) and transcriptional silencing events (Matzke et al. 2004).

Ironically, though transgene silencing presents a problem for crop improvement, it has nevertheless provided the research community with a powerful tool to unravel fascinating details about the different ways plants use epigenetic strategies for gene regulation, stress response, genome organisation and other vital biological aspects, all of which could prove essential for the development of reliable crop plants.

3.4 Enhanced Biodiversity via Epigenetic Mechanisms in Agro-Systems

An essential prerequisite for plant breeding is phenotypic variation in natural populations, upon which selection could be applied toward the development of elite cultivars. Until recently, phenotypic variation was almost exclusively attributed to DNA sequence divergence and hence the accumulation of favourable alleles in a few improved lines. This could theoretically shrink the spectrum of genetic variability and inevitably lead to restricted potential for further improvement. Nevertheless, advances in the field of gene regulation have revealed a suite of epigenetic mechanisms controlling the temporal, spatial and abundance patterns of gene expression that could have morphological, physiological and ecological consequences. This means that genetic variation alone is not responsible for every aspect of evolutionary change and so selection experiments could have a positive result even in seemingly narrow genetic pools, such as inbred lines.

The two sources of variability, namely genetic and epigenetic, have been compared by Tsaftaris and Polidoros (2000), who concluded that epialleles usually emerge as a response to developmental or environmental cues, they happen more frequently and could be reversible, unlike spontaneous DNA mutations that are usually random and more stable. Fusions of the two phenomena are also possible, as demonstrated by the case of transposable elements (TEs) released through epigenetic chromatin modifications, but causing a genetic mutation by the destruction of a gene locus. The implications for the arising variability in plant populations could be significant in terms of developmental regulation, adaptation to different environmental conditions, response to abiotic and biotic stress and even product quality.

Examples illustrating the involvement of epigenetic mechanisms in developmental regulation and environmental adaptation could include the case of *Stellaria longipes*, where Tatra et al. (2000) suggested that DNA methylation plays a key role in the shade-avoidance response by demonstrating that stem elongation in differential-light-condition treated plants and in different ecotypes was associated with methylation levels. In other words, plants treated with low (as opposed to high)

Red/Far Red light ratios as well as those belonging to the prairie (rather than alpine) ecotype were taller, which coincided with lower levels of methylation.

Another notable example involves the regulation of flowering time by the *FLOWERING LOCUS C* (*FLC*), a repressor of flowering, which is silenced through the formation of heterochromatin, via reduced acetylation at H3K9 and H3K14 and increased methylation of H3K9 and H3K27, after prolong exposure to low temperatures (Bastow et al. 2004; Sung and Amasino 2004). In the same sense, reduced DNA methylation in transgenic Arabidopsis plants overexpressing a chromatin remodeling gene (*AtCHR12*) was postulated to be responsible for the growth arrest of primary buds as well as growth reduction of the primary stem that was particularly obvious under drought and heat stress; when an *AtCHR12*-knockout mutant, however, was tested under stress, it exhibited a growth arrest phenotype less pronounced than wild type plants (Mlynarova et al. 2007). Intriguingly, stress response pathways could be interconnected with developmental ones, as demonstrated for instance through the plant hormone- and stress-related histone deacetylases (HDACs) *HDA19* and *HDA6*. In more detail, mutants with reduced expression of *HDAC19* showed delayed flowering, whereas mutants with reduced expression of *HDAC6* displayed both downregulation (i.e. reduction) of JA (jasmonic acid)-responsive genes, which are associated with stress responses, as well as upregulation (i.e. induction) of the *FLC* locus and so delayed flowering (Long et al. 2006; Tian and Chen 2001; Wu et al. 2008).

Histone deacetylases HDA6 and HDA19 have been involved in histone deacetylation following biotic and abiotic stress stimuli in Arabidopsis. Thus, *HDA6* is induced by jasmonic acid (JA) and ethylene (Zhou et al. 2005) and takes part in transcriptional gene silencing (TGS) (Probst et al. 2004) and RNA-directed DNA methylation (Aufsatz et al. 2002). Similarly, JA and ethylene induce the expression of *HDA19*, which is also true for wounding and infection by a fungus (*Alternaria brassicicola*). Not surprisingly, increased expression levels of *ERF1* (ethylene response factor) and *PR* (pathogenesis related) genes coincided with the overexpression of *HDA19* and reduced acetylation levels in transgenic plants (Zhou et al. 2005). Members of HDAC families of proteins in rice are also differentially regulated by abiotic stress factors, such as cold, osmotic stress and salt, as well as by JA and salicylic acid (Fu et al. 2007).

Besides histone modifications, DNA methylation levels have been involved in stress responses, too. For example in tobacco, abiotic stress factors, like aluminium, paraquat, salt and low temperatures, induced demethylation and transcriptional activation of a glycerophosphodiesterase-like protein, although a similar reaction was not evident after challenge with a pathogen (Choi and Sano 2007). Reduction in DNA methylation levels was also observed correlating with the cold stress-induced expression of the *ZmM1* locus in maize (Steward et al. 2002), as well as in *Bryonia dioica* after the application of mechanical stress (Galaud et al. 1993). In contrast, hypermethylation was recorded following drought stress in pea (Labra et al. 2002), similarly to the case of tobacco cell-suspension culture after the application of osmotic stress (Kovarik et al. 1997).

Notably, the observation that specific histone modification-dependent pathways could possibly facilitate the methylation process of the methylated loci in Arabidospis, lead to the speculation that histone marks could be turned into the more stable DNA methylation marks (Zhu 2008). At the same time, it is fascinating that a direct correlation has been revealed between the ability of genomic sequences to produce small RNAs and DNA methylation (Lister et al. 2008). Recently, Zheng et al. (2008) demonstrated that a small RNA-binding protein (ROS3) mediates sequence-specific demethylation performed by the DNA glycosylase ROS1 (repressor of silencing 1), which has been previously reported to have a critical role in erasing DNA methylation and preventing TGS of target genes in Arabidopsis (Agius et al. 2006).

Endogenous siRNAs that are regulated by abiotic stress stimuli, like dehydration, salinity, cold stress or the plant stress hormone abscisic acid, have been identified in Arabidopsis (Sunkar and Zhu 2004). Moreover, a microRNA (miR389) in Arabidopsis has been shown to repress the expression of two genes involved in plant resistance during oxidative stress conditions, i.e. high light, heavy metal and methyl viologen (Sunkar et al. 2006). In detail, oxidative stresses reduce the expression levels of miR398 normally targeting *CSD1* and *CSD2*, allowing the accumulation of mRNA from the two genes. These genes are involved in detoxifying superoxide radicals, which are collected as a result of different environmental stresses and could otherwise cause loss of crop productivity through effects in cellular function, such as nucleic acid damage or protein oxidation (Bartels and Sunkar 2005). More importantly, mutant plants able to accumulate large amounts of CSD2 via overexpression of a miR398-resistant form of CSD2 exhibited enhanced tolerance to oxidative stresses (Sunkar et al. 2006).

Apart from abiotic stress tolerance, though, technologies based on RNA silencing can also be applied in challenges by viruses. In other words, transgenic plants displaying resistance via RNA degradation utilizing small RNAs would not involve the transgenic production of functional viral genes or proteins; nevertheless, degradation is sequence-sensitive and overall the methodology doesn't overcome either the obstacle of transformation success rate, or the fact that there is still public skepticism toward the use of transgenic crops (Fermin et al. 2004).

Additionally, RNA interference (RNAi) mechanisms both via miRNAs and siRNAs could be applied in the development of improved products. One such example is the development through mutational breeding of the rice cultivar LGC1 that carries a dominant mutation on *low glutelin content1* (*Lgc1*) locus, which reduces the content of glutelin in rice grains making them suitable for patients with kidney disease (Kusaba et al. 2003). Glutelin is a major seed storage protein encoded by a multigene family whose members are classified into two subfamilies, GluA and GluB. *Lgc1* suppresses the production of GluB protein expression via RNA silencing, although suppression of GluA could further reduce protein production. Recently, the LGC1 cultivar was crossed with another mutant line induced by gamma-irradiation, that was deficient through deletion in protein 'globulin', resulting in reducing the total amount of digestible protein to about half compared to regular cultivars and 15% less than in LGC1 (Nishimura et al. 2009).

Another example is the development of transgenic coffee plants (*Coffea arabica*) in which expression of one of the three genes involved in caffeine biosynthesis – namely theobromine synthase (CaMXMT1) – is repressed by RNA interference (RNAi); as a result caffeine content was reduced by up to 70% (Ogita et al. 2003). Furthermore, transgenic tomatoes with improved carotenoid or flavonoid content, both of which are beneficial for human health, were constructed using RNAi technology to suppress the expression of a gene involved in photomorphogenesis regulation, DET1 (DE-ETIOLATED 1) (Davuluri et al. 2005). Earlier studies have indicated that DET1 plays an important role throughout plant development and may control light-regulated gene expression (Davuluri et al. 2004).

More examples could include the production of high-amylose wheat through the RNAi down-regulation of two starch-branching enzymes (SBE) (SBEIIa and SBEIIb) in wheat endosperm, raising amylose content in starch by 70% (Regina et al. 2006). Also, notable is the design of an RNAi construct that was able to reduce the levels of the carcinogen nornicotine in cured leaves of tobacco plants (*Nicotiana tabacum*) by 3.6 fold compared to commercial varieties (Gavilano et al. 2006). Finally, equally fascinating was the genetic transformation of onion plants (*Allium cepa*) so as to suppress the eye-irritating lachrymatory factor synthase gene using RNAi silencing (Eady et al. 2008).

Undoubtedly, genome plasticity plays a key role in the adjustment of growth and development in order for the plant to utilize the available resources effectively and is especially critical under stress conditions. Consequently, it would be expected for the majority of the stress-induced modifications to be "reset" to original levels once the causal agent has ceased to exist, while some of the modifications could be more stable.

Vernalization, or the ability to grow after prolonged exposure to low temperatures, presents an example of plants receiving an environmental cue and adjust to it via epigenetic modification, without this modification becoming heritably stable; in this case propagation of the particular epigenetic state is stable through mitosis but not meiosis (review by Grant-Downton and Dickinson 2006). At the other end, heritable epigenetic alleles have been identified in the cases of the *Lcyc* gene in *Linaria vulgaris* causing a flower morphological change, and the *LGC-1* gene in rice resulting to lower protein content in the seed, both of which have been stable through a number of generations (Cubas et al. 1999; Kusaba et al. 2003).

Intriguingly, tobacco plants challenged by tobacco mosaic virus (TMV), generated progeny that displayed hypomethylation in several leucine-rich repeat (LRR)-containing loci, which are involved in disease resistance. Additionally, they showed a higher frequency of recombination in the also hypomethylated LRR-containing *N* gene, that is necessary for resistance to TMV (Mlynarova et al. 2007). In a similar fashion, progenies of rice that have been treated with an inhibitor of DNA cytosine methylation (namely, 5-aza-deoxycytidine) were hypomethylated. One of these progenies in particular exhibited complete absence of methylation in the promoter of an *Xa21*-like gene (*Xa21G*), which is imperative in disease resistance against *Xanthomonas oryzae*. This resulted in constitutive expression of *Xa21G* and enhanced resistance to the pathogen race PR2 (Akimoto et al. 2007).

4 Conclusion

Overall, it has become apparent that a series of molecular channels are responsible for extensive biodiversity in nature. Usually, the genetic changes are more permanent but frequently random, whereas the epigenetic alterations seem to be more targeted but less stable. Nevertheless, examples of more stable epigenetic changes have been presented. In this case, stable, epigenetic adaptation could help plants respond more effectively in subsequent stress situations. At the same time, however, it raises questions as to the negative impact on crop yield in case it prevents the plant from growing to its full potential. The implications are far reaching, taking into account the farmer's practice of retaining seed from previous crops that could have suffered from different stress conditions, as well as breeding-for-stress-environments programmes. Therefore, the continuing studies in the field of genetic and epigenetic variation including the driving forces and consequences both in short and long term of such phenomena is essential for the critical analysis of the processes involved and the utilization of the resulting knowledge in order to facilitate sustainable crop management and improve plant breeding.

Undoubtedly, both genetic and epigenetic mechanisms have an essential role to play both in evolution and dispersion, as well as conservation and sustenance in nature. More importantly, however, biodiversity can serve the purpose of sustainable agriculture, on condition that its stewardship is implemented in a friendly system to ensure viability to perpetuity. Plant breeding enables the management of existing and newly developed genetic diversity toward new cultivars that could entirely meet the needs of a sustainable agriculture. Up until now, lack of appropriate cultivars has been a crucial obstacle for sustainable agriculture to be widely adopted. The producers' belief that conventional systems give higher yield combined with the increasing pressure on agricultural systems to supply more food to a growing population, are insuperable barriers for sustainable agricultural systems to have a broad acceptance.

The molecular mechanisms that have been discussed here highlight nature's ability to perpetuate the existence of biodiversity. In this sense, proper management of such biodiversity is essential in order to develop agricultural systems that are less reliant on financial subsidies, with reduced economic fragility and able to better support self-reliance in local communities, leading at the same time to the production of high-quality food. The on-going climate changes on the one hand result in newly developed genetic variation and on the other hand impose the need to develop new cultivars.

Plant breeding, orientated towards the major principles of sustainable agriculture, has the potential to overcome the particular impediment of suitable for sustainable agriculture cultivars. Availability of ample supply of new cultivars adaptable to low-inputs, improved for resource-use efficiency, tolerant to biotic and abiotic stresses, and self-reproducible combined with sustainable agriculture strategies might prove a fruitful parameter to the future expansion of sustainable agricultural practices.

References

Adams KL, Cronn R, Percifield R, Wendel JF (2003) Genes duplicated by polyploidy show unequal contributions to the transcriptome and organ-specific reciprocal silencing. Proc Natl Acad Sci USA 100:4649–4654. doi:10.1073/pnas.0630618100

Adams KL, Percifield R, Wendel JF (2004) Organ-specific silencing of duplicated genes in a newly synthesized cotton allotetraploid. Genetics 168:2217–2226. doi:10.1534/genetics.104.033522

Agius F, Kapoor A, Zhu JK (2006) Role of the *Arabidopsis* DNA glycosylase/lyase ROS1 in active DNA demethylation. Proc Natl Acad Sci USA 103:11796–11801. doi:10.1073/pnas.0603563103

Ahloowalia BS, Maluszynski M, Nichterlein K (2004) Global impact of mutation-derived varieties. Euphytica 135:187–204. doi:10.1023/B:EUPH.0000014914.85465.4f

Akimoto K, Katakami H, Kim HJ, Ogawa E, Sano CM, Wada Y, Sano H (2007) Epigenetic inheritance in rice plants. Ann Bot 100:205–217. doi:10.1093/aob/mcm110

Alvarez-Venegas R, Pien S, Sadder M, Witmer X, Grossniklaus U, Arramova Z (2003) ATX-1 an *Arabidopsis* homologue of trithorax activates flower homeotic genes. Curr Biol 13:627–637. doi:10.1016/S0960-9822(03)00243-4

Aufsatz W, Mette MF, Van Der Winden J, Matzke M, Matzke AJ (2002) HDA6, a putative histone deacetylase needed to enhance DNA methylation induced by double-stranded RNA. EMBO J 21:6832–6841. doi:10.1093/emboj/cdf663

Baroux C, Spillane C, Grossniklaus U (2002) Evolutionary origins of the endosperm in flowering plants. Genome Biol 3:1026.1–1026.5. doi:10.1186/gb-2002-3-9-reviews1026

Bartel DP (2004) MicroRNAs. Cell 116:281–297. doi:10.1016/S0092-8674(04)00045-5

Bartels D, Sunkar R (2005) Drought and salt tolerance in plants. Crit Rev Plant Sci 24:23–58. doi:10.1080/07352680590910410

Bastow R, Mylne J, Lister C, Lippman Z, Martiennsen R, Dean C (2004) Vernalization requires epigenetic silencing of FLC by histone methylation. Nature 427:164–167. doi:10.1038/nature02269

Baumbusch LO, Thorstensen T, Kraus V, Fischer A, Numann K, Assalkhou R, Schulz I, Reuter G, Aalen RB (2001) The *Arabidopsis thaliana* genome contains at least 29 active genes encoding SET domain proteins that can be assigned to four evolutionary conserved classes. Nucleic Acids Res 29:4319–4333. doi:10.1093/nar/29.21.4319

Bennetzen JL (2005) Transposable elements, gene creation and genome rearrangement in flowering plants. Curr Opin Genet Dev 15:621–627. doi:10.1016/j.gde.2005.09.010

Bennetzen JL, Ma J, Devos KM (2005) Mechanisms of recent genome size variation in flowering plants. Ann Bot 95:127–132. doi:10.1093/aob/mci008

Bernstein E, Caudy AA, Hammond SM, Hannon GJ (2001) Role for a bidentate ribonuclease in the initiation step of RNA interference. Nature 409:363–366. doi:10.1038/35053110

Bhasin M, Reinherz EL, Reche PA (2006) Recognition and classification of histones using support vector machine. J Comput Biol 13:102–112. doi:10.1089/cmb.2006.13.102 DOI:dx.doi.org

Blanc G, Wolfe KH (2004) Functional divergence of duplicated genes formed by polyploidy during Arabidopsis evolution. Plant Cell 16:1679–1691. doi:10.1105/tpc.021410

Budar F, Thia-Toong L, Van Montagu M, Hernalsteens J-P (1986) Agrobacterium-mediated gene transfer results mainly in transgenic plants transmitting T-DNA as a single Mendelian factor. Genetics 114:303–313, PMID: 17246346, PMCID: PMC1202937

Capy P, Gasperi G, Biemont C, Bazin C (2000) Stress and transposable elements: co-evolution or useful parasites? Heredity 85:101–106. doi:10.1046/j.1365-2540.2000.00751.x

Castiglione MR, Cremonini R, Frediani M (2002) DNA methylation patterns on plant chromosomes. Caryologica 55:275–282. doi:10.1007/s00709-010-0116-x

Chandler V, Stam M (2004) Chromatin conversations: mechanisms and implications of paramutation. Nat Rev Genet 5:532–544. doi:10.1038/nrg1378

Chen X, Liu X, Wu D, Shu QY (2006) Recent progress of rice mutation breeding and germplasm enhancement in China. Plant Mutat Rep 1:4–6

Chinnusamy V, Zhu J-K (2009) Epigenetic regulation of stress responses in plants. Curr Opin Plant Biol 12:1–7. doi:10.1016/j.pbi.2008.12.006

Choi CS, Sano H (2007) Abiotic-stress induces demethylation and transcriptional activation of a gene encoding a glycerophosphodiesterase-like protein in tobacco plants. Mol Genet Genomics 277:589–600. doi:10.1007/s00438-007-0209-1

Choi Y, Gehring M, Johnson L, Hannon M, Harada JJ, Goldberg RB, Jacobsen SE, Fischer RL (2002) DEMETER, a DNA glycosylase domain protein, is required for endosperm gene imprinting and seed viability in Arabidopsis. Cell 110:33–42. doi:10.1016/S0092-8674(02)00807-3

Cosgrove MS, Boeke JD, Wolberger C (2004) Regulated nucleosome mobility and the histone code. Nat Struct Mol Biol 11:1037–1043. doi:10.1038/nsmb851 DOI:dx.doi.org

Cubas P, Vincent C, Coen E (1999) An epigenetic mutation responsible for natural variation in floral symmetry. Nature 401:157–161. doi:10.1038/43657

Cullis CA (2005) Mechanisms and control of rapid genomic change in flax. Ann Bot 95:201–206. doi:10.1093/aob/mci013

Davuluri GR, Tuinen A, Mustilli AC, Manfredonia A, Newman R, Burgess D, Brummell DA, King SR, Palys J, Uhlig J, Pennings HMJ, Bowler C (2004) Manipulation of *DET1* expression in tomato results in photomorphogenic phenotypes caused by post-transcriptional gene silencing. Plant J 40:344–354. doi:10.1111/j.1365-313X.2004.02218.x

Davuluri GR, vanTuinen A, Fraser PD, Manfredonia A, Newman R, Burgess D, Brummell DA, King SR, Palys J, Uhlig J, Bramley PM, Pennings HM, Bowler C (2005) Fruit-specific RNAi-mediated suppression of *DET1* enhances carotenoid and flavonoid content in tomatoes. Nat Biotechnol 23:890–895. doi:10.1038/nbt1108

De Chiara TM, Robertson EJ, Efstratiadis A (1991) Parental imprinting of the mouse insulin-like growth factor II gene. Cell 64:849–859. doi:10.1016/0092-8674(91)90513-X

Dou Y, Milne TA, Tackett AJ, Smith ER, Fukuda A, Wysocka J, Ailis DC, Chait BT, Hess JL, Roeder R (2005) Physical association and coordinate function of H3K4 methylotransferase MLL1 and the H4K16 acetylotransferase MOF. Cell 121:873–885. doi:10.1016/j.cell.2005.04.031

Dudley JW, Lambert RJ (2004) 100 generations of selection for oil and protein in corn. Plant Breed Rev 24:79–110

Eady CC, Kamoi T, Kato M, Porter NG, Davis S, Shaw M, Kamoi A, Imai S (2008) Silencing onion lachrymatory factor synthase causes a significant change in the sulfur secondary metabolite profile. Plant Physiol 147:2096–2106. doi:10.1104/pp. 108.123273

Elbasir SM, Lendeckel W, Tuschl T (2001) RNA interference is mediated by 21- and 22-nucleotide RNAs. Genes Dev 15:188–200. doi:10.1101/gad.862301

Elliot FC (1958) Plant breeding and cytogenetics. McGraw Hill Book Company, Inc., New York

Fasoula DA (1990) Correlations between auto-, allo- and nill-competition and their implications in plant breeding. Euphytica 50:57–62. doi:10.1007/BF00023161

Fasoula VA, Boerma HR (2005) Divergent selection at ultra-low plant density for seed protein and oil content within soybean cultivars. Field Crops Res 91:217–229. doi:10.1016/j.fcr.2004.07.018 DOI:dx.doi.org

Fasoula VA, Boerma HR (2007) Intra-cultivar variation for seed weight and other agronomic traits within three elite soybean cultivars. Crop Sci 47:367–373. doi:10.2135/cropsci2005.09.0334

Fasoula VA, Fasoula DA (2002) Principles underlying genetic improvement for high and stable crop yield potential. Field Crop Res 75:191–209. doi:10.1016/S0378-4290(02)00026-6 DOI:dx.doi.org

Fasoulas AC (2000) Building up resistance to Verticillium wilt in cotton through honeycomb breeding. In: Gillham FM (ed) New frontiers in cotton research. Proceedings of the 2nd world cotton research conference, 6–12 Sept 1998, Athens, pp 120–124

Felsenfeld G, Groudine M (2003) Controlling the double helix. Nature 421:448–453. doi:10.1038/nature01411DOI:dx.doi.org

Fermin G, Tennant P, Gonsalves C, Lee D, Gonsalves D (2004) Comparative development and impact of transgenic papayas in Hawaii, Jamaica, and Venezuela. Methods Mol Biol 286: 399–430. doi:10.1385/1-59259-827-7:399

Fire A, Xu SQ, Montgomery MK, Kostas SA, Driver SE, Mello CC (1998) Potent and specific genetic interference by double-stranded RNA in *Caenorhabditis elegans*. Nature 391:806–811. doi:10.1038/35888

Floyd SK, Bowman JL (2005) MicroRNAs: micro-managing the plant genome. In: Meyer P (ed) Plant epigenetics. Blackwell Scientific Publications, Oxford, pp 244–279

Fransz PF, Alonso-Blanco C, Liharska TB, Peeters AJM, Zabel P, Hans de Jong J et al (1996) High-resolution physical mapping in Arabidopsis thaliana and tomato by fluorescence in situ hybridization to extended DNA fibres. Plant J 9:421–430. doi:10.1046/j.1365-313X.1996.09030421

Freese E (1959) The difference between spontaneous and base-analogue induced mutations of phage T4. Proc Natl Acad Sci USA 45:622–633. doi:10.1073/pnas.45.4.622

Fu W, Wu K, Duan J (2007) Sequence and expression analysis of histone deacetylases in rice. Biochem Biophys Res Commun 356:843–850. doi:10.1016/j.bbrc.2007.03.010

Galaud JP, Gaspar T, Boyer N (1993) Inhibition of internode growth due to mechanical stress in *Bryonia dioica*: relationship between changes in DNA methylation and ethylene metabolism. Physiol Plant 87:25–30. doi:10.1007/BF02912989

Gavilano LB, Coleman NP, Burnley L-E, Bowman ML, Kalengamaliro NE, Hayes A, Bush L, Siminszky B (2006) Genetic Engineering of *Nicotiana tabacum* for Reduced Nornicotine Content. J Agric Food Chem 54:9071–9078. doi:10.1021/jf0610458

Gethi JG, Labate JA, Lamkey KR, Smith ME, Kresovich S (2002) SSR variation in important U.S. maize inbred lines. Crop Sci 42:951–957

Grant-Downton RT, Dickinson HG (2005) Epigenetics and its implications for plant biology. 1. The epigenetic network in plants. Ann Bot 96:1143–1164. doi:10.1093/aob/mci273

Grant-Downton RT, Dickinson HG (2006) Epigenetics and its implications for plant biology. 2. The 'epigenetic epiphany': epigenetics, evolution and beyond. Ann Bot 97:11–27. doi:10.1093/aob/mcj001

Griffiths AJF, Miller JH, Suzuki DT, Lewontin RC, Gelbart WM (1993) An introduction to genetic analysis. W.H. Freeman & Company, New York

Gustafsson A, Gadd I (1965) Mutations and crop improvement. II. The genus *Lupinus*(Leguminosae). Hereditas 53:15–39. doi:10.1111/j.1601-5223.1965.tb01977

Haig D, Westoby M (1989) Parent specific gene expression and the triploid endosperm. Am Nat 134:147–155. doi:10.1086/284971

Hall IM, Grewal SI (2003) Structure and function of heterochromatin: implications for epigenetic gene silencing and genome organization. In: Hannon GJ (ed) RNAi: a guide to gene silencing. Cold Spring Harbor Laboratory Press, Cold Spring Harbor, pp 205–232

Hamilton AJ, Baulcombe DC (1999) A species of small antisense RNA in post-transcriptional gene silencing in plants. Science 286:950–952. doi:10.1126/science.286.5441.950

Hashida SN, Uchiyama T, Martin C, Kishima Y, Sano Y, Mikami T (2006) The temperature-dependent change in methylation of the *Antirrhinum* transposon Tam3 is controlled by the activity of its transposase. Plant Cell 18:104–118. doi:10.1105/tpc.105.037655

Henderson IR, Jacobsen SE (2007) Epigenetic inheritance in plants. Nature 44:418–424. doi:10.1038/nature05917

Hollick JB, Dorweiler JE, Chandler VL (1997) Paramutation and related allelic interactions. Trends Genet 13:302–308. doi:10.1016/S0168-9525(97)01184-0

Jacobsen SE, Meyerowitz EM (1997) Hypermethylated SUPERMAN epigenetic alleles in Arabidopsis. Science 277:1100–1103. doi:10.1126/science.277.5329.1100

Jenuwein T, Allis CD (2001) Translating the histone code. Science 293:1074–1080. doi:10.1126/science.1063127 DOI:dx.doi.org

Jiang C-X, Wright RJ, El-Zik KM, Paterson AH (1998) Polyploid formation created unique avenues for response to selection in *Gossypium* (cotton). Proc Natl Acad Sci USA 95:4419–4424, PMID: 9539752, PMCID: PMC22504

Jiang N, Bao Z, Zhang X, Eddy SR, Wessler SR (2004) Pack-MULE transposable elements mediate gene evolution in plants. Nature 431:569–573. doi:10.1038/nature02953

John MC, Amasino RM (1989) Extensive changes in DNA methylation patterns accompany activation of a silent T-DNA ipt gene in *Agrobacterium tumefaciens*-transformed plant cells. Mol Cell Biol 9:4298–4303, PMID: 2479825, PMCID: PMC362510

Jones PA, Rideout WM, Shen J-C, Spruck CH, Tsai YC (1992) Methylation, mutation and cancer. BioEssays 14:33–36. doi:10.1002/bies.950140107

Jost JP, Siegmann M, Sun L, Leung R (1995) Mechanisms of DNA demethylation in chicken embryos. Purification and properties of a 5-methylcytosine-DNA glycosylase. J Biol Chem 270:9734–9739. doi:10.1074/jbc.270.17.9734

Kankel MW, Ramsey DE, Stokes TL, Flowers SK, Haag JR, Jeddeloh JA, Riddle NC, Verbsky ML, Richards EJ (2003) Arabidopsis *MET1* cytosine methyltransferase mutants. Genetics 163:1109–1122, PMID: 12663548, PMCID: PMC1462485

Kashkush K, Feldman M, Levy AA (2002) Gene loss, silencing and activation in a newly synthesized wheat allotetraploid. Genetics 160:1651–1659, PMID: 11973318, PMCID: PMC1462064

Kazazian HH Jr (2004) Mobile elements: drivers of genome evolution. Science 303:1626–1632. doi:10.1126/science.1089670

Kermicle JL, Alleman M (1990) Gametic imprinting in maize in relation to the angiosperm life cycle. Development 108 (Suppl):9–14. PMID: 2090436

Khoshoo TN (1959) Polyploidy in gymnosperms. Evolution 13:24–39, http://www.jstor.org/stable/2405943

Kidwell MG, Lisch D (2002) Transposable elements as sources of genomic variation. In: Craig NL, Craigie R, Gellert M, Lambowitz AM (eds) Moblie DNA II. American Society for Microbiology Press, Washington, DC, pp 59–90

Kinoshita T, Yadegari R, Harada JH, Goldberg RB, Fisher RL (1999) Imprinting of the *MEDEA* Polycomb gene in the Arabidopsis endosperm. Plant Cell 11:1945–1952. doi:10.1105/tpc.11.10.1945

Kinoshita T, Miura A, Choi Y, Kinoshita Y, Cao X, Jacobsen SE, Fischer RL, Kakutani T (2004) One-way control of *FWA* imprinting in *Arabidopsis* endosperm by DNA methylation. Science 303:521–523. doi:10.1126/science.1089835

Kovarik A, Koukalova B, Bezdek M, Opatrn Z (1997) Hypermethylation of tobacco heterochromatic loci in response to osmotic stress. Theor Appl Genet 95:301–306. doi:10.1007/s001220050563

Kusaba M, Miyahara K, Iida S, Fukuoka H, Takano T, Sassa H, Nishimura M, Nishio T (2003) *Low glutelin content 1*: a dominant mutation that suppresses the *glutelin* multigene family via RNA silencing in rice. Plant Cell 15:1455–1467. doi:10.1105/tpc.011452

Labra M, Ghiani A, Citterio S, Sgorbati S, Sala F, Vannini C, Ruffini-Castiglione M, Bracale M (2002) Analysis of cytosine methylation pattern in response to water deficitin pea roottips. Plant Biol 4:694–699. doi:10.1055/s-2002-37398

Lee Y, Jeon K, Lee JT, Kim S, Kim VN (2002) MicroRNA maturation: stepwise processing and subcellular localization. EMBO J 21:4663–4670. doi:10.1093/emboj/cdf476

Lippman Z, Gendrel A-V, Black M, Vaughn M, Dedhla N, McComble R, Lavine K, Mittal V, May B, Kasschau KD, Carrington JC, Doerge RW, Colot V, Martienssen R (2004) Role of transposable elements in heterochromatin and epigenetic control. Nature 430:471–476. doi:10.1038/nature02651

Lister R, O'Malley RC, Tonti-Filippini J, Gregory BD, Berry CC, Millar AH, Ecker JR (2008) Highly integrated single-base resolution maps of the epigenome in *Arabidopsis*. Cell 133:523–536. doi:10.1016/j.cell.2008.03.029

Long JA, Ohno C, Smith ZR, Meyerowitz EM (2006) TOPLESS regulates apical embryonic fate in *Arabidopsis*. Science 312:1520–1523. doi:10.1126/science.1123841

Louwers M, Haring M, Stam M (2005) When alleles meet: paramutation. In: Meyer P (ed) Plant epigenetics. Blackwell Scientific Publications, Oxford, pp 134–173

Luger K, Mder AW, Richmond RK, Sargent DF, Richmond TJ (1997) Crystal structure of the nucleosome core particle at 2.8 A resolution. Nature 389:251–260. doi:10.1038/38444

Ma X-F, Fang P, Gustafson JP (2004) Polyploidization-induced genome variation in triticale. Genome 47:839–848. doi:10.1139/G04-051

Martinez J, Patkaniowska A, Urlaub H, Luhrmann R, Tuschl T (2002) Single-stranded antisense siRNAs guide target RNA cleavage in RNAi. Cell 110:563–574. doi:10.1016/S0092-8674(02)00908-X

Matzke M, Aufsatz W, Kanno T, Daxinger L, Papp I, Mette MF, Matzke AJM (2004) Genetic analysis of RNA-mediated transcriptional gene silencing. Biochim Biophys Acta 1677:129–141

McClintock B (1984) The significance of responses of the genome to challenge. Science 226:792–801. doi:10.1126/science.15739260

McCormick S (2004) Control of male gametophyte development. Plant Cell 16(Suppl):S142–S153. doi:10.1105/tpc.016659

Meister G, Tuschl T (2004) Mechanisms of gene silencing by double-stranded RNA. Nature 431:343–349. doi:10.1038/nature02873

Mlynarova L, Nap JP, Bisseling T (2007) The SWI/SNF chromatin-remodeling gene *AtCHR12* mediates temporary growth arrest in *Arabidopsis thaliana* upon perceiving environmental stress. Plant J 51:874–885. doi:10.1111/j.1365-313X.2007.03185.x

Mochida K, Yamazaki Y, Ogihara Y (2003) Discrimination of homoeologous gene expression in hexaploid wheat by SNP analysis of contigs grouped from a large number of expressed sequence tags. Mol Gen Genet 270:371–377. doi:10.1007/s00438-003-0939-7

Morales-Ruiz T, Ortega-Galisteo AP, Ponferrada-Marn MI, Martnez-Macas MI, Ariza RR, Roldán-Arjona T (2006) *DEMETER* and *REPRESSOR OF SILENCING 1* encode 5-methylcytosine DNA glycosylases. Proc Natl Acad Sci USA 103:6853–6858. doi:10.1073/pnas.0601109103

Morel JB, Godon C, Mourrain P, Beclin C, Boutet S, Feuerbach F, Proux F, Vaucheret H (2002) Fertile hypomorphic *ARGONAUTE (ago1)* mutants impaired in post -transcriptional gene silencing and virus resistance. Plant Cell 14:629–639. doi:10.1105/tpc.010358

Morgante M, Brunner S, Pea G, Fengler K, Zuccolo A, Rafalski A (2005) Gene duplication and exon shuffling by helitron-like transposons generate intraspecies diversity in maize. Nat Genet 37:997–1002. doi:10.1038/ng1615

Muskens MWM, Vissers APA, Mol JNM, Kooter JM (2000) Role of inverted DNA repeats in transcriptional and post-transcriptional gene silencing. Plant Mol Biol 43:243–260. doi:10.1023/A:1006491613768

Mysore KS, Nam J, Gelvin SB (2000) An *Arabidopsis* histoneH2Amutant is deficient in *Agrobacterium* T-DNA integration. Proc Natl Acad Sci USA 97:948–953, PMID: 10639185, PMCID: PMC15436

Napoli C, Lemieux C, Jorgensen R (1990) Introduction of a chimeric chalcone synthase gene in to petunia results in reversible co-suppression of homologous genes in trans. Plant Cell 2:279–289. doi:10.1105/tpc.2.4.279

Nishimura M, Morita R, Kusaba M (2009) Utilization and molecular characterization of seed protein composition mutants in rice plants. JARQ 43:1–5

Ogita S, Uefuji H, Yamaguchi Y, Koizumi N, Sano H (2003) Producing decaffeinated coffee plants. Nature 423:823. doi:10.1038/423823a

Olufowote JO, Xu Y, Chen X, Park WD, Beachell HM, Dilday RH, Goto M, McCouch SR (1997) Comparative evaluation of within-cultivar variation of rice (Oryza sativa L.) using microsatellite and RFLP markers. Genome 40:370–378. doi:10.1139/g97-050

Ossowski S, Schneeberger K, Lucas-Lledó JI, Warthmann N, Clark RM, Shaw RG, Weigel D, Lynch M (2010) The rate and molecular spectrum of spontaneous mutations in Arabidopsis thaliana. Science 327:92–94. doi:10.1126/science.1180677

Otto SP, Whitton J (2000) Polyploidy incidence and evolution. Annu Rev Genet 34:401–437. doi:10.1146/annurev.genet.34.1.401

Papp I, Mette MF, Aufsatz W, Daxinger L, Schauer SE, Ray A, van der Winden J, Matzke M, Matzke AJM (2003) Evidence for nuclear processing of plant microRNA and short interfering RNA precursors. Plant Physiol 132:1382–1390. doi:10.1104/pp.103.021980

Parizotto EA, Dunoyer P, Rahm N, Himber C, Voinnet O (2004) In vivo investigation of the transcription, processing, endonucleolytic activity, and functional relevance of the spatial distribution of a plant miRNA. Genes Dev 18:2237–2242. doi:10.1101/gad.307804

Paterson AH, Bowers JE, Peterson DG, Estill JC, Chapman BA (2003) Structure and evolution of cereal genomes. Curr Opin Genet Dev 13:644–650. doi:10.1016/j.gde.2003.10.002

Patterson GI, Thorpe CJ, Chandler VL (1993) Paramutation, an allelic interaction, is associated with a stable and heritable reduction of transcription of the maize *b* regulatory gene. Genetics 135:881–894, PMID: 7507455, PMCID: PMC 1205727

Paull TT, Rogakou EP, Yamazaki V, Kirchgessner CU, Gellert M, Bonner WM (2000) A critical role for histone H2AX in recruitment of repair factors to nuclear foci after DNA damage. Curr Biol 10:886–895. doi:10.1016/S0960-9822(00)00610-2

Probst AV, Fagard M, Proux F, Mourrain P, Boutet S, Earley K, Lawrence RJ, Pikaard CS, Murfett J, Furner I, Vaucheret H, Mittelsten SO (2004) *Arabidopsis* histone deacetylase HDA6 is required for maintenance of transcriptional gene silencing and determines nuclear organization of rDNA repeats. Plant Cell 16:1021–1034. doi:10.1105/tpc.018754

Prymakowska-Bosak M, Przewloka MR, Slusarczyk J, Kuras M, Lichota J, Kilianczyk B, Jerzmanowski A (1999) Linker histones play a role in male meiosis and the development of pollen grains in tobacco. Plant Cell 11:2317–2329. doi:10.1105/tpc.11.12.2317

Rapp RA, Wendel JF (2005) Epigenetics and plant evolution. New Phytol 168:81–91. doi:10.1111/j.1469-8137.2005.01491

Rasmusson DC, Phillips RL (1997) Plant breeding progress and genetic diversity from de novo variation and elevated epistasis. Crop Sci 37:303–310

Regina A, Bird A, Topping D, Bowden S, Freeman J, Barsby T, Kosar-Hashemi B, Li Z, Rahman S, Morell M (2006) High-amylose wheat generated by RNA interference improves indices of large-bowel health in rats. Proc Natl Acad Sci USA 103:3546–3551. doi:10.1073/pnas.0510737103

Reinhart BJ, Weinstein EG, Rhoades MW, Bartel B, Bartel DP (2002) MicroRNAs in plants. Genes Dev 16:1616–1626. doi:10.1101/gad.1004402

Reuter G, Fischer A, Hofmann I (2005) Heterochromatin and the control of gene silencing in plants. In: Meyer P (ed) Plant epigenetics. Blackwell Scientific Publications, Oxford, pp 134–173

Rhoades MW, Reinhart BJ, Lim LP, Burge CB, Bartel B, Bartel DP (2002) Prediction of plant microRNA targets. Cell 110:513–520. doi:10.1016/S0092-8674(02)00863-2

Rideout WM 3rd, Coetzee G, Olumi A, Jones P (1990) 5-Methylcytosine as an endogenous mutagen in the human LDL receptor and p53 genes. Science 249:1288–1290. doi:10.1126/science.1697983

Roth BM, Pruss GJ, Vance VB (2004) Plant viral suppressors of RNA silencing. Virus Res 102:97–108. doi:10.1016/j.virusres.2004.01.020

Rusche LN, Kirchmaier AL, Rine J (2003) The establishment, inheritance, and function of silenced chromatin in *Saccharomyces cerevisiae*. Annu Rev Biochem 72:481–516. doi:10.1146/annurev.biochem.72.121801.161547

Schwartz YB, Pirrotta V (2008) Polycomb complexes and epigenetic states. Curr Opin Cell Biol 20:266–273. doi:10.1016/j.ceb.2008.03.002

Schwarz DS, Hutvagner G, Du T, Xu Z, Aronin N, Zamore PD (2003) Asymmetry in the assembly of the RNAi enzyme complex. Cell 115:199–208. doi:10.1016/S0092-8674(03)00759-1

Sheldon CC, Burn JE, Perez PP, Metzger J, Edwards JA, Peacock WJ, Dennis ES (1999) The FLF MADS box gene: a repressor of flowering in Arabidopsis regulated by vernalization and methylation. Plant Cell 11:445–458, PMID: 10072403, PMCID: PMC144185

Shen JC, Rideout WM III, Jones PA (1994) The rate of hydrolytic deamination of 5-methylcytosine in double-stranded DNA. Nucleic Acids Res 22:972–976. doi:10.1093/nar/22.6.972

Slotkin RK, Martienssen R (2007) Transposable elements and the epigenetic regulation of the genome. Nat Rev Genet 8:272–285. doi:10.1038/nrg2072

Slotkin RK, Freeling M, Lisch D (2005) Heritable transposon silencing initiated by a naturally occurring transposon inverted duplication. Nat Genet 37:641–644. doi:10.1038/ng1576

Smith WK, Gorz HJ (1965) Sweetclover improvement. Adv Agron 17:163–231

Soltis DE, Soltis PS (1995) The dynamic nature of polyploid genomes. Proc Natl Acad Sci USA 92:8089–8091

Stam M, Belele CL, Ramakrishna W, Dorweiler JE, Bennetzen JL, Chandler VL (2002) The regulatory regions required for B′ paramutation and expression are located far upstream of the maize *b1* transcribed sequences. Genetics 162:917–930, PMID: 12399399, PMCID: PMC 1462281

Stebbins GL (1950) Variation and evolution in plants. Columbia University Press, Columbia

Steward N, Ito M, Yamaguchi Y, Koizumi N, Sano H (2002) Periodic DNA methylation in maize nucleosomes and demethylation by environmental stress. J Biol Chem 277:37741–37746. doi:10.1074/jbc.M204050200

Sung S, Amasino RM (2004) Vernalization and epigemetics: how plants remember winter. Curr Opin Plant Biol 7:4–10. doi:10.1016/j.pbi.2003.11.010

Sunkar R, Zhu JK (2004) Novel and stress-regulated microRNAs and other small RNAs from *Arabidopsis*. Plant Cell 16:2001–2019. doi:10.1105/tpc.104.022830

Sunkar R, Kapoor A, Zhu J-K (2006) Post transcriptional induction of two cu/zn superoxide dismutase genes in *Arabidopsis* is mediated by down regulation of miR398 and important for oxidative stress tolerance. Plant Cell 18:2051–2065. doi:10.1105/tpc.106.041673

Sweredoski M, DeRose-Wilson L, Gaut BS (2008) A comperative computational analysis of nonautonomous *Helitron* elements between maize and rice. BMC Genomics 9:467. doi:10.1186/1471-2164/9/467

Takeda S, Sugimoto K, Otsuki H, Hirochika H (1999) A 13-bp *cis*-regulatory element in the LTR promoter of the tobacco retrotransposon *Tto1* is involved in responsiveness to tissue culture, wounding, methyl jasmonate and fungal elicitors. Plant J 18:383–393. doi:10.1046/j.1365-313X.1999.00460.x

Tatra GS, Miranda J, Chinnappa CC, Reid DM (2000) Effect of light quality and 5-azacytidine on genomic methylation and stem elongation in two ecotypes of *Stellaria longipes*. Physiol Plant 109:313–321. doi:10.1034/j.1399-3054.2000.100313.x

Thomas BC, Pedersen B, Freeling M (2006) Following tetraploidy in an Arabidopsis ancestor, genes were removed preferentially from one homeolog leaving clusters enriched in dose-sensitive genes. Genome Res 16:934–946. doi:10.1101/gr.4708406

Tian L, Chen ZF (2001) Blocking histone deacetylation in *Arabidopsis* induces pleiotropic effects on plant gene regulation and development. Proc Natl Acad Sci USA 98:200–205. doi:10.1073/pnas.011347998

Tokatlidis IS (2000) Variation within maize lines and hybrids in the absence of competition and relation between hybrid potential yield per plant with line traits. J Agric Sci 134:391–398. doi:10.1017/S0021859699007637

Tokatlidis IS, Tsialtas JT, Xynias IN, Tamoutsidis E, Irakli M (2004) Variation within a bread wheat cultivar for grain yield, protein content, carbon isotope discrimination and ash content. Field Crops Res 86:33–42. doi:10.1016/S0378-4290(03)00169-2

Tokatlidis IS, Xynias IN, Tsialtas JT, Papadopoulos II (2006) Single-plant selection at ultra low density to improve stability of a bread wheat cultivar. Crop Sci 46:90–97. doi:10.2135/cropsci2005.0125

Tokatlidis IS, Tsikrikoni C, Tsialtas JT, Lithourgidis AS, Bebeli PJ (2008) Variability within cotton cultivars for yield, fibre quality and physiological traits. J Agric Sci 146:483–490. doi:10.1017/S0021859608007867

Tokatlidis IS, Papadopoulos II, Baxevanos D, Koutita O (2010) Genotype×environment effects on single-plant selection at low density for yield and stability in climbing dry bean populations. Crop Sci 50:775–783. doi:10.2135/cropsci2009.08.0459

Tokatlidis IS, Tsikrikoni C, Lithourgidis AS, Tsialtas JT, Tzantarmas C (2011) Intra-cultivar variation in cotton: response to single-plant yield selection at low density. J Agric Sci 149: 197–204. doi:10.1017/S0021859610000596

Tranel PJ, Wright TR (2002) Resistance of weeds to ALS-inhibiting herbicides: what have we learned? Weed Sci 50:700–712. doi:10.1614/0043-1745(2002) 050[0700:RROWTA]2.0.CO;2

Tsaftaris AS, Kafka M (1998) Mechanisms of heterosis in crop plants. J Crop Prod 1:95–111. doi:10.1300/J144v01n01_05

Tsaftaris AS, Polidoros AN (2000) DNA methylation and plant breeding. Plant Breed Rev 18:87–176

Tsaftaris AS, Polidoros AN, Kapazoglou A, Tani E, Kovacevic NM (2008) Epigenetics and plant breeding. Plant Breed Rev 30:49–178

Tschiersch B, Hofmann A, Krauss V, Dorn R, Korge G, Reuter G (1994) The protein encoded by the *Drosophila* position-effect variegation suppressor gene *Su(var)3–9* combines domains of antagonistic regulators of homeotic gene complexes. EMBO J 13:3822–3831

Turner BM (2000) Histone acetylation and an epigenetic code. BioEssays 22:836–845. doi:10.1002/1521-1878(200009)22:9<836::AID-BIES9>3.0.CO;2-X

van der Krol AR, Mur LA, Beld M, Mol JNM, Stuitje AR (1990) Flavonoid genes in petunia: addition of a limited number of gene copies may lead to a suppression of gene expression. Plant Cell 2:291–299. doi:10.1105/tpc.2.4.291

Van Holde KE (1988) Chromatin, Springer series in molecular biology. Springer, New York

Vance V, Vaucheret H (2001) RNA silencing in plants-defense and counterdefense. Science 292:2277–2280. doi:10.1126/science.1061334

Vaucheret H, Vazquez F, Crata P, Bartel DP (2004) The action of *ARGONAUTE 1* in the miRNA pathway and its regulation by the miRNA pathway are crucial for plant development. Genes Dev 18:1187–1197. doi:10.1101/gad.1201404

Voinnet O (2001) RNA silencing as a plant immune system against viruses. Trends Genet 17: 449–459. doi:10.1016/S0168-9525(01)02367-8

Wang Q, Dooner HK (2006) Eukaryotic transposable elements and genome evolution special feature: remarkable variation in maize genome structure inferred from haplotype diversity at the bz locus. Proc Natl Acad Sci USA 103:17644–17649. doi:10.1073/pnas.0603080103

Wang J, Tian L, Lee H-S, Wei NE, Jiang H, Watson B, Madlung A, Osborn TC, Doerge RW, Comai L, Jeffrey Chen ZJ (2006) Genomewide nonadditive gene regulation in Arabidopsis allotetraploids. Genetics 172:507–517. doi:10.1534/genetics.105.047894

Wassenegger M, Heimes S, Riedel L, Sanger HL (1994) RNA-directed de novo methylation of genomic sequences in plants. Cell 76:567–576. doi:10.1016/0092-8674(94)90119-8

Wendel JF (2000) Genome evolution in polyploids. Plant Mol Biol 42:225–249. doi:10.1023/A:1006392424384

Wessler SR (1996) Plant retrotransposons: turned on by stress. Curr Biol 6:959–961. doi:10.1016/S0960-9822(02)00638-3

Weterings K, Russell SD (2004) Experimental analysis of the fertilization process. Plant Cell 16(Suppl):S107–S118. doi:10.1105/tpc.016873

Wolfe KH (2001) Yesterday's polyploids and the mystery of diploidization. Nature Rev Genet 2:323–341. doi:10.1038/35072009

Woodcock CL (2005) A milestone in the odyssey of higher-order chromatin structure. Nat Struct Mol Biol 12:639–640. doi:10.1038/nsmb0805-639

Wu K, Zhang L, Zhou C, Yu CW, Chaikam V (2008) HDA6 is required for jasmonate response, senescence and flowering in *Arabidopsis*. J Exp Bot 59:225–234. doi:10.1093/jxb/erm300

Xiao W, Gehring M, Choi Y, Margossian L, Pu H, Harada JJ, Goldberg RB, Pennell RI, Fischer RL (2003) Imprinting of the MEA Polycomb gene is controlled by antagonism between MET1 methyltransferase and DME glycosylase. Dev Cell 5:891–901. doi:10.1016/S1534-5807(03)00361-7

Yadegari R, Drews GN (2004) Female gametophyte development. Plant Cell 16(Suppl): S133–S141. doi:10.1105/tpc.018192

Ye J, Ai X, Eugeni EE, Zhang L, Carpenter LR, Jelinek MA, Freitas MA, Parthun MR (2005) Histone H4 lysine 91 acetylation. Mol Cell 18:123–130. doi:10.1016/j.molcel.2005.02.031 DOI:dx.doi.org

Zhang YX, Gentzbittel L, Vear F, Nicolas P (1995) Assessment of inter- and intra-inbred line variability in sunflower (Helianthus annuus) by RFLPs. Genome 38:1040–1048. doi:10.1139/g95-138

Zhang X, Yazaki J, Sundaresan A, Cokus S, Chan SW-L, Chen H, Henderson IR, Shinn P, Pellegrini M, Jacobsen SE, Ecker JR (2006) Genome-wide high-resolution mapping and functional analysis of DNA methylation in Arabidopsis. Cell 126:1189–1201. doi:10.1016/j.cell.2006.08.003

Zhang K, Sridhar VV, Zhu J, Kapoor A, Zhu JK (2007) Distinctive core histone post-translational modification patterns in *Arabidopsis thaliana*. PLoS ONE 2:e1210. doi:doi:10.1371/journal.pone.0001210

Zheng X, Pontes O, Zhu J, Miki D, Zhang F, Li WX, Iida K, Kapoor A, Pikaard CS, Zhu JK (2008) ROS3 is an RNA-binding protein required for DNA demethylation in *Arabidopsis*. Nature 455:1259–1262. doi:10.1038/nature07305

Zhou C, Zhang L, Duan J, Miki B, Wu K (2005) *HISTONE DEACETYLASE 19* is involved in jasmonic acid and ethylene signaling of pathogen response in *Arabidopsis*. Plant Cell 17:1196–1204. doi:10.1105/tpc.104.028514

Zhu JK (2008) Epigenome sequencing comes of age. Cell 133:395–397. doi:10.1016/j.cell.2008.04.016

Zhu J, Kapoor A, Sridhar VV, Agius F, Zhu J-K (2007) The DNA glycosylase/lyase ROS1 functions in pruning DNA methylation patterns in *Arabidopsis*. Curr Biol 17:54–59. doi:10.1016/j.cub.2006.10.059

Zilberman D, Gehring M, Tran RK, Ballinger T, Henikoff S (2007) Genome-wide analysis of Arabidopsis thaliana DNA methylation uncovers an interdependence between methylation and transcription. Nat Genet 39:61–69. doi:10.1038/ng1929

Integrated Agri-Aquaculture Systems

Ariel D. Zajdband

Abstract The integration of aquaculture within wider farming systems has been promoted as a way of increasing food production, conserving the environment and ensuring food security. Such systems are known as integrated agri-aquaculture systems (IAAS), which are characterized by their capacity to generate synergies between farm enterprises. Agri-aquaculture systems are generally family farming systems, comprised of three major sub-systems: aquaculture, agriculture and household. In this chapter, most of the common positive interactions of agri-aquaculture systems are reviewed, such as: (1) the use of animal manure as pond fertilizer, (2) the use of crop by-products as supplementary feed for fish, (3) the use of pond sediments as terrestrial crop fertilizers, and (4) the use of aquaculture wastewater for crop irrigation. Other indirect positive interactions between farm components on agri-aquaculture systems are also reviewed, such as the control of pests and weeds by fish in rice fields, the control of malaria, and the use of on-farm substrates for periphyton growth. This review shows how synergies are usually tested through controlled experiments that do not provide sound information about the behavior of the system as a whole. When integrated farming systems are analyzed, trade-off becomes evident among farming components, and therefore the opportunities for positive interactions that they generate. These trade-offs make difficult the inclusion of different kinds of synergies within the same farm. Approaches for the analysis of agri-aquaculture systems must involve broader scales, and as such, resilience theory provides a useful framework for further studies.

Keywords Integrated farming • Fish culture • Synergies • Malaria

A.D. Zajdband (✉)
School of Agronomy, University of Buenos Aires/CONICET,
Av. San Martín 4453 (1414DSE), Ciudad de Buenos Aires, Argentina
e-mail: zajdband@agro.uba.ar

E. Lichtfouse (ed.), *Genetics, Biofuels and Local Farming System*, Sustainable Agriculture Reviews 7, DOI 10.1007/978-94-007-1521-9_4,

1 Introduction

Agriculture is facing a great triple challenge that involves producing more food for a growing population, conserving the environment and ensuring food security (Godfray et al. 2010). It is estimated that food production will need to double by 2030 to feed the projected global human population (Tilman 1999). Increased income is also changing diets towards consumption of more animal products (Pingali 2007). At the same time, opportunities for the expansion of agricultural systems are declining (Rudel et al. 2009), and one billion people are currently estimated to be undernourished, even when food production is enough to satisfy the existing demand (FAO 2009b). Therefore, producing more food has proven to be not enough to solve the problem of hunger (Pretty et al. 2003).

During the last 40 years, the industrialization of agriculture has been responsible for an increase in food production (Hazell and Wood 2008). However, the Green Revolution also has resulted in negative environmental impacts and has been unable to solve the problem of malnutrition (Matson et al. 1997). The achievement of sustainability goals such as food production and reducing negative environmental impact has been difficult because of the existence of trade-offs between them (Naylor 2008). One alternative is to design agricultural systems based on the management of agroecosystem complexity (Pretty 2008), that enable the production of low-cost food while minimizing environmental impact, based on synergies that contrast with trade-offs in industrial farming practices (Cassman et al. 2005).

Traditional farming systems have proven to be capable of producing food in a environmentally sound manner and especially in supplying food to most vulnerable groups such as the rural poor (Herrero et al. 2010). These diversified systems often exist in marginal environments, where farmers manage agroecosystem complexity based on traditional ecological knowledge (Altieri 1999). This diversity reduces risks and allows integration between components, reducing the need for external inputs and energy sources. For instance, in crop-livestock systems, the animal component serves multiple purposes, such as providing animal manure and draft power for crop cultivation, from which products and wastes are used to feed livestock (Devendra and Thomas 2002a). Furthermore, traditional mixed farming systems have been proposed as a way to increase food security and mitigate climate change (FAO 2009a).

Aquaculture is being recognized as an important way of increasing food production (FAO 2007a; Fedoroff et al. 2010; Godfray et al. 2010). Aquaculture is the farming of aquatic organisms such as animals and plants (FAO 2008). Fish protein accounts for 15.3% of total world animal protein, being especially important in the poorest countries, where its contribution to total animal protein consumption reaches 18.5% (FAO 2009c). Fish plays an important role in food security by providing many nutrients, including high-quality protein, omega-3 polyunsaturated fatty acids, and micronutrients (Li and Hu 2009). During the last several decades, global demand for fish has increased as a consequence of population growth and increases in income, and fish has played an important role in the rapid growth of the consumption of animal products in developing countries. However, the stagnation of wild fish catch has created a gap between the supply and the increased demand for fish. This difference has been filled by aquaculture, which has been responsible for most of the net growth in fish production during the last decade (Delgado et al. 2003). Aquaculture is the fastest growing animal food production sector since 1970, contributing at present to almost half of global fish supply (FAO 2009c).

The aquaculture sector involves a wide variety of production systems that differ in their ability to increase food production and achieve food security while conserving the environment (Lazard et al. 2010). A wide classification system defines two types of production systems usually known as industrial and rural aquaculture (Diana 2009; Naylor et al. 2000). The first is generally dedicated to the cultivation of high-value, carnivorous species in monocultures such as salmon and shrimp, with an intense use of external inputs, for export and urban markets. Industrial aquaculture systems are promoted as a way of generating income and employment in developing countries (Hishamunda and Ridler 2002, 2006). However, they also are causing different social and environmental problems (Adger 2000; Bergquist 2007; Irz et al. 2007; Naylor and Burke 2005; Naylor et al. 2009; Primavera 2005; Stonich and Vandergeest 2001; Tacon et al. 2010). Although industrial systems can be improved to produce food while reducing environmental negative impacts (Chopin et al. 2010; Nobre et al. 2010; Troell et al. 2009), they are not able to produce cheap fish for the poorest and most food-insecure groups (Rivera-Ferré 2009).

The development of rural or small-scale aquaculture is the only way to increase the availability of fish for the rural poor (Ahmed and Lorica 2002; Brummett and Williams 2000; FAO 2007b). These systems are mostly focused on the production of different, relatively low-value species for household consumption or local markets. Although the contribution of small-scale aquaculture systems is underestimated in statistics (FAO 2003), it accounts for the majority of global aquaculture production (Muir 2005). These farming systems are utilized by families that use extensive and semi-intensive practices to raise herbivorous/ omnivorous species (Demaine 2009). Small-scale aquaculture is generally a component in wider farming systems where other activities take place.

The introduction of aquaculture into existing agricultural systems is being promoted as a sustainable alternative for the future of food production (Cassman et al. 2005; IAASTD 2009; Pretty 2008). The integration of aquaculture and agriculture enables the generation of synergies between farm components. According to Edwards (1998), synergies occur when *"an output from one sub-system in an integrated farming system which may otherwise may have been wasted becomes an input to another sub-system resulting in a greater efficiency of output of desired products from the land/water area under the farmer's control"*. Integrated agri-aquaculture systems (IAAS) are defined as the concurrent or sequential linkage between two or more agricultural activities, of which at least one is aquaculture (Little and Edwards 2003). Generally, IAAS have been named on the basis of their interacting components, i.e. rice/fish, pig/fish, poultry/fish or multi-component farms with three or more linked elements. Typically, animal manure and crop wastes are used as fertilizers and feed, respectively, for fish ponds (Prein 2002). The management of these synergies would reduce the need for external inputs, thus increasing total farm productivity and profitability in an ecologically sound manner through the increase in resource-use efficiency (FAO/IIRR/WorldFish Center 2001).

IAAS are also promoted as an efficient way to enhance food security (Karapanagiotidis et al. 2009; Kawarazuka 2010; Prein and Ahmed 2000). Generally, these farms produce low-value fish that is fed with what would otherwise be on-farm wastes. Moreover, these species often are raised in polyculture, taking advantage of the synergies between them that enable higher yields per hectare (Milstein 1992). Cultured species are complemented with a non-cultured component known as "self-recruiting species". The sale of the former generally is a source of income, and the latter are an important component of farmers' food consumption (Kadir et al. 2007; Thilsted et al. 1997). The introduction of aquaculture can also contribute to the alleviation of poverty through employment generation and the supply of low-cost fish for the poor (Edwards 1999).

IAAS have been described as more sustainable when compared with other food production systems (Prein 2007). Integrated farming systems are usually compared to less diverse and more open systems in terms of nutrients and energy, such as monoculture systems (Gomiero et al. 1997; Kautsky et al. 1997). It is considered that the diversity of enterprises in IAAS offers lower risks (Prein et al. 1998; Pullin 1998), so the introduction of aquaculture can contribute to increased farm resilience (Bailey 2008). Furthermore, the increased degree of internal recycling and the decreased dependence on external inputs is often considered as more sustainable (Cavalett et al. 2006; Dalsgaard et al. 1995; Dalsgaard and Oficial 1997; Pullin et al. 2007). In many cases, the integration of aquaculture into agricultural production on family farms has enhanced income generation (Berg 2002; Dey et al. 2007; Hishamunda et al. 1998; Kumaresan et al. 2009; Middendorp and Verreth 1986; Pant et al. 2005; Shang and Costa-Pierce 1983), improving social status and food consumption (Ahmed and Garnett 2010). The resilience of IAAS has been demonstrated in many cases around the world, where these systems have persisted for long periods of time (Costa-Pierce 2002; De Silva and Davy 2010).

While IAAS are traditional and have a long history in some regions, in other regions they are only just being developed. The traditional rice-fish systems in China or VACs (VACs = "vuon, ao, chuong", garden, pond and pigsty) in Vietnam have persisted for centuries, showing a high resilience in the face of different crisis (Lu and Li 2006; Mohanty et al. 2004; Weimin 2010). These cases have supported the idea that IAAS are only a viable option in South East Asia, where are traditional food production systems. This was used to explain the failure of many projects that tried to introduce aquaculture as a new activity in other regions such as Africa and Latin America (FAO 1997). However, traditional IAAS also exist in other regions, such as the *Chinampas* systems in Mexico (Smardon 2006), and are being introduced successfully as new practices in Latin America (Pilarski et al. 2004; Zajdband 2009) and Africa (Brummett 1999). Finally, fish culture is not widespread in Asia and there it has the potential even in this region for aquaculture expansion (Fernando and Halwart 2000), even when agricultural expansion opportunities are over (Devendra and Thomas 2002b). In spite of the differences among IAAS, they share the common feature of being located in rural areas of developing countries.

The aim of this paper is to review the synergies between aquaculture and other farm components and to discuss the opportunity that these systems have to maximize these synergies as a way to meet the challenge of producing more food, conserving the environment and enhancing food security.

2 The Aquaculture Sub-System

Aquaculture production systems are generally are classified in terms of their intensity, which determines different levels of resource use. The degree of intensification, which depends on a wide variety of factors such as capital, labor and mechanization, generally is defined in aquaculture according to feeding practices. In the case of IAAS, aquaculture can be considered as semi-intensive, as it relies on fertilization to produce natural feed and/or supplementary feeds. It represents an intermediate level between extensive systems, where fish only rely on natural feed, and the intensive systems, where fish is only supplied with an artificial external diet (Edwards 1993). Aquaculture operations can be classified as land-based, such as ponds or tanks, or flushed through water-based systems such as cages or long-lines (Muir 2005). While most small-scale aquaculture is land-based, taking place in ponds and rice fields, in some cases it is performed in water-based units, as when fish is raised in cages that are placed in artificial or natural water bodies within the farm (Pant et al. 2004; Phillips and De Silva 2006).

Ponds or rice fields for aquaculture are usually considered as ecosystems of their own. The biological components of these ecosystems can be classified as autotrophs and heterotrophs. The first group is constituted of primary producers, which are able to synthesize organic matter; these include algae, macrophytes, and certain non-photosynthetic bacteria. The natural productivity of the water body supports the entire food chain. In contrast, the heterotrophs rely on the consumption of organic matter produced by the autotrophs. This group involves fungi, bacteria, mollusks, crustaceans, and fish. The balance between respiration and photosynthesis of the biological components primarily determines the daily dissolved oxygen and pH fluctuations.

Primary productivity in ponds is mainly regulated by the availability of light, suitable water temperatures, and nutrients such as nitrogen, phosphorus, and carbon. Sufficient solar radiation and appropriate temperatures are functions of weather, location, and pond turbidity. Although most of the elementary nutrients are needed, nitrogen and phosphorus are of the greatest importance to primary production. Optimal concentrations of these nutrients in water bodies for aquaculture production are close to 4 and 1 kg (ha day)$^{-1}$, respectively. Although phosphorus is most commonly considered the major limiting nutrient in freshwater environments, natural pond productivity is usually limited by nitrogen in tropical environments (Knud-Hansen et al. 1991). In sediments with high clay content or those that contain acid-sulphate soils, available phosphorus can also easily limit algal productivity, since it is rapidly fixed by various cations and bound up in sediments (Knud-Hansen 1998).

Nutrient recycling of the organic matter deposited in the pond bottom enhances primary productivity. In ponds, water turnover is normally small, and water losses are caused mainly by evaporation and seepage (Avnimelech et al. 2008). Organic residues such as decaying algae settle to the pond bottom, a place with limited oxygen supply due to water stratification. Water stratification in tropical ponds is disturbed during the night due to cooling of surface waters, or during the day due to strong winds or rain (Diana et al. 1997). Stocking of fish species that browse in the sediment for food, such as common carp, may also lead to the mixing of sediment and thus help oxidize the pond bottom (Phan-Van et al. 2008; Ritvo et al. 2004). The oxygenation of the pond bottom allows the decomposition of organic matter by aerobic bacteria, prevents the release of reduced compounds into pond water that are toxic to fish, and provides a good habitat for benthic organisms (Boyd and Bowman 1997).

2.1 Stocked Species

Cultured aquatic species in IAAS are generally low-value herbivorous/omnivorous fish such as carps and tilapia. Species selection depends on different factors such as availability, the required equipment and facilities, and husbandry methods (Edwards 1998). The shortage of quality fish seed has been one of the most important constraints in the development of aquaculture (Muir 2005). Seed, i.e. newly hatched small fish, can be collected from the wild or from hatcheries that generally use artificial methods to stimulate fish spawning. Carp and tilapia seed can be produced without sophisticated knowledge and technology in rural areas (Little et al. 2007a). Furthermore, these species feed low in the food chain, which allow them to be raised by enhancing pond primary productivity through fertilization and the supply of on-farm wastes as supplementary feeds – as opposed to carnivorous species that require addition of high value protein feed. Although carps and tilapia present many advantages, in most regions they are non-native species that could potentially negatively impact local biodiversity (De Silva et al. 2009).

Different species usually are raised together in the same water body, allowing for an efficient use of various feeding and spatial niches. Generally, the most traditional polyculture is the stocking of different Chinese carp species. Increasing the number of cultured species results in the spreading of risk and often also allows for higher yields as a consequence of the synergistic interactions between species. However, interactions do not always result in better outcomes. The stocking of different species must be planned with knowledge of their complementarities or lack thereof. This property is determined not only by the species, but also the stocking density. The species ratio in the mixture must reflect the relative amount of different natural-food niches in the water body (Milstein 1992). However, fish feeding habits are opportunistic, and it is known that some cultured species can shift their feeding habits as a result of changes in the endogenous food supply or the presence of other species (Rahman et al. 2008).

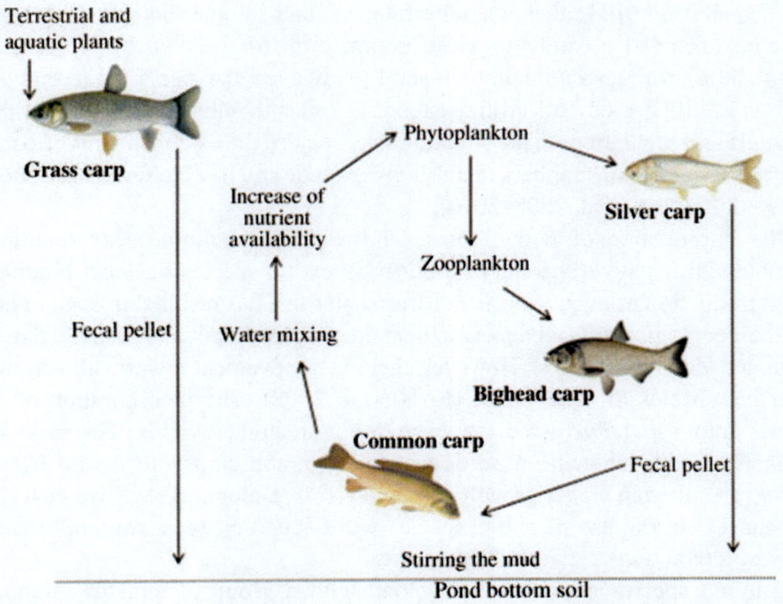

Fig. 1 Polyculture of Chinese carp in ponds and the synergies between them. Each species occupies different food and spatial niches, allowing efficient utilization of resources. Furthermore, the activity of some species enhances the growth of others, as in the case of common carp and silver carp

In the traditional Chinese-carp polyculture, each species has a different food and spatial niche and also present synergistic interactions (Fig. 1). The major Chinese carp are the silver carp (*Hypophthalmichthys molitrix*), bighead carp (*Aristichthys nobilis*), grass carp (*Ctenopharyngodon idella*), and common carp (*Cyprinus carpio*). The silver carp feeds on phytoplankton in the water column, the bighead consumes zooplankton, the grass carp eats plants, and common carp eat benthos/detritus (Kestemont 1995). There are two main synergistic interactions among carp species reared in polyculture. First, silver carp fecal pellets rich in partially digested phytoplankton are eaten by common carp that otherwise could not utilize these algae. Second, common carp stir up mud, recirculating nutrients into the water column that interfere with the development of filamentous algae and higher plants, thereby increasing nutrient availability for phytoplankton production and hence food for silver carp (Milstein 1992).

Relatively high value species also can be found in IAAS, even though their cultivation often requires the use of more external inputs (Demaine 2009). In this case, aquaculture is expected to contribute to poverty alleviation more through employment and foreign exchange, as high value species are generally exported, instead of supplying cheap protein for the poor. This is a common practice in Bangladesh, where freshwater prawn has become one of the most important sources of income for family farmers, after being introduced into fish polyculture

(Ahmed et al. 2010). Prawn is a bottom feeder, and so other bottom feeder species must be excluded in order to avoid competition for food niches (Hossain and Islam 2006). Prawn cultivation generally involves the use of external inputs such as certified seed and artificial feed, in order to satisfy international market demands and regulations. This situation may require the establishment of contracts between farmers and suppliers that usually exclude small-scale farmers (Belton and Little 2008; Lebel et al. 2008, 2010).

The introduction of native species into fish polyculture seeks to minimize the potential negative impacts of cultivating exotic species on local biodiversity. The introduction of exotic species of freshwater fish has resulted in social benefits, and their ecological impact appear to be moderate (Arthur et al. 2010; Gozlan 2008; Tapia and Zambrano 2003). However, there is a movement towards the cultivation of native species in aquaculture (Ross et al. 2008). The incorporation of native species into fish polyculture can even result in higher yields. For instance, in Southern Brazil, replacing a portion of the common carp with *jundiá* (*Rhamdia quelens*) resulted in higher growth rates and yields (Bolognesi Da Silva et al. 2006). The success in the use of native species will depend on seed availability and the study of interactions between these species.

Cultured species often share the pond with a group of aquatic animals that occur naturally without regular stocking called 'self-recruiting-species' (SRS) or 'small-indigenous species' (SIS) (Amilhat et al. 2009; Wahab et al. 2004). In the past, SRS were considered as harmful, and poisons were used to kill these fish species (Kadir et al. 2006). Fortunately, the importance of SRS is nowadays widely recognized, especially due to their enhancement of household food security. SRS are often consumed whole, including bones, and they constitute a rich source of vitamins and micronutrients (Morales et al. 2006; Thilsted et al. 1997). Studies are being conducted with the aim of understanding the interactions of these species with large carp and prawns, whose sale is a source of income for family farmers (Kadir et al. 2006, 2007; Kunda et al. 2008, 2009; Milstein et al. 2008, 2009; Wahab et al. 2004, 2008).

3 The Agricultural Sub-System

By definition, IAAS involve an agricultural component that generally exists before the introduction of aquaculture. Agro-ecological conditions such as climatic, soil-related and biotic factors determine the type of crop and animal system that has been developed at each location. Most IAAS are crop-dominated, but they generally also include livestock husbandry and non-cultivated areas such as woodlands. For instance, a typical farming system in South East Asia consists of rice as the main crop, with other enterprises occurring on the farm such as annual and perennial crops; a mixed garden around the homestead for fruits and vegetables, grass-fed cattle, and scavenging pigs or poultry (Devendra and Thomas 2002b; Edwards et al. 1988). These activities usually co-exist in small-scale farms that often occupy less than 5 ha, and that are located in risk-prone, marginal environments with infertile soils. The diversity of activities in these farms is usually a strategy of risk spreading, which guides many decisions in family farming systems (Altieri 2002).

The diversity of activities within the agricultural sub-system often shapes the way in which aquaculture is introduced. Conversely, management of agricultural activities may change as a result of the introduction of fish farming. For instance, large variations in dissolved oxygen concentrations in rice-based systems may require the stocking of fish species that are adapted to such conditions, such as bottom-feeder carps (Mohanty et al. 2004). In turn, the stocking of fish into rice fields may reduce the use of chemical pesticides (Berg 2002; Weimin 2010). Furthermore, the introduction of fish farming may result in the development of new agricultural activities, such as the cultivation of fruit and vegetables in dikes (Ahmed and Garnett 2010; Edwards 2008) or the introduction of ducks (Soliman et al. 2000).

Agricultural activities play an important role in providing food for household consumption and products for sale in local markets. Although most studies on IAAS have been focused on the positive effects of fish on nutrition and income (Prein and Ahmed 2000), aquaculture is sometimes not the most profitable enterprise or the most efficient way of enhancing household nutrition (Hishamunda et al. 1998; Wetengere 2009). The production and sale of products such as rice, vegetables, and fruits within IAAS sometimes have stronger nutritional and economic benefits than fish farming (Bosma et al. 2007; Ruddle and Prein 1997). However, as mentioned above, family farming systems support a wide diversity of activities for multiple purposes and to spread risk, as they are not driven solely by profit generation.

Agricultural crops and livestock are multipurpose, with importance beyond providing food and fiber. In integrated systems such as IAAS, each farm component is not only valued for the amount of product that it can produce, but also for its role in supplying inputs to other farm enterprises. For instance, grass-fed livestock such as cattle or buffalo are used for draft power in Asia or for milk production in India, but they are also important in providing manure for crop fertilization (Little and Edwards 2003). Cattle may also fulfill other roles, acting as a kind of savings account, and even as a display of status (Moll 2005). Furthermore, some crops or non-cultivated species can be used for farmers for social, ritualistic, or medicinal purposes (Tipraqsa et al. 2007).

4 The Household Subsystem

The integration of agricultural activities and aquaculture generally takes place within family farming systems. In these systems, household members manage the farm and provide most of the labor, although sometimes external workers are hired (Brookfield and Parsons 2007). Therefore, in IAAS, family labor is not taken into account in the estimation of production costs (Hishamunda et al. 1998). In addition to providing farm labor, household members are also consumers of different farm products. However, the proportion of laboring persons with respect to those who do not work i.e. 'consumers', varies, as when new members are incorporated through births, young people start working on the farm, or adults age. Typically, the household subsystem passes through different phases according to the ratio of workers to consumers (Bosma et al. 2007). Unfortunately, household dynamics have received little attention in the analysis of IAAS. In fact, the poor understanding of the socio-cultural environment has been indicated as a major impediment for the introduction of aquaculture in regions such as Africa and Latin America (Martinez-Espinosa 1992; Ruddle 1996).

The study of household subsystems in IAAS is generally motivated by the question of why they did or did not adopt IAAS. Most analyses seek to find a set of common features to describe the types of farmer that have adopted IAAS. In Vietnam, Bosma (2007) found that adoption mainly depends on the availability of family labor, farmers' wellbeing, and the area of the homestead. Farmers' wellbeing seems to be an important determinant of IAAS adoption in different regions. The introduction of aquaculture usually involves high initial costs that constrain its adoption by the poorest farmers (Kipkemboi et al. 2007; Lewis 1997; Little et al. 1996; Nhan et al. 2007; Tipraqsa et al. 2007). Furthermore, in some cases where IAAS are also adopted by the poor, wealthier farmers benefit the most (Ahmed et al. 2010; Little et al. 2007b). However, there are cases where IAAS have improved food production, and increased and diversified the incomes of the poorest farmers (Ahmed and Garnett 2010; Dey et al. 2010; Duc 2009). In fact, once aquaculture is adopted, the degree of farmer satisfaction with aquaculture is more related to the relative income provided by fish culture, than the farm absolute income (Duc 2008). Despite the importance of intrinsic socio-economic variables in explaining the adoption of IAAS, they are not always useful to explain the difference in fish yields after the introduction of aquaculture (Veerina et al. 1999).

IAAS can contribute to improved food security of poor households mainly by increasing available food supply and by generating income for the purchase of additional food (Ahmed and Lorica 2002). For instance, in Southwest Bangladesh, the sale of relatively high-value prawn for export is an important source of income for farmers, and other products such as fish, rice and vegetables are consumed by households and are sold in local markets (Ahmed and Garnett 2010). Although employment generation is often cited as another way of enhancing household food security, small-scale aquaculture mainly uses family labor, and thus it usually does not create new employment opportunities. However, if practices are intensified, they can require hired workers, which can be off-farm employment opportunities for members of poor households (Murshed-e-Jahan et al. 2010).

5 Synergies in Integrated Agri-Aquaculture Systems

The integration of aquaculture into existing farming systems aims to maximize the positive interactions among activities. Aquaculture systems, especially in nutrient-poor environments, require a supply of nutrients to support significant yields

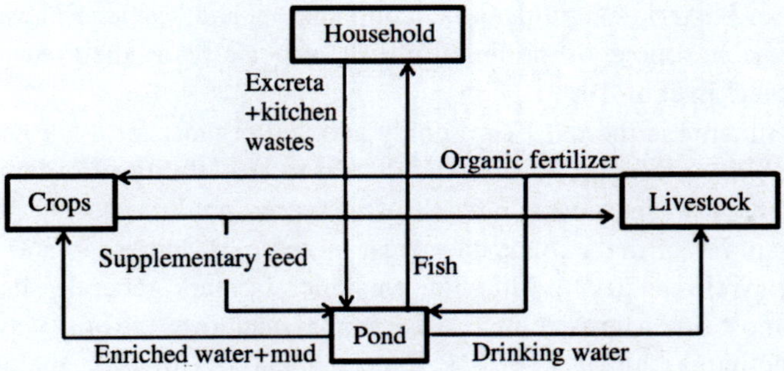

Fig. 2 Possible on-farm interactions between IAAS subsystems (Adapted from Edwards et al. 1988). Focus on the central role of aquaculture must be noted – for example, interactions also exist between households and livestock or crops, but they are not usually considered when IAAS are analyzed

(Hambrey 2004). On small-holder farms, these inputs are available mainly in the form of wastes from crops or animals. As part of an integrated farming system, aquaculture also plays a role as a supplier of inputs to other farm components. By-products of aquaculture such as enriched waste water and sediments can be used as inputs for other agricultural activities (Little and Edwards 2003) (Fig. 2).

Different methods were used to analyze the interactions among farm components (Prein 2007). These methods consider each interaction as a flow of nutrients, energy or mass, and thus, a budget analysis is usually performed (Dashu and Jianhuo 1996; Ruddle et al. 1986). In the 1990s, bio-resource flow diagrams were widely used as a tool for the study of interactions (Lightfoot et al. 1994), but they did not consider the external flows that enter or leave the farm, such as inorganic fertilizers or products for sale. Therefore, other approaches were developed later on that take into account all flows. Generally, these approaches involve a set of indicators and often require the use of ECOPATH software (Dalsgaard et al. 1995; Dalsgaard and Oficial 1997; Phong et al. 2010).

The common synergies found in IAAS are reviewed below. While the first two categories of positive interactions (5.1 and 5.2) involve the use of wastes from the agricultural sub-system as aquaculture inputs, the last two categories are related to the use of aquaculture wastes as inputs for agricultural activities. Although it is recognized that more interactions are usually present in IAAS, the focus is on the synergies that involve the aquaculture sub-system.

5.1 Organic Fertilization of Ponds

In IAAS, ponds are fertilized with on-farm crop and animal wastes. The purpose of pond fertilization in aquaculture is to stimulate pond primary productivity, and thus, to provide natural foods for cultured organisms. Pond fertilization stimulates the production of bacteria, phytoplankton, zooplankton, and benthos. These organisms serve as a rich source of protein for fish, especially in their early stages of development (Lin et al. 1997).

Animal manure is the most commonly adopted organic fertilizer because of its wide availability on farms. In fact, the increased availability of organic fertilizers during the last few decades as a result of increased poultry and pig production is one important factor that explains the great increase in Chinese inland aquaculture production (Weimin 2010). Although wealthier farmers generally use inorganic fertilizers, most small-farmers have to use animal manure as the only way to enhance pond productivity (Ahmed et al. 2010; Pant et al. 2005). However, another approach is the combination of both organic and mineral fertilizers (Bhakta et al. 2004). Manure includes farm animal feces, with or without urine, spilled feed, and bedding material. Organic fertilizers also include green manures such as legumes with high-nitrogen content and fresh or composted agricultural by-products such as pressmud, a sugarcane by-product (Keshavanath et al. 2006). A large variety of animal manures have been used to fertilize fish ponds. The quality of the manure as fertilizer depends on its composition, which may vary over time. The application of manure to the pond must follow a set of stages that enable efficient use of this resource. This will be discussed in further detail below.

5.1.1 Factors Affecting the Quantity and Quality of Animal Manure

The amount and nutrient composition of animal manure are highly variable. These depend on a wide variety of factors that can be classified as (1) animal-specific; (2) animal feed and feeding; (3) housing, bedding materials and waste collection; and (4) waste transportation, processing and storage. Good-quality manure can be defined as manure that supplies adequate amounts of the nutrients that most commonly limit pond primary productivity, such as nitrogen and phosphorus. The majority of published trials rarely report either manure nutrient composition or any of the other factors listed above.

Animal-Specific Factors

Animal manure is composed of the undigested fraction of the diet consumed by livestock and a small proportion of microbial debris and endogenous substances. The ability to digest and utilize consumed food differs among animal species and even among individuals of the same species with different size and growth rates. Animals that have been successfully employed in integrated livestock-aquaculture farming systems can be classified into monogastrics and ruminants. Monogastric animals such as poultry and pigs are usually fed with high-quality feeds and tend to have more nutrient-dense and valuable manure than ruminants, particularly those ruminants raised in nutrient-poor tropical environments; however, the range of waste quality varies enormously among ruminants. Manure from monogastrics usually presents a lower carbon to nitrogen ratio and higher phosphorus content. Ruminant manure also may contain tannins which stain pond water brown, increasing water turbidity and reducing phytoplankton growth (Hassan et al. 1997; Tersiyski et al. 2007). As ruminant manure has low nutrient density and a high concentration of tannins, its potential as pond fertilizer is limited.

In general, the selection of a particular fertilizer is more related to its availability and to cultural values than to its intrinsic value. Although monogastrics produce the best manure for pond fertilization, pig manure often is not used for religious reasons (Wetengere 2009). Nightsoil can be considered as monogastric manure used for pond fertilization, the adoption of which depends in part on cultural values (Ellis and Wang 1997; Knudsen et al. 2008). In contrast, ruminant fecal waste is still one of the most commonly used fertilizers in fish ponds in developing countries because of its availability. The use of rabbit manure is another example of locally available manure sources (Breine et al. 1996).

Animal Feeding and Nutrition

The composition of animal feed and how it is supplied affects manure characteristics. The quantity of manure is determined by the total amount of food consumed and the digestibility of that food. Any improvement in animal-nutrient retention will result in a reduction of manure quality. Clearly, the wastes from poorly fed animals are often low quality (Markewich et al. 2010). Furthermore, wastes from animals that have been in close contact with pesticides or drugs must be avoided. Feeding methods are also important, as animals that are fed *ad-libitum* will produce more manure and will also tend to spill more uneaten food that, if collected, can be used by fish (Little and Edwards 2003).

Housing, Bedding Materials and Waste Collection

The environment in which livestock live affects the loss of nutrients from manure. The main factors that can be controlled are related to the type of housing, the nature and management of bedding materials, and the collection of manure. The most common losses of nutrients in animal wastes are generally in the form of nitrogen, by leaching and volatilization. In contrast, phosphorus is mostly bound to solids, and thus, losses are often minimal.

Livestock can be classified into three groups: (1) those that are completely confined; (2) those that pass part of the day in a housing facility, and (3) those that are completely free. As it is much easier to collect manure from a confined animal, most experiments on organic fertilization use manure collected in these systems. However, the efficiency of collection will depend on other factors such as construction characteristics of the facilities and the bedding material. Livestock facilities can also be located above or close to the ponds, allowing fresh manure to be applied directly. If not, manure must be collected and then processed and/or stored.

The location of livestock facilities adjacent to or above the pond allows for the use of fresh animal manure and wastewater from cleaning facilities (Vu et al. 2007). This practice: (1) conserves the nutritional content of manure; (2) allows for the direct consumption of the uneaten livestock feed; (3) reduces the need for labor for collection, storage and transportation, and (4) usually saves land area for other purposes.

If the facility is not above the water body, then manure has to be collected. Livestock facilities should reduce the exposure of manure to environmental factors such as direct sunlight, wind, and rain that reduce their quality. These facilities must also be designed to allow the easy collection of wastes. Urine from ruminants is often overlooked, but it contains a considerable amount of nitrogen. Construction details such as hard flooring and roofing can prevent nitrogen leaching and volatilization, respectively.

Bedding material can be used to absorb moisture, reduce odor and nitrogen loss, and provide a soft floor surface, improving animal living conditions. Bedding high in carbon such as straw particularly improves the absorption of nitrogen. The mixing of the manure with bedding increases dry matter, ash, fiber and crude protein content of the litter (Ogunwande et al. 2008).

Manure from animals raised in scavenging systems can also be used. The collection of this manure is more difficult, and its quality is always worse due to the loss of urinary nitrogen. In some cases, collection can be facilitated by confinement of animals during nighttime. Under free-grazing conditions, livestock tend to deposit excreta in agriculturally non-productive areas where they tend to spend the most time, such as beneath trees, around sources of water, and gateways. Cattle manure plays an important role in fertilization of rangelands. Therefore, their re-distribution on IAAS implies a trade-off between rangeland fertility and pond productivity (Rufino et al. 2006).

Waste Transportation, Processing, and Storage

After collection, waste is transported to be processed and stored. Processing of manure can consist of composting or fermentation, i.e. controlled aerobic or anaerobic microbial decomposition, respectively. Manure processing aims to reduce the time between manure application and the release of nutrients in the pond water, since processed manure contains nutrients in a more available form. Another goal of processing is to facilitate the destruction of pathogens and parasites that can be transferred to fish, resulting in a more stable and drier product that can be stored with little odor (Bernal et al. 2009). Furthermore, processing enables the use of wastes that cannot be used in a raw state such as rice straw and stems, the addition of which may reduce nitrogen loss (Mahimairaja et al. 1995; Phuong et al. 2006).

Nutrient losses can also occur during waste processing and storage. In both aerobic and anaerobic processing, carbon losses are most significant. Some nitrogen may be lost during processing, although the final processed product generally presents higher nitrogen and phosphorus concentrations than at the beginning of processing (Goyal et al. 2005; Parkinson et al. 2004). The stabilized product can then be stored. During storage, nutrient losses will depend mostly on the exposure of processed manure to climatic factors such as temperatures, wind, and rain (Sagoo et al. 2007; Sommer et al. 2007).

5.1.2 Application of Animal Manure to Ponds

Animal manure can be applied to ponds at a fixed rate or a rate determined during the fish cultivation period from pond measurements. Generally, experiments are based on fixed fertilization rates (Garg and Bhatnagar 2000; Kaggwa et al. 2008); however, the requirements of the water body tend to vary during the production cycle. To maximize fertilization efficiency, the nutrient demand of the aquatic organisms should be synchronized with the nutrient supply from manure. If not, fish production is constrained due to either lack or overload of nutrients. Furthermore, fertilization management should also consider previous pond history, since older ponds will require relatively low fertilizer doses (Das and Jana 2003).

5.2 Supplementary Feeding

Supplementary feeding is a common practice that seeks to increase the stocking density of the water body, allowing higher yields per hectare. When the protein content of natural feed exceeds fish protein-requirements for growth, fish may use protein as an energy source. In this case, supplementation with high-energy feed can be used as a strategy to increase protein-use efficiency, which enables a higher stocking density. After this period of supplementation, growth of fish biomass will lead to higher protein requirements that cannot be met by the natural feed supply. Therefore, supplementation with high protein food may be needed (De Silva 1993).

Increase in fish yields due to the use of supplemental feeding is widely recognized, especially in herbivorous/omnivorous species. The relative utilization of supplementary feed varies among fish species, being mainly related to the digestibility of the ingredients and the digestive physiology of the fish. Herbivorous and omnivorous fishes appear to digest starchy components of plant materials more effectively than carnivorous fishes. Therefore, the former group of species is commonly found in IAAS, as most on-farm wastes that are used as supplementary feeds are crop by-products, which cannot be efficiently used by carnivorous species. For instance, common carp and rohu (*Labeo rohita*) fed with fish meal, rice bran, and mustard oil cake showed 1.5 and 2.1 times higher yield, respectively, than in treatments without supplementary feed (Rahman et al. 2006).

5.2.1 Food Ingredients

Any edible food that can be consumed directly by fish may be considered as a supplementary feed. Therefore, a wide variety of feed ingredients are used to prepare supplemental feeds. As with animal manure, the selection of ingredients by farmers is primarily determined by availability rather than by nutritional value, and the nutritional value of each ingredient varies widely. Ingredients can be classified into: (1) cereal grains and by-products; (2) oilseeds and by-products; (3) grain legumes; (4) root crops; (5) fruits; (6) grass and plant feedstuffs; (7) aquatic plants; (8) exogenously supplied live invertebrate organisms; (9) animal by-products, and (10) kitchen wastes.

Most ingredients of plant origin often contain anti-nutritional factors that can affect fish growth performance, interfering with food utilization and affecting animal health. One important group of anti-nutritional factors involves substances that affect protein utilization and digestion such as protease inhibitors and tannins. These factors are present in many plant-derived ingredients, particularly legumes. For instance, legumes such as *Sesbiana aculeata* can be used as an ingredient of supplementary feed, and it composes up to 12% of carp diets in untreated form. However, higher levels of inclusion are not recommended due to the presence of various anti-nutritional factors such as tannins and saponins (Hossain et al. 2001). Anti-nutritional factors may be destroyed or deactivated through processing techniques such as heating, solvent extraction and enzyme treatment. Nevertheless, before deciding on treatment procedures to reduce anti-nutritional factors, the different tolerance limits of different fish species must be considered. For example, tilapia seems to be more tolerant than carp to the increased presence of these substances (Francis et al. 2001).

Cereals include barley, maize, millet, oats, rice, rye, sorghum, and wheat. Cereal grains and by-products usually are used as dietary energy sources for aquatic cultured organisms. Cereal grains are rich sources of carbohydrates and poor sources of protein. Fagbenro (1999) found that although the composition and proportion of protein and carbohydrates in different cereal by-products are not uniform, their apparent digestibilities in common carp diets are similar, and thus, they can be used interchangeably in diet formulations. In one experiment, common carp and rohu fed on diets containing *Amaranthus* seeds, replacing rice bran and groundnut oil cake at different levels, show better growth than the control (Virk and Saxena 2003).

Both oilseeds and their by-products can also be used as feed. In oilseeds, the major food reserves are lipids, instead of starch as in cereals. Oilseeds include soybean, cotton, groundnut, sunflower, rape, flax, coconut, sesame, palm kernel, and mustard. By-products from the extraction of oil from these seeds are used as supplementary feed primarily in the form of cakes and meals, which are relatively rich sources of protein. However, their amino acid content is usually unbalanced. For example, Singh et al. (2003) found that the performance of mrigal (*Cirrhinus mrigala*) fingerlings, i.e. young or small fish, was better when groundnut, a legume, was used when compared with canola, sunflower, mustard or sesame cake as a protein source in the diet. Oilseeds and their byproducts also are being increasingly used in the replacement of fish meal as protein sources. Fagbenro (1998) found that the replacement of 67% of fish-meal protein with soybean, sunflower seed meal, or winged-bean meal did not affect growth, feed conversion efficiency or protein utilization by Nile tilapia (*Oreochromis niloticus*).

Legumes used as feed include chick pea, cowpea, lupin, and pigeon pea. Groundnut and soybean are oleaginous legume crops and thus are grown primarily for processing into edible oils and protein concentrates. Grain legumes are good sources of protein, energy and B vitamins such as thiamine and riboflavin, and often are considered natural supplements to cereal grains, because of their amino acid composition. Garg et al. (2002) have studied the effect of four different raw and processed leguminous seeds, namely soybean, *Vigna radiata*, cowpea and *Cyamopsis tetragonaloba*, on the growth of mrigal and rohu fingerlings. The processing of beans affected growth and other variables such as feed conversion ratio, protein efficiency ratio and gross energy retention.

Root crops are plant species with extensive carbohydrate food reserves in their underground stems, i.e., tubers or roots, and include taro, carrot, sweet potato, cassava, and potato. Root crops and tubers are poor sources of protein, vitamins, calcium and phosphorus, but are rich dietary sources of potassium and digestible carbohydrates. In one trial, increased cassava incorporation in tilapia diet as an energy source resulted in better growth and food utilization (Wee and Ng 1986).

Many plant fruits and their processed by-products may be considered for fish feeding. Fruits used as supplementary feed in aquaculture include banana, tangerine, papaya, orange and avocado. Although fruits generally are regarded as poor dietary sources of protein, they are rich sources of digestible carbohydrates and therefore energy. Ulloa and Verreth (2003) recommend the inclusion of coffee pulp in diets for *Oreochromis aurea,* limited to no more than 130 $g.kg^{-1}$ when fish are raised in earthen ponds and natural food is available. Ulloa et al. (2004) also found that green-banana meal may be a good energy source in omnivorous fish diets because of its high energy content.

Grass and forage crops constitute a valuable source of dietary carbohydrates, protein, vitamins and minerals. Results of proximate analysis have indicated the high potential of some of these plant materials, such as cassava and mulberry leaves, as supplementary feed because of their high protein and energy content (Dongmeza et al. 2009; Ng and Wee 1989). Grass carp was successfully grown using different forages usually available in IAAS such as maize leaves (Dongmeza et al. 2010), and a mixture of annual ryegrass (*Lolium multiflorum*) and napier grass (*Pennisetum purpureum*) (Casaca 2008). The inclusion of grass into tilapia supplementary feeds was also found as a viable option for small-holder farmers (Chikafumbwa et al. 1993; Poot-López et al. 2010).

Aquatic plants are a valuable source of nutrients for herbivorous fish species. In general, aquatic macrophytes are poor sources of dietary protein and lipids, but are regarded as good sources of digestible carbohydrates and minerals. However, they may contain considerable amounts of crude fiber and anti-nutritional factors such as tannins. Kalita et al. (2007) found that *Ipomoea reptans* and *Lemna minor*, two aquatic weed species, have the potential to become nutrient sources for fish diets since their concentrations of anti-nutritional factors were within tolerable limits for fish. Bairagi et al. (2002) found that fermented *Lemna* leaf meal can be incorporated into rohu diets at percentages up to 30%.

Macro-invertebrate animals such as insect larvae and oligochaete worms, either alive or in processed form, can be used as feed for fish. For instance, the replacement of fish meal with fermented silkworm-pupae silage in the formulation of carp diets resulted in higher body-weight gain and feed-conversion rate (Rangacharyulu et al. 2003). Freshwater snail (*Pila globosa*) is mixed with cooked rice, rice bran, oil cake, and fish meal to prepare home-made feed for the culture of freshwater prawns in rice fields in Bangladesh. However, snails have become extinct in most prawn farming areas due to excessive harvesting (Ahmed et al. 2008a).

Slaughterhouse by-products and kitchen wastes also are used as supplementary feeds (Little and Edwards 2003; Prein 2002). The majority of animal by-products has a well-balanced essential amino-acid profile and is good dietary sources of protein, lipid, energy, minerals and vitamins. Kitchen wastes are used widely in ponds, having the advantage of being processed before application. However, in order to improve palatability, it is advised to be at least sun-dried before being used as feed, since fish prefer feed to be dried (Negesse et al. 2009).

5.2.2 Food Processing and Storage

Feed preparation involves a wide range of processes from feed mixing to cooking and pelleting. The cooking of feed is not always possible because it consumes labor and firewood, two often scarce resources on small-holder farms. Therefore, feed ingredients can be simply mixed in mash, balls or pellets to provide a more balanced diet to fish. In tropical environments, sun-drying is a low-cost way to remove water from fresh feed, allowing long-term storage (Gabriel et al. 2007). Cooking does have several benefits such as (1) destruction of bacteria that potentially could contaminate the feed, (2) gelatinization of starch, increasing digestibility, (3) binding together of feed ingredients, reducing dispersion losses, and (4) preservation of feed for storage.

Feed must be stored to allow a regular food supply for cultured fish. The problem of feed availability during part of the year leads to inconsistency in the feeding regime. Without proper storage, feed quality can deteriorate, increasing the risk of fish diseases. The most important considerations when storing feed are temperature and moisture. As small-holder farms often lack cold-storage facilities, ensiling, i.e. acidification, can be a suitable method to preserve different feed resources in long-term storage (Vidotti et al. 2002).

5.3 Pond Mud for Terrestrial-Crop Fertilization

Sediment accumulated in aquaculture ponds can be used as a fertilizer for terrestrial crops in agricultural systems, reducing the need for external fertilizer inputs. Sedimentation is a continuous process in aquaculture ponds and results from both external and internal sources of solids (Boyd 1995). The accumulation of sediments affects the soil quality of the bottom of the pond and reduces pond volume (Avnimelech and Ritvo 2003). However, the removal of sediment is a limited practice in pond culture since it is labor-intensive and can result in significant losses of volatile nutrients, especially nitrogen in the form of ammonia. An alternative practice is the planting of rooting aquatic plants that are able to extract nutrients from mud substrates (Yi et al. 2002).

The quality of sediment as fertilizer for terrestrial crops depends mainly on the level of inputs used in aquaculture. The main sources of sediments are the solids that enter the pond in run-off due to watershed erosion, the erosion of pond levees, and organic inputs used for aquaculture. The latter are the main determinants of the nutrient composition of pond sediments. Uneaten feed, dead plankton and fish excrement are high in nitrogen and accumulate on the pond bottom, which serves as a nutrient trap. Highly insoluble phosphorus compounds also end up in the sediment. However, increased phosphorus inputs, increased sediment pH, and the bioturbation of the sediment by benthivorous fish enable a greater release of this nutrient to the water (Ritvo et al. 2004).

The removal of sediment from old ponds can improve the physical and chemical characteristics of pond-bottom soils. The deposition of organic matter increases microbial activity, thereby increasing oxygen demand, which can result in the depletion of dissolved oxygen in the pond bottom. The removal of the sediment reduces this oxygen demand, and thus, the risk of developing an anaerobic zone, which is not favorable for fish growth due to the release of toxic substances such as nitrites and hydrogen sulfides. Other benefits derived from the removal of the sediment are an increase in bottom compaction that facilitates many operations such as draining, harvesting, and dry-out between crops. However, the removal of the nutrients contained in sediment could also reduce productivity in fertilized ponds (Avnimelech and Ritvo 2003; Yuvanatemiya and Boyd 2006).

Sediments with high organic-matter and nutrient content have the potential to be used as terrestrial crop fertilizers. Organic matter can maintain or improve physical soil properties such as aeration, water-holding capacity, and aggregate stability. Furthermore, nutrient-rich sediments can be utilized to meet crop nutrient requirements (Mizanur et al. 2004). In contrast, sediments with low nutrient content have limited use as crop fertilizers. Accumulated pond sediments from one hectare of a semi-intensive tilapia culture cycle in Egypt potentially could meet the nitrogen and potassium fertilizer requirements for 0.35–1.2 and 0.7–1.5 ha of maize, respectively, but less than 1% of the phosphorus fertilizer requirement for the same areas (Muendo 2006).

5.4 *Pond Water for Irrigation*

Effluents from aquaculture may be used for terrestrial-crop irrigation. In fact, the construction of ponds on farms may mainly occur for irrigation purposes instead of fish farming (Fernando and Halwart 2000; Nhan et al. 2007). Pond-outlet water often is nutrient-rich, since it contains both dissolved and suspended inorganic and organic matter from fish culture such as fertilizers and feeds, or other external nutrients, such as matter derived from soil erosion, run-off and leaching. However, aquaculture effluents also may contain chemical residues such as antibiotics or pesticides. Typically, pond-effluent composition varies with the season, being richer in summer, when fertilization and feeding rates are higher.

The irrigation of terrestrial crops with aquaculture effluents have been conducted with success. However, in most of these experiments, fish were fed with completely artificial diets, which led to high nutrient concentrations in water (Al-Jaloud et al. 1996; Hussain and Al-Jaloud 1995, 1998; Lin and Yi 2003; Sikawa and Yakupitiyage 2010; Silva Castro et al. 2006). The effluents in these experiments differ significantly from effluents from ponds located in low-input systems. As such, while wastewater from fish ponds can be used as a source of irrigation in areas subject to seasonal drought, it does not always have high value as fertilizer (Costa-Pierce 1987). Pond water also can be used for other purposes such as drinking water for cattle and household bathing water; however, little research could be found on these issues (Edwards et al. 1988; Jewel et al. 2003).

6 Indirect Positive Interactions

Certain positive interactions are not considered when farms are analyzed only in terms of direct flows among activities. These interactions among activities result in mutual benefits for aquaculture and other related farm components, but they are the result of more complex processes than the simple transfer of mass, nutrients, or energy.

6.1 Fish Control of Pests and Weeds in Rice Fields

The introduction of fish into rice fields can help control rice pests and weeds. Although fish are an ancient component of rice fields in Asia (Liu and Cai 1998; Weimin 2010), the introduction of high-yielding varieties of rice and the use of fertilizers and pesticides have reduced the opportunity for the development of integrated rice-fish systems (Berg 2001). In the last decade, however, there has been a shift towards the use of integrated pest-management techniques, where fish play a fundamental role. The most important factors for effective pest control by fish are related to the feeding habit of the stocked species and their distribution within the rice fields (Halwart and Gupta 2004; Halwart et al. 1996). Stocking fish into rice fields allows for interaction between fish and some of the major pests of rice cultivation, such as insects, snails and weeds. Fish stocking often is part of a broader integrated pest-management strategy that reduces farmers' inclination to spray agrochemicals (Ahmed and Garnett 2010; Horstkotte-Wesseler 1999). Nevertheless, stocking fish into rice fields does not always demonstrate a clear influence on yields (Yang et al. 2006). While there is evidence that rice yields increase in integrated systems when compared with rice monoculture (Frei and Becker 2005; Gurung and Wagle 2005), other experiences do not find a clear effect of fish stocking on rice yields (Berg 2002; Frei et al. 2007a; Rothuis et al. 1998; Vromant and Chau 2005).

The ability of stocked fish to control arthropod populations depends on fish size and the vertical location of insects eggs. Small fish may control only arthropod species that lay their eggs in the water or have an aquatic stage during their life cycle. This is the case with the rice caseworm, (*Nymphula depunctinalis*) which lays? its eggs on floating rice leaves (Vromant et al. 1998). In contrast, plant suckers such as the brown (*Nilavaparta lugens*), green (*Nephotettix virescens*) and white (*Cofana spectra*) planthopper cannot be controlled by small fish due to their location far from water, laying eggs on rice leaves above water surfaces and having no aquatic phase (Frei et al. 2007b; Vromant et al. 2002). Rice planthoppers can only be controlled through the stocking of relatively large fish into rice fields (Yan et al. 1995).

The stocking of fish into rice fields also can be used as a control strategy for the golden apple snail (*Pomacea canaliculata*), one of the most important rice pests in Asia. Common carp is the only fish species recommended for the biological control of the golden apple snail in rice. However, small fish can consume only immature snails, and the stocking of bigger fish, i.e. with a body weight of more than one kilogram, is not practical. Thus, other control methods are needed to keep pest populations at tolerable level (Sin 2006).

Fish in rice fields also could control weeds either directly, by feeding on plants and seeds or uprooting plants, or indirectly, by increasing water turbidity. Direct control of weeds can be performed by different species such as grass carp, tilapia, and silver barb (*Puntius gonionotus*). Only bottom-feeding species such as common carp can uproot weeds and increase water turbidity through soil perturbation (Rothuis et al. 1999). The stocking of fish in rice fields can decrease the workload of manual weeding, and even lead to reduced herbicide application (Frei et al. 2007b).

6.2 Malaria Control

The stocking of fish into artificial water bodies constructed for irrigation purposes can be an effective strategy to meet the challenge of controlling malaria. Irrigation is currently viewed as the most effective way to increase food production in arid and semi-arid areas in Africa (You et al. 2010). However, the construction of artificial water bodies may provide breeding sites for mosquitoes, which are malaria vectors (Ijumba and Lindsay 2001). Malaria was responsible for more than 800,000 deaths in 2008, mainly among children younger than 5 years old living in sub-Saharan Africa (WHO 2009). Abandoned ponds can be suitable habitats for mosquitoes (Howard and Omlin 2008); however, the use of these ponds for aquaculture production can reduce mosquito populations. The removal of vegetation on the water surface can increase pond primary productivity and control mosquito populations by reducing shelter for mosquito larvae (Edwards et al. 1988). The stocking of fish can reduce mosquito pressure because different fish species can directly consume mosquito larvae (Aditya et al. 2010; Coates 1984; Frei et al. 2007b). The intake of mosquito larvae seems to be related to fish species and feeding regime, but not correlated with body size (Jianguo and Dashu 1995). Therefore, the selection of a suitable fish species and feeding practice are fundamental to fightingmalaria. For instance, the culture of common carp and *Poecilia reticulata*, a larvivorous fish, has resulted in a decreased incidence of malaria in Indonesia (Nalim 1994).

6.3 Periphyton-Based Aquaculture

Woody branches, bamboo poles, or any other hard substrates from the farm can be fixed in aquaculture water bodies to enhance the growth of periphyton. This technique is a simple way of producing natural food for cultured fishes. It is based on the growth of periphyton, which are microbial communities that grow attached to substrate. Periphyton are composed of sessile autotrophic and heterotrophic biota such as bacteria, fungi, protozoa, phytoplankton, and zooplankton. In most experiments, fish production was greater in ponds with additional substrates compared to controls without substrates, and this difference was attributed to periphyton (van Dam et al. 2002). The growth of these rich-protein communities can replace or complement supplemental feed in fish ponds (Azim et al. 2003).

Different substrates have been assessed for periphyton growth. Among these substrates, bamboo (*Bambusa sp.*) poles have shown the best performance, and thus are the most used in experiments (Azim et al. 2001, 2002; Uddin et al. 2006; Wahab et al. 1999). Rai et al. (2008) found no significant differences between bamboo poles and rice straw in fish growth, but rice straw led to the highest economic return due to its low cost. The substrates on which periphyton grow may be found on smallholder farms as wastes, without opportunity costs. Hence, the use of on-farm wastes as periphyton substrates may be the best choice from an economic point of view. Rice leaves also act as a substrate for periphyton growth in rice-fish systems, but their importance has not been recognized until recently (Saikia and Das 2009).

7 Managing Synergies at the Agroecosystem Level

As agroecosystems, IAAS present emergent behavior that has to be considered in their design and promotion. Agroecosystems are ecological systems modified by human beings to produce food, fiber or other agricultural products (Conway 1987). IAAS are usually presented, at the system level, as a way of increasing productivity while reducing the need for external inputs through the generation of synergies between farm components. However, the system is often analyzed through reductionist approaches that try to understand the system by studying each isolated interaction. When the system is analyzed as a whole, it becomes clear that it is difficult to include different kinds of synergies in the same system because of the existence of trade-offs between them. Assessments at the farm level usually are not able to detect these situations, or to find ways to overcome them.

7.1 Competition Between Activities and Trade-Offs Between Synergies

In IAAS, competition may exist among farm activities for scarce nutrients. Although nutrients in animal manure and crop by-products are presented as wastes from the point of view of each activity (Edwards 1998), there often are different farm components that compete for their use. Therefore, the aquaculture component becomes another destination that competes for those resources. For instance, ruminant manure may be used as a source of nutrients and energy for pond aquaculture, but it also has the capacity to improve soil fertility for terrestrial crops or grasslands. Ruminant manure can further be used by farmers for fuel or house building. In the same way, pig manure can be used to fertilize rice and vegetables. Supplementary feeding ingredients such as rice bran may have other potential uses, such as ruminant feed. Therefore, the introduction of aquaculture may result in the reduction of staple-crop outputs due to the reallocation of feeds or fertilizers.

Further trade-offs between aquaculture and other farming activities may be due to competition for important production factors such as land and labor. For instance, the construction of fish refuges in rice fields may reduce the area available for rice cultivation by 10–20%. Furthermore, stocking of fish usually involves spacing plants further apart to allow for more light penetration, increasing the natural productivity of the water, but decreasing rice productivity and yields (Sin 2006). In Bangladesh, some farmers do not cultivate rice during the rainy season to avoid competition with prawns for living space (Ahmed et al. 2008b). The introduction of fish culture also can increase the labor demand for households. Most work in aquaculture is concentrated in a few tasks in which cooperative forms of work are common, such as pond preparation, stocking, and harvesting, which reduces the individual household burden. However, certain tasks, such as the removal of sediments, imply much more significant amounts of labor, which limits the adoption of these practices. As among farm sub-systems, there also are trade-offs among synergies. When the scale of analysis is broadened, it becomes clear that many synergies cannot occur together, especially in low-input systems. For instance, there is often a trade-off between the quality of feed and natural pond productivity. The quality of livestock feed is one main factor that determines the quality of animal manure for fertilization. Since small-scale farms are resource limited, available animal feed can either be given to fish or to terrestrial livestock. If feed is given to terrestrial livestock, they may then produce better quality manure. Another trade-off occurs between the use of pond sediment as fertilizer for terrestrial crops, and the use of wastewater for irrigation. If sediment is removed, when the pond is refilled, the "new" sediment will absorb and retain a significant amount of nutrients added to the pond, especially phosphorus. Therefore, fewer nutrients will be available for enriching the water column, resulting in a lower nutrient content available for primary productivity and in wastewater that may be used for irrigation.

The existence of trade-offs between farm components and their interactions are often overlooked in the study of IAAS. Research on the integration of aquaculture and other farm components is based on classic research methods, which tend to focus on single processes. In these controlled experiments, trade-offs cannot be found, since the experiments only focus on one positive interaction between two farm components. Furthermore, sometimes on-station research is carried out using wastes from industrial systems such as feedlots, where genetically-improved livestock races are raised and fattened on high-quality artificial diets in closed facilities. These conditions contrast sharply with those commonly found on IAAS, where few animals from autochthonous races are raised in scavenging systems (Altieri 2002). Therefore, it is impossible to scale-up the results obtained in these controlled experiments to field conditions, especially resource-limited farming systems (Little and Edwards 2003).

On-farm research is able to deal with farm complexity, but it also presents the problem of uncontrolled confounding variables. A reductionist approach is necessary to isolate interactions and to focus on a few variables. Since it is impossible to study the effect of every combination of variables, mechanistic research can help in the design of integrated systems; however, on-farm adaptive experimentation and monitoring are the only way to identify patterns of response at the system level (Shennan 2008).

7.2 Is More Integration Always Better?

It is generally established that more integration among activities will result in higher production while minimizing wastes, improving IAAS' sustainability. In fact, the degree of integration has been used as an indicator of stability (Dalsgaard and Oficial 1997). It is assumed that the lack of integration among activities is due to farmers' limited understanding about their own farming systems (Bosma 2007; Nhan et al. 2006). Integration in IAAS often has been measured as the number of flows among farm components. These studies assign a value of 1 to each individual flow, and thus the sum of flows on each farm indicates the degree of farm integration (Bosma et al. 2006). In IAAS in Malawi, the level of integration is primarily governed by the number of enterprises, their relative size, and their distance to each other, the pond, and the homestead (Dey et al. 2010). In Thailand, the distance to market seems to drive the degree of integration, as the relative inaccessibility to markets reduces the availability of external inputs, increasing the need of using on-farm wastes (Pant et al. 2005). Although the number of flows is a useful indicator in some cases, it may also be important to analyze the quality of each interaction and the structure of the system as a whole.

The integration of farm components has many benefits; however, a high degree of connectivity among them should be avoided. A high degree of connectivity may lead to strong interdependence, which can affect resilience of the system as a whole. In China, mulberry trees were replaced with sugarcane when silk prices declined in 1929, resulting in the demise of the traditional mulberry-dike-fish-pond landscape (Weng 2007; Yee 1999). However, in this case, fish culture was retained, and new activities were introduced to these Chinese farms, such as cultivation of vegetables, fruits and flowers, in addition to raising poultry, i.e. ducks and chicken (Lo 1996). In other cases, changes in agricultural practices such as the introduction of high yielding, short duration rice varieties led to the exclusion of fish culture from rice-based farms (Berg 2002). As such, a high degree of dependence is risky because failure in one component may result in collapse of the whole system. Instead of extreme connectivity, some degree of separation, i.e. modularity, among system components is more robust (Webb and Bodin 2008). Although there is no optimal degree of modularity, if farm components are not completely dependent on one another, failure can be isolated and remains at the sub-system level.

7.3 Farms Are More Than Flows of Mass, Energy, and Nutrients

The structure of IAAS usually is represented as farm components related through flows of mass, nutrients, and/or energy. The use of bio-resource flow diagrams presents some advantages, since it enables the participation of farmers with a wide variety of educational backgrounds, and structures information, providing a brief summary of the whole farm (Dalsgaard and Oficial 1997). Nevertheless, these diagrams are unable to integrate all system dynamics. As demonstrated above, considering the farm only in terms of direct flows between activities disregards some important positive interactions such as the control of pests and human vector diseases, and the use of on-farm substrates for periphyton growth. Moreover, thinking of a farm in this way also carries other incorrect assumptions or omissions that are widespread in analyses of IAAS, such as that: (1) no losses occur during the flow from one activity to another, (2) integrated systems are closed in terms of nutrients and energy, and (3) people are considered only in terms of flows.

The representation of flows among activities does not consider losses that occur during the transfer of resources. Between the production of "waste" from one activity, and its use, there are intermediary stages, e.g. collection, processing, and storage, that lead to losses. These losses occur both because of mismatches in space and time. The availability of some resources is affected greatly by seasonality, potentially causing a temporal mismatch. The efficient use of on-farm inputs depends on their collection, processing, and storage, allowing for synchronization between collection, application, and crop/animal uptake. As pond inputs are not available throughout the year, their storage allows for their use when needed (Prein 2002).

IAAS usually are characterized as nearly closed systems in terms of nutrients and energy. Bio-resource flow diagrams depict the flows that occur among farm components, reinforcing this vision. However, IAAS are far from closed ecological cycles (Edwards 1993, 2008). Not only do farms often export their products and import external inputs, they also are open systems that interact with their environment. These relationships are numerous and important for understanding nutrient dynamics within a farm. For example, nitrogen inputs come from fixation and purchased external inputs. Nitrogen can also be exported or lost from the system through volatization, run-off, leaching, predation of fish or crops by wild birds, and the sale of harvested products (Guo and Bradshaw 1993). The use of external inputs is more important in resource-poor environments, where inputs are required to create minimal pond fertility levels (Demaine 2009).

In bio-resource flow diagrams, households are considered as just another farm component that can provide or consume farm products and wastes; however, people are the managers of these systems, with the capacity to modify interactions between farm components. Therefore, not only must people be considered as more than just another activity, but the system structure and interactions must be considered as dynamic. Systems can be changed by people in response to events such as extreme weather or market changes. Farmers do not manage the system only in terms of flows of mass, nutrients, and energy, with the aim of maximizing their internal recycling. Farm components have different significance for people, such as the complex role of traditional mixed crop-livestock systems, where livestock plays many roles (Devendra and Thomas 2002a). On IAAS, ponds may play different roles, such as water for bathing or washing, or savings accounts. (Little et al. 2007b). Farmers also have their own cultural values that explain the use or non-use of some inputs, such as pig manure.

8 Future Challenges for Integrated Agri-Aquaculture Systems

Aquaculture must play an important role in the future of food production, and IAAS offer many advantages, especially when compared with more intensive forms of aquaculture. However, integrated aquaculture is not a one-size-fits-all practice, especially in marginal environments, where integration does not always offer the most efficient way to increase farm productivity (Michielsens et al. 2002). Therefore, IAAS should not be promoted as a panacea. Socio-economic and environmental conditions should be evaluated before IAAS are promoted in a given region (Hambrey et al. 2008). This implies that approaches for analyzing IAAS must involve scales broader than the farm level to understand their contribution to sustainable rural development (Bondad-Reantaso et al. 2009).

The potential contribution of IAAS to poverty reduction and food security is related to their ability to become a source of income for poor farmers or hired workers, and to increase the availability of cheap animal protein for poor consumers (Ahmed and Lorica 2002). However, the trend towards introduction of relatively high-value, export-oriented species such as freshwater shrimp is reducing opportunities for the poor to benefit from the introduction of IAAS. These systems exclude the poorest farmers as producers and reduce the availability of cheap protein for consumers.

The cultivation of high-value species often involves vertical integration between a large firm and a group of farmers. Vertical integration is purported to provide producers with the benefits of a secure market and the opportunity to modernize. In contracts between small-scale farmers and large firms, firms supply inputs and transfer technology to farmers, while ensuring them a market for their final product (Hishamunda and Ridler 2002). Nevertheless, in practice, vertical integration often results in highly unfavorable contracts for farmers (Belton and Little 2008; Lebel et al. 2010) that may exclude the poorest (Ahmed and Garnett 2010). Furthermore, international standards for high-value species may create barriers for farmers in entering wider markets (Bostock et al. 2010). The increasing adoption of high-value species cultivation further reduces the availability of other, cheaper species such as carp, which are more affordable to the poor consumers. The introduction of aquaculture and its integration within farming systems must be focused on the poor to be an effective way of enhancing food security.

Negative environmental consequences derived from the adoption of IAAS mainly occur outside the farm, involving broader spatial and temporal scales than those generally utilized in the assessment of IAAS. The most important environmental impact of IAAS is related to the use of alien species and water pollution. The introduction of alien species can result in escapees of these fish. If non-native species escape from ponds, they can contribute to habitat deterioration, increased competition for resources, and disease transmission (De Silva et al. 2009). However, negative effects from this phenomenon appear to be moderate and must be balanced against the economic benefit derived from the introduction of alien species (Gozlan et al. 2010). The adoption of IAAS also can contribute to increased water pollution as a consequence of nutrient discharge from ponds. Nhan et al. (2008) have shown that excreta-fed aquaculture practiced in Vietnam is not sustainable in the long-run if farmers continue to adopt it. Both factors may contribute to biodiversity loss, due to the adoption of IAAS. Other mechanisms can reduce local biodiversity after the introduction of aquaculture as well, such as the overharvesting of prawn fry and snails, which are used for stocking and feeding, respectively, from natural environments (Ahmed and Garnett 2010). Other environmental impacts associated with IAAS are related to the emission of greenhouse gases such as methane and nitrous oxide (Datta et al. 2009).

The theory of resilience provides a dynamic framework for the study of IAAS, as both social and ecological components must be considered through different scales (Gunderson and Holling 2002). IAAS tend to be considered as resilient systems because of their diversity (Bailey 2008). However, diversity is only one mechanism of system robustness. It will be important to assess IAAS in terms of the other mechanisms of system robustness, which are (1) the redundancy of system components, (2) the control of flows within the system, and (3) modular structure (Webb and Bodin 2008). Furthermore, the persistence of IAAS in some regions such as rice-fish culture in China has reinforced the idea that IAAS are resilient systems (De Silva et al. 2010). However, high system resilience, i.e. a system's ability to resist external disturbances and persist, does not necessarily mean that the system is sustainable (Holling 2001). For instance, a model in which large firms take advantage of their dominant position can be resilient, even though this occurs at the expense of farmers' wellbeing. In contrast, sustainability is related to enhancing the resilience of the system in desirable states and reducing the resilience of the system in undesirable states (Perrings 1998).

9 Conclusion

Synergies between aquaculture and agricultural activities may enhance nutrient recycling, increasing resource use efficiency. The use of wastes as inputs for other agricultural activities usually is tested in isolation. When analyses are performed at the farm level, it is realized that many positive interactions may not be able to coexist on the same farm. Further studies of IAAS must incorporate scales broader than the farm to allow for understanding of the interactions between the farm and broader processes occurring on larger scales. Resilience theory provides a useful framework for further studies on IAAS. Resilience of IAAS depends not only on farm dynamics, but also the broader social-ecological system.

Acknowledgements The author would like to thank Laura J. Hess for her invaluable help, as well as the two anonymous reviewers for their useful comments.

References

Adger NW (2000) Social and ecological resilience: are they related? Prog Hum Geogr 24(3):347–364

Aditya G, Pal S, Saha GK (2010) An assessment of fish species assemblages in rice fields in West Bengal, India: implications for management. J Appl Icthtyol 26:535–539. doi:10.1111/j.1439-0426.2010.01460.x

Ahmed N, Garnett ST (2010) Sustainability of freshwater prawn farming in rice fields in southwest Bangladesh. J Sustain Agric 34:659–679. doi:10.1080/10440046.2010.493397

Ahmed M, Lorica MH (2002) Improving developing country food security through aquaculture development – lessons from Asia. Food Policy 27(2):125–141. doi:10.1016/S0306-9192(02)00007-6

Ahmed N, Demaine H, Muir JF (2008a) Freshwater prawn farming in Bangladesh: history, present status and future prospects. Aquacult Res 39(8):806–819. doi:10.1111/j.1365-2109.2008.01931.x

Ahmed N, Brown JH, Muir JF (2008b) Freshwater prawn framing in *gher* systems in southwest Bangladesh. Aquac Econ Manage 12:207–223

Ahmed N, Ellison EH, Muir JF (2010) Rice fields to prawn farms: a blue revolution in southwest Bangladesh? Aquac Int 18:555–574. doi:10.1007/s10499-009-9276-0

Al-Jaloud AA, Hussian G, Karimulla S, Al-Hamidi AH (1996) Effect of irrigation and nitrogen on yield and yield components of two rapeseed cultivars. Agric Water Manage 30(1):57–68

Altieri MA (1999) The ecological role of biodiversity in agroecosystems. Agric Ecosyst Environ 74:19–31

Altieri MA (2002) Agroecology: the science of natural resource management for poor farmers in marginal environments. Agric Ecosyst Environ 1971:1–24

Amilhat E, Lorenzen K, Morales EJ, Yakupitiyage A, Little DC (2009) Fisheries production in Southeast Asian farmer-managed aquatic systems (FMAS). I. Characterisation of systems. Aquaculture 296:219–226. doi:10.1016/j.aquaculture.2009.08.014

Arthur RI, Lorenzen K, Homekingkeo P, Sidavong K, Sengvilaikham B, Garaway CJ (2010) Assessing impacts of introduced aquaculture species on native fish communities: Nile tilapia and major carps in SE Asian freshwaters. Aquaculture 299:81–88. doi:10.1016/j.aquaculture.2009.11.022

Avnimelech Y, Ritvo G (2003) Shrimp and fish pond soils: processes and management. Aquaculture 220:549–567

Avnimelech Y, Verdegem MCJ, Kurup M, Keshavanath P (2008) Sustainable land-based aquaculture: rational utilization of water, land and feed resources. Mediterr Aquac J 1(1):45–55

Azim ME, Wahab MA, van Dam AA, Beveridge MCM, Verdegem MCJ (2001) The potential of periphyton-based culture of two Indian major carps, rohu *Labeo rohita* (Hamilton) and *gonia Labeo gonius* (Linnaeus). Aquac Res 32(3):209–216. doi:10.1046/j.1365-2109.2001.00549.x

Azim ME, Wahab MA, Verdegem MCJ, van Dam AA, van Rooij JM, Beveridge MCM (2002) The effects of artificial substrates on freshwater pond productivity and water quality and the implications for periphyton-based aquaculture. Aquat Living Resour 15:231–241

Azim ME, Verdegem MCJ, Mantigh I, van Dam AA, Beveridge MCM (2003) Ingestion and utilization of periphyton grown on artificial substrates by Nile tilapia, *Oreochromis niloticus L.* Aquac Res 34:85–92

Bailey C (2008) Human dimensions of an ecosystem approach to aquaculture. In: Soto D, Aguilar-Manjarrez J, Hishamunda N (eds) Building an ecosystem approach to aquaculture. FAO, Rome, pp 37–46

Bairagi A, Ghosh KS, Sen SK, Ray AK (2002) Duckweed (*Lemna polyrhiza*) leaf meal as a source of feedstuff in formulated diets for rohu (*Labeo rohita Ham.*) fingerlings after fermentation with a fish intestinal bacterium. Bioresour Technol 85:17–24

Belton B, Little DC (2008) The development of aquaculture in central Thailand: domestic demand versus export-led production. J Agrarian Change 8(1):123–143

Berg H (2001) Pesticide use in rice and rice–fish farms in the Mekong Delta, Vietnam. Crop Prot 20:897–905

Berg H (2002) Rice monoculture and integrated rice-fish farming in the Mekong Delta, Vietnam-economic and ecological considerations. Ecol Econ 41:95–107

Bergquist DA (2007) Sustainability and local people's participation in coastal aquaculture: regional differences and historical experiences in Sri Lanka and the Philippines. Environ Manage 40:787–802. doi:10.1007/s00267-006-0108-y

Bernal MP, Alburquerque JA, Moral R (2009) Composting of animal manures and chemical criteria for compost maturity assessment: a review. Bioresour Technol 100(22):5444–5453. doi:10.1016/j.biortech.2008.11.027

Bhakta JN, Sarkar D, Jana S, Jana BB (2004) Optimizing fertilizer dose for rearing stage production of carps under polyculture. Aquaculture 239:125–139. doi:10.1016/j.aquaculture.2004.03.006

Bolognesi Da Silva L, Barcellos LJG, Quevedo RM, de Souza SMG, Kreutz LC, Ritter F, Finco JA, Bedin AC (2006) Alternative species for traditional carp polyculture in southern South America: initial growing period. Aquaculture 255(1–4):417–428. doi:10.1016/j.aquaculture.2005.12.024

Bondad-Reantaso MG, Bueno P, Demaine H, Pongthanapanich T (2009) Development of an indicator system for measuring the contribution of small-scale aquaculture to sustainable rural development. In: Bondad-Reantaso MG, Prein M (eds) Measuring the contribution of small-scale aquaculture: an assessment. FAO Fisheries and Aquaculture Technical Paper 534. FAO, Rome, pp 161–179

Bosma RH (2007) Using fuzzy logic models to reveal farmers' motives to integrate livestock, fish, and crops. Ph.D. thesis, Wageningen University, Wageningen

Bosma RH, Phong LT, Kaymak U, Van Den Berg J, Udo HMJ, Van Mensvoort MEF, Tri LQ (2006) Assessing and modelling farmers' decision making on integrating aquaculture into agriculture in the Mekong Delta. NJAS 53(3–4):281–300

Bosma RH, Udo HMJ, Verreth JAJ, Visser LE, Nam CQ (2007) Agriculture diversification in the Mekong Delta: farmers' motives and contributions to livelihoods. Asian J Agric Dev 2(1–2): 49–66

Bostock J, McAndrew B, Richards R, Jauncey K, Telfer T, Lorenzen K, Little D, Ross L, Handisyde N, Gatward I, Corner R (2010) Aquaculture: global status and trends. Phil Trans R Soc B 365:2897–2912. doi:10.1098/rstb.2010.0170

Boyd CE (1995) Bottom soils, sediment, and pond aquaculture. Chapman & Hall, New York

Boyd CE, Bowman JR (1997) Pond bottom soils. In: Egna HS, Boyd CE (eds) Dynamics of pond aquaculture. CRC Press, Boca Raton, pp 135–162

Breine JJ, Teugels GG, Podoor N, Ollevier F (1996) First data on rabbit dung as a water fertilizer in tropical fish culture and its effect on the growth of *Oreochromis niloticus* (Teleostei, Cichlidae). Hydrobiologia 321(2):101–107. doi:10.1007/BF00023167

Brookfield H, Parsons H (2007) Family farms: survival and prospect: a worldwide analysis. Routledge, London

Brummett RE (1999) Integrated aquaculture in sub-Saharan Africa. Environ Dev Sustain 1:315–321

Brummett RE, Williams MJ (2000) The evolution of aquaculture in African rural and economic development. Ecol Econ 33(2):193–203. doi:10.1016/S0921-8009(99)00142-1

Casaca JM (2008) Policultivos de peixes integrados à produção vegetal: avaliação econômica e sócio ambiental (peixe-verde). Ph.D. thesis, UNESP, Jaboticabal

Cassman KG, Wood S, Choo PS, Cooper HD, Devendra C, Dixon J, Gaskell J, Kahn S, Lal SR, Lipper L, Pretty J, Primavera J, Ramankutty N, Viglizzo E, Wiebe K (2005) Cultivated systems. In: Hassan R, Scholes R, Ash N (eds) Ecosystems and human well-being: current state and trends, vol 1, Millennium ecosystem assessment series. Island Press, Washington, DC, pp 745–794

Cavalett O, Queiroz JF, Ortega E (2006) Emergy assessment of integrated production systems of grains, pig and fish in small farms in the South Brazil. Ecol Modell 193:205–224

Chikafumbwa FJKT, Costa-Pierce BA, Jamu DM, Kadongola WK, Balarin JD (1993) Investigations on the use of on-farm resources as pond inputs to culture *Tilapia rendalli* and *Oreochromis shiranus* on smallholder farms in rural Malawi. Aquaculture 117:261–271

Chopin T, Troell M, Reid GK, Knowler D, Robinson SMC, Neori A, Buschmann AH, Pang SJ, Fang J (2010) Integrated multi-trophic aquaculture. In: Advancing the aquaculture Agenda: workshop proceedings. OECD, Paris, pp 195–218, http://www.sourceoecd.org/agriculture/9789264088719

Coates D (1984) A survey of the fish fauna of Sudanese irrigation systems with reference to the use of fishes in the management of ecological problems (the control of aquatic weeds, malaria and infective schistosomiasis). Fish Manage 15(3):81–96

Conway G (1987) The properties of agroecosystems. Agric Syst 24:95–117

Costa-Pierce BA (1987) Initial feasibility of integrated aquaculture on lava soils in Hawaii. Aquacult Eng 6:171–182

Costa-Pierce BA (2002) Ecological aquaculture: the evolution of the blue revolution. Blackwell, Oxford

Dalsgaard JPT, Oficial RT (1997) A quantitative approach for assessing the productive performance and ecological contributions of smallholder farms. Agric Syst 55(4):503–533

Dalsgaard JPT, Lightfoot C, Christensen B (1995) Towards quantification of ecological sustainability in farming systems analysis. Ecol Eng 4:181–189

Das SK, Jana BB (2003) Pond fertilization regimen: state of the art. J Appl Aquac 13(1–2):35–66

Dashu N, Jianhuo W (1996) Material cycles and economic returns in ricefish ecosystem. In: Mackay KT (ed) Rice-fish culture in China. IDRC, Ottawa, pp 177–182

Datta A, Nayak DR, Sinhababu DP, Adhya TK (2009) Methane and nitrous oxide emissions from an integrated rainfed rice-fish farming system of Eastern India. Agric Ecosyst Environ 129(1–3):228–237. doi:10.1016/j.agee.2008.09.003

De Silva SS (1993) Supplementary feeding in semi-intensive aquaculture systems. In: New MB, Tacon AGJ, Csavas I (eds) Farm-made aquafeeds. Proceedings of the FAO/AADCP regional expert consultation on farm-made aquafeeds, 14–18 Dec 1992, Bangkok. FAO-RAPA/AADCP, Bangkok, pp 24–60

De Silva SS, Davy FB (2010) Success stories in Asian aquaculture. Springer, Dordrecht

De Silva SS, Davy FB, Phillips MJ (2010) Synthesis and lessons learned. In: De Silva SS, Davy FB (eds) Success stories in Asian aquaculture. Springer, Ottawa, pp 187–199

De Silva SS, Nguyen TTT, Turchini GM, Amarasinghe US, Abery NW (2009) Alien species in aquaculture and biodiversity: a paradox in food production. Ambio 38(1):24–28

Delgado CL, Wada N, Rosegrant MW, Meijer S, Ahmed M (2003) Fish to 2020: supply and demand in changing global markets. IFPRI/World Fish Center, Washington, DC

Demaine H (2009) Rural aquaculture: reflections ten years on. In: Bondad-Reantaso MG, Prein M (eds) Measuring the contribution of small-scale aquaculture: an assessment. FAO, Rome, pp 45–58

Devendra C, Thomas D (2002a) Crop–animal interactions in mixed farming systems in Asia. Agric Syst 71:27–40

Devendra C, Thomas D (2002b) Smallholder farming systems in Asia. Agric Syst 71:17–25

Dey MM, Kambewa P, Prein M, Jamu D, Paraguas FJ, Pemsl DE, Briones RM (2007) WorldFish centre- impact of the development and dissemination of integrated aquaculture–agriculture technologies in Malawi. In: Waibel H, Zilberman D (eds) International research on natural resource management: advances in impact assessments. FAO/CAB International, Wallingford, pp 118–146

Dey MM, Paraguas FJ, Kambewa P, Pemsl DE (2010) The impact of integrated aquaculture–agriculture on small-scale farms in Southern Malawi. Agric Econ 41:67–79. doi:10.1111/j.1574-0862.2009.00426.x

Diana JS (2009) Aquaculture production and biodiversity conservation. Bioscience 59(1):27–38. doi:10.1525/bio.2009.59.1.7

Diana JS, Szyper JP, Batterson TR, Boyd CE, Piedrahita RH (1997) Water quality in ponds. In: Egna HS, Boyd CE (eds) Dynamics of pond aquaculture. CRC Press, Boca Raton, pp 53–71

Dongmeza E, Steinbronn S, Francis G, Focken U, Becker K (2009) Investigations on the nutrient and antinutrient content of typical plants used as fish feed in small scale aquaculture in the mountainous regions of Northern Vietnam. Anim Feed Sci Technol 149:162–178

Dongmeza E, Francis G, Steinbronn S, Focken U, Becker K (2010) Investigations on the digestibility and metabolizability of the major nutrients and energy of maize leaves and barnyard grass in grass carp (*Ctenopharyngodon idella*). Aquac Nutr 16(3):313–326. doi:10.1111/j.1365-2095.2009.00667.x

Duc NM (2008) Farmers' satisfaction with aquaculture – a logistic model in Vietnam. Ecol Econ 68:525–531. doi:10.1016/j.ecolecon.2008.05.009

Duc NM (2009) Economic contribution of fish culture to farm income in Southeast Vietnam. Aquac Int 17:15–29. doi:10.1007/s10499-008-9176-8

Edwards P (1993) Environmental issues in integrated agriculture-aquaculture and wastewater-fed fish culture system. In: Pullin RSV, Rosenthal H, Maclean JL (eds) Environment and aquaculture in developing countries. ICLARM Conference Proceedings 31. ICLARM, Manila, pp 139–170

Edwards P (1998) A systems approach for the promotion of integrated aquaculture. Aquac Econ Manage 2(1):1–12

Edwards P (1999) Towards increased impact of rural aquaculture. Discussion paper prepared for the first meeting of the APFIC ad-hoc working group of experts on rural aquaculture. FAO Regional Office for Asia and the Pacific (RAP), Bangkok

Edwards P (2008) From integrated carp polyculture to intensive monoculture in the Pearl River Delta, South China. Aquac Asia 13(2):3–7

Edwards P, Pullin RSV, Gartner JA (1988) Research and education for the development of integrated crop-livestock-fish farming systems in the tropics. ICLARM studies and reviews 16. ICLARM, Manila

Ellis EC, Wang SM (1997) Sustainable traditional agriculture in the Tai Lake Region of China. Agric Ecosyst Environ 61:177–193

Fagbenro O (1998) Apparent digestibility of legume seeds by Nile tilapia. Aquac Int 6:83–87

Fagbenro O (1999) Apparent digestibility of various cereal grain by-products in common carp diets. Aquac Int 7:277–281

FAO (1997) Expert consultation on small-scale rural aquaculture. FAO Fisheries Report 458. FAO Committee on Fisheries, Rome, 28–31 May 1996

FAO (2003) Report of the second session of the sub-committee on aquaculture. FAO Fisheries Report 716. FAO Committee on Fisheries, Trondheim, 7–11 Aug 2003

FAO (2007a) The role of aquaculture in sustainable development. Thirty-fourth Session, 17–24 Nov. FAO, Rome

FAO (2007b) The state of world fisheries and aquaculture. FAO, Rome

FAO (2008) Glossary of aquaculture. FAO, Rome

FAO (2009a) Foods security and agricultural mitigation in developing countries: options for capturing synergies. FAO, Rome

FAO (2009b) The state of food insecurity in the world 2009: economic crises – impacts and lessons learned. FAO, Rome

FAO (2009c) The state of world fisheries and aquaculture. FAO, Rome

FAO/IIRR/WorldFish Center (2001) Integrated agriculture-aquaculture: a primer. FAO Fisheries Report 407, Rome

Fedoroff NV, Battisti DS, Beachy RN, Cooper PJM, Fischhoff DA, Hodges CN, Knauf VC, Lobell D, Mazur BJ, Molden D, Reynolds MP, Ronald PC, Rosegrant MW, Sanchez PA, Vonshak A, Zhu J-K (2010) Radically rethinking agriculture for the 21st century. Science 327(5967): 833–834. doi:10.1126/science.1186834

Fernando CH, Halwart M (2000) Possibilities for the integration of fish farming into irrigation systems. Fish Manage Ecol 7:45–54

Francis G, Makkar HPS, Becker K (2001) Antinutritional factors present in plant-derived alternate fish feed ingredients and their effects in fish. Aquaculture 199(3–4):197–227. doi:10.1016/S0044-8486(01)00526-9

Frei M, Becker K (2005) Integrated rice-fish culture: coupled production saves resources. Nat Resour Forum 29:135–143

Frei MMA, Razzak MM, Hossain M, Oehme SD, Becker K (2007a) Performance of common carp, *Cyprinus* carpio L. and Nile tilapia, *Oreochromis niloticus* (L.) in integrated ricefish culture in Bangladesh. Aquaculture 262:250–259. doi:10.1016/j.aquaculture.2006.11.019

Frei M, Khan MAM, Razzak MA, Hossain MM, Dewan S, Becker K (2007b) Effects of a mixed culture of common carp, *Cyprinus carpio L.*, and Nile tilapia, *Oreochromis niloticus (L.)*, on terrestrial arthropod population, benthic fauna, and weed biomass in rice fields in Bangladesh. Biol Control 41:207–213. doi:10.1016/j.biocontrol.2007.02.001

Gabriel UU, Akinrotimi OA, Bekibele DO, Onunkwo DN, Anyanwu PE (2007) Locally produced fish feed: potentials for aquaculture development in subsaharan Africa. Afr J Agric Res 2(7):287–295

Garg SK, Bhatnagar A (2000) Effect of fertilization frequency on pond productivity and fish biomass in still water ponds stocked with *Cirrhinus mrigala* (Ham.). Aquac Res 31(5): 409–414. doi:10.1046/j.1365-2109.2000.00422.x

Garg SK, Kalla A, Bhatnagar A (2002) Evaluation of raw and hydrothermically processed leguminous seeds as supplementary feed for the growth of two Indian major carp species. Aquac Res 33(3):151–163

Godfray HCJ, Beddington JR, Crute IR, Haddad L, Lawrence D, Muir JF, Pretty J, Robinson S, Thomas SM, Toulmin C (2010) Food security: the challenge of feeding 9 billion people. Science 327(812):812–818. doi:10.1126/science.1185383

Gomiero T, Giampietro M, Bukens SGF, Paoletti MG (1997) Biodiversity use and technical performance of freshwater fish aquaculture in different socioeconomic contexts: China and Italy. Agric Ecosyst Environ 2:169–185

Goyal S, Dhull SK, Kapoor KK (2005) Chemical and biological changes during composting of different organic wastes and assessment of compost maturity. Bioresour Technol 96(14): 1584–1591. doi:10.1016/j.biortech.2004.12.012

Gozlan RE (2008) Introduction of non-native freshwater fish: is it all bad? Fish Fish 9:106–115

Gozlan RE, Britton JR, Cowx I, Copp GH (2010) Current knowledge on non-native freshwater fish introductions. J Fish Biol 76:751–786. doi:10.1111/j.1095-8649.2010.02566.x

Gunderson LH, Holling CS (2002) Panarchy: understanding transformations in human and natural systems. Island Press, Washington, DC

Guo JY, Bradshaw AD (1993) The flow of nutrient and energy through a Chinese farming system. J Appl Ecol 30:86–94

Gurung TB, Wagle SK (2005) Revisiting underlying ecological principles of rice-fish integrated farming for environmental, economical and social benefits. Our Nat 3:1–12

Halwart M, Gupta MV (eds) (2004) Culture of fish in rice fields. FAO/The WorldFish Center, Rome

Halwart M, Borlinghaus M, Kaule G (1996) Activity pattern of fish in rice fields. Aquaculture 145:159–170

Hambrey J (2004) A brief review of small-scale aquaculture in Asia, its potential for poverty alleviation, with a consideration of the merits of investment and specialization. In: Halwart M, Moehl JF (eds) Regional technical expert workshop on cage culture in Africa. Fisheries proceedings 6. FAO, Rome

Hambrey J, Edwards P, Belton B (2008) An ecosystem approach to freshwater aquaculture: a global review. In: Soto D, Aguilar-Manjarrez J, Hishamunda N (eds) Building an ecosystem approach to aquaculture. FAO Fisheries and Aquaculture Proceedings 14. FAO/Universitat de les Illes Balears Experts Workshop, 7–11 May 2007, Mallorca Spain. FAO, Rome, pp 117–221

Hassan S, Edwards P, Little DC (1997) Comparison of Tilapia monoculture and carp polyculture in fertilized earthen ponds. J World Aquac Soc 28(3):268–274

Hazell P, Wood S (2008) Drivers of change in global agriculture. Phil Trans R Soc B 363:495–515. doi:10.1098/rstb.2007.2166

Herrero M, Thornton PK, Notenbaert AM, Wood S, Msangi S, Freeman HA, Bossio D, Dixon J, Peters M, van de Steeg J, Lynam J, Parthasarathy Rao P, Mcmillan S, Gerard B, McDermott J, Serand C, Rosegrant M (2010) Smart investments in sustainable food production: revisiting mixed crop-livestock systems. Science 327:822–825. doi:10.1126/science.1183725

Hishamunda N, Ridler NB (2002) Macro policies to promote sustainable commercial aquaculture. Aquac Int 10:491–505

Hishamunda N, Ridler NB (2006) Farming fish for profits: a small step towards food security in sub-Saharan Africa. Food Policy 31:401–414

Hishamunda N, Jolly CM, Engle C (1998) Evaluation of small-scale aquaculture with intra-rural household trade as an alternative enterprise for limited resource farmers: the case of Rwanda. Food Policy 23(2):143–154

Holling CS (2001) Understanding the complexity of economic, ecological, and social systems. Ecosystems 4:390–405. doi:10.1007/s10021-001-0101-5

Horstkotte-Wesseler G (1999) Socioeconomics of rice-aquaculture and IPM in the Philippines: synergies, potential and problems. ICLARM Technical Report 57. ICLARM, Manila

Hossain MA, Islam MS (2006) Optimization of stocking density of freshwater prawn *Macrobrachium rosenbergii* (de Man) in carp polyculture in Bangladesh. Aquac Res 37:994–1000. doi:10.1111/j.1365-2109.2006.01518.x

Hossain MA, Focken U, Becker K (2001) Evaluation of an unconventional legume seed, Sesbania aculeata, as a dietary protein source for common carp, *Cyprinus carpio* L. Aquaculture 198:129–140

Howard AFV, Omlin FX (2008) Abandoning small-scale fish farming in western Kenya leads to higher malaria vector abundance. Acta Trop 105:67–73. doi:10.1016/j.actatropica.2007.09.010

Hussain G, Al-Jaloud AA (1995) Effect of irrigation and nitrogen on water use efficiency of wheat in Saudi Arabia. Agric Water Manage 27:143–153

Hussain G, Al-Jaloud AA (1998) Effect of irrigation and nitrogen on yield, yield components and water use efficiency of barley in Saudi Arabia. Agric Water Manage 36:55–70

IAASTD (2009) Agriculture as a crossroads: global report. Island Press, Washington, DC

Ijumba JN, Lindsay SW (2001) Impact of irrigation on malaria in Africa: paddies paradox. Med Vet Entomol 15:1–11

Irz X, Stevenson JR, Tanoy A, Villarante P, Morissens P (2007) The equity and poverty impacts of aquaculture: insights from the Philippines. Dev Policy Rev 25(4):495–516

Jewel MAS, Affan MA, Khan S (2003) Fish mortality due to cyanobacterial bloom in an aquaculture pond in Bangladesh. Pak J Biol Sci 6(12):1046–1050

Jianguo W, Dashu N (1995) A comparative study of the ability of fish to catch mosquito larva. In: Mackay KT (ed) Rice-fish culture in China. IDRC, Ottawa, pp 217–222

Kadir A, Kundu RS, Milstein A, Wahab MA (2006) Effects of silver carp and small indigenous species on pond ecology and carp polycultures in Bangladesh. Aquaculture 261(3):1065–1076. doi:10.1016/j.aquaculture.2006.09.010

Kadir A, Wahab MA, Milstein A, Hossain MA, Seraji MTI (2007) Effects of silver carp and the small indigenous fish mola *Amblypharyngodon mola* and punti *Puntius sophore* on fish polyculture production. Aquaculture 273:520–531. doi:10.1016/j.aquaculture.2007.07.012

Kaggwa R, van Dam A, Balirwa J, Kansiime F, Denny P (2008) Increasing fish production from wetlands at Lake Victoria, Uganda using organically manured seasonal wetland fish ponds. Wetlands Ecol Manage 17(3):257–277. doi:10.1007/s11273-008-9105-7

Kalita P, Mukhopadhyay PK, Mukherjee AK (2007) Evaluation of the nutritional quality of four unexplored aquatic weeds from northeast India for the formulation of cost-effective fish feeds. Food Chem 103:204–209. doi:10.1016/j.foodchem.2006.08.007

Karapanagiotidis IT, Yakupitiyage A, Little DC, Bell MV, Mente E (2009) The nutritional value of lipids in various tropical aquatic animals from rice-fish farming systems in northeast Thailand. J Food Compos Anal 23(1):1–8. doi:10.1016/j.jfca.2009.08.001

Kautsky N, Berg H, Folke C, Larsson J, Troell M (1997) Ecological footprint for assessment of resource use and development limitations in shrimp and tilapia aquaculture. Aquac Res 28: 753–766

Kawarazuka N (2010) The contribution of fish intake, aquaculture, and small-scale fisheries to improving nutrition: a literature review. The WorldFish Center Working Paper No. 2106. The WorldFish Center, Penang

Keshavanath P, Shivanna, Gangadhara B (2006) Evaluation of sugarcane by-product pressmud as a manure in carp culture. Bioresour Technol 97:628–634. doi:10.1016/j.biortech.2005.03.019

Kestemont P (1995) Different systems of carp production and their impact on the environment. Aquaculture 129:347–372

Kipkemboi J, van Dam AA, Ikiara MM, Denny P (2007) Integration of smallholder wetland aquaculture–agriculture systems (fingerponds) into riparian farming systems on the shores of Lake Victoria, Kenya: socio-economics and livelihoods. Geogr J 173:257–272

Knud-Hansen CF (1998) Pond fertilization: ecological approach and practical application. Pond Dynamics/Aquaculture CRSP, Corvalis, 135 p

Knud-Hansen CF, Batterson TR, McNabba CD, Harahat IS, Sumantadinata K, Eidman HM (1991) Nitrogen input, primary productivity and fish yield in fertilized freshwater ponds in Indonesia. Aquaculture 94:49–63

Knudsen LG, Phuc PD, Hiep NT, Samuelsen H, Jensen PK, Dalsgaard A, Raschid-Sally L, Konradsen F (2008) The fear of awful smell: risk perceptions among farmers in Vietnam using wastewater and human excreta in agriculture. Southeast Asian J Trop Med Public Health 39(2):341–353

Kumaresan A, Pathak KA, Bujarbaruah KM, Vinod K (2009) Analysis of integrated animal-fish production system under subtropical hill agro ecosystem in India: growth performance of animals, total biomass production and monetary benefit. Trop Anim Health Prod 41:385–391

Kunda M, Azim ME, Wahab MA, Dewan S, Roos N, Thilsted SH (2008) Potential of mixed culture of freshwater prawn (*Macrobrachium rosenbergii*) and self-recruiting small species mola (*Amblypharyngodon mola*) in rotational ricefish/prawn culture systems in Bangladesh. Aquac Res 39(5):506–517. doi:10.1111/j.1365-2109.2008.01905.x

Kunda M, Azim ME, Wahab MA, Dewan S, Majid MA, Thilsted SH (2009) Effects of including catla and tilapia in a freshwater prawnmola polyculture in a rotational ricefish culture systems. Aquac Res 40(9):1089–1098. doi:10.1111/j.1365-2109.2009.02204.x

Lazard J, Baruthio A, Mathé S, Rey-Valette H, Chia E, Clément O, Aubin J, Morissens P, Mikolasek O, Legendre M, Levang P, Blancheton J-P, René F (2010) Aquaculture system diversity and sustainable development: fish farms and their representation. Aquat Living Resour 23:187–193. doi:10.1051/alr/2010018

Lebel L, Lebel P, Garden P, Giap D, Khrutmuang S, Nakayama S (2008) Places, chains, and plates: governing transitions in the shrimp aquaculture production-consumption system. Globalizations 5(2):211–226. doi:10.1080/14747730802057589

Lebel L, Mungkung R, Gheewala SH, Lebel P (2010) Innovation cycles, niches and sustainability in the shrimp aquaculture industry in Thailand. Environ Sci Policy 13(4):291–302. doi:10.1016/j.envsci.2010.03.005

Lewis D (1997) Rethinking aquaculture for resource-poor farmers: perspectives from Bangladesh. Food Policy 22(6):533–546

Li D, Hu X (2009) Fish and its multiple human health effects in times of threat to sustainability and affordability: are there alternatives? Asia Pac J Clin Nutr 18(4):553–563

Lightfoot C, Prein M, Lopez T (1994) Bioresource flow modeling with farmers. ILEIA Newsl 10(3):22–23

Lin CK, Teichert-Coddington DR, Green BW, Veverica KL (1997) Fertilization regimes. In: Egna HS, Boyd CE (eds) Dynamics of pond aquaculture. CRC Press, Boca Raton, pp 73–107

Little DC, Edwards P (2003) Integrated livestock-fish farming systems. FAO Inland Water Resources and Aquaculture Service Animal Production Service, Rome

Little DC, Surintaraseree P, Innes-Taylor N (1996) Fish culture in rainfed rice fields of northeast Thailand. Aquaculture 140:295–321

Little DC, Barman BK, Haque MM, Wahab MA (2007a) Descentralised Nile tilapia seed production. In: van der Zijpp AJ, Verreth JAJ, Le Quang Tri, van Mensvoort MEF, Bosma RH, Beveridge MCM (eds) Fishponds in farming systems. Proceedings of a symposium held in Can Tho City, 28–30 Apr 2006, organized by Can Tho University. Vietnam and Wageningen University, Wangeningen, pp 49–58

Little DC, Karim M, Turongrouang D, Morales EJ, Murray FJ, Barman BK, Hague MM, Kundu N, Belton B, Faruque ASG, Azim ME, Islam FU, Pollock L, Verdegem MCJ, Young JA, Leschen W, Wahab MA (2007b) Livelihood impacts of ponds in Asia- opportunities and constraints. In: van der Zijpp AJ, Verreth JAJ, Le Quang Tri, van Mensvoort MEF, Bosma RH, Beveridge MCM (eds) Fishponds in farming systems. Proceedings of a symposium held in Can Tho City, 28–30 Apr 2006, organized by Can Tho University. Vietnam and Wageningen University, Wangeningen, pp 177–202

Liu J, Cai Q (1998) Integrated aquaculture in Chinese lakes and paddy fields. Ecol Eng 11:49–59

Lo CP (1996) Environmental impact on the development of agricultural technology in China: the case of the dike-pond ('jitang') system of integrated agriculture-aquaculture in the Zhujiang Delta of China. Agric Ecosyst Environ 60:183–195

Lu J, Li X (2006) Review of ricefish-farming systems in China – one of the Globally Important Ingenious Agricultural Heritage Systems (GIAHS). Aquaculture 260:106–113. doi:10.1016/j.aquaculture.2006.05.059

Mahimairaja S, Bolan NS, Hedley MJ (1995) Denitrification losses of N from fresh and composted manures. Soil Biol Biochem 27(9):1223–1225

Markewich HA, Pell AN, Mbugua DM, Cherney DJR, van Es HM, Lehmann J, Robertson JB (2010) Effects of storage methods on chemical composition of manure and manure decomposition in soil in small-scale Kenyan systems. Agric Ecosyst Environ 139(1–2):134–141. doi:10.1016/j.agee.2010.07.010

Martinez-Espinosa M (1992) Rural aquaculture: from myth to reality. FAO Aquac Newsl 2:13–15

Matson PA, Parton WJ, Power AG, Swift MJ (1997) Agricultural intensification and ecosystem properties. Science 277:504–509

Michielsens CGJ, Lorenzen K, Philips MJ, Gauthier R (2002) Asian carp farming systems: towards a typology and increased resource use efficiency. Aquac Res 33:403–413

Middendorp AJ, Verreth JAJ (1986) The potential of and constraints to fish culture in integrated farming systems in the Lam Pao Irrigation Project, Northeast Thailand. Aquaculture 56(1): 63–78

Milstein A (1992) Ecological aspects of fish species interactions in polyculture ponds. Hydrobiologia 231:177–186

Milstein A, Kadir A, Wahab MA (2008) The effects of partially substituting Indian carps or adding silver carp on polycultures including small indigenous fish species (SIS). Aquaculture 279(1–4):92–98. doi:10.1016/j.aquaculture.2008.04.009

Milstein A, Wahab MA, Kadir A, Sagor MFH, Islam MA (2009) Effects of intervention in the water column and/or pond bottom through species composition on polycultures of large carps and small indigenous species. Aquaculture 286:246–253. doi:10.1016/j.aquaculture.2008.09.036

Mizanur R, Yakupitiyage A, Ranamukhaarachchi SL (2004) Agricultural use of fishpond sediment for environmental amelioration. Thammasat Int J Sci Technol 9(4):1–10

Mohanty RK, Verma HN, Brahmanand PS (2004) Performance evaluation of ricefish integration system in rainfed medium land ecosystem. Aquaculture 230:125–135. doi:10.1016/S0044-8486(03)00423-X

Moll HAJ (2005) Costs and benefits of livestock systems and the role of market and nonmarket relationships. Agricultural Economics 32:181–193

Morales EJ, Little DC, Imminka A, Demaine H, Yakupitayage A, Amilhat E, Lorenzen K (2006) Project report: contribution of self-recruiting species produced in farmer-managed aquatic systems in rural areas of Southeast Asia to food consumption. J Food Compos Anal 19:759–760. doi:10.1016/j.jfca.2006.03.003

Muendo PN (2006) The role of fish ponds in the nutrient dynamics of mixed farming systems. Ph.D. thesis, Wageningen University, Wageningen

Muir J (2005) Managing to harvest? Perspectives on the potential of aquaculture. Phil Trans R Soc B 360:191–218. doi:10.1098/rstb.2004.1572

Murshed-e-Jahan K, Ahmed M, Belton B (2010) The impacts of aquaculture development on food security: lessons from Bangladesh. Aquac Res 41:481–495. doi:10.1111/j.1365-2109.2009.02337.x

Nalim S (1994) The impact of fish in enhancing rice field ecosystems. In: Dela Cruz CR (ed) Role of fish in enhancing ricefield ecology and in integrated pest management. ICLARM conference proceedings 43. ICLARM, Manila

Naylor RL (2008) Managing food production systems for resilience. In: Chapin FS, Kofinas GP, Folke C (eds) Principles of natural resource stewardship: resilience-based management in a changing world. Springer, New York, pp 259–280

Naylor RL, Burke M (2005) Aquaculture and ocean resources: raising tigers of the sea. Annu Rev Environ Resour 30:185–218

Naylor RL, Goldburg RJ, Primavera JH, Kautsky N, Beveridge MCM, Clay J, Folke C, Lubchenco J, Mooney H, Troell M (2000) Effect of aquaculture on world fish supplies. Nature 405:1017–1024. doi:10.1038/35016500

Naylor RL, Hardy RW, Bureau DP, Chiu A, Elliott M, Farrell AP, Forster I, Gatlin DM, Goldburg RJ, Hua K, Nichols PD (2009) Feeding aquaculture in an era of finite resources. Proc Natl Acad Sci USA 106(36):15103–15110. doi:10.1073/pnas.0905235106

Negesse T, Makkar HPS, Becker K (2009) Nutritive value of some non-conventional feed resources of Ethiopia determined by chemical analyses and an in vitro gas method. Anim Feed Sci Technol 154:204–217. doi:10.1016/j.anifeedsci.2009.09.010

Ng WK, Wee KL (1989) The nutritive value of cassava leaf meal in pelleted feed for Nile tilapia. Aquaculture 83(1–2):45–58. doi:10.1016/0044-8486(89)90059-8

Nhan DK, Milstein A, Verdegem MCJ, Verreth JAV (2006) Food inputs, water quality and nutrient accumulation in integrated pond systems: a multivariate approach. Aquaculture 261:160–173. doi:10.1016/j.aquaculture.2006.07.015

Nhan DK, Phong LT, Verdegem MJC, Duong LT, Bosma RH, Little DC (2007) Integrated freshwater aquaculture, crop and livestock production in the Mekong delta, Vietnam: determinants and the role of the pond. Agric Syst 94:445–458. doi:10.1016/j.agsy.2006.11.017

Nhan DK, Verdegem MCJ, Binh NT, Duong LT, Milstein A, Verreth JAJ (2008) Economic and nutrient discharge tradeoffs of excreta-fed aquaculture in the Mekong Delta, Vietnam. Agric Ecosyst Environ 124:259–269. doi:10.1016/j.agee.2007.10.005

Nobre AM, Robertson-Andersson D, Neori A, Sankar K (2010) Ecologicaleconomic assessment of aquaculture options: comparison between abalone monoculture and integrated multi-trophic aquaculture of abalone and seaweeds. Aquaculture 306:116–126. doi:10.1016/j.aquaculture.2010.06.002

Ogunwande GA, Osunade JA, Adekalu KO, Ogunjimi LAO (2008) Nitrogen loss in chicken litter compost as affected by carbon to nitrogen ratio and turning frequency. Bioresour Technol 99(16):7495–7503. doi:10.1016/j.biortech.2008.02.020

Pant J, Demaine H, Edwards P (2004) Assessment of the aquaculture subsystem in integrated agriculture–aquaculture systems in Northeast Thailand. Aquac Res 35:289–298

Pant J, Demaine H, Edwards P (2005) Bio-resource flow in integrated agriculture-aquaculture systems in a tropical monsoonal climate: a case study in Northeast Thailand. Agric Syst 83:203–219

Parkinson R, Gibbs P, Burchett S, Misselbrook T (2004) Effect of turning regime and seasonal weather conditions on nitrogen and phosphorus losses during aerobic composting of cattle manure. Bioresour Technol 91:171–178. doi:10.1016/S0960-8524(03)00174-3

Perrings C (1998) Resilience in the dynamics of economy-environment systems. Environ Resour Econ 11(3–4):503–520

Phan-Van M, Rosseau D, De Pauw N (2008) Effects of fish bioturbation on the vertical distribution of water temperature and dissolved oxygen in a fish culture- integrated waste stabilization pond system in Vietnam. Aquaculture 281(1–4):28–33. doi:10.1016/j.aquaculture.2008.04.033

Phillips MJP, De Silva SS (2006) Finfish cage culture in Asia: an overview of status, lessons learned and future developments. In: Halwart M, Moehl JF (eds) FAO Regional Technical Expert Workshop on Cage Culture in Africa. Entebbe, Uganda, 20–23 October 2004. FAO, Rome, pp 49–72

Phong LT, van Dam AA, Udo HMJ, van Mensvoort MEF, Tri LQ, Steenstra FA, van der Zijpp AJ (2010) An agro-ecological evaluation of aquaculture integration into farming systems of the Mekong Delta. Agric Ecosyst Environ 138(3–4):232–241. doi:10.1016/j.agee.2010.05.004

Phuong ND, Tuan VD, Toan TD (2006) Farmers practices in organic and inorganic fertilization on crops, trees and vegetables. In: Porphyre V, Coi NQ (eds) Pig production development, animal waste management and environmental protection: a case study in Thai Binh Province, Northern Vietnam. PRISE Publications, Hanoi, pp 145–163

Pilarski F, Tomazelli Jnior O, Casaca JM, Mello Garcia FR, Tomazelli IB, dos Santos IR (2004) Integrated fish/pig systems: environmental feature and fish quality. R Bras Zootec 33(2):267–276. doi:10.1590/S1516-35982004000200001

Pingali P (2007) Westernization of Asian diets and the transformation of food systems: implications for research and policy. Food Policy 32(3):281–298. doi:10.1016/j.foodpol.2006.08.001

Poot-López GR, Hernández JM, Gasca-Leyva E (2010) Input management in integrated agriculture–aquaculture systems in Yucatan: tree spinach leaves as a dietary supplement in tilapia culture. Agric Syst 103:98–104. doi:10.1016/j.agsy.2009.11.003

Prein M (2002) Integration of aquaculture into cropanimal systems in Asia. Agric Syst 71:127–146

Prein M (2007) Comparative analysis of material flows in low input carp and poultry farming: an overview of concepts and methodology. In: Bartley DM, Brugre C, Soto D, Gerber P, Harvey B (eds) Comparative assessment of the environmental costs of aquaculture and other food production sectors: methods for meaningful comparisons. FAO/WFT Expert Workshop, 24–28 Apr 2006, Vancouver. FAO fisheries proceedings No. 10. FAO, Rome, pp 183–200

Prein M, Ahmed M (2000) Integration of aquaculture into smallholder farming systems for improved food security and household nutrition. Food Nutr Bull 21(4):466–472

Prein M, Lightfoot C, Pullin RSV (1998) ICLARM's approach to the integration of aquaculture into sustainable farming systems. In: ADB/NACA. Report on a regional study and workshop on aquaculture sustainability and the environment. Network of Aquaculture Centres in Asia, Bangkok, Thailand and the Asian Development Bank, Manila, pp 257–265

Pretty J (2008) Agricultural sustainability: concepts, principles and evidence. Phil Trans R Soc B 363:447–465. doi:10.1098/rstb.2007.2163

Pretty JN, Morison JIL, Hine RE (2003) Reducing food poverty by increasing agricultural sustainability in developing countries. Agric Ecosyst Environ 95(1):217–234. doi:10.1016/S0167-8809(02)00087-7.

Primavera JH (2005) Mangroves, fishponds, and the quest for sustainability. Science 310:57–59

Pullin RSV (1998) Aquaculture, integrated resources management and the environment. In: Mathias JA, Charles AT, Baotong H (eds) Integrated fish farming. Proceedings of a workshop on integrated fish farming, 11–15 October 1994, Wuxi, Jiangsu Province. CRC Press, Boca Raton, pp 19–43

Pullin R, Froese R, Pauly D (2007) Indicators for the sustainability of aquaculture. In: Bert TM (ed) Ecological and genetic implications of aquaculture activities. Springer, Dordrecht, pp 53–72

Rahman MM, Verdegem MCJ, Nagelkerke LAJ, Wahab MA, Milstein A, Verreth JAJ (2006) Growth, production and food preference of rohu *Labeo rohita* (H.) in monoculture and in polyculture with common carp *Cyprinus carpio* (L.) under fed and non-fed ponds. Aquaculture 257:359–372. doi:10.1016/j.aquaculture.2006.03.020

Rahman MM, Jo Q, Gong YG, Miller SA, Hossai MY (2008) A comparative study of common carp (*Cyprinus carpio L.*) and calbasu (*Labeo calbasu* Hamilton) on bottom soil resuspension, water quality, nutrient accumulations, food intake and growth of fish in simulated rohu (*Labeo rohita* Hamilton) ponds. Aquaculture 285(1–4):78–83. doi:10.1016/j.aquaculture.2008.08.002

Rai S, Yi Y, Wahab MA, Bart AN, Diana JS (2008) Comparison of rice straw and bamboo stick substrates in periphyton-based carp polyculture systems. Aquac Res 39:464–473. doi:10.1111/j.1365-2109.2008.01898.x

Rangacharyulu PV, Giri SS, Paul BN, Yashoda KP, Jagannatha Rao R, Mahendrakar NS, Mohanty SN, Mukhopadhyay PK (2003) Utilization of fermented silkworm pupae silage in feed for carps. Bioresour Technol 86:29–32

Ritvo G, Kochba M, Avnimelech Y (2004) The effects of common carp bioturbation on fish pond bottom soil. Aquaculture 242(1):345–356

Rivera-Ferré MG (2009) Can export-oriented aquaculture in developing countries be sustainable and promote sustainable development? The shrimp case. J Agric Environ Ethics 22(4):301–321. doi:10.1007/s10806-009-9148-7

Ross LG, Martinez-Palacios CA, Morales EJ (2008) Developing native fish species for aquaculture: the interacting demands of biodiversity, sustainable aquaculture and livelihoods. Aquac Res 39(7):675–683. doi:10.1111/j.1365-2109.2008.01920.x

Rothuis AJ, Nhan DK, Richter CJJ, Ollevier F (1998) Rice with fish culture in the semi-deep waters of the Mekong Delta, Vietnam: a socio-economical survey. Aquac Res 29:47–57

Rothuis AJ, Vromant N, Xuan VT, Richter CJJ, Ollevier F (1999) The effect of rice seeding rate on rice and fish production, and weed abundance in direct-seeded ricefish culture. Aquaculture 172:255–274. doi:10.1016/S0044-8486(98)00396-2

Ruddle K (1996) Households and community factors affecting development of small-scale fish farming in Africa. In: Bailey C, Jentoft S, Sinclair P (eds) Aquaculture development: social dimensions of an emerging industry. Westview Press, Boulder

Ruddle K, Prein M (1997) Assessing potential nutritional and household economic benefits of developing integrated farming systems. In: Mathias J, Charles AT, Hu B (eds) Integrated fish farming. Proceedings of the international workshop held in Wuxi, Peoples Republic of China, 11–15 October 1994. CRC Press, Boca Raton/New York, pp 111–121

Ruddle K, Deng H, Liang G (1986) Energy exchanges and the energy efficiency of household ponds in the dike-pond system of the Zhujiang Delta, China. Bull Natl Mus Ethnol 11(1):323–343

Rudel TK, Schneider L, Uriarte M, Turner BL II, DeFriesc R, Lawrence D, Geoghegan J, Hecht S, Ickowitzf A, Lambin EF, Birkenholtz T, Baptista S, Grau R (2009) Agricultural intensification and changes in cultivated areas, 1970–2005. Proc Natl Acad Sci USA 106(49):20675–20680. doi:10.1073/pnas.0812540106

Rufino MC, Rowe EC, Delve RJ, Giller KE (2006) Nitrogen cycling efficiencies through resource-poor African croplivestock systems. Agric Ecosyst Environ 112(4):261–282. doi:10.1016/j.agee.2005.08.028

Sagoo EJR, Williams BJ, Chambers LO, Boyles RM, Chadwick DR (2007) Integrated management practices to minimise losses and maximise the crop nitrogen value of broiler litter. Biosystems Eng 97(4):512–519. doi:10.1016/j.biosystemseng.2007.03.032

Saikia SK, Das DN (2009) Potentiality of periphyton-based aquaculture technology in rice-fish environment. J Sci Res 1(3):624–634. doi:10.3329/jsr.v1i3.2114

Shang YC, Costa-Pierce B (1983) Integrated aquaculture-agriculture farming systems: some economic aspects. J World Maric Soc 14:523–530

Shennan C (2008) Biotic interactions, ecological knowledge and agriculture. Phil Trans R Soc B 363:717–739

Sikawa DC, Yakupitiyage A (2010) The hydroponic production of lettuce (*Lactuca sativa* L) by using hybrid catfish (*Clarias macrocephalus* x *C. gariepinus*) pond water: potentials and constraints. Agric Water Manage 97(9):1317–1325. doi:10.1016/j.agwat.2010.03.013

Silva Castro R, Borges Azevedo CMS, Bezerra-Neto F (2006) Increasing cherry tomato yield using fish effluent as irrigation water in Northeast Brazil. Sci Hortic 110:44–50

Sin TS (2006) Evaluation of different species of fish for biological control of golden apple snail *Pomacea canaliculata* (Lamarck) in rice. Crop Prot 25:1004–1012

Singh K, Garg SK, Kalla A, Bhatnagar A (2003) Oilcakes as protein sources in supplementary diets for the growth of *Cirrhinus mrigala* (Ham.) fingerlings: laboratory and field studies. Bioresour Technol 86:283–291

Smardon RC (2006) Heritage values and functions of wetlands in Southern Mexico. Landsc Urban Plan 74:296–312. doi:10.1016/j.landurbplan.2004.09.009

Soliman AK, El-Horbeety AA, Essa MAR, Kosba MA, Kariony IA (2000) Effects of introducing ducks into fish ponds on water quality, natural productivity and fish production together with the economic evaluation of the integrated and non-integrated systems. Aquac Int 8:315–326

Sommer SG, Petersen SO, Srensen P, Poulsen HD, Mller HB (2007) Methane and carbon dioxide emissions and nitrogen turnover during liquid manure storage. Nutr Cycl Agroecosyst 78(1):27–36. doi:10.1007/s10705-006-9072

Stonich SC, Vandergeest P (2001) Violence, environment, and industrial shrimp farming. In: Pelusso NL, Watts M (eds) Violent environments. Cornell University Press, Ithaca, pp 260–289

Tacon AGJ, Metian M, Turchini GM, De Silva SS (2010) Responsible aquaculture and trophic level implications to global fish supply. Rev Fish Sci 18(1):94–105

Tapia M, Zambrano L (2003) From aquaculture goals to real social and ecological impacts: carp introduction in rural Central Mexico. Ambio 32(4):252–257

Terziyski D, Grozev G, Kalchev R, Stoeva A (2007) Effect of organic fertilizer on plankton primary productivity in fish ponds. Aquac Int 15:181–190. doi:10.1007/s10499-007-9086-1

Thilsted S, Roos N, Hassan N (1997) The rôle of small indigenous fish species in food and nutrition security in Bangladesh. NAGA The ICLARM Quarterly JulyDecember, pp 82–84

Tilman D (1999) The ecological consequences of changes in biodiversity: a search for general principles. Ecology 80(5):1455–1474

Tipraqsa P, Craswell ET, Noble AD, Schmidt-Vogt D (2007) Resource integration for multiple benefits: multifunctionality of integrated farming systems in Northeast Thailand. Agric Syst 94:694–703. doi:10.1016/j.agsy.2007.02.009

Troell M, Joyce A, Chopin T, Neori A, Buschmann AH, Fang J-G (2009) Ecological engineering in aquaculture – potential for integrated multi-trophic aquaculture (IMTA) in marine offshore systems. Aquaculture 297:1–9. doi:10.1016/j.aquaculture.2009.09.010

Uddin S, Ekram-Ul-Azim M, Wahab A, Verdegem MCJ (2006) The potential of mixed culture of genetically improved farmed tilapia (*Oreochromis niloticus*) and freshwater giant prawn (*Macrobrachium rosenbergii*) in periphyton-based systems. Aquac Res 37:241–247. doi:10.1111/j.1365-2109.2005.01424

Ulloa JB, Verreth JAJ (2003) Growth of Oreochromis aureus fed with diets containing graded levels of coffee pulp and reared in two culture systems. Aquaculture 217:275–283

Ulloa JB, van Weerd JH, Huisman EA, Verreth JAJ (2004) Tropical agricultural residues and their potential uses in fish feeds: the Costa Rican situation. Waste Manage 24:87–97. doi:10.1016/j.wasman.2003.09.003

van Dam AA, Beveridge MCM, Azim ME, Verdegem MCJ (2002) The potential of fish production based on periphyton. Rev Fish Biol Fish 12:1–31

Veerina SS, Nandeesha MC, De Silva SS, Ahmed M (1999) An analysis of production factors in carp farming in Andhra Pradesh, India. Aquac Res 30:805–814

Vidotti RM, Carneiro DJ, Viegas E (2002) Growth rate of Pacu, *Piaractus mesopotamicus*, fingerlings fed diets containing co-dried fish silage as replacement of fish meal. J Appl Aquac 12(4):77–88. doi:10.1300/J028v12n04_07

Virk P, Saxena K (2003) Potential of Amaranthus seeds in supplementary feed and its impact on growth in some carps. Bioresour Technol 86(1):25–27. doi:10.1016/S0960-8524(02)00139-6

Vromant N, Chau NTH (2005) Overall effect of rice biomass and fish on the aquatic ecology of experimental rice plots. Agric Ecosyst Environ 111:153–165. doi:10.1016/j.agee.2005.06.015

Vromant N, Rothuis AJ, Cuc NTT, Ollevier F (1998) The effect of fish on the abundance of the rice caseworm *Nymphula depunctalis* (Guenée) (Lepidoptera: Pyralidae) in direct seeded, concurrent rice fish yields. Biocontrol Sci Technol 8:539–546. doi:10.1080/09583159830054

Vromant N, Nhan DK, Chau NTH, Ollevier F (2002) Can fish control planthopper and leafhopper populations in intensive rice culture? Biocontrol Sci Technol 12:695–703. doi:10.1080/0958315021000039879

Vu TKV, Tran MT, Dang TTS (2007) A survey of manure management on pig farms in Northern Vietnam. Livest Sci 112:288–297

Wahab MA, Azim ME, Ali MH, Beveridge MCM, Khan S (1999) The potential of periphyton-based culture of the native major carp calbaush, *Labeo calbasu* (Hamilton). Aquac Res 30:409–419. doi:10.1046/j.1365-2109.1999.00337.x

Wahab MA, Alim MA, Milstein A (2004) Effects of adding the small fish punti (*Puntius sophore*) and/or mola (*Amblypharyngodon mola*) to a polyculture of large carp. Aquac Res 35(2):124–133. doi:10.1111/j.1365-2109.2004.00990.x

Wahab MA, Kunda M, Azim ME, Dewan S, Thilsted SH (2008) Evaluation of freshwater prawn-small fish culture concurrently with rice in Bangladesh. Aquac Res 39(14):1524–1532. doi:10.1111/j.1365-2109.2008.02025.x

Webb C, Bodin O (2008) A network perspective on modularity and control of flow in robust systems. In: Norberg J, Cumming GS (eds) Complexity theory for a sustainable future. Columbia University Press, New York, pp 95–118

Wee KL, Ng LT (1986) Use of cassava as an energy source in a pelleted feed for the tilapia, *Oreochromis niloticus* L. Aquac Res 17(2):129–138. doi:10.1111/j.1365 -2109.1986.tb00094.x

Weimin M (2010) Recent developments in rice-fish culture in China: a holistic approach for livelihood improvement in rural areas. In: De Silva SS, Davy B (eds) Success stories in Asian aquaculture. Springer, Dordrecht, pp 27–40

Weng Q (2007) A historical perspective of river basin management in the Pearl River Delta of China. J Environ Manage 85:1048–1062. doi:10.1016/j.jenvman.2006.11.008

Wetengere K (2009) Socio-economic factors critical for adoption of fish farming technology: the case of selected villages in Eastern Tanzania. Int J Fish Aquac 1(3):28–37

WHO (2009) World malaria report 2009. World Health Organization, Geneva

Yan YS, Shang W, Fu WH, An KD, Rong KM, Zhai WQ (1995) Ability of fish to control rice diseases, pests, and weeds. In: Mackay KT (ed) Rice-fish culture in China. IDRC, Ottawa, pp 223–229

Yang Y, Hang H, Xiao-jun HU, Dai Q, Zang Y (2006) Characteristics of growth and yield formation of rice in rice-fish farming system. Agric Sci China 5(2):103–110

Yee AWC (1999) New developments in integrated dike-pond agriculture-aquaculture in the Zhujiang Delta, China: ecological implications. Ambio 28(6):529–533

Yi Y, Lin CK, Diana JS (2002) Recycling pond mud nutrients in integrated lotus-fish culture. Aquaculture 212:217–230

You L, Ringler C, Nelson G, Wood-Sichra U, Robertson R, Wood S, Guo Z, Zhu T, Sun Y (2010) What is the irrigation potential for Africa? A combined biophysical and socioeconomic approach. IFPRI Discussion Paper 993. IFPRI, Washington, DC

Yuvanatemiya V, Boyd CE (2006) Physical and chemical changes in aquaculture pond bottom soil resulting from sediment removal. Aquacult Eng 35:199–205. doi:10.1016/j.aquaeng.2006.02.001

Zajdband AD (2009) Integrando actividades en la Provincia de Misiones. Rev LEISA de Agroecol 25(1):31–33

Small RNA-Mediated Defensive and Adaptive Responses in Plants

Afsar Raza Naqvi, Nirupam Roy Choudhury, and Qazi Mohd. Rizwanul Haq

Abstract Agriculture is the backbone of economy for most countries, especially the developing countries with high population density. However, the change in climate and its influence on the environment has led to low productivity of several staple crops that can be attributable to prevailing biotic and abiotic factors. In the long run, these factors are likely to have more adverse effects on the crop yield than what is currently being encountered. Therefore, scientists across the world develop strategies to tackle future problems leading to food insecurity. Plants are exposed to a variety of stresses under natural conditions. To encounter these challenges plants have efficiently evolved with several endogenous mechanisms to defend themselves against such harmful situations.

Small ribonucleic acids (sRNA) have recently been identified as critical molecules that regulate wide variety of biological phenomenon, both in plants and animals. Several lines of evidence demonstrates that small RNAs profiles changes rapidly in response to adverse stimuli. In certain cases, specific small RNAs were found to be associated with a particular stress, thus, indicating the

A.R. Naqvi
Department of Biosciences, Jamia Millia Islamia (A Central University), New Delhi 110 025, India

Plant Molecular Biology Group, International Centre for Genetic Engineering and Biotechnology, Aruna Asaf Ali Marg, New Delhi 110 067, India
e-mail: afsarrazaa@yahoo.co.in

N.R. Choudhury
Plant Molecular Biology Group, International Centre for Genetic Engineering and Biotechnology, Aruna Asaf Ali Marg, New Delhi 110 067, India
e-mail: nirupam@icgeb.res.in

Q.M. Rizwanul Haq (✉)
Department of Biosciences, Jamia Millia Islamia (A Central University), New Delhi 110 025, India
e-mail: qmrhaque@gmail.com

E. Lichtfouse (ed.), *Genetics, Biofuels and Local Farming System*, Sustainable Agriculture Reviews 7, DOI 10.1007/978-94-007-1521-9_5,

direct involvement of small RNA-mediated pathways in plethora of stresses. We review here the advancements of small RNA biology, including microRNA and short-interfering RNA, and aspects related to their possible role in counter defense of biotic and abiotic stress responses. The review has been broadly classified into two sections describing role of small RNAs assigned in response to various biotic and abiotic stresses. The possible evidence of cross-talk among diverse range of stresses is also discussed. Future investigations may explore putative novel pathways downstream to this small RNA misexpression and consequently open avenues to design strategies to raise transgenic crops resistant to multiple stresses.

Keywords Argonaute • Cross-talk • Dicer • MiRNAs • Plant defense responses • SiRNAs • Stress

1 Introduction

Plants products, including food and medicines, have allowed human survival on this planet over the centuries. However, with the prevailing conditions of deforestation, soil-erosion, floods and other factors, there is significant reduction in the crop productivity worldwide. This is further worsened with an emergence of huge variety of plant pathogens that can infect wide variety of hosts. With an ever-increasing human population there is threat of food insecurity, if the situation continues (Ortiz 1998). To meet the future demands, plant biotechnology is aimed at raising transgenic crops that can cope with adverse biotic and abiotic stresses and are conducive to sustainable agriculture and remain high-yielding even in the absence of petrochemical applications (Izquierdo 1999). Plant breeders have attempted to generate new varieties with better adaptability in presence of disease causing pathogens and natural adversities. Further, the crop improvement by genetic approaches assisted with conventional plant breeding can contribute significantly to achieving these goals with a realistic interpretation of the global environmental and economic situation. The dissections of pathways compromised or induced in response to variety of stresses have lead plant biologists to adopt strategies that can confer resistance against such unpleasant environmental conditions.

Plants being sessile, incessantly encounter wide spectrum of stresses, both of biotic and abiotic in origin. Along the evolution, plants have designed and adopted strategies to cope with these adversities. Discovered in the last decade, the small RNA-mediated pathways have been demonstrated to be able to defend and regulate almost all kinds of stress responses. Besides their role in defense related pathways, small RNA have also been implicated in chromatin remodeling, DNA elimination in *Tetrahymena thermophilus* (scan siRNAs) and defense against mobile elements (Mette et al. 2000; Pal-Bhadra et al. 2002; Zilberman et al. 2003; Mochizuki and Gorovsky 2004). These small RNAs present across the eukaryotic kingdom are critical regulatory species that spatiotemporally orchestrate the transcriptome. Deep sequencing of small RNA pool in various organisms led to the discovery of

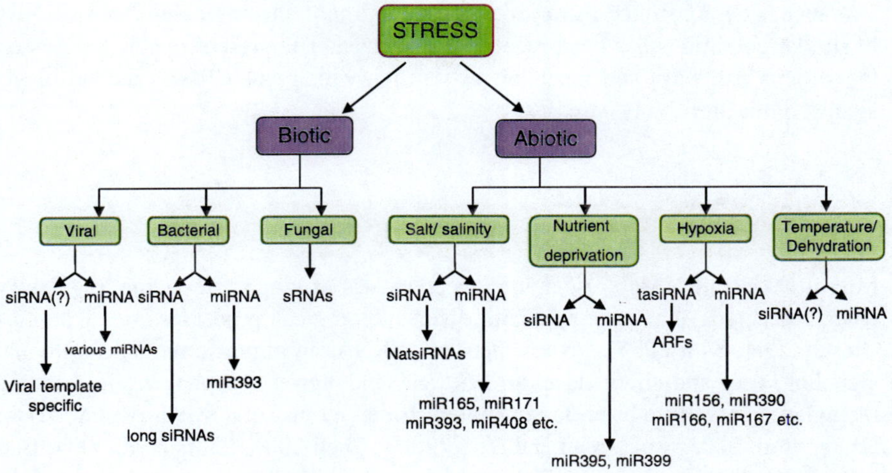

Fig. 1 Involvement of small RNAs in biotic and abiotic stresses. Hosts' RNAi responses are induced in response to majority of the abiotic or biotic stresses studied so far. Most of the stresses mentioned in the figure triggers the generation of, both, short interfering RNA (*siRNA*) and microRNA pathway. With the deep sequencing technology the role of these components is becoming more evident

new classes namely, piwi-interacting RNAs (piRNAs), 21-U RNAs, nat-siRNAs, etc., suggesting that they are inevitable in various biologically important pathways (reviewed by Naqvi et al. 2009).

Major classes of small RNAs constituting the RNA interference (RNAi) pathway include short interfering RNAs (siRNAs) and microRNAs (miRNAs) that differ in their biogenesis and downstream effect. The biogenesis of small RNAs requires endogenously transcribed RNAs or exogenous RNAs having potential to assume secondary structures or forming dsRNAs. These are recognized and processed by RNase III protein (Dicer) to generate duplex small RNAs of ~21–30 nt length (Bartel 2004). Argonaute, an RNA binding protein with inherent endonucleolytic activity, is recruited to these duplex small RNAs. Of the two strands, passenger strand is released and degraded while the surviving strand (guide strand) remains bound to proteins. The effector proteins loaded onto guide strand together constitutes si- or mi- RNA Induced Silencing Complex (si-/ mi-RISC). The small RNAs with associated proteins mediate their effect by binding to the complementary sites on the target message(s) that leads to either target cleavage and/or translation repression and is termed post-transcriptional gene silencing (PTGS). Alternately, siRNAs can also be incorporated into RNA Induced Transcriptional Silencing (RITS) complex, protein machinery distinct from siRISC (Pal-Bhadra et al. 2002). These siRNAs guide RITS complex to genomic sequences that modulate the local chromatin status thereby affecting transcription through such locus, a phenomenon known as transcriptional gene silencing (TGS).

As evident from Fig. 1, generation of both the classes of small RNAs are triggered by abiotic and biotic stresses. While the role of miRNAs in conferring stress tolerance is demonstrated in various plants, the importance of endogenous siRNAs in similar conditions has been recently recognized. The present review focuses on the various pathways and mechanisms through which small RNAs defend plants against biotic and abiotic stresses.

2 MicroRNAs

MicroRNAs (miRNAs) are genome encoded ~21 nt long, non-coding, regulatory RNA molecules that participate in diverse biological processes in eukaryotes (Bartel 2004; Kim 2005). Involvement of miRNAs in organ formation, hormonal signaling, root and shoot development, leaf and flower morphogenesis, genome methylation, etc. have been demonstrated (Emery et al. 2003; Mallory et al. 2004; Lauter et al. 2005; Aukerman and Sakai 2003; Chen 2004; Laufs et al. 2004; Bao et al. 2004; Guo et al. 2005). Their role in response to biotic and abiotic stresses has also been recognized and is still expanding rapidly (Navarro et al. 2006; Jones-Rhoades and Bartel 2004).

The miRNA genes are transcribed from their own promoters by RNA pol II activity (Fig. 2). The transcription leads to the generation of primary-miRNA

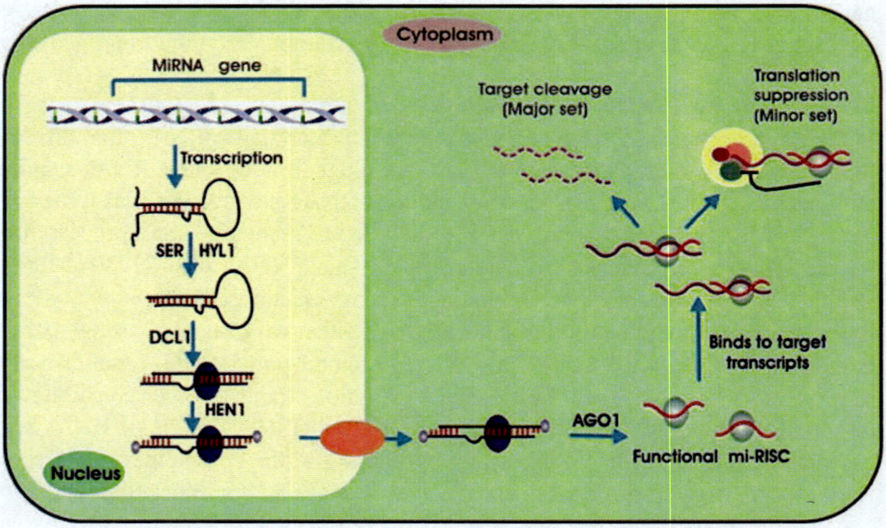

Fig. 2 MicroRNA (miRNA) biogenesis in plants: transcription of miRNA locus gives rise to a pri-miRNA that eventually assumes hairpin structure and is sequentially processed by a set of RNase III-type proteins to finally generate duplex miRNAs. These miRNAs are methylated by HUA-1 enhancer (HEN1) before leaving the nucleus. Argonaute binds to duplex miRNAs and, presumably in conjunction with unknown factors, release single stranded mature miRNAs. miRNAs in association with RISC bind to target transcripts and leads to target cleavage while a minor set of same miRNA might also be involved in translation suppression of similar targets. RISC: RNA induced silencing complex

(pri-miRNA) that is a prerequisite for miRNA formation. The pri-miRNAs possess typical stem-loop structure with characteristic 5′ cap and 3′ poly-A tail (Lee et al. 2004, Kurihara and Watanabe 2004). The sizes of cloned pri-miRNAs are typically found to be ~1 kb long and can vary for different family members. These pri-miRNAs are cleaved to pre-miRNAs by Dicer like-1 (DCL1), Hyponasty-like 1 (HYL1) and SERRATE proteins that remove the flanking sequences of the duplex stem (Han et al. 2004; Park et al. 2005; Song et al. 2007). Pre-miRNAs, while still in the nucleus, are cleaved to ~21 nt long duplex RNAs by Dicer like-1 action (Bartel 2004; Park et al. 2005). Recently, Dicer like-4 (DCL4) has also been demonstrated to generate few miRNAs namely, miR822 and miR839 (Rajagopalan et al. 2006). Unlike those in animals, plant miRNA duplexes are modified at their 3′-terminal by Hua-enhancer (HEN1), a nuclear methylase, which accounts for the increased stability of plant miRNAs (Yu et al. 2005; Li et al. 2005). After being modified, the duplex miRNA is transported out to cytoplasm by HASTY, a plant analog of Exportin-5 in animals (Bollman et al. 2003; Shibata et al. 2006).

In cytoplasm, these miRNA duplex molecules recruit Argonaute1 (Fagard et al. 2000; Morel et al. 2002), another crucial protein of RNA interference (RNAi) machinery. After sensing thermodynamic stabilities of both the ends of duplex, one of the components of RISC, initiates the unwinding from the strand with lower thermodynamic stability (Khvorova et al. 2003). The guide strand is retained within the complex machinery following the helicase activity of RISC, while the other strand (passenger strand) is degraded by various exonucleases. Recent studies have shown that mature miRNAs are also degraded by a set of exonucleases encoded by *SMALL RNA DEGRADING NUCLEASE* (*SDN*) genes (Vanitharani and Chen 2008). These proteins specifically recognize single stranded small RNAs (upto ~27 nt) *via* 2′-O-methyl modifications on small RNAs and degrade them to maintain the steady-state levels of mature miRNAs. In *Arabidopsis*, there are three SDN genes and knockout of all these genes leads to severe developmental defects in plants indicating that these are indispensable in maintaining the miRNA flux. These findings have led to better insights about regulatory mechanisms that release the pressure exerted by miRNAs on their target function(s).

The guide strand along with AGO1 and other accessory proteins are termed mi-RNA Induced Silencing Complex (mi-RISC). This active RNA-protein machinery, when encounters any transcript bearing complementary sequence, leads to the degradation of the target. However, a recent report suggests that while a major subset of a given miRNA molecule is engaged in degrading their respective targets, a small proportion of the same miRNA pool might participate in translation suppression of the target as well (Fig. 2; Brodersen et al. 2008). In this respect, plants resemble animal miRNA pathway where translation repression is more commonly observed as a consequence of miRNA binding to the target transcript (Pillai et al. 2005).

It can be speculated that besides miRISC, other miRNA-protein complex(s), presumably lacking the component possessing endonucleolytic activity, is (are) likely to participate in the process. Interestingly, AGO1 and DCL1, the integral components of miRNA pathway, are themselves regulated by miR168 and miR162, respectively (Xie et al. 2003; Vaucheret et al. 2006). This negative feedback inhibition of the global miRNA biogenesis might be a necessity to introduce another layer of fine regulation within the pathway.

Many reports on cloning and deep sequencing of small RNAs suggest that miRNAs have a fundamental role in maintaining the physiology of plants under stress. Here, we discuss various stresses of both biotic and abiotic nature and the roles played by miRNA in combating such adverse situations.

2.1 MicroRNAs in Bacterial Infection

Plants can succumb to a variety of bacterial infections, generally by those residing in the rhizosphere. In response to such stresses plants elicit several defense related pathways that include miRNA-mediated RNAi. Subtle and temporal changes in the endogenous miRNA levels can eventually reduce the levels of certain transcripts that might participate in defense responses. These changes introduced in the hosts' transcriptome thus prevent pathogen to successfully inhabit the host. Recently it was observed that bacterial infections modulate host miRNA levels (Navarro et al. 2006, 2008). In animals, several miRNAs are known to regulate host immunity against bacterial pathogens. However, our understanding of how these molecules contribute to plants innate immunity has just started to emerge. In an attempt to elucidate the effect of bacterial infection on host transcriptome, Navarro et al. (2006) observed that when *Arabidopsis* plants were treated with protein of bacterial origin (Flg22), a specific class of transcripts is down-regulated. These transcripts turned out to be the components of auxin signaling pathway namely, Transport Inhibitor Response 1 (TIR1), Auxin F-box protein (AFB2 and AFB3) and possess binding sites for miR393. Analysis of cleaved target products in various *dcl* mutant backgrounds (*dcl1, dcl2 and dcl3*) revealed that these targets are cleaved by miR393 in DCL1 dependent manner. Following Flg22 treatment *Arabidopsis* seedlings showed ~2 fold up-regulation of miR393 levels while no changes were observed in case of $Flg^{A.tum}$ (a non-effector protein). To investigate whether the changes in miR393 levels were induced at the transcription levels, ~1.5 kb sequence encompassing all required *cis*- acting elements for miR393 transcription was cloned upstream to Green Fluorescent Protein (GFP). When *Arabidopsis* GFP-promoter construct transformants were treated with Flg22, they exhibited enhanced GFP expression but this was not observed in $Flg^{A.tum}$ treated plants. This indicates that the promoter of miR393 is activated by bacterial effectors.

Further, in response to Flg22 treatment it was also observed that *Arabidopsis* miR393 over-expressing lines show significantly reduced levels of TIR1 (a miR393 target). This reduction in TIR1 levels renders stability to Aux/ Indole-3-acetic acid (IAA) proteins that repress auxin signaling by forming heterodimers with Auxin Response Factors (ARFs) and modulate the transcription of various primary auxin-response genes to which ARFs bind. These findings were further corroborated by analyzing the levels of three primary auxin-responsive genes namely, GH3-like, BDL/IAA12 and AXR3/ IAA17 that repressed in response to Flg22 treatment. Moreover, *Arabidopsis* lines over-expressing AFB1 (refractory to miR393) were sensitive to bacterial infection as evidenced by high bacterial titers. The studies clearly demonstrated that down-regulation of auxin signaling, resulting in ARF inactivation, is a part of a pathogen-induced immune response. Since miRNAs can directly regulate the expression of various ARFs, it is apparent that miRNAs contribute to plants basal defense.

This hypothesis gained further support from the studies performed by the same group where they comprehensively demonstrated that *Arabidopsis* plants with defective miRNA biogenesis promoted the growth of avirulent bacterial strains (Navarro et al. 2008). *Arabidopsis dcl1* and *hen1* mutants, when challenged with non-pathogenic *Pseudomonas syringae* and *E. coli* strains, exhibited increased growth, similar to the virulent *Pseudomonas syringae* pv . *tomato* DC3000 (Pto DC3000). This strongly suggests the fundamental role of miRNAs in conferring robust immunity against bacterial and other pathogens as well. To elucidate how bacterial effectors interfere with the host miRNA pathway, the influence of infection on the miRNA transcription, biogenesis/ stability and their activity was studied individually. The expression levels of Pathogen Associated Molecular Pattern (PAMP) -sensitive and -insensitive miRNAs in response to virulent and non-virulent bacterial strains were analyzed. Intriguingly, the suppression of certain PAMP-sensitive pri-miR393 and pri-miR396b was observed in response to pathogens, while PAMP insensitive pri-miR166 and pri-miR173 levels remained unaltered (Navarro et al. 2008). When E3 ligase activity of *Pseudomonas* effector protein (AvrPto) were abolished by introducing mutation (Y525A), the effector suppressed the pri-miRNA levels to a similar extent, suggesting that the effectors function is not required to suppress miRNA pathway. It appears that the sensitivity towards certain pathogens has allowed plants to induce specific pathways and miRNA represents one of them. The transcriptional suppression of miRNA-genes might be utilized by bacteria to establish/sustain inside host. To confirm that the changes in pri-miRNA levels were due to altered transcription of miRNA genes, they used previously described promoter-GFP constructs, wherein GFP induction was observed in response to effectors treatment.

To study the influence of the bacterial effector on miRNA biogenesis/ stability transgenic expressing WT AvrPto were analyzed. These transgenic accumulate miRNA precursors while mature miR393 were concomitantly decreased and is probably achieved by conferring stability to pre-miRNAs. It is known that AvrPto interacts with BAK-1(BRI-1 associated receptor kinase) and various PRRs (Pathogen Recognition Receptor). BAK-1 acts as ligand for BRI-1 (Brassinosteroid Insensitive- 1) and FLS (Flagellin sensing) interaction and its sequestration by AvrPto results in termination of the signal leading to pathogen triggered immunity (PTI). Therefore, any mutation rendering AvrPto non-functional can provide evidence for its direct influence on hosts' miRNA biogenesis. This was evaluated by studying $AvrPto_{Y89D}$ mutant that is unable to interact with BAK-1, while $AvrPto_{G2A}$ leads to mutated myristoylation and cannot localize itself to the membrane.

The mutants were unable to alter the levels or stability of the pri- or pre-miRNAs. Together, these results demonstrate that interaction of AvrPto with host proteins, but not its activity, is required for suppression of transcription of certain miRNAs.

In order to investigate the role of bacterial effectors on miRNA activity, the suc-sul lines were used. These suc-sul lines are stable *Arabidopsis* transgenics expressing hairpin RNA (bearing homology to endogenous SULPHUR transcript) restricted to phloem tissues. Consequently, the endogenous SULPHUR mRNA is degraded due to generation of siRNAs and the plants exhibit veinal chlorosis. Of various *Arabidopsis* plants over-expressing bacterial effectors, HopT1-1 showed reduced chlorosis. Since the silencing is achieved by AGO1 activity, it is plausible that the HopT1-1 protein interferes with the miRISC function. The miRNA target transcript levels were not affected. However, there was significant reduction in the corresponding protein levels. These observations suggest that probably HopT1-1 blocks the AGO1-mediated target translation inhibition. These bacterial effectors can be treated as Bacterial Suppressors of RNA silencing (BSRs) and in this respect are akin to VSRs.

To demonstrate that compromised miRNA pathway promotes growth of avirulent bacterial strains, *Arabidopsis* plants infected with *Turnip mosaic virus* were challenged with non-virulent bacterial strain. This virus encodes potent VSR namely, P1-HCPro and severely compromise the host RNAi machinery (both siRNA and miRNA pathways). Consequently, this leads to an enhanced growth of the non-virulent strains. These observations supports the hypothesis that phytopathogens (here viruses and bacteria) have evolved with similar defense strategies to modulate host RNAi responses. It would be interesting to study and elucidate if other plant pathogens like fungi or worms also employ such mechanisms to gain entry into host and improved pathogenesis.

The above findings corroborated with the miRNA expression profiling studies of *Pst DC3000 hrcC* treated *Arabidopsis* plants where a set of miRNAs were deregulated (Fahlgren et al. 2007). Three miRNAs namely, miR393, miR167 and miR160 were up regulated while miR825 was found to be down regulated under similar infection. Intriguingly, all the three PAMP-sensitive miRNAs target genes were involved in auxin signaling. This further supports that the repression of auxin pathway is a prerequisite to counteract bacterial infection. miR825, which is down-regulated under infection, targets three genes namely, zinc finger homeobox gene, remorin and frataxin-related gene that participate in gene regulation, biotic stress responses and iron regulator, respectively. How these genes confer resistance to host or is being used by pathogen to modulate host transcriptome is not yet understood.

In summary, these studies highlight that certain molecules act as receptors to pathogens and transduce signals to modulate expression of specific miRNA(s) which, in turn, would bring about changes in the host transcriptome. Other pathways involving miRNAs might also operate in similar fashion contributing to plant resistance.

2.2 MicroRNA in Viral Infection

The initial reports on viral encoded protein mediated inhibition of host RNAi response provided strong evidence of the importance of this phenomenon in conferring immunity against pathogen invasion (Anandalakshmi et al. 1998; Kasschau et al. 2003) and since then a huge number of viruses have been demonstrated to employ similar mechanism(s) to interfere with hosts RNAi pathways for their survival and tropism. Plant viruses have been demonstrated to counteract the host miRNA pathway utilizing specific self encoded protein or RNA structure called viral suppressors of RNAi (VSRs) leading to developmental defects in plants (Kasschau et al. 2003; Dunoyer et al. 2004). These VSRs are believed to alter the host miRNA pathway by disturbing the biogenesis at one or more of the crucial steps. As demonstrated in the transgenic plants expressing viral proteins like HC-Pro, P19, P21, etc., the global miRNA levels are generally deregulated (Chapman et al. 2004; Chen et al. 2004; Chellappan et al. 2005). Stable transgenic overexpressing VSRs showed marked reduction in miRNA-mediated target cleavage. Thus, certain miRNAs can be potentially utilized as biomarker(s) for specific viral disease and provide opportunity to employ these RNAs to design antiviral strategies (Naqvi et al. unpublished data). Interestingly, it has also been shown that the viral infection can lead to de-regulation of AGO1, an important biogenesis factor governing miRNA flux (Vaucheret et al. 2006). This hints on the other possible modes of VSR action. Moreover, as siRNA and miRNA pathways share common intermediates, it is possible that a VSR alone can modulate both the pathways. The action of few known VSRs with their important functions is listed in Table 1.

In animals, upregulation of specific miRNA have been demonstrated to confer resistance against viruses (Lecellier et al. 2005; Jopling et al. 2005). Besides, many animal viruses are shown to encode miRNAs that help them to establish infection (Pfeffer et al. 2004). However, in plant viruses none of the above mechanisms has been observed till date. Interestingly, artificially designed miRNAs (amiRNAs) against certain plant viruses confer strong resistance against them without exhibiting any off-target effects (Alvarez et al. 2006; Schwab et al. 2006). Different amiRNA based strategies have been devised to make plants immune to viral attacks. For example, endogenous miRNA precursor sequence was used to incorporate sequence of *Rice dwarf virus* and the rice plants were transformed with such construct (Ma et al. 2004), where transformants were found to be resistant to the virus. In another study, *Plum pox virus* genome was modified to incorporate sequences that were complementary to hosts' endogenous miRNAs (Simón-Mateo and Garcia 2006). The *Nicotiana* plants transformed with engineered PPV genome bearing target sequences (derived from *Arabidopsis*) for various miRNAs namely, miR159, miR167 and miR171 exhibited robust resistance against *Plum pox virus*. In yet another example, similar strategy was designed to confer resistance against multiple viruses. Here, miRNA-precursor was engineered to generate two different amiRNAs. These miRNAs were designed to target *Turnip yellow mosaic virus* and

Table 1 Few viral suppressors of RNAi with their functions highlighting influence on host RNAi pathways

Viral suppressor	Origin (virus genus)	Function(s) attributed	Refs.
P1/HC Pro	Potyvirus	Assembly and/or targeting of RISC	Anandalakshmi et al. (1998) and Kasschau et al. (2003)
2b	Cucumovirus	Binds to dsRNA; interfere with the spread of silencing	Brigneti et al. (1998) and Qi et al. (2004)
P19	Tombusvirus	Binds to siRNAs	Silhavy et al. (2002) and Qi et al. (2004)
P25	Potato Virus X	Interfere with movement of silencing signal	Voinnet et al. (2000)
Coat protein	Carmovirus	Interfere with the processing of dsRNA	Qu et al. (2003) and Qi et al. (2004)
FHV-B2	Nodavirus	Binds to dsRNA	Fenner et al. (2006)
AC4	Begomovirus	miRNA binding	Chellappan et al. (2005) and Vanitharani et al. (2004)
AC2	Begomovirus	Interfere with general methylation	Trinks et al. (2005) and Bisaro (2006)
βC1	Begomovirus	Interacts with certain host TFs to mediate their effects	Cui et al. (2005)
VA1 RNA	Adenovirus	Sequesters dsRNA processing proteins of RNAi	Lu and Cullen (2004) and Andersson et al. (2005)

Turnip mosaic virus, and thereby leading to simultaneous elimination of viruses (Niu et al. 2006). *Arabidopsis* plants transformed with this construct were found to exhibit resistance when challenged with either of these viruses.

Overall, these studies reveal the potential of employing miRNA-mediated strategy to fight various viral pathogens.

2.3 *MicroRNA in Abiotic Stresses*

Abiotic stress can profoundly affect the plants' yield and to deal with such adverse climatic stresses plants have to adjust their physiology accordingly. This can be achieved by induction and/or repression of a large array of transcripts. MiRNAs being the regulators of huge set of transcription factors (AP2, MYB, TCP, etc.,) and defense related genes (PPR, WRKY, etc.,) are critical molecules to study under stressed physiologies. Recent advancements in deep sequencing supported by microarray based profiling have allowed us to gain better insights into complex miRNA-regulated circuits that are activated in response to stress. Notable factors that modulate plants growth and yield include nutrient deficiency (phosphate deficiency, sulphate deficiency), salt stress, hypoxia, etc. The role of miRNAs under various abiotic stresses is discussed below:

2.3.1 Phosphate Deficiency

Phosphate is a critical macro-nutrient for plant growth and is evidently being regulated by miRNA (Sunkar and Zhu 2004). MiR399 binds at the multiple locations on the 5′-UTR of ubiquitin conjugating enzyme *UBC24*, an important component of protein degradation machinery (Fujii et al. 2005). Undetectable under normal conditions, miR399 is induced by low-inorganic phosphate (Pi) conditions that bring about the repression of *UBC24* transcripts. This leads to the upregulation of a phosphate transporter, *At*PT1 that helps in assimilation of inorganic phosphate. The increased Pi intake helps plant to cope with low-phosphate stress. This was supported by introducing mutations in the miR399 binding sites on the *UBC24* transcripts. Further, studies by Aung et al. (2006) clearly revealed that miR399 was present in the tissue domains where *UBC24* was found, supporting the notion that the co-localization of miRNA and its target(s) is required for better coordination. After analyzing the Pi levels in roots and shoots of wild-type and miR399 over-expressing transgenics an interesting observation was made that the shoots of the transgenics plants accumulate large amount of Pi while the corresponding levels in the roots were comparable. These findings suggest remobilization of Pi from shoots to roots. However, further study is required to shed light on the finer details of miRNA-mediated phosphate level regulation pathway.

2.3.2 Sulfate Starvation

Similar to inorganic phosphate, sulfate metabolism is also regulated by miRNA. Induced under low- sulfate stress miR395 targets ATP sulfurylases (*APS1*, *APS3* and *APS4*), the enzyme(s) that participates in catalysis of inorganic sulfate assimilation (Jones-Rhoades and Bartel 2004; Sunkar and Zhu 2004). Such induction of miR395 was restricted only to sulfate deficiency, as other salt stresses did not affect the level of miR395. MiR395 also binds to *AST68*, a low affinity sulfate transporter that acts as cargo for transporting sulfate from root to shoot. It is noteworthy that miR395 regulates two different groups of genes controlling the sulfate metabolism thereby signifying the crucial role of miRNAs in abiotic stress responses.

2.3.3 Oxidative Stress

Plants have evolved extensively to deal with the oxidative stress which severely affects their existence and a large number of genes have been implicated in the process. Copper/Zinc Superoxide Dismutase (CSD1 and CSD2) represent a class of genes that confer oxidative tolerance by scavenging free radicals. These genes were found to be targeted by miR398 (Jones-Rhoades and Bartel 2004). Under normal conditions, miR398 is expressed at high levels that lead to the cleavage of CSD1 and CSD2. High oxidative stress inhibits the expression of miR398 and consequently increases CSD1 and CSD2 levels that confer oxidative stress tolerance to plants (Sunkar and Zhu 2004). This finding was supported by raising transgenic plants with mutated miR398 which were able to withstand a high degree of oxidative stress.

2.3.4 Salt/Temperature/Drought Stress

Salt, drought and temperature represent a major class of abiotic stress. There exists enough literature supporting the role of various transcription factors during above mentioned stresses. Transcription factors bind to well defined motifs present in the promoter region of many genes thereby regulating their expression. Majority of these transcription factors are shown to be controlled by miRNAs, but to discern how miRNA expression gets altered under these conditions remains a major challenge. Since miRNAs are known to be expressed at specific developmental time point and in a tissue specific manner, it is conceivable that certain miRNAs might be selectively produced following a particular stress. Thus, it is imperative to gain information about such miRNAs as they would prove beneficial in raising transgenic crop plants that can cope with extreme environmental conditions. Advent of high throughput sequencing has made it possible to obtain such miRNA sequences that would help us providing insights into these complex regulatory pathways. Studies on *Arabidopsis* and rice have clearly shown that expression of many miRNAs is significantly altered after stress treatment (Sunkar et al. 2006; Liu et al. 2008; Zhou et al. 2008), some of these are listed in Table 2.

2.3.5 Hypoxia

Hypoxia is a common stress associated with low oxygen availability that consequently reduces the overall energy status inside the cell. Analyses with the small RNA population from root revealed that the expression levels of both miRNAs and tasiRNAs are profoundly altered in treated tissues compared to healthy counterpart (Moldovan et al. 2010). There are about 25 miRNAs (belonging to 19 families) that are induced in response to hypoxia while 5 miRNAs (belonging to 4 families) were significantly down-regulated. Few miRNAs namely, miR156, miR166 and miR172, in view of their role in plant development, are believed to control root growth and after sensing hypoxia they can arrest root growth (Schwarz et al. 2008; Aukerman and Sakai 2003; Boualem et al. 2008). Only few miRNAs that were highlighted in hypoxic conditions were common to other abiotic stresses. Intriguingly, majority of hypoxia induced miRNAs overlapped with those altered following viral infection indicating the involvement of similar pathways during these stresses. For instance, of the 20 miRNAs altered in viral infection (Tobacco mosaic virus-Cg), 15 miRNAs were observed to be deregulated under low oxygen condition (Tagami et al. 2007), suggesting a possible cross-talk among these stresses. To test whether these dysregulated miRNAs are specific to hypoxia,

Table 2 List of microRNAs (miRNA) altered during salt, drought and temperature stresses

miRNA	Stress (D/S/T/H)[a]	Up/down	Target	References[b]
miR156	D/H	Up	Squamosa Binding Protein TFs	b, d
miR158	D/H	Up	Unknown protein	b, d
miR159	D/H	Up	MYB TFs	b, d
miR164	T	Up	NAC TFs	C
miR165	D/S/T	Up	Class III HD-ZIP TFs	b, c
miR167	D/S/H	Up	ARF6 and ARF8	b, c
miR168	D/S	Up	AGO1	b, c
miR169	D/T/H	Up/down (H)	CBF HAP-2 like TFs	a, b, c, d
miR171	D/S	Up	SCL TFs	B
miR172	T/H	Up	Apetala TFs	c, d
miR319	D/T	Up	TCP TFs	b, c
miR389	D/S/T	Down	Unknown protein	A
miR393	D/S/T	Up	F-box protein; bHLH TFs	b, c
miR394	D	Up	F-box protein	B
miR396	D/S/T	Up	GRL TFs	b, c
miR397	T	Up	Laccases	C
miR398	T	Up	Superoxide dismutases	C
miR402	T	Up	ROS-1 like, DNA glycosylase	a, b, c
miR408	S	Up	Plantacyanin	a, b, c

[a]The list was compiled from data on rice and Arabidopsis. Abbreviations used: Drought (D), Salt (S), Temperature (T) and Hypoxia (H)

[b]References (a) Sunkar and Zhu (2004), (b) Liu et al. (2008), (c) Zhou et al. (2008) and (d) Moldovan et al. (2010)

expression of selected miRNAs were studied in presence of inhibitors of mitochondrial respiration. Most of the miRNAs exhibited similar expression levels suggesting that the fluctuations observed are genuine consequences of cellular energy burden. The promoter analysis of several dysregulated miRNAs revealed that HD-ZIP I motif was common to all miRNA genes which, incidentally, gets down-regulated in response to hypoxia (Zimmermann et al. 2004). Together, these observations support the direct regulation of several miRNAs by this class of transcription factors.

Moreover, two miRNAs, namely miR173 and miR390, are known to trigger tasiRNA biogenesis (also discussed in 'tasiRNA' section on Page 24). Both of these miRNAs were observed to be up-regulated following hypoxia and so were the levels of their corresponding tasiRNAs (representing TAS1a, TAS2 and TAS3 classes). As many of the targets regulated by these tasiRNAs are involved in plant defense and development, it appears that an intricate network is activated in response to hypoxia. However, the underlying mechanism(s) behind the adaptive response remains to be unraveled.

A list of miRNAs that are demonstrated to be deregulated under salt, drought, temperature and hypoxia stress is presented in Table 2. Two important interpretations can be made from Table 2 – firstly, most of the miRNAs are induced in response to stress, however, a few miRNAs e.g., miR389 are down-regulated following stress. It can be speculated that either the promoter elements of these miRNAs are inducible and hypersensitive to most of the stresses. Secondly, a significant fraction of these miRNAs are deregulated in more than one stress, suggesting that these miRNAs are engaged in multiple functions. The analysis of the putative promoter sequences has revealed that miRNA targeted transcription factors have motifs to bind to the *cis*-elements of miRNA genes. For instance, the promoter of miR169 that targets CCAAT- binding factors (CBF transcription factors) have Dehydration response element (DRE), a motif recognized by CBFs (Zhou et al. 2008). It is thus apparent that miRNAs and transcription factors regulate each other thereby indirectly participating in their own feedback regulation. However, detailed studies on the promoters of individual miRNAs might shed more light on these issues.

2.3.6 Mechanical Stress

In view of the critical regulatory roles played by miRNAs in diverse stress responses, Lu et al. (2005) addressed a question whether these small molecules can sense adverse mechanical stimuli as well. Intriguingly, they observed modest changes in the expression levels of various miRNAs when the xylem tissues were subjected to mechanical stress (tension and compression stress). Based on the mechanical stress applied they classified the miRNA expression patterns in five different groups.

Group A: Suppressed similarly by tension and compression and is represented by miR156, miR162, miR475, miR480 and miR481.

Group B: Upregulated in tension and compression stressed tissues and include miR408.

Group C: Preferentially upregulated in compressed tissues. This group includes miR159, miR476 and miR479.

Group D: Down-regulated only in compression stressed tissue and is represented by miR160 and miR172.

Group E: miR168 was specifically induced under tension stressed tissues.

The above studies suggest that like other stresses, mechanical stress profoundly induces changes in the expression levels of miRNAs and that they are specific to particular stress namely, compression or tension.

3 Short Interfering RNAs (siRNAs)

Besides miRNAs, plants have other related pathway, namely siRNA-mediated RNAi to avoid pathogen invasion. These responses are believed to be ancient innate immunity-like cellular responses that were employed to take care of the entry of undesired nucleic acid molecules including transposons, viral or bacterial genome etc. (Mette et al. 2000; Zilberman et al. 2003; Mochizuki and Gorovsky 2004). Plants have evolved intricate RNAi mechanisms to combat adverse conditions. The activation of siRNA pathway requires a long dsRNA formation that activates RNase

Table 3 *Arabidopsis* Argonautes and their assigned functions

Argonaute	Function assigned	Size of small RNA associated	Loss-of-function	References
AGO1	Main Slicer, required in miRNA and tasiRNA biogenesis	21 nt	Dwarf and sterile plants, defective in virus defense	Morel et al. (2002) and Baumberger and Baulcombe (2005)
AGO2	TasiRNA biogenesis	21 nt	No obvious developmental defects	Lobbes et al. (2006) and Mi et al. (2008)
AGO3	Likely association with tasiRNAs	21 nt	No obvious developmental defects	Lobbes et al. (2006) and Mi et al. (2008)
AGO4	RNA-mediated DNA and Histone modification	24 nt	No obvious developmental defects	Zilberman et al. (2003) and Qi et al. (2006)
AGO5	TasiRNA biogenesis	21–24 nt	No obvious developmental defects	Mi et al. (2008) and Takeda et al. (2008)
AGO6	Transcriptional gene silencing	24 nt	No obvious developmental defects	Zheng et al. (2007)
AGO7 (ZIPPY)	TasiRNA biogenesis	21 nt	Regulates vegetative phase changes	Fahlgren et al. (2006)
AGO8	Probable involvement in DNA methylation and chromatin silencing	24 nt	No obvious developmental defects	Zheng et al. (2007) and Takeda et al. (2008)
AGO9	Probably participate in TGS	24 nt	No obvious developmental defects	Zheng et al. (2007) and Takeda et al. (2008)
AGO10 (PIN-HEAD/ ZWILLE)	miRNA-directed translation inhibition	21 nt	Pleiotropic developmental defects	Mallory et al. (2009)

III like endonucleases present throughout the cell. These mainly include DCLs that release 21–24 nt long siRNA duplexes with 2nt 3′ overhangs (Elbashir et al. 2001). Duplex siRNAs with DCL attracts another important protein component of RNAi machinery namely, AGOs. *Arabidopsis* genome codes for ten AGO proteins and each AGO protein has specific function to perform, however they exhibit redundancy in their function (Refer to Table 3). The inherent helicase activity of RISC results in unwinding of siRNA duplex. The siRNA strand retained with AGO is termed guide strand, while the other strand removed from the complex is referred to as passenger strand. The passenger strand is soon degraded by various endogenous RNases, as exemplified by Exoribonuclease-4 (XNR4) and SDN exonucleases. The slicer activity of AGO is induced as soon as the single stranded siRNA guide molecules recognizes any transcript bearing sequence complementarity, which eventually leads to the breakdown of target messages. In plants, siRNA pathways are known to contribute to overcome stresses induced by viral and bacterial infection and abiotic stresses like salt, hypoxia, etc.

3.1 siRNAs in Viral Infection

Although the phenomenon of recovery of plants following severe viral infection has been known since long, Hamilton and Baulcombe (1999) were the first to demonstrate the involvement of long dsRNA in inducing homologous gene silencing. Both DNA and RNA viruses, during their entire life cycle, generates dsRNA structures that can induce RNAi machinery of the hosts. Similar to miRNA pathway, the viral gene silencing requires production of dsRNA as a prerequisite (Fig. 3). In case of RNA viruses, the replication involves dsRNA forms as intermediates, while most of the DNA viruses generate dsRNA either by producing overlapping transcripts from opposite strands or the viral RNA itself adopts a structure that is recognized by RNAi specific proteins e.g., Dicers. Besides, certain DNA viruses can also attract the RNAi machinery thereby modifying their genome and consequently effecting the viral replication (Fig. 3). Thus, viruses are inducer and targets of RNAi machinery.

The RNAi mechanisms operative in response to viral entry have been studied extensively in plants. Various classes of siRNAs are reported and their generation depends on downstream dicer (DCL) requirement. These are of different lengths and require specific dicer for their biogenesis (Henderson et al. 2006). Unlike animal system, plants possess multiple dicers (e.g., *Arabidopsis* has four, rice has six) that mediate cleavage of various kinds of long dsRNA molecules. Each dicer is known to have specific substrate (Table 4). Studies on *Arabidopsis* have clearly demonstrated that DCL2 generates a 22nt long natural antisense siRNAs (nat-siRNA) and in few cases of viral siRNA, DCL3 participates in biogenesis of 24 nt heterochromatic siRNAs while DCL4 releases 21 nt tasiRNA and viral siRNA from long dsRNAs (Blevins et al. 2006; Deleris et al. 2006). The notable findings on the activity of viral siRNA biogenesis came from the various studies focusing on the functional hierarchy of dicers (Xie et al. 2005; Gasciolli et al. 2005; Henderson et al. 2006). While studying on three different RNA viruses, Deleris et al. (2006) showed that DCL4 is the major dicer participating in antiviral defense mechanism in *Arabidopsis*. Interestingly, under *dcl4* background it was observed that DCL2 can substitute for DLC4 function, suggesting that DCL4 and DCL2 are predominantly involved in silencing of foreign nucleic acid. However, DCL4 is the primary player

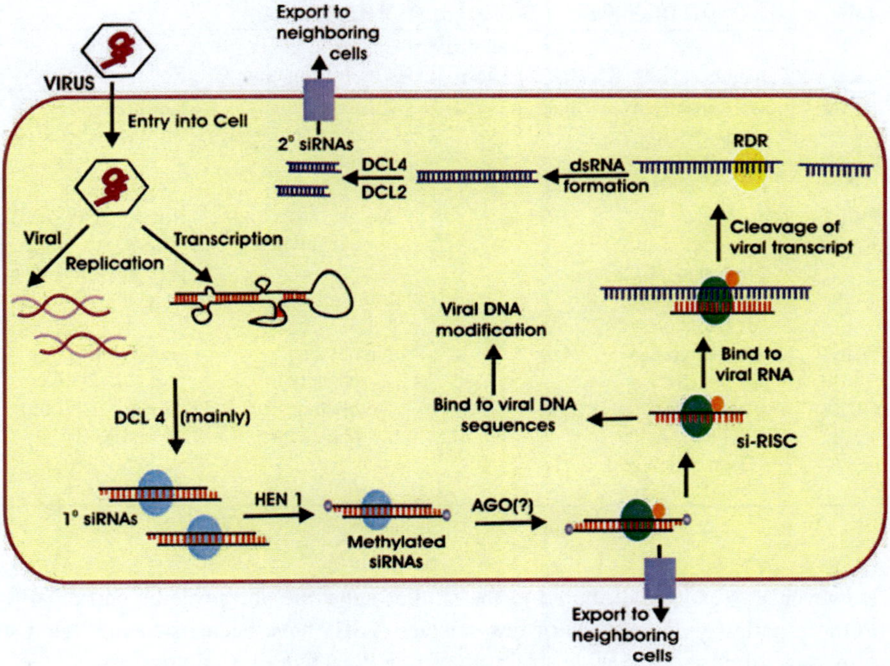

Fig. 3 Biogenesis of viral siRNAs: viral entry into the host activates siRNA-mediated RNAi pathway that recognizes dsRNA substrates. Dicer-like 4 (*DCL4*) initiates the cleavage of viral dsRNAs that can be methylated by host HEN1. These siRNAs will cleave viral transcripts, which can further be recognized by RNA dependent RNA polymerase (*RDR*) proteins to generate dsRNAs, which in turn, are cropped by DCL4/DCL2 to produce secondary siRNAs for robust silencing. Both, primary and secondary siRNAs can emanate to achieve systemic silencing. siRNAs may also recognize viral genome sequences and bring about DNA modification that interferes with the viral replication

leading to the degradation of viral nucleic acid. This suggests that in order to make RNAi-mediated immune system more efficient, the important players have evolved with functional hierarchy.

Viruses have evolved to escape hosts' RNAi-mediated defense mechanisms by employing self genome encoded proteins called Viral RNAi suppressors (VSRs; Voinnet 2005; Li and Ding 2006). These proteins contribute significantly to the pathogenicity of viruses and act by interfering with the RNAi related pathways (siRNA and miRNA pathways). Transgenic expressing mutated VSRs exhibit less pronounced developmental defects and much reduced miRNA accumulation, supporting the importance of these proteins in viral pathogenesis (van Wezel et al. 2002). It is worth noting that these VSRs can even act in a cross-kingdom manner i.e., a VSR from plant virus can efficiently work in animals and *vice versa*, suggesting that VSRs have evolved specifically to act against RNAi pathway(s).

Table 4 Arabidopsis Dicer-like proteins (DCLs) and their functions

Dicer	Function	Associated small RNA species	Loss-of-function	References
DCL1	miRNA, tasiRNA and nat-siRNA biogenesis	21 nt	Leaf developmental defects	Kurihara and Watanabe (2004) and Borsani et al. (2005)
DCL2	Nat-siRNAs generation and viral nucleic acid degradation	22–23 nt	No obvious phenotype	Borsani et al. (2005) and Katiyar-Agarwal et al. (2006)
DCL3	DNA methylation and TGS	24 nt	No obvious phenotype	Henderson et al. (2006)
DCL4	TasiRNA biogensis and Degradation of viral RNAs and	22 nt	No obvious phenotype	Allen et al. (2005) and Xie et al. (2005)

However, this can be attributed to the similar substrate and proteins participating in these pathways. Functions of few selected VSRs have been listed in Table 1 to provide a glimpse of the wide spectrum of functions that they can perform.

3.2 Short Interfering RNAs in Bacterial Infection

Short-interfering RNAs (siRNAs) have been known much before miRNAs but recently there has been rapid improvement in our understanding about their functional roles in plants. Described below are various classes of siRNAs with their assigned functions.

3.2.1 Long siRNAs

One class of small RNAs which has been discovered recently are long siRNAs (lsi-RNAs) that are >30 nt long molecules and are specifically induced either under bacterial infection (bearing *avrRpt2* effector) or growth conditions (Katiyar-Agarwal et al. 2006). About six different lsiRNAs were predicted by bioinformatics analyses which were further validated by molecular analysis. It was found that lsiRNA 1, 4, 5 and 6 are generated by NAT (Natural Antisense Transcript) pairs encoding proteins and expressed in different tissues or cell types (Fig. 4). Biogenesis of lsiRNAs requires proteins involved in both miRNA and siRNA pathways. For example, the generation of lsiRNAs depends on DCL1 but not DCL2 or 3, and show marked reduction under HYL1 and HST mutant background. Moreover,

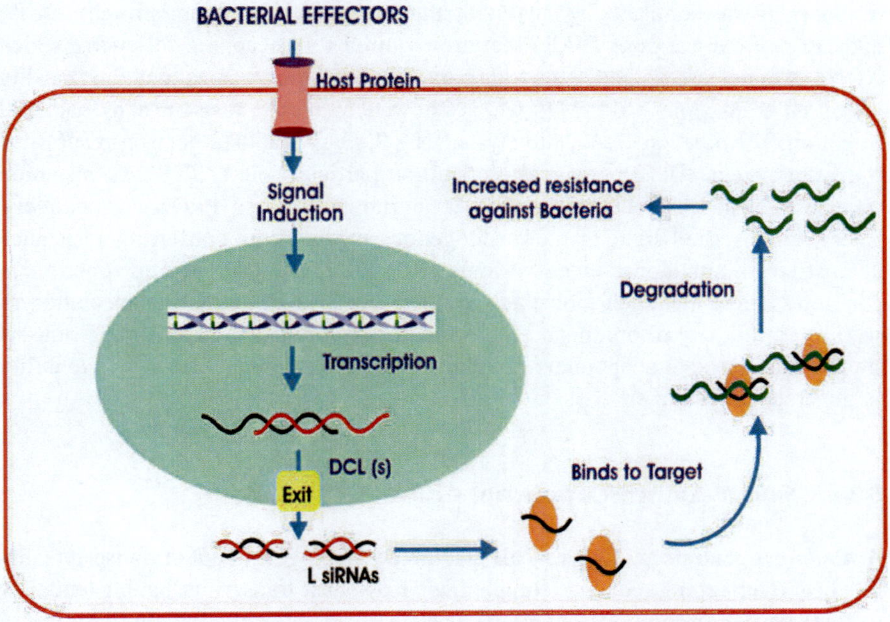

Fig. 4 Biogenesis of siRNAs induced by bacterial infection: bacterial effectors recognize host surface proteins to transduce signals inside the host cell that consequently induce transcription from a specific locus, a NAT pair in this case. The activated Natural Antisense Transcript (*NAT*) pairs, in turn, induce the RNAi machinery to generate >30 nt long siRNAs that exclusively participate in cleavage of the constitutive component of NAT pair. This helps plants to activate certain pathways that are otherwise suppressed thereby gaining immunity against specific bacterial pathogen

when components of tasiRNA biogenesis, namely, AGO7, RNA Dependent RNA Polymerase (RDR6) and Nuclear RNA Polymerase D (NRPD1a and 1b), were mutated, a significant reduction was observed in the lsiRNA levels. Similar mutations in AGO4, AGO2, Suppressor of Gene Silencing (SGS3) or SDE3 did not show significant changes in the lsiRNA production (Katiyar-Agarwal et al. 2007). The studies with HEN1 mutant elucidate that 2′-OH or 3′-OH of 3′-terminal ribose is protected by HEN1 leading to reduced siRNA levels as a result of this mutation. Based on the above observations, it appears that the requirement of DCL1 is likely to be associated with the recruitment of downstream proteins since it is known that DCL1 is a prerequisite for *trans*-acting siRNA (tasiRNA; discussed below in Page 25) biogenesis. It would be interesting to know if DCL1 could manipulate its own activity alone or if it requires some other factors for the purpose.

*At*lsiRNA-1 is highly induced by avirulent *Pst avrRpt2*, while AtlsiRNA-2, -3 and -4 are modestly induced and relatively weak induction is observed with *At*lsiRNA-5. *At*lsiRNA-1 targets the antisense *At*RAP transcript in *cis,* which is a negative regulator of plant defense response. Interestingly, it has been discovered that lsiRNAs act by decapping of target transcript, a mode previously not observed in plants. It was confirmed by analyzing the mutants for decapping protein (DCP1 and DCP2) and Varicose (VAR) that are required for decapping following which XRN4 exonuclease degrades the transcript. The inference was validated by raising knockout transgenic of *AtRAP,* which showed increased resistance to bacterial infection. The other lsiRNAs, namely, lsiRNA-2, -3, -4 and -5 target iron-responsive transporter gene (IRT2), a signal recognition particle gene (AT7SL-1), an anion transporter gene (ANTR2), and a protoporphyrinogen oxidase (PPOX), respectively. These studies shed light on new endogenous mechanisms conferring resistance against stress situations. Since a single siRNA can achieve specific function, it can be assumed that such molecules could be decisive factors in the speciation as well as complexity of organism. Therefore, the potential of a plant to fine tune its transcriptome and to adapt under diverse physiologies probably lies in its capability to generate dissimilar siRNAs.

3.2.2 Natural Antisense Transcript-siRNA (NAT-siRNA)

Arabidopsis genome analysis reveals that there are a huge amount of transcript pairs that bear overlapping sequence complementarity. These transcripts, having tendency to form partial or complete complementary dsRNAs, are called Natural Antisense Transcripts (NATs). The NATs are endogenous protein coding or non-coding transcripts that can bind to other cellular RNAs with complementary sequence. There are about 2,000 nat-siRNA pairs within the *Arabidopsis* genome. One component of the NAT pair is stress (abiotic or biotic) inducible and downregulates the constitutive transcript of the NAT pair. On the basis of their origin NATs can be classified as: (1) *Cis*-NATs: Transcripts that are generated from opposite strands of the same genomic locus and have a long perfect complementary overlap between the sense and antisense transcripts are termed *cis-* NATs, and (2) *Trans*-NATs: These are pair of RNA transcribed from different genomic loci and they often have short and imperfect complementarity.

More than hundred *cis*-NATs have been predicted to bind other transcripts in *trans*, albeit with weak complementarity, thereby increasing the complexity of the regulatory network (Wang et al. 2005). Trans NATs are further classified into immediate and distal NATs depending on the location of the complementary transcript originating from the genome. Although ~2,000 immediate *trans*-NATs have been discovered, there is no report on distal *trans*-NATs (Wang et al. 2006).

The RNAi machinery of the cell recognizes dsRNA (formed from NAT pairing) as substrate and cleaves them into small RNAs. The biogenesis of nat-siRNA requires a set of proteins namely, DCL1, HYL1, HEN1, DCL2, RDR6, SGS3 and NRPD1a that generate 21–24 nt long nat-siRNAs (Fig. 5). The mature species resides in a complex of one of the AGO proteins that preferentially guides cleavage of the constitutive transcript. Although many studies have shown the presence of endogenous siRNAs, none has demonstrated their functional significance. Till date, only two nat-siRNAs have been validated experimentally and these have

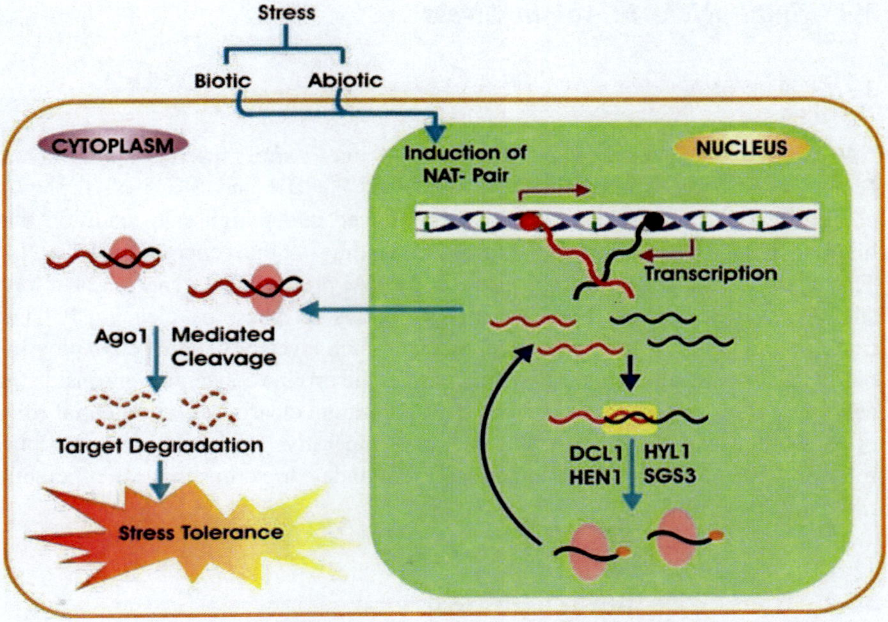

Fig. 5 NAT-siRNA biogenesis: induced by various abiotic or biotic stimuli, Natural Antisense Transcript (*NAT*) pair shares a region of sequence that leads to the formation of full or partially complementary dsRNA structure. Such structures are recognized by Dicer-like protein (*DCL*)/ Suppressor of gene silencing (*SGS3*) to produce 21–24 long nat-siRNAs that cleave the constitutive component of NAT pair. The biogenesis also requires few components of miRNA pathway like HEN1 and HYL1

been demonstrated to play critical roles under abiotic and biotic stresses (Borsani et al. 2005; Katiyar-Agarwal et al. 2006).

Study by Katiyar-Agarwal et al. (2006) demonstrated the involvement of nat-siRNA in conferring resistance against biotic stress. It was observed that a bacterially induced transcript *ATGB2* recognize the cognate antisense *PPRL* transcripts to generate dsRNA like structure, thereby, triggering small RNA production. This nat-siRNA, named nat-siRNAATGB2, was identified from the small RNA database and found to originate from NAT pair of Rab-2 like gene *ATGB2* and a pentatricopeptide repeats protein-like gene (*PPRL*). The *PPRL* gene is a negative regulator of effector-triggered immunity (ETI). The nat-siRNAATGB2 binds to its target gene, *PPRL*, and relieves negative influence, thereby allowing host to elicit rapid response against pathogen. Importantly, the induction of this nat-siRNA requires bacterial *RPS2* gene and another race specific gene *NDR1*. These molecules bind to *PPRL* transcripts which are negative regulators of *RPS2* and degrade it to confer resistance against bacteria. These findings demonstrate how critically the biogenesis of these nat-siRNAs is regulated and highlights their importance in conferring tolerance to plants against various stresses.

3.3 Small RNAs in Abiotic Stress

3.3.1 NAT-siRNAs

NAT-siRNAs have been shown to participate in abiotic stress response, too. Borsani et al. (2005) demonstrated the involvement of P5CDH and SRO5 NAT pair in conferring salt stress tolerance. SRO5 is induced under high salt condition that binds to P5CDH (constitutively expressed) leading to the generation of dsRNA. This consequently activates a specific set of RNAi proteins that generate 24 nt long siRNA molecules. These 24 nt nat-siRNA species set phase by cleaving P5CDH transcript and thereby lowering down its expression levels. The reduced expression of P5CDH consequently leads to an increase in the proline concentrations inside the cell that confer better salt resistance. These findings clearly reveal potential roles of endogenous small RNAs. It thus becomes imperative to characterize additional roles of these siRNAs in order to employ them to develop crops that can efficiently resist diverse stresses.

3.3.2 Trans Acting siRNAs (tasiRNAs)

These are 21 nt long non-coding, endogenously expressed regulatory small RNAs found exclusively in plants (Vazquez et al. 2004; Allen et al. 2005). The discovery of tasiRNAs began with *rdr6* mutant screening where a non-protein coding transcript accumulate in *rdr6* mutant background that was later on found to be the parent strand for these classes of siRNAs. TasiRNAs generation depends on the miRNA/DCL1-mediated cleavage of TAS mRNA, one of the cleaved products (either 5′ or 3′) used as a template for RDR6 and SGS3 converting ssRNA to dsRNA in primer-independent fashion (Fig. 6). How RDR6 and associated proteins recognize the particular cleaved end to bind still remains to be elucidated. The dsRNA thus formed are diced by DCL4 to produce phased 21 nt long tasiRNAs that bind to transcripts different from the parent mRNA cleaved by miRNA (Xie et al. 2005). In contrast to other classes of siRNA, they functionally resemble miRNAs in this aspect. Similar to miRNAs, tasiRNAs can negatively regulate their generation by cleaving the precursor transcripts. TasiRNAs generated from the minus strand can prevent the accumulation of TAS precursors. These mature tasiRNAs are loaded onto AGO1- or AGO7-RISC complexes and processed subsequently.

Both tasiRNA and miRNAs have been demonstrated to target different members of the same gene family. However, in case of TAS4 siRNAs, they bind to the parent transcript. This interesting observation came from the tasiRNA target analysis where both, parent miRNA (miR828) and tasiRNAs from TAS4 locus targets the same MYB transcription factors (Rajagopalan et al. 2006). The binding site within the MYB transcript is different for miRNA and tasiRNA. This example clearly demonstrates that both miRNA and siRNA can participate in conjunction in achieving proper regulation of certain transcripts. Five different TAS loci have

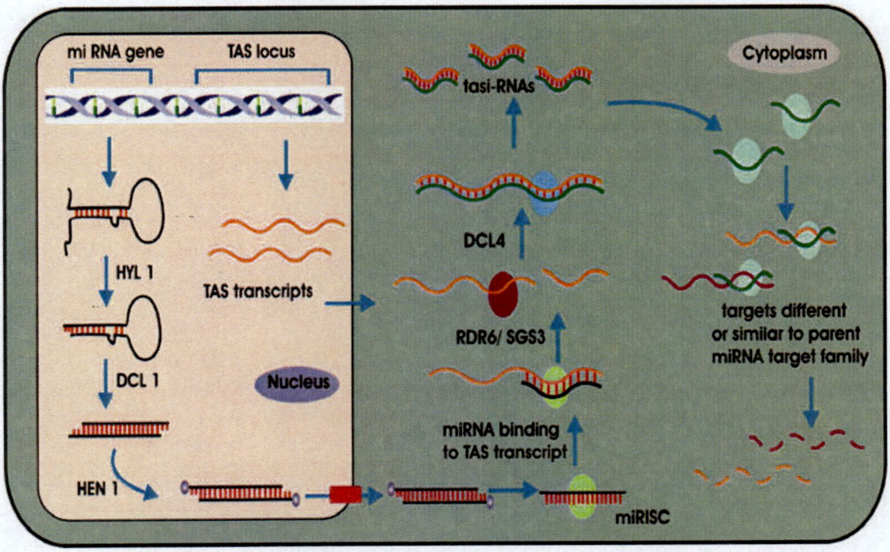

Fig. 6 TasiRNA biogenesis: tasiRNA biogenesis is dependent on miRNAs. Mature miRNAs generated after processing of hairpin transcripts are finally loaded on to the protein machinery called mi-RISC. The miRNAs then guide this complex to degrade their target transcripts. Few of such transcripts are recognized by RNA dependent RNA polymerase (*RDR6*)/Suppressor of gene silencing (*SGS3*) proteins that convert single stranded cleaved products to double stranded RNAs that are further diced by Dicer-like 4 (*DCL4*) to produce ~21 nt long ta-siRNAs. These siRNAs get associated with si-RISC, bind to and degrade mRNAs that are similar to or different from the parent miRNA targets

been found to regulate several developmentally important genes that participate in leaf development, auxin signaling, defense responses, etc. (Table 5). TasiRNAs target many genes including those demonstrated to play critical roles in plants defense mechanisms. Although there is no report showing the direct involvement of tasiRNAs in abiotic and biotic stresses, following examples relate them with stress physiology: (a) Auxin Responsive Factors are well known class of transcription factors that are required for plants basal defense and during various abiotic stresses (also mentioned above). (b) Pentatricopeptide Repeat proteins constitute another category of transcripts regulated by tasiRNAs (from *TAS2* locus) and adjust plants' hypersensitive responses against various pathogens effectors.

It appears that certain transcripts that escaped miRNA-mediated regulation are likely evolved in a manner to be a part of RNAi-mediated regulatory network encompassing as much protein coding transcript. MiRNA-triggered siRNA biogenesis clearly suggests that RNAi machinery has extended its regulatory roots and the overlapping functional domains of small RNAs are evident in the form of highly controlled expression of vast array of genes. Recently, it has been demonstrated that tasiRNA pathway can be hypothetically utilized to knockdown any gene of interest when flanked by miRNA (that activates tasiRNA pathway) at

Table 5 List of TAS loci with their parent miRNA and target genes

TAS locus	miRNA(s) required	Genes targeted	Reference
TAS1a	miR173	FAD-binding domain-containing protein	Vazquez et al. (2004)
TAS1b	miR173	FAD-binding domain-containing protein	Vazquez et al. (2004)
TAS1c	miR173	FAD-binding domain-containing protein	Vazquez et al. (2004)
TAS2	miR161	PPR subfamily proteins	Allen et al. (2005) and Yoshikawa et al. (2005)
TAS3	miR390, miR160, miR167	ARF family proteins	Fahlgren et al. (2006) and Montgomery et al. (2008)
TAS4	miR828	MYB75, MYB90, MYB115	Rajagopalan et al. (2006)

both the ends (Felippes and Weigel 2009). The work demonstrated that besides the miRNA binding site, other sequences are also required for the tasiRNA generation. Heterologous transcripts containing trigger miRNA-binding sequences lead to tasiRNA generation, suggesting that such sequences act as *cis* sequences necessary for tasiRNA formation. This study has opened avenues to unravel functionality of the gene of interest.

4 Cross Talk Among Stresses

Several studies on high-throughput sequencing together with miRNA profiling following diverse stresses revealed that a huge subset of miRNAs were overlapping with each other. Interestingly, majority of these miRNAs are conserved across the plant species suggesting their putative roles in combating adverse stimuli. Most of these dysregulated miRNAs control transcription factors which, in turn, regulate expression of several downstream genes. As is evident from Fig. 7a, a significant population of these miRNAs is common to abiotic and biotic stresses. Moreover, among abiotic stresses, too, there are a number of miRNAs (miR165, miR167, miR169, miR393 etc.) responding to multiple stresses, while certain miRNAs (miR164, miR394 etc.) are specifically induced following a particular type of stress. Both these observations clearly point towards an obvious cross-talk between signaling cascades triggered following stress sensitization.

From Fig. 7b, it is obvious that virus induced miRNA profile of *Arabidopsis* plants is significantly similar to that of hypoxia treated plants (Tagami et al. 2007; Moldovan et al. 2010). There are seven miRNAs common to both salt and virus stressed samples while drought and cold stress share 3 similar miRNAs with virus infected samples (Liu et al. 2008; Zhou et al. 2008). It can be assumed

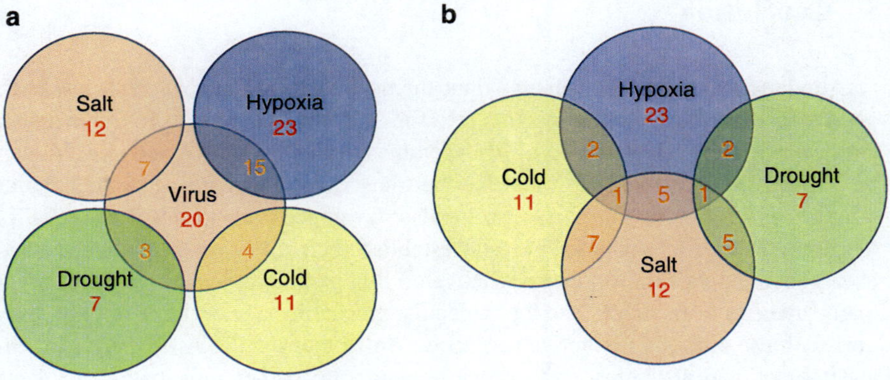

Fig. 7 Cross talk among various stress components: (**a**) overlap of microRNA profile among biotic and abiotic stresses. Tobacco Mosaic Virus-Cg responsive miRNAs share greater overlap with that of hypoxia induced miRNAs. (**b**) Among abiotic stresses, miRNAs induced in salt treated are common to those highlighted in drought and cold stress, while only few hypoxia-responsive miRNAs were common to other abiotic stresses

that virus infected, hypoxic and salt treated plants trigger overlapping pathways that converge early during the execution of the response(s). It can be noted that different tissues (leaves in case of viral infection and roots in hypoxia) were used for miRNA profiling; still significant similarity was observed in the pattern of miRNA deregulation. However, this does not appear to occur in case of drought and cold stressed samples, with less than half of the miRNAs being represented compared to biotic stressed sample. Here, abiotic and biotic stress responses are presumably divergent and therefore, require unique pathways to be induced in order to sustain stable cellular *millieu*. When similar analysis was performed among various abiotic stresses, we observed strikingly different results. We observed 5 miRNAs (less than 50%) that were common to hypoxia and salt treated tissues while, only 2 miRNAs were similarly altered during cold, drought and hypoxia response (Fig. 7). Intriguingly, when salt stress responsive miRNAs were compared with cold and drought treated samples, approximately 65% of the total deregulated miRNAs were shared between these stresses. This further supports the notion that plants response to virus infection and hypoxia is mechanistically different than salt, cold or drought stress. Overall, these observations signify that the extent of cross-talk is likely to vary among different stresses.

Nonetheless, these unique dysregulated sets of miRNAs provide an opportunity to dissect the functions of genes that probably act as key factors in conferring tolerance towards particular stress. MicroRNAs common to different stresses can also be exploited in generating transgenic that can survive under wide range of stress conditions and thus could be potential contributors in multiple stress tolerance. In future it will be important to map how a single stimulus leads to deregulation of several genes simultaneously and how the intricate networking among these genes achieve adaptive response.

5 Conclusion

Our understanding on how plants adapt themselves to the environment has been greatly increased by gaining insights into the finer details of small RNA-mediated pathways. Both the components of RNAi pathways namely, siRNA and miRNA, can be utilized simultaneously to fine tune the transcriptome according to the cellular requirements. Bacteria and viruses have also developed arms in the form of VSRs to counteract the RNAi pathways and establish their infection. Although studies have generated vast important knowledge on this paradigm, we are yet to uncover many layers to fully understand the molecular mechanisms underlying this complex arms defense and other adaptive responses. Since many miRNA and siRNAs still await functional attributes, it becomes necessary to gather information on these small RNA-mediated pathways to bridge the prevailing gaps. In a given time point, plants may encounter variable stresses, therefore, in near future it would be crucial to study the impact of endogenous small RNAs in response to multiple stresses. In future, it will be interesting to know what kind of stimuli would induce a particular set of small RNA(s). The post-genome era would witness miRNAs as promising tools in understanding mechanisms and development of strategies and approaches to unleash the complicated networks operating in living systems.

Acknowledgements Financial assistance to ARN by CSIR, Government of India, is duly acknowledged.

References

Allen E, Xie Z, Gustafson AM, Carrington JC (2005) microRNA-directed phasing during trans-acting siRNA biogenesis in plants. Cell 121:207–221. doi:10.1016/j.cell.2005.04.004

Alvarez JP, Pekker I, Goldshmidt A, Blum E, Amsellem Z, Eshed Y (2006) Endogenous and synthetic microRNAs stimulate simultaneous, efficient, and localized regulation of multiple targets in diverse species. Plant Cell 18:1134–01151. doi:10.1105/tpc.105.040725

Anandalakshmi R, Pruss GJ, Ge X, Marathe R, Mallory AC, Smith HT, Vane VB (1998) A viral suppressor of gene silencing in plants. Proc Natl Acad Sci USA 95:13079–13084. doi:10.1073/pnas.1733874100

Andersson MG, Haasnoot PC, Xu N, Berenjian S, Berkhout B, Akusjarvi G (2005) Suppression of RNA interference by adenovirus virus-associated RNA. J Virol 79:9556–9565. doi:10.1128/JVI.79.15.9556–9565.2005

Aukerman MJ, Sakai H (2003) Regulation of flowering time and floral organ identity by a microRNA and its APETALA2-like target genes. Plant Cell 15:2730–2741. doi:10.1105/tpc.016238

Aung K, Lin S, Wu C, Huang Y, Su C, Chiou T (2006) pho2, a phosphate overaccumulator, is caused by a nonsense mutation in a microRNA399 target gene. Plant Physiol 141:1000–1011. doi:10.1104/pp. 106.078063

Bao N, Lye KW, Barton MK (2004) MicroRNA binding sites in arabidopsis class III HD-ZIP mRNAs are required for methylation of the template chromosome. Dev Cell 7:653–662. doi:10.1016/j.devcel.2004.10.003

Bartel DP (2004) MicroRNAs: genomics, biogenesis, mechanism, and function. Cell 116:281–297. doi:10.1016/S0092-8674(04)00045-5

Baumberger N, Baulcombe DC (2005) Arabidopsis ARGONAUTE1 is an RNA Slicer that selectively recruits micro RNAs and short interfering RNAs. Proc Natl Acad Sci USA 102:11928–11933. doi:10.1073/pnas.0505461102

Bisaro DM (2006) Silencing suppression by geminivirus proteins. Virology 344:158–168. doi:10.1016/j.virol.2005.09.041

Blevins T, Rajeswaran R, Shivaprasad PV, Beknazariants D, Si-Ammour A, Park HS, Vazquez F, Robertson D, Meins FJ, Hohn T, Pooggin MM (2006) Four plant Dicers mediate viral small RNA biogenesis and DNA virus induced silencing. Nucleic Acids Res 34:6233–6246. doi:10.1093/nar/gkl886

Bollman KM, Aukerman MJ, Park MY, Hunter C, Berardini TZ, Poethig RS (2003) HASTY, the Arabidopsis ortholog of exportin 5/MSN5, regulates phase change and morphogenesis. Development 130:1493–1504. doi:10.1242/10.1242/dev.00362

Borsani O, Zhu J, Verslues PE, Sunkar R, Zhu JK (2005) Endogenous siRNAs derived from a pair of natural cisantisense transcripts regulate salt tolerance in Arabidopsis. Cell 123:1279–1291. doi:10.1016/j.cell.2005.11.035

Boualem A, Laporte P, Jovanovic M, Laffont C, Plet J, Combier JP, Niebel A, Crespi M, Frugier F (2008) MicroRNA 166 controls root and nodule development in *Medicago trancatula*. Plant J 54:876–887. doi:10.1111/j.1365-313X.2008.03448.x

Brigneti G, Voinnet O, Li WX, Ji LH, Ding SW, Baulcombe DC (1998) Viral pathogenicity determinants are suppressors of transgene silencing in Nicotiana benthamiana. EMBO J 17:6739–6746. doi:10.1093/emboj/17.22.6739

Brodersen P, Sakvarelidze-Achard L, Bruun-Rasmussen M, Dunoyer P, Yamamoto YY, Sieburth L, Voinnet O (2008) Widespread translational inhibition by plant miRNAs and siRNAs. Science 320:1185–1190. doi:10.1126/science.1159151

Chapman EJ, Prokhnevsky AI, Gopinath K, Dolja VV, Carrington JC (2004) Viral RNA silencing suppressors inhibit the microRNA pathway at an intermediate step. Genes Dev 18:1179–1186. doi:10.1101/gad.1201204

Chellappan P, Vanitharani R, Fauquet CM (2005) MicroRNA-binding viral protein interferes with Arabidopsis development. Proc Natl Acad Sci USA 102:10381–10386. doi:10.1073/pnas.0504439102

Chen X (2004) A microRNA as a translational repressor of APETALA2 in Arabidopsis flower development. Science 303:2022–2025. doi:10.1126/science.1088060

Chen J, Li WX, Xie D, Peng JR, Ding SW (2004) Viral virulence protein suppresses RNA silencing-mediated defense but upregulates the role of microRNA in host gene expression. Plant Cell 16:1302–1313. doi:10.1105/tpc.018986

Cui X, Li G, Wang D, Hu D, Zhou X (2005) A Begomovirus DNA beta-encoded protein binds DNA, functions as a suppressor of RNA silencing, and targets the cell nucleus. J Virol 79:10764–10775. doi:10.1128/JVI.79.16.10764-10775.2005

Deleris A, Gallego-Bartolome J, Bao J, Kasschau KD, Carrington JC, Voinnet O (2006) Hierarchial action and inhibition of plant Dicer-like proteins in antiviral defense. Science 303:68–71. doi:10.1126/science.1128214

Dunoyer P, Lecellier CH, Parizotto EA, Himber C, Voinnet O (2004) Probing the microRNA and small interfering RNA pathways with virus-encoded suppressors of RNA silencing. Plant Cell 16:1235–1250. doi:10.1105/tpc.020719

Elbashir SM, Lendeckel W, Tuschl T (2001) RNA interference is mediated by 21- and 22-nucleotide RNAs. EMBO J 17:6739–6746. doi:10.1093/emboj/20.23.6877

Emery JF, Floyd SK, Alvarez J, Eshed Y, Hawker NP, Izhaki A, Baum SF, Bowman JL (2003) Radial patterning of Arabidopsis shoots by class III HD-ZIP and KANADI genes. Curr Biol 13:1768–1774. doi:10.1016/S0960-9822(03)00718-8

Fagard M, Boutet S, Morel JB, Bellini C, Vaucheret H (2000) AGO1, QDE-2, and RDE-1 are related proteins required for post-transcriptional gene silencing in plants, quelling in fungi, and RNA interference in animals. Proc Natl Acad Sci USA 97:11650–11654. doi:10.1073/pnas.200217597

Fahlgren N, Montogomery TA, Howell MD, Allen E, Dvorak SK, Alexander AL, Carrington JC (2006) Regulation of AUXIN RESPONSE FACTOR3 by TAS3 ta-siRNA affects developmental timing and patterning in Arabidopsis. Curr Biol 16:939–944. doi:10.1016/j.cub.2006.03.065

Fahlgren N, Howell MD, Kasschau KD, Chapman EJ, Sullivan CM, Cumbie JS, Givan SA, Law TF, Grant SR, Dangl JL, Carrington JC (2007) High-throughput sequencing of Arabidopsis microRNAs: evidence for frequent birth and death of MIRNA genes. PLoS ONE 2:e219. doi:10.1371/journal.pone.0000219

Felippes FF, Weigel D (2009) Triggering the formation of tasiRNAs in Arabidopsis thaliana: the role of microRNA miR173. EMBO Rep 10:264–270. doi:10.1038/embor.2008.247

Fenner BJ, Thiagarajan R, Chua HK, Kwang J (2006) Betanodavirus b2 is an RNA interference antagonist that facilitates intracellular viral RNA accumulation. J Virol 80:85–94. doi:10.1128/JVI.80.1.85-94.2006

Fujii H, Chiou TJ, Lin SI, Aung K, Zhu JK (2005) A miRNA involved in phosphate-starvation response in Arabidopsis. Curr Biol 15:2038–2043. doi:10.1016/j.cub.2005.10.016

Gasciolli V, Mallory AC, Bartel DP, Vaucheret H (2005) Partially redundant functions of arabidopsis DICER-like enzymes and a role for DCL4 in producing trans-acting siRNAs. Curr Biol 15:1494–1500. doi:10.1016/j.cub.2005.07.024

Guo HS, Xie Q, Fie JF, Chua NH (2005) MicroRNA directs mRNA cleavage of the transcription factor NAC1 to downregulate auxin signals for Arabidopsis lateral root development. Plant Cell 17:1376–1386. doi:10.1105/tpc.105.030841

Hamilton A, Baulcombe D (1999) A species of small antisense RNA in posttranscriptional gene silencing in plants. Science 286:950–952. doi:10.1126/science.286.5441.950

Han MH, Goud S, Song L, Fedoroff N (2004) The arabidopsis double-stranded RNA-binding protein HYL1 plays a role in microRNA-mediated gene regulation. Proc Natl Acad Sci USA 101:1093–1098. doi:10.1073/pnas.0307969100

Henderson IR, Zhang X, Lu C, Johnson L, Meyers BC, Green PJ, Jacobsen SE (2006) Dissecting Arabidopsis thaliana DICER function in small RNA processing, gene silencing and DNA methylation patterning. Nat Genet 38:721–725. doi:10.1038/ng1804

Izquierdo J (1999) Biotechnology can help crop production to feed and increasing world population?: positive and negative aspects need to be balanced, a perspective from FAO. Presented at International Symposium on Plant Genetic Engineering, 6–10 Dec 1999, Havana (in print, Elsevier)

Jones-Rhoades MW, Bartel DP (2004) Computational identification of plant microRNAs and their targets, including a stress-induced miRNA. Mol Cell 14:787–799. doi:10.1016/j.molcel.2004.05.027

Jopling CL, Yi M, Lancaster AM, Lemon SM, Sarnow P (2005) Modulation of hepatitis C virus RNA abundance by liver-specific MicroRNA. Science 309:1577–1581. doi:10.1126/science.1113329

Kasschau KD, Xie Z, Allen E, Llave C, Chapman EJ, Krizan KA, Carrington JC (2003) P1/HC-Pro, a viral suppressor of RNA silencing, interferes with Arabidopsis development and miRNA function. Dev Cell 4:205–217. doi:10.1016/S1534-5807(03)00025-X

Katiyar-Agarwal S, Morgan R, Dahlbeck D, Borsani O, Villegas A Jr, Zhu JK, Staskawicz BJ, Jin H (2006) A pathogen-inducible endogenous siRNA in plant immunity. Proc Natl Acad Sci USA 103:18002–18007. doi:10.1073/pnas.0608258103

Katiyar-Agarwal S, Gao S, Vivian-Smith A, Jin H (2007) A novel class of bacteria-induced small RNAs in Arabidopsis. Genes Dev 21:3123–3134. doi:10.1101/gad.1595107

Khvorova A, Reynolds A, Jayasena SD (2003) Functional siRNAs and miRNAs exhibit strand bias. Cell 115:209–216. doi:10.1016/S0092-8674(03)00801-8

Kim VN (2005) MicroRNA biogenesis: coordinated cropping and dicing. Nat Rev Mol Cell Biol 6:376–385. doi:10.1038/nrm1644

Kurihara Y, Watanabe Y (2004) Arabidopsis micro-RNA biogenesis through Dicer-like 1 protein functions. Proc Natl Acad Sci USA 101:12753–12758. doi:10.1073/pnas.0403115101

Laufs P, Peaucelle A, Morin H, Traas J (2004) MicroRNA regulation of the CUC genes is required for boundary size control in Arabidopsis meristems. Development 131:4311–4322. doi:10.1242/dev.01320

Lauter N, Kampani A, Carlson S, Goebel M, Moose SP (2005) MicroRNA172 down-regulates glossy15 to promote vegetative phase change in maize. Proc Natl Acad Sci USA 102:9412–9417. doi:10.1073/pnas.0503927102

Lecellier CH, Dunoyer P, Arar K, Lehmann-Che J, Eyquem S, Himber C, Saïb A, Voinnet O (2005) A cellular microRNA mediates antiviral defense in human cells. Science 308:557–560. doi:10.1126/science.1108784

Lee Y, Kim M, Han J, Yeom KH, Lee S, Baek SH, Kim VN (2004) MicroRNA genes are transcribed by RNA polymerase II. EMBO J 23:4051–4060. doi:10.1038/sj.emboj.7600385

Li F, Ding SW (2006) Virus counterdefense: diverse strategies for evading the RNA-Silencing immunity. Annu Rev Microbiol 60:503–531. doi:10.1146/annurev.micro.60.080805.142205

Li J, Yang Z, Yu B, Liu J, Chen X (2005) Methylation protects miRNAs and siRNAs from a 3′-end uridylation activity in arabidopsis. Curr Biol 15:1501–1507. doi:10.1016/j.cub.2005.07.029

Liu HH, Tian X, Li YJ, Wu CA, Zheng CC (2008) Microarray-based analysis of stress-regulated microRNAs in Arabidopsis thaliana. RNA 14:836–843. doi:10.1261/rna.895308

Lobbes D, Rallapalli G, Schmidt DD, Martin C, Clarke J (2006) SERRATE: a new player on the plant microRNA scene. EMBO Rep 7:1052–1058. doi:10.1038/sj.embor.7400806

Lu S, Cullen BR (2004) Adenovirus VA1 noncoding RNA can inhibit small interfering RNA and microRNA biogenesis. J Virol 78:12868–12876. doi:10.1128/JVI.78.23.12868-12876.2004

Lu S, Sun Y-S, Shi R, Clark C, Li L, Chiang VL (2005) Novel and mechanical stressresponsive microRNAs in Populus trichocarpa that are absent from Arabidopsis. Plant Cell 17:2186–2203. doi:10.1105/tpc.105.033456

Ma ZL, Yang HY, Wang R, Tian B (2004) Construct hairpin RNA to fight against rice dwarf virus. Acta Bot Sin 46:332–336

Mallory AC, Reinhart BJ, Jones-Rhoades MW, Tang G, Zamore PD, Barton MK, Bartel DP (2004) MicroRNA control of PHABULOSA in leaf development: importance of pairing to the microRNA 5′ region. EMBO J 23:3356–3364. doi:10.1038/sj.emboj.7600340

Mallory AC, Hinze A, Tucker MR, Bouché N, Gasciolli V, Elmayan T, Lauressergues D, Jauvion V, Vaucheret H, Laux T (2009) Redundant and specific roles of the ARGONAUTE proteins AGO1 and ZLL in development and small RNA-directed gene silencing. PLoS Genet 5:e1000646. doi:10.1038/sj.emboj.7600340

Mette MF, Aufsatz W, van der Winden J, Matzke MA, Matzke AJ (2000) Transcriptional silencing and promoter methylation triggered by double stranded RNA. EMBO J 19:5194–5201. doi:10.1093/emboj/19.19.5194

Mi S, Cai T, Hu Y, Chen Y, Hodges E, Ni F, Wu L, Li S, Zhou H, Long C, Chen S, Hannon GJ, Qi Y (2008) Sorting of small RNAs into Arabidopsis argonaute complexes is directed by the 5′ terminal nucleotide. Cell 133:116–127. doi:10.1016/j.cell.2008.02.034

Mochizuki K, Gorovsky MA (2004) Conjugation-specific small RNAs in tetrahymena have predicted properties of scan (scn) RNAs involved in genome rearrangement. Genes Dev 18:2068–2073. doi:10.1101/gad.1219904

Moldovan D, Spriggs A, Yang J, Pogson BJ, Dennis ES, Wilson IW (2010) Hypoxia responsive miRNAs and *trans*-acting small interfering RNAs in *Arabidopsis*. J Exp Bot 61:165–177. doi:10.1093/jxb/erp296

Montgomery TA, Howell MD, Cuperus JT, Li D, Hansen JE, Alexander AL, Chapman EJ, Fahlgren N, Allen E, Carrington JC (2008) Specificity of ARGONAUTE7-miR390 interaction and dual functionality in TAS3 trans-acting siRNA formation. Cell 133:128–141. doi:10.1016/j.cell.2008.02.033

Morel JB, Godon C, Mourrain P, Beclin C, Boutet S, Feuerbach F, Proux F, Vaucheret H (2002) Fertile hypomorphic ARGONAUTE (ago1) mutants impaired in posttranscriptional gene silencing and virus resistance. Plant Cell 14:629–639. doi:10.1105/tpc.010358

Naqvi AR, Islam MN, Choudhury NR, Haq QM (2009) The fascinating world of RNA interference. Int J Biol Sci 5:97–117, PMCID: PMC2631224

Navarro L, Dunoyer P, Jay F, Arnold B, Dharmasiri N, Estelle M, Voinnet O, Jones JDG (2006) A plant miRNA contributes to antibacterial resistance by repressing auxin signaling. Science 312:436–439. doi:10.1126/science.1126088

Navarro L, Jay F, Nomura K, He SY, Voinnet O (2008) Suppression of the microRNA pathway by bacterial effector proteins. Science 321:964–967. doi:10.1126/science.1159505

Niu QW, Lin SS, Reyes JL, Chen KC, Wu HW, Yeh SD, Chua NH (2006) Expression of artificial microRNAs in transgenic Arabidopsis thaliana confers virus resistance. Nat Biotechnol 24:1420–1428. doi:10.1038/nbt1255

Ortiz R (1998) Critical role of plant biotechnology for the genetic improvement of food crops-perspective for the next millennium. Elec J Biotechnol 1(3). http://www.ejb.ucv.cl/content/vol1/issue3/full/7/

Pal-Bhadra M, Bhadra U, Birchler JA (2002) RNAi related mechanisms affect both transcriptional and post-transcriptional transgene silencing in Drosophila. Mol Cell 9:315–327. doi:10.1016/S1097-2765(02)00440-9

Park MY, Wu G, Gonzalez-Sulser A, Vaucheret H, Poethig RS (2005) Nuclear processing and export of microRNAs in arabidopsis. Proc Natl Acad Sci USA 102:3691–3696. doi:10.1073/pnas.0405570102

Pfeffer S, Zavolan M, Grässer FA, Chien M, Russo JJ, Ju J, John B, Enright AJ, Marks D, Sander C, Tuschl T (2004) Identification of virus-encoded microRNAs. Science 304:734–736. doi:10.1126/science.1096781

Pillai RS, Bhattacharyya SN, Artus CG, Zoller T, Cougot N, Basyuk E, Bertrand E, Filipowicz W (2005) Inhibition of translational initiation by Let-7 microRNA in human cells. Science 309:1573–1576. doi:10.1126/science.1115079

Qi Y, Zhong X, Itaya A, Ding B (2004) Dissecting RNA silencing in protoplasts uncovers novel effects of viral suppressors on the silencing pathway at the cellular level. Nucleic Acids Res 32:e179. doi:10.1093/nar/gnh180

Qi Y, He X, Wang XJ, Kohany O, Jurka J, Hannon GJ (2006) Distinct catalytic and non-catalytic role of ARGONAUTE4 in RNA-directed DNA methylation. Nature 443:1008–1012. doi:10.1038/nature05198

Qu F, Ren T, Morris TJ (2003) The coat protein of turnip crinkle virus suppresses posttranscriptional gene silencing at an early initiation step. J Virol 77:511–522. doi:10.1128/JVI.77.1.511-522.2003

Rajagopalan R, Vaucheret H, Trejo J, Bartel DP (2006) A diverse evolutionary fluid set of microRNAs in Arabidopsis thaliana. Genes Dev 20:3407–3425. doi:10.1101/gad.1476406

Schwab R, Ossowski S, Riester M, Warthmann N, Weigel D (2006) Highly specific gene silencing by artificial microRNAs in Arabidopsis. Plant Cell 18:1121–1133. doi:10.1105/tpc.105.039834

Schwarz S, Grande AV, Bujdoso N, Saedler H, Huijser P (2008) The microRNA regulated SBP-box genes *SPL9* and *SPL15* control shoot maturation in Arabidopsis. Plant Mol Biol 67:183–195. doi:10.1007/s11103-008-9310-z

Shibata S, Sasaki M, Miki T, Shimamoto A, Furuichi Y, Katahira J, Yoshihiro Y (2006) Exportin-5 orthologues are functionally divergent among species. Nucleic Acids Res 34:4711–4721. doi:10.1093/nar/gkl663

Silhavy D, Molnar A, Lucioli A, Szittya G, Hornyik C, Tavazza M, Burgyán J (2002) A viral protein suppresses RNA silencing and binds silencing-generated, 21- to 25-nucleotide double-stranded RNAs. EMBO J 21:3070–3080. doi:10.1093/emboj/cdf312

Simón-Mateo C, García JA (2006) MicroRNA-guided processing impairs Plum pox virus replication, but the virus readily evolves to escape this silencing mechanism. J Virol 80:2429–2436. doi:10.1128/JVI.80.5.2429-2436.2006

Song L, Han MH, Lesicka J, Fedoroff N (2007) Arabidopsis primary microRNA processing proteins HYL1 and DCL1 define a nuclear body distinct from the cajal body. Proc Natl Acad Sci USA 104:5437–5442. doi:10.1073/pnas.0701061104

Sunkar R, Zhu JK (2004) Novel and stress regulated microRNAs and other small RNAs from Arabidopsis. Plant Cell 16:2001–2019. doi:10.1105/tpc.104.022830

Sunkar R, Kapoor A, Zhu JK (2006) Posttranscriptional induction of two Cu/Zn superoxide dismutase genes in Arabidopsis is mediated by downregulation of miR398 and important for oxidative stress tolerance. Plant Cell 18:2051–2065. doi:10.1105/tpc.106.041673

Tagami Y, Inaba N, Kutsuna N, Kurihara Y, Watanabe Y (2007) Specific enrichment of miRNAs in *Arabidopsis thaliana* infected with tobacco mosaic virus. DNA Res 14:227–233. doi:10.1093/dnares/dsm022

Takeda A, Iwasaki S, Watanabe T, Utsumi M, Watanabe Y (2008) The mechanism selecting the guide strand from small RNA duplexes is different among argonaute proteins. Plant Cell Physiol 49:493–500. doi:10.1093/pcp/pcn043

Trinks D, Rajeswaran R, Shivaprasad PV, Akbergenov R, Oakeley EJ, Veluthambi K, Hohn T, Pooggin MM (2005) Suppression of RNA silencing by a geminivirus nuclear protein, AC2, correlates with transactivation of host genes. J Virol 79:2517–2527. doi:10.1128/JVI.79.4.2517-2527.2005

van Wezel R, Dong X, Liu H, Tien P, Stanley J, Hong Y (2002) Mutation of three cysteine residues in Tomato yellow leaf curl virus-China C2 protein causes dysfunction in pathogenesis and posttranscriptional gene-silencing suppression. Mol Plant Microbe Interact 15:203–208. doi:10.1094/MPMI.2002.15.3.203

Vanitharani R, Chen X (2008) Degradation of microRNAs by a family of exoribonucleases in Arabidopsis. Science 321:1490–1492. doi:10.1126/science.1163728

Vanitharani R, Chellappan P, Pita JS, Fauquet CM (2004) Differential roles of AC2 and AC4 of cassava geminiviruses in mediating synergism and suppression of posttranscriptional gene silencing. J Virol 78:9487–9498. doi:10.1128/JVI.78.17.9487-9498.2004

Vaucheret H, Mallory AC, Bartel DP (2006) AGO1 homeostasis entails coexpression of MIR168 and AGO1 and preferential stabilization of miR168 by AGO1. Mol Cell 22:129–136. doi:10.1016/j.molcel.2006.03.011

Vazquez F, Vaucheret H, Rajagopalan R, Lepers C, Gasciolli V, Mallory AC, Hilbert JL, Bartel DP, Crete P (2004) Endogenous trans-acting siRNAs regulate the accumulation of Arabidopsis mRNAs. Mol Cell 16:69–79. doi:10.1016/j.molcel.2004.09.028

Voinnet O (2005) Induction and suppression of RNA silencing: insights from viral infections. Nat Rev Genet 6:206–220. doi:10.1038/nrg1555

Voinnet O, Lederer C, Baulcombe DC (2000) A viral movement protein prevents spread of the gene silencing signal in Nicotiana benthamiana. Cell 103:157–167. doi:10.1016/S0092-8674(00)00095-7

Wang XJ, Gaasterland T, Chua NH (2005) Genome-wide prediction and identification of cis-natural antisense transcripts in Arabidopsis thaliana. Genome Biol 6:R30. doi:10.1186/gb-2005-6-4-r30

Wang H, Chua NH, Wang XJ (2006) Prediction of trans-antisense transcripts in Arabidopsis thaliana. Genome Biol 7:R92. doi:10.1186/gb-2006-7-10-r92

Xie Z, Kasschau KD, Carrington JC (2003) Negative feedback regulation of Dicer-like1 (DCL1) in arabidopsis by microRNA-guided mRNA degradation. Curr Biol 13:784–789. doi:10.1016/S0960-9822(03)00281-1

Xie Z, Allen E, Wilken A, Carrington JC (2005) Dicer-like 4 functions in trans-acting small interfering RNA biogenesis and vegetative phase change in Arabidopsis thaliana. Proc Natl Acad Sci USA 102:12984–12989. doi:10.1073/pnas.0506426102

Yoshikawa M, Peragine A, Park MY, Poethig RS (2005) A pathway for the biogenesis of trans-acting siRNAs in Arabidopsis. Genes Dev 19:2164–2175. doi:10.1101/gad.1352605

Yu B, Yang Z, Li J, Minakhina S, Yang M, Padgett RW, Steward R, Chen X (2005) Methylation as a crucial step in plant microRNA biogenesis. Science 307:932–935. doi:10.1126/science.1107130

Zheng X, Zhu J, Kapoor A, Zhu JK (2007) Role of Arabidopsis AGO6 in siRNA accumulation, DNA methylation and transcriptional gene silencing. EMBO J 26:1691–1701. doi:10.1038/sj.emboj.7601603

Zhou X, Wang G, Sutoh K, Zhu JK, Zhang W (2008) Identification of cold-inducible microRNAs in plants by transcriptome analysis. Biochim Biophys Acta 1799:780–788. doi:10.1016/j.bbagrm.2008.04.005

Zilberman D, Cao X, Jacobsen SE (2003) ARGONAUTE4 control of locus-specific siRNA accumulation and DNA and histone methylation. Science 299:716–719. doi:10.1126/science.1079695

Zimmermann P, Hirsch-Hoffmann M, Henning L, Gruissem W (2004) GENEVESTIGATOR. Arabidopsis microarray database and analysis toolbox. Plant Physiol 136:2621–2632. doi:10.1104/pp. 104.046367

Beneficial Soil Microorganisms, an Ecological Alternative for Soil Fertility Management

Claudio Altomare and Ivanka Tringovska

Abstract The new challenges that will be faced by agriculture in the twenty-first century impose the adoption of strategies able to increase food production without further increase the area of arable land and with low environmental impact. Soil microorganisms are a major component of the natural fertility of soils. They can promote plant growth, increase crop productivity and contribute significantly to the mineral nutrition of crop plants. This review examines the up-to-date knowledge about the potential and existing uses of beneficial microbes as biofertilizers and gives an outline of their modes of action.

Plant growth promoting microorganisms (PGPM) influence plant nutrition and growth through various mechanisms including nitrogen fixation, breakdown of organic matter, solubilization of sparingly soluble minerals, release of chelating compounds and biologically active substances such as phytohormones, vitamins and enzymes, and increase of the root system efficiency in nutrient uptake. Non symbiotic soil- or endophytic bacteria belonging to the genera *Azospirillum*, *Azotobacter*, *Acetobacter*, *Gluconacetobacter*, *Azoarcus*, *Bacillus*, *Paenibacillus*, *Burkholderia*, *Herbaspirillum*, *Clostridium*, *Klebsiella*, *Enterobacter*, *Citrobacter*, and *Pseudomonas* are able to fix atmospheric N_2 and have been found to be responsible for supply of biologically fixed nitrogen to crop plants. They may also improve plant growth through production of bioactive metabolites and indirect mechanisms, such as suppression of phytopathogens or induction of resistance to pathogens in plants.

C. Altomare (✉)
Institute of Sciences of Food Production, National Research Council,
Via Giovanni Amendola 122/O, 70126 Bari, Italy
e-mail: claudio.altomare@ispa.cnr.it

I. Tringovska
Maritsa Vegetable Crops Research Institute, 32 Brezovsko shosse Street, Plovdiv 4003, Bulgaria
e-mail: dwdt@abv.bg

E. Lichtfouse (ed.), *Genetics, Biofuels and Local Farming System*, Sustainable Agriculture Reviews 7, DOI 10.1007/978-94-007-1521-9_6,

Numerous species of soil and rhizosphere microorganisms may solubilize insoluble mineral phosphates, mainly through acidification and production of organic acids, and thus mobilize the enormous reserves of phosphorus (P) that are stored in most soils and are otherwise unavailable to plants. Generally, fungi exhibit greater P-solubilizing ability than bacteria. Members of the genera *Aspergillus*, *Penicillium*, and *Trichoderma* are particularly efficient P-solubilizers. Among bacteria, good results have been obtained with *Bacillus* spp. and *Pseudomonas* spp., especially in combination with P-solubilizing fungi and arbuscular mycorrhizal fungi (AMF). Under field conditions, the combined use of P-solubilizing microorganisms with mineral fertilizers such as rock-phosphate has often given successful results.

Also, the biological activities of microorganisms in the rhizosphere can mediate the solubility, and hence the availability at root surface of micro-nutrients, of which most soils are defective due to the fast depletion resulting from intensive farming. While other elements are also involved, Fe, Mn and Zn deficiencies have the greatest impact on the yields and quality of agricultural produce. PGPM and AMF have the capability to alter soil pH and modify the equilibrium of many chemical and biochemical reactions, such as precipitation/dissolution, adsorption/desorption, complexation/dissociation, and oxidation/reduction of metal cations and thus regulate the plant uptake. Beside enhancing plant nutrition under limited or deficient conditions, they may also reduce detrimental effects of excess of micronutrients, which may occur in acid or polluted soils.

In conclusion, in order to reduce the environmental and economic costs of the massive use of synthetic fertilizers and to obtain safer food, the use of PGPM as biofertilizers appears to be a concrete perspective. All the evidence summarized in this review clearly shows that beneficial soil microorganisms, either alone or in combination with mineral or organic fertilizers, may be utilized to increase crop productivity and maintain the fertility of soils without threatening the environment. These multifunctional agents are a renewable resource with low environmental impact. Therefore, beneficial soil microbes should be further studied and exploited for the development of sustainable agriculture.

Keywords Nitrogen • Phosphorus • Iron • Manganese • Zinc • Biofertilization • Soil microorganisms • Plant growth promotion

Abbreviations

AM	arbuscular mycorrhizae
AMF	arbuscular mycorrhizal fungi
BNF	biological nitrogen fixation
B	boron
Ca	calcium
Co	cobalt
Cu	copper

Fe	iron
HYVs	high-yield varieties
K	potassium
Mg	magnesium
Mo	molybdenum
Mn	manganese
N	nitrogen
Na	sodium
P	phosphorus
PGPM	plant growth promoting microorganisms
PGPR	plant growth-promoting rhizobacteria
S	sulfur
Zn	zinc

1 Introduction

In the twenty-first century, agriculture has to face complex and, in several ways, new challenges. The human population of the world has doubled within the last 45 years, and it is expected to grow at an average rate of 1.1–1.3% per year, increasing from about 6 billion people today to 8.3 billion people in 2030 (UN 2004). Currently, about 800 million people in the world are suffering from chronic malnutrition due to shortage of suitable foods. In developing countries, food demand will grow at a rate higher than 2.5% per year, due to a constant increase of per capita consumption. It is foreseen that an extra billion tonnes of cereals, by far the world's most important food source for direct human consumption and meat production, will be needed by 2030 (OECD-FAO 2007).

In developed countries, the total area of arable land is slowly declining as a result of expansion of urban settlements, industrial activities, and development of infrastructure, especially transport systems. In contrast the arable area is likely to increase in sub-Saharan Africa and Latin America. In Latin America a considerable part of the extra area will come from forest clearance (FAO 2002) and will probably have a substantial environmental impact. Global climate change will also contribute to the reduction of arable land. More sites will be affected by drought, flooding, salt water intrusion or sea surges (Fischer et al. 2002). The decline of per capita available arable land is aggravated by the degradation of soils. Throughout the world, current erosion rates are higher than ever. According to studies of the International Food Policy Research Institute (http://www.ifpri.org), each year an estimated 10 million hectares of arable land worldwide are abandoned due to soil erosion and diminished production. Another 10 million hectares are critically damaged each year by salinization, largely due to irrigation and/or improper drainage methods. The total loss accounts annually to more than 1.3% of the total arable land. Finally, in both developed and developing countries, an increasing share of arable land is discontinued from food production and dedicated to non-food use for biofuels production. Large increases in production volumes of biofuels have been occurring in Brazil, United States (US), European Union (EU), China, India, Indonesia and Malaysia, with annual global production of bioethanol projected to increase to 120 billion litres by 2020, and that of biodiesel to 12 billion litres (IEA 2004). In the EU, about 3 billion litres of biofuel (mostly biodiesel from rapeseed oil) were produced in 2004. The EU grew oilseeds for biofuel on an estimated 7.5 million hectares (Schnepf 2006). In 2000, about 6% of USA corn production went into ethanol, which increased to 14% of corn crop for biofuels in 2005, and further to 20% in 2006 (OECD-FAO 2007). This has resulted in food shortage and increasing prices, particularly in developing countries.

In this context, increasing food production without further increasing the area of arable land will require a rational exploitation of soil fertility and its careful management. In the second half of the twentieth century, the Green Revolution allowed food production to keep pace with the world population growth. During the Green Revolution, crop productivity was boosted by improvements in technology and changes in farming systems (Khush 1999). The development of improved crop varieties that were able to use higher nitrogen than the older varieties boosted production even further. These varieties are generally referred to as "high-yielding varieties" (HYVs). In general, HYVs require high levels of fertilization and agricultural maintenance, such as intensive pest control and controlled water supply. Under these conditions, HYVs significantly outperform traditional varieties (Khush 1999). However, without these inputs, traditional varieties may outperform HYVs (Third World Network Dossier 1990). While HYVs have enabled the increase of agricultural production, their higher demands for fertilizers, pesticides and water have social-, environmental- and health-related drawbacks (Shiva 1991; Pepper 2008). The production of chemical fertilizers needs great quantities of fossil resources, making them very expensive and ultimately not affordable for many farmers, especially in the developing countries. Moreover, the massive use of chemical fertilizers and other agrochemicals may lead to ground water contamination, while use of HYVs can lead to depletion of soil nutrients, eventually resulting in reduction of crop yield. Given the above premises, it is clear that agricultural production is not sustainable and a change is needed. Therefore the main goal for agricultural research should be the development of sustainable tools to maintain and where possible increase food production in an environmental and economically sound way.

It has long been known that soil microorganisms are one major component of the natural fertility of soils and that they can influence, directly or indirectly, plant growth, development, plant health, and nutrient and water uptake (Kennedy 1999). Beneficial soil microorganisms, either alone or in combination with mineral or organic fertilizers, may be utilized to increase crop productivity and maintain the fertility of soils without threatening the environment. This review summarizes the up-to-date knowledge about the existing uses of beneficial microbes for improvement of plant nutrition, growth and productivity, and investigate the potentials. We will review cases of soil bacteria and fungi, including mycorrhizal fungi, used as biofertilizers and plant growth enhancers. We will examine how these microorganisms can influence plant nutrition, with particular reference to nitrogen, phosphorus and major microelements, namely iron, manganese and zinc. Also, we will outline the chemical and biochemical mechanisms that influence solubility of nutrients in the rhizosphere and ultimately control their uptake by plants and microorganisms. *Rhizobia* have not been included in this review, since the unique association between legume plants and symbiotic nitrogen fixing bacteria known as rhizobia has been already extensively reviewed elsewhere (Atkins 1984; Spaink 2000; Jones et al. 2007).

2 Biofertilizers and Plant Growth Promoting Microorganisms

The management of soil microbial community is an important aspect of successful and sustainable agriculture. The use of beneficial microorganisms as bioinoculants for enhancement of plant growth and productivity has been intensively studied (Harman and Björkman 1998; Lucy et al. 2004; Artursson et al. 2006; Berg 2009), especially in cereal and vegetable crops. These studies have led to the concept of "biofertilizer", which in this review is defined in accordance with Vessey (2003) as: "a substance which contains living microorganisms which, when applied to seed, plant surfaces, or soil, colonizes the rhizosphere or the interior of the plant and promotes growth by increasing the supply or availability of primary nutrients to the host plant". This definition includes microorganisms that increase the growth of plants by enriching the soil nutrient pool, such as N_2-fixing rhizobia or free living bacteria, or make nutrients more available, e.g. by solubilization of phosphates or micronutrients, or increase plant access to nutrients by increasing root surface area. In addition, other mechanisms including production of biologically active substances, such as amino acids, vitamins and phytohormones, can play a role in plant growth promotion by biofertilizers (Arshad and Frankenberger 1991; González-López et al. 2005). Beneficial soil microorganisms can improve soil fertility and plant growth also through indirect mechanisms, such as amelioration of soil structure (Ingham 1998; Piotrowski et al. 2004; Rillig and Mummey 2006), enhancement of plant resistance to abiotic stresses (Feng et al. 2002; Mayak et al. 2004; Vivas et al. 2003), suppression of phytopathogenic fungi and bacteria or induction of plant resistance to pathogenes (Brundrett et al. 1996; Duffy et al. 2004; Nagesh and Parvatha-Reddy 2004; Siddiqui and Shaukat 2004).

Kennedy et al. (2004) reviewed the use of plant growth-promoting rhizobacteria (PGPR) as biofertilizers in open field crops. The authors concluded that PGPR increase crop yield by modifying the soil-plant processes by means of which nitrogen (N) and other nutrients are retained in the plant-soil system. As a result, the need for external input of N and phosphorus (P) through chemical fertilizers is reduced. Biofertilization through microorganisms can replace or reduce the use of chemical fertilizers (Şahin et al. 2004). In principle, biofertilizers are expected to be less expensive and more environmentally-friendly than chemical fertilizers. Application of biofertilizers to main food crops such as wheat, maize and rice could bring several benefits. These include higher incomes due to higher yields, reduced fertilization costs, reduced emission of greenhouse gases and reduced leaching of NO_3–N to groundwater. Biofertilizers might be particularly important in agricultural systems where chemical fertilizer inputs are either impractical (rangelands), undesirable (organic farming), or not possible (subsistence agriculture) (Pacovsky 1990). Worldwide, various biofertilizers are currently commercially available for agricultural application.

It is known that the efficacy of plant growth promotion by microbial inoculants is affected by soil nutritional condition. Better stimulatory effect on plant growth is usually observed in nutrient deficient soil than in nutrient rich soil (Burdman et al. 2000). Nevertheless, it has been demonstrated that even in the presence of optimum levels of N in soil, inoculation with PGPR may improve the growth and yield of inoculated plants (Shaharoona et al. 2006).

It was found that the PGPR strains *Pseudomonas alcaligenes* PsA15, *Bacillus polymyxa* BcP26 and *Mycobacterium phlei* MbP18 had a much better stimulatory effect on plant growth and nutrient uptake of maize in a nutrient deficient calcisol soil than in a relatively rich loamy soil (Egamberdiyeva 2007). The efficacy of Pseudomonads in improving plant growth and yields of wheat decreases with increasing rates of synthetic fertilizers added to the soil (Shaharoona et al. 2008). Field grown wheat and barley inoculated with *Azospirillum brasilense* or *Bacillus* sp. showed increased grain yield at all levels of nitrogen fertilizer, compared to non-inoculated control plots. However, this effect diminished at high fertilizer levels and the highest benefit from bacterial inoculation occurred at low levels of N-fertilization (Ozturk et al. 2003). Similar results have been obtained with sweet potato (Saad et al. 1999; Yasmin et al. 2007), carrots (Wange 1996), tomatoes (Barakrt and Gabr 1998; Adesemoye et al. 2009), cabbage (Verma et al. 1997). Arbuscular mycorrhizal fungi (AMF) form symbiotic mutualistic associations with plants, which affect beneficially plant physiology and nutrition and result in increased plant growth and tolerance to stresses and higher yields of many crops (Vosátka and Albrechtová 2009). It is known that AMF can work efficiently only at low levels of P in soil solution (Peters and Habte 2001; Olsson et al. 2002; Aikio and Ruotsalainen 2002; Schroeder and Janos 2004; Javaid 2009). It has been demonstrated that high P-levels significantly decrease mycorrhizal colonization of tomato, corn and cilantro. At low P-levels colonization of roots is more successful and the plant response is higher (Schroeder and Janos 2004). In sweet basil, more biomass and higher concentrations of phytochemicals were obtained after inoculation with AMF at low P fertilization levels (Toussaint et al. 2007). Tawaraya et al. (1996) found that root exudates from P-deficient or P-sufficient onions stimulated hyphal growth of AMF, whereas those from P-rich roots had no effect. Thus, the P content of the host plant has an influence on the quantity and/or quality of root exudates, which in turn affect hyphal growth of AMF.

Inoculation with plant-growth promoting microorganisms, and in particular with AMF, is also a promising way to help plants to overcome abiotic stresses. Under drought or salt stress, mycorrhizal plants show improved growth and nutrient acquisition, especially of immobile elements such as P, Cu, and Zn, than non-mycorrhizal.

This indicates that greater nutrient acquisition in response to AMF colonization could be a plant strategy for salt stress tolerance (Al-Karaki and Al-Raddad 1997; Subramanian and Charest 1997; Al-Karaki 2000, 2006). Mycorrhizal tomato plants, grown under field and drought-stress conditions showed improved drought resistance and fruit quality as a consequence of enhanced nutritional status, especially with regard to P and N (Subramanian et al. 2006). Similar results were obtained with drought-stressed maize where mycorrhizal plants had significantly greater shoot biomass and higher N, P, K, Mg, Mn, and Zn content in grains than non-mycorrhizal plants (Subramanian and Charest 1997, 1999; Subramanian et al. 2006). Improved growth of plants subjected to water or salt stress has been reported for several other crops, including chile pepper (Mena-Violante et al. 2006), clover (Shokri and Maadi 2009), cotton (Tian et al. 2004), lettuce (Ruiz-Lozano et al. 1995, 1996), soybean (Porcel and Ruiz-Lozano 2004) watermelon (Kaya et al. 2003), wheat (Al-Karaki and Al-Raddad 1997) and zucchini (Colla et al. 2008).

The efficacy of microbial plant-growth promotion is often improved by combining different microorganisms or different strains of one particular microorganism. This multi strain approach has often resulted in more pronounced effects than single strain inoculation. For instance, the yield response of wheat to inoculation with a mixture of *A. brasilense, Azotobacter chroococcum, Bacillus polymyxa,* and *Enterobacter cloacae* was higher than in response to a single strain-inoculation (de Freitas 2000). Sugar beet root and barley yields were higher with dual inoculation of N_2-fixing bacteria of the genus *Bacillus* or with a combination of N_2-fixing and P-solubilizing bacteria than after a single inoculation with N_2-fixing bacteria (Şahin et al. 2004). A combination of N_2-fixing and P-solubilizing bacteria has been shown to improve potato yield and quality (Mahendran et al. 1996; Mahendran and Chandramani 1998; Mahendran and Kumar 1998). Yield and white sugar content of sugar beet were increased by combining N_2-fixing and P-solubilizing strains of *Bacillus*. The plants inoculated with these bacteria performed better than non-inoculated plants fertilized with 120 kg ha^{-1} urea (Şahin et al. 2004). Higher protein content in chickpeas has been obtained by triple inoculation with *Mesorhizobium ciceri, A. chroococcum,* and *Pseudomonas* sp. or *Bacillus* sp., compared to the non-inoculated control (Wani et al. 2007). In pot experiments, Zaidi et al. (2003) observed that the highest chickpea grain and straw yields, as well as the highest uptake of N and P, were obtained with a combined inoculation of *Rhizobium* sp., *Pseudomonas striata* and *Glomus fasciculatum*. The authors concluded that combinations of N_2-fixing, P-solubilizing microorganisms and AMF can increase the yield of chickpea crop. Wani et al. (2007) obtained similar results with field grown chickpea inoculated with asymbiotic N_2-fixing *Azotobacter* in combination with the P-solubilizing bacteria *Bacillus* or *Pseudomonas*. The co-inoculation resulted in enhanced plant growth, nodulation, nutrient uptake and yield. Triple inoculation of wheat plants with *Glomus intraradices*, *Pseudomonas* spp., and *Aspergillus awamori* resulted in higher grain yields and shoot dry matter, as well as higher P concentrations in both grains and shoots than either single or dual inoculation (Babana and Antoun 2006). In a greenhouse trial, Wu et al. (2005) found that the dual inoculation of maize plants with beneficial bacteria and AMF could, at least to some extent, compensate the nutrient deficiency in soils and assure normal plant development. Also, the production of plant-growth promoting metabolites by some free-living bacteria is enhanced in the presence of AM fungi. A synergistic effect on plant growth was observed in wheat after dual inoculation with *A. chroococcum* and *G. fasciculatum* (Behl et al. 2003). *Glomus mosseae* in association with indol-acetic-acid (IAA) and gibberellins producing strains of *Bacillus pumillus* or *B. licheniformis* enhanced the yield of alfalfa plants and increased the efficiency of N utilization, as determined by the amount of biomass produced per unit of nitrogen in plant tissues (Medina et al. 2003).

Another approach to improve the efficacy of microbial plant-growth promotion is the combined application of biofertilizers and mineral or organic fertilizers. In field trials, dual inoculation of bean plants with the N_2-fixer *Azospirillum brasilense* Sp7 and the biocontrol and growth-promoting fungus *Trichoderma harzianum* Rifai 1295-22, combined with rock phosphate applied at $1\,kg\,ha^{-1}$ significantly ($P < 0.05$) increased seed yield and total seed N and P content (Öğüt et al. 2005). Nuruzzaman et al. (2003) examined the effect of the N_2-fixers *A. chroococcum* and *Azospirillum brasilense*, cow dung and urea on okra vegetative growth in field situations. The application of cow dung plus *A. chroococcum*, *A. brasilense,* or both, had a positive effect on plant growth and was comparable with the application of 60% urea. Inoculation with N_2-fixing bacteria has been reported to allow the reduction of chemical fertilizers, especially N fertilizers, by 20–50%. Though, to achieve superior results it needs to be combined with organic fertilizers (Bashan et al. 2004; Das et al. 2004). Enrichment of vermicompost with the N_2-fixing bacteria *A. chroococcum* and *Azospirillum lipoferum* has resulted in an increase of its N content (Kumar and Singh 2001). Compost enriched with N_2-fixing bacteria *Azotobacter vinelandii, Beijerinckia derxii* and *Azospirillum sp.* has promoted the growth of tomato plants (Meunchang et al. 2006). These data suggest that N_2-fixers used in combination with organic fertilizers may be a feasible alternative to synthetic N fertilizers in organic and low-input farming. Phosphate fertilization reduce mycorrhizal infection and root colonization (Amijee et al. 1989; de Miranda et al. 1989; Koide 1991). However, long term field trials with clover and alfalfa have shown that, while mineral fertilization reduces the growth of AMF, the development of AMF mycelium is increased by organic fertilization (Joner 2000; Gryndler et al. 2006).

Several microbial biocontrol agents have been reported to promote plant growth and increase yields. In some cases, the plant growth promotion has been related to suppression of minor plant pathogens (Kloepper and Schroth 1981a, b). These are pathogens that can be detrimental to the plant but do not cause apparent symptoms of disease. For this reason, some authors (including Vessey 2003) separate biofertilizers from "biopesticides" and include the biocontrol agents that also promote plant growth in the latter. According to other authors, there is no clear separation of growth promotion and biological control induced by bioinoculants (Kloepper 1993). As a matter of fact, often the boundaries between biofertilization and biocontrol are vague. This is not only because plant growth promotion by biocontrol agents may occur even in gnotobiotic conditions (that is in the absence of any plant pathogen) (Windham et al. 1986), but also because biocontrol can occur by means of the very same mechanisms that are thought to be involved in plant growth promotion. Examples of this are siderophores secretion by fluorescent pseudomonads (Haas and Défago 2005) and reduction of manganese oxides by soil bacteria that control the "take-all" disease of grains (Marschner et al. 1991; Huber and McCay-Buis 1993) (see farther in this review for more details). These mechanisms operate concurrently on both the plant pathogen, through inhibition of growth or virulence, and plant nutrition through increase of nutrient solubility and uptake. Recent insights in the modes of action of biocontrol agents have made more and more apparent that biological control is not a simple fight between good and bad microbes, but it is actually a three-way interaction in which plants play a major role (Harman et al. 2004; Preston 2004) and, therefore, the outcome is greatly influenced by plant vigour and nutritional status.

In conclusion, soil microorganisms applied as biofertilizers have proven to be able to promote plant growth and increase crop productivity. They operate in different ways, including enrichment of soil nutrient pool, release of nutrients from soil or organic matter, production of hormone-like metabolites and enhancement of plant health. In many cases, the combination of different biofertilizers that operate through different modes of action has been reported to result in synergistic beneficial effects on plant nutrition and growth.

3 Improvement of Plant Nutrition by Soil Microorganisms

3.1 Nitrogen

Nitrogen (N) is an extremely important and often limiting nutrient in agroecosystems. It is necessary for plants for synthesis of proteins, nucleic acids, chlorophyll, coenzymes (vitamins) and other compounds. In plants about 85% of N is incorporated into proteins, about 5% into nucleic acids DNA and RNA and the remaining portion into low-molecular-weight, water-soluble, organic compounds such as amides, amines and amino acids (Barker and Bryson 2007). Plants need N in all stages of their development, but the highest demand occurs during the stage of intensive growth. The N content in plants varies widely with and within species. The concentration of total N in plant tissue is 1–7% of dry weight, depending on the plant part. Fruits, seeds and leaves contain more N than stems and roots. N is very mobile within plant and is translocated from older to younger growing plant parts (Barker and Bryson 2007).

In soil, N is present mainly in organic forms, humic and nonhumic fractions. It is concentrated mainly in the top layers of soil. Humic substances or humus are more stable than nonhumic ones and slowly mineralizable, hence they contribute little to plant N nutrition. Conversely, low molecular substances such as amino acids and amino sugars are readily available to plants. Inorganic forms of N are mostly nitrate and ammonium, and plants take these ions up directly. Nitrate ions are not adsorbed to soil particles and a significant part can be lost through leaching. Ammonium is adsorbed nonspecifically to colloids in the soil. This ammonium is protected from nitrification and is available to plants at the same time (Barker and Bryson 2007; Mengel 2008).

Soil microorganisms play a major role in the nutrient cycle of N. They transform inorganic and organic N in forms that can be readily used by plants. The four processes involved in the cycling of N are N_2-fixation, ammonification, nitrification and denitrification. The rate and direction of these processes depend on the structure of microbial consortia and their activity in soil. Although the processes of ammonification, nitrification and denitrification are important factors for mobilization of N from soil organic matter, N_2-fixation is considered as a characteristic trait of plant growth promoting microorganisms and therefore this phenomenon is reviewed further in more detail.

Diverse microorganisms inhabitant of rhizosphere and bulk soil contribute significantly to soil N pools by transformation of atmospheric N_2 to organic N compounds that serve as a nutrient source for other organisms. The process is known as *biological nitrogen fixation* (BNF) (Howard and Rees 1996). Global terrestrial BNF is between 100 and 290 million tonnes of N year^{-1} (Cleveland et al. 1999), 40–48 million tonnes year^{-1} of which are fixed by agricultural crops in fields (Galloway et al. 1995; Jenkinson 2001). It was estimated that cyanobacteria (symbiotic N_2-fixers) contribute 7–80 kg N ha^{-1} year^{-1}, free-living bacteria 15 kg N ha^{-1} year^{-1}, and associative (endophytic) bacteria – 36 kg N ha^{-1} year^{-1} (Elkan 1992).

Literature shows successful examples of non-symbiotic BNF especially for cereal crops, where endophytic diazotrophs have been found to be responsible for supply of biologically fixed N to their host plant. It was found that cereals could obtain up to 30% of their N from BNF when fertilized with high rates of phosphorus and potassium and with micro elements. The largest effect was observed on sugar cane, which can obtain up to 150 kg N ha^{-1} from BNF (Döbereiner 1997; Kennedy et al. 2004; Pedraza 2008). Nevertheless, despite of reports of significant contribution of N_2-fixing bioinoculants to the N content of plants there is some controversy with regard to the actual contribution of N_2-fixing bacteria to the N budget of plant (Hoefsloot et al. 2005). It is now generally accepted that the beneficial effects on plant growth are at least partially due to phytohormone production by these microorganisms. Also, it has been evidenced that production of biologically active substances such as amino acids, vitamins and phytohormones, influences morphology, geometry and physiology of the root system, as well as the ability of free living N_2-fixers to mobilize sparingly soluble substances (Arshad and Frankenberger 1991; Shantharam and Mattoo 1997; Bashan 1999; de Freitas 2000; Cocking 2003; Bashan et al. 2004; González-López et al. 2005; Hoefsloot et al. 2005; Furina and Bonartseva 2007), and all these mechanisms may concur to promote plant growth. Although BNF may contribute to a fraction of crop N requirements, the reduction of losses from the soil ecosystem may be a much more significant contribution to the N economy of crop production (Kennedy et al. 2004). The N captured in dead microbial biomass may either be rapidly mineralised or initially immobilised into microbial biomass and later released by mineralisation. Usually both processes, mineralization and immobilization, occur simultaneously (Powlson et al. 2001).

Besides N-fixing bacteria, other soil microorganisms may play a role in plant acquisition of N. For instance, mycorrhizal fungi are of importance especially with regard to phosphorus nutrition of plants but several reports show that they can also contribute significantly to plant N-nutrition. Their contribution involves direct and indirect effects. Hyphae of mycorrhizal fungi are thinner than plant roots and can penetrate decomposing organic matter and better absorb recently mineralized N. In addition, these fungi can have an indirect effect on soil organic matter breakdown by stimulating decomposing bacteria (Azcon et al. 2001; Hodge 2001; Hodge et al. 2001; Hodge 2003; Artursson et al. 2006). Finally, they change the ecological conditions of soil and the composition of the soil microbial community and thus may help to mitigate ecosystem N losses (Medina et al. 2003; Cavagnaro et al. 2006).

3.1.1 Non Symbiotic Nitrogen-Fixing Bacteria

Non symbiotic N_2-fixing soil bacteria inhabit bulk soil, root surface, root interior and even plant aerial tissues. In the vicinity of plants they utilize root exudates as an energy source and thus N_2-fixation in the rhizosphere is more efficient than in bulk soil (Bürgmann et al. 2005). These bacteria use root exudates containing carbohydrates, organic acids, amino acids and amides, vitamins and other compounds as carbon sources (Lynch and Whipps 1990; Marschner 1995; Kuiper et al. 2002).

Among the N_2-fixing bacteria, the most extensively studied genus is *Azospirillum*. Azospirillum spp. are gram-negative facultative endophytic diazotrophs. Most species prefer rhizosphere and root habitats, where they are commonly found, but a fraction of the population persists in bulk soil (Baldani et al. 1997; Bashan 1999; Steenhoudt and Vanderleyden 2000; Tejera et al. 2005). Frequently *Azospirillum* invades root tissues without developing any nodules (de Oliveira Pinheiro et al. 2002). Strains of *Azospirillum* species have been isolated from the rhizosphere of all the major cereal crops and grasses (Grifoni et al. 1995; Kahinidi et al. 1997) or directly from the soil (Egorenkova et al. 2000). The ability of *Azospirillum* spp. to fix atmospheric N_2 varies widely between members of this group. Strains of *A. lipoferum* have been recognized as the most active N_2-fixers (Han and New 1998). Indeed, *Azospirillum* spp. have been reported to be able to affect plant growth and yield trough different mechanisms, including improvement of root development, increase of water and nutrients uptake by roots and, to a lesser extent, BNF (El Zemrany et al. 2007; Okon and Itzigsohn 1995). Some examples of plant growth promotion achieved by inoculation of *Azospirillum* spp. and the mechanisms of action that are thought to be involved are shown in Table 1.

Beside *Azospirillum*, other N_2-fixing bacteria have the potential for development of biofertilizers. They belong to the genera *Azotobacter, Acetobacter, Gluconacetobacter, Azoarcus, Bacillus, Paenibacillus, Burkholderia, Herbaspirillum, Clostridium, Klebsiella, Enterobacter, Citrobacter, Pseudomonas* and have

Table 1 Examples of effective association between *Azospirillum* spp. and crop plants. BNF: biological nitrogen fixation

Bioinoculant(s)	Crop	Effect	Mechanism	Reference
A. brasilense 42 M or *A. lipoferum Sp 242*	Maize	Higher fresh weight and higher N content in the leaves	BNF	Ribaudo et al. (2001)
A. amazonense	Rice	Increased number of panicles, grain dry matter and nitrogen accumulation at grain maturation	BNF	Rodrigues et al. (2008)
Mixed inoculum of *A. lipoferum, A. brasilense, Azoarcus, Zoogloea* and *Pseudomonas*	Rice	Higher biomass and nitrogen content	BNF	Malik et al. (1997)
A. brasilense BNM-10	Wheat	Increased grain yield, protein concentration in grains and NO_3 concentration in plant tissues	Increase in the uptake of inorganic N by roots or by a better soil exploration, but not BNF	Saubidet et al. (2002)

been reviewed in detail by Chan et al. (1994), Baldani et al. (1997), Mrkovački and Milic (2001), Cocking (2003), Choudhury and Kennedy (2004), Kennedy et al. (2004) and Pedraza (2008). Table 2 shows some examples of successful association between diazotrophic bacteria other than *Azospirillum* and crop plants.

In the majority of studies concerned with the use on N_2-fixing bacteria as biofertilizer, combined inocula of selected microorganisms have shown a more pronounced effect on plant growth and nutrition than the single components of the mixture. Studies with mixed inocula of *Azospirillum lipoferum, Azospirillum brasilense, Azoarcus* sp., *Pseudomonas* sp., *Zoogloea* sp. to sorghum plants indicated that nearly 29% of plant N came from the atmosphere. *Azoarcus* sp. alone supplied through BNF the 10.7% of N to shoots and 2.0% to roots (Stein et al. 1997). Inoculation of micropropagated sugarcane plants with a mixture of *Gluconacetobacter diazotrophicus* (syn. *Acetobacter diazotrophicus*), *Herbaspirillum seropedicae, Herbaspirillum rubrisubalbicans, Azospirillum amazonense* and *Burkholderia* sp. contributed to approximately 30% of the total N accumulation in plants (Oliveira et al. 2002). In addition, N_2-fixers can interact profitably with other beneficial microorganisms. For example, *Azospirillum* increases nodulation in the legume-rhizobia symbiotic association (Dobbelaere et al. 2003).

Intensive agriculture relies heavily on synthetic N fertilizers, but more than half of the applied N is lost through denitrification, ammonia volatilization, leaching and runoff. This leads to considerable environmental risks, including nitrate occurrence in groundwater, greenhouse effect and destruction of the ozone layer, and represents an economic loss for the farmers. Alternative and renewable sources of N are non symbiotic soil bacteria, able to fix atmospheric N_2. They are widely distributed in

Table 2 Examples of N_2-fixing bacteria that increase crop production

Bioinoculant(s)	Crop	Effect	Reference
Azotobacter chroococcum	Wheat	Increased yield with less N fertilizers	Narula et al. (2005)
	Sugar beet	Increase of dry weight and N content	Mrkovačkiki et al. (1997)
Azoarcus spp.	Sorghum	Increased total N content	Stein et al. (1997)
Burkholderia vietnamiensis	Rice	Increased grain yield	Govindarajan et al. (2008)
Bacillus spp.	Rice	Significant increase in the root and shoot of plants	Beneduzi et al. (2008)
Paenibacillus polymyxa; Pseudomonas putida; Bacillus sp.	Wheat and spinach	Increased N concentration in plant tissues; stimulated plant growth	Çakmakçi et al. (2007b)
Pseudomonas fluorescens	Black pepper	Increased dry matter; increased nutrient mobilization in the rhizosphere, enhanced plant vigor; N-uptake increased by 65% over control	Diby et al. (2005)
Gluconacetobacter diazotrophicus plus *Herbaspirillum* sp.	Sugarcane	Increased biomass and leaf N-content without N-fertilization or with a reduced dose of N fertilizer	Muthukumarasamy et al. (2006)

bulk soil, root surface, root interior and plant aerial tissues. These microorganisms can promote plant growth by supplying host plant with biologically fixed N or by production of biologically active substances (amino acids, vitamins and phytohormones) that influence root system development. The efficacy of biological nitrogen fixation and contribution to N budget of the plants can be enhanced by using combined inocula. The application of non symbiotic N_2-fixing bacteria in agriculture can improve N management in more "low-input" direction, and can constitute an integral part of sustainable agriculture.

3.2 Phosphorus

Phosphorus (P) is an essential nutrient for plant growth, due to its specific and important functions in plant structure and metabolism. P is involved in energy transfer processes as a component of adenosine triphosphate (ATP) and its derivatives, adenosine diphosphate (ADP) and adenosine monophosphate (AMP). A considerable amount of P in cells is a part of phosphorylated intermediates in pathways of the energetic metabolism (Calvin cycle, photorespiratory pathway and glycolysis), amino acid and nucleotide metabolism, and nitrogen and sulfur assimilation. P occurs as phospholipids in cell membranes, in phosphoproteins and enzymes. Phytic acid is a common phosphorus reserve in seeds. Inorganic forms of P are located in cell vacuoles as storage P. Inorganic and organic forms of

P serve as buffers in the maintenance of cellular pH (Sanchez 2007; White and Hammond 2008). Total phosphorus content in plant tissue ranges from about 0.1 to 1% (Sanchez 2007). Due to its crucial role in plant organism, this nutrient is needed in large amounts already at very early stage of plant's growth. Functions influenced by P nutrition include root development, absorption of water and nutrients, flower blooming and seed production, ability to tolerate biotic and abiotic stress, N_2-fixing capacity of legumes-root nodule bacteria associations (Marschner 1995). P is easily moveable in plant and usually is concentrated in rapidly growing parts, younger tissues, flowers and seeds.

In soil, P exists in organic and inorganic forms. Organic forms of P are inositol phosphate (soil phytate), phosphomonoesters, phosphodiesters including phospholipids and nucleic acids, and phophotriesters (Rodríguez and Fraga 1999; Soon 2008). Inositol phosphate is the most stable form of organic phosphorus, while nucleic acids and phospholipids are much more readily mineralized and bioavailable (Soon 2008). It has been estimated that about 5% of soil P is comprised in soil living biomass (White and Hammond 2008).

Inorganic P exists in soil as free phosphates, labile P and non-labile P. Free phosphates are in soil solution. Labile P is adsorbed on the surface of soil particles and can be released by anion exchange. A part of surface P may become non-labile, due to the processes of precipitation or diffusion into the solid phase (Soon 2008; White and Hammond 2008). Inorganic P forms relatively insoluble phosphate complexes with iron (Fe) and aluminum (Al) in acidic soils (especially those with pH lower than 5.0), or with calcium (Ca) in alkaline soils (pH above 7.0). The biggest reserves of mineral P are rocks, such as apatites $[Ca_5(PO_4)_3(OH, F)]$ and other Ca-P, Al-P and Fe-P minerals, some of which are found in soil as reaction products of phosphate fertilizers. The majority of these minerals are of little significance for the mineral nutrition of plants during a cultivation cycle because of their low solubility (Soon 2008).

P is available to plants only as orthophosphate ions, both $H_2PO_4^-$ and HPO_4^{2-}, depending on soil pH. In some cases, plants may absorb soluble organic phosphates, such as nucleic acids. All other P sources must be solubilized or mineralized to release P in forms that plants can take up. The rate of conversion between different P fractions varies widely and is influenced by many factors, such as soil structure, organic matter, soil pH, temperature, soil moisture, quantity of P fertilizers administered, total P concentration in soil, vegetation and composition of the community of soil micro-organisms (White and Hammond 2008). Despite the large quantity of phosphorus in soils (in average 600 mg kg^{-1} of soil, according to Fernández et al. 2007), this element is a major plant growth-limiting nutrient, since most of it is fixed as insoluble phosphates. An input of P with fertilizers is often required to sustain crop production. However, the supplied P reacts rapidly with cations in soil to form insoluble salts. Hence, in order to prevent P deficiency and crop losses, farmers often tend to over-fertilize.

In most soils, the enormous P reserves can support crop requirements for a long time, provided that it is mobilized through appropriate soil management practices, such as application of slowly-acting natural fertilizers (e.g. rock phosphates), addition of organic matter, and exploitation of P-solubilizing capability of soil microorganisms. Numerous species of soil or rhizosphere microorganisms are directly or indirectly involved in the fluxes between the different P pools. They can solubilize insoluble mineral phosphate complexes and mineralize organic phosphates, making P available to plants (Rodríguez and Fraga 1999; Gyaneshwar et al. 2002). In general, natural soils support diverse groups of P-solubilizing microorganisms, which possess an enormous potential for provision of P in forms available to plants and could serve as efficient biofertilizers.

Besides solubilization of mineral phosphates and mineralization of organic P compounds, soil microorganisms can also influence diffusion of phosphates from adsorbing surfaces to soil solution. Soil bacteria produce mucilages (mainly polysaccharides) that are powerful surfactants that alter the interaction of soil solids with water and ions. The occupation of adsorption sites by microbial polysaccharides and their effect on stability of soil aggregates and water balance facilitate the diffusion of phosphate ions from bulk soil through the rhizosphere (Deubel and Merbach 2005). Also, many rhizosphere bacteria and fungi produce plant hormones that influence root architecture, development of root hairs and affinity of roots for phosphate, thus indirectly affecting the P uptake by plants (Deubel and Merbach 2005).

One of the most important benefits that come to higher plants from the symbiotic association with mycorrhizal fungi is the improved uptake of nutrients, especially of P (Allen et al. 2003). Mycorrhizal fungi possess high-affinity P uptake mechanisms. In addition, they produce organic acids and can determine modifications of the root architecture, such as increase of total root surface and volume, number of tips, degree of root branching etc. (Lin et al. 2002; Villegas and Fortin 2001; Gamalero et al. 2002, 2004; Martin and Stutz 2004). In this way arbuscular mycorrhizae (AM) increase the volume of soil explored by the roots. The external mycorrhizal hyphae, which are quite thinner than plant roots, can access to sites normally not penetrable by roots or root hairs. Thus, acting as extensions of plant roots, external hyphae increase the volume of root-exploitable soil and extend the area of nutrient-uptaking surface (Brundrett et al. 1996; Ness and Vlek 2000).

The effect of microorganisms on phosphorus bioavailability in soils has been reviewed recently by Deubel and Merbach (2005), Vassilev et al. (2006b) and Khan et al. (2007).

3.2.1 Mechanisms of Microbial Mobilization of P in Soil

P-mobilization is a complex process and there are multiple mechanisms involved in microbial mobilization of mineral and organic P in soil. Summarized information about the mechanisms of phosphate solubilization can be found in several reviews (Rodríguez and Fraga 1999; Gyaneshwar et al. 2002; Vessey 2003; Bashan et al. 2004; Khan et al. 2007).

The processes of acidification, chelation, and exchange reactions can be involved in solubilization of inorganic P, but the prevalence of one or another of these mechanisms is organism-dependent (Plante 2007; Vassilev et al. 2006a). The most common mechanism of solubilization of mineral phosphate is soil acidification. At high pH values, P-solubilization often results from the release of organic acids or protons by the edaphic biological systems and subsequent drop of the pH in the soil matrix (Nautiyal et al. 2000; El-Komy 2005; Pandey et al. 2006; Barroso and Nahas 2007). Different organic acids (namely gluconic, citric, lactic, succinic, propionic, 2-ketogluconic, isovaleric, isobutyric, acetic, oxalic, malonic, and tartaric acids) are released by soil microorganisms and are thought to be responsible for solubilization of sparingly soluble phosphates (Reyes et al. 1999a, b, 2001; Rodríguez and Fraga 1999; Gyaneshwar et al. 2002; Chen et al. 2006). However, in some cases a lack of correlation between the amount of organic acids released in the medium and the quantity of mineral P solubilized has been reported (Deubel and Merbach 2005; Rudresh et al. 2005b). This supports the hypothesis that mechanisms other than production of organic acids may also be involved in the acidic dissolution of mineral P. Illmer and Schinner (1995) and Illmer et al. (1995) found significant correlation between P mobilization and microbial biomass. They concluded that solubilization was dependant on the release of protons associated to microbial biomass production or biomass-associated processes, such as respiration or NH_4^+ assimilation. According to Rodríguez and Fraga (1999) and Gyaneshwar et al. (2002), another possible mechanism of inorganic P solubilization by soil microorganisms is production of carbonic acid, derived from respiratory CO_2. However, Plante (2007) found poor correlation between CO_2 or HCO_3^- levels in soil and dissolution of apatite.

Although acidification is an important mechanism of P-solubilization, it is not the only possible one. Chelation by organic acids or other metabolites, such as siderophores, that complexate metal ions (Ca, Fe and Al) associated with phosphates, may also play a role (Illmer et al. 1995; Altomare et al. 1999; Rodríguez and Fraga 1999; Shekhar et al. 2000; Gyaneshwar et al. 2002; Welch et al. 2002; Hamdali et al. 2008). Deubel and Merbach (2005) found that a significant amount of P was released from $Ca_3(PO_4)_2$ through the chelating activity of succinic, lactic and citric acids and suggested that carboxylic anions solubilize more phosphorus and play a greater role than acidification under natural soil (buffered) conditions. Gluconic acid forms an insoluble complex with Ca^{2+} (Lin et al. 2006), while citric and oxalic acids form stable complexes with free Fe^{3+} and Al^{3+} (Illmer and Schinner 1995; Deubel and Merbach 2005) and thus liberate phosphates from complexed metals into soil solution. Among the above mentioned organic acids, gluconic acid is considered as a major factor of solubilization of P from $Ca_3(PO_4)_2$ (Lin et al. 2006) and citric acid from $AlPO_4$ (Illmer et al. 1995).

Solubilization of inorganic phosphate by microorganisms could be stimulated by soil nutritional status and organic energy sources in soils. (Kim et al. 1998; Reyes et al. 1999a, b; Nautiyal et al. 2000). For instance, Kim et al. (1998) investigated the effect of different C sources on the survival of the P solubilizer *Enterobacter agglomerans* and soil microbial activity. The authors found that organic P incorporated into soils could provide the necessary energy for solubilization of insoluble P–forms and concluded that the key for P mobilization is the application of appropriate energy sources that stimulate the microbial community. Some *in vitro* tests aiming to investigate the mechanisms of inorganic phosphate solubilization by microorganisms showed the important role of the type of N, P and C sources (Reyes et al. 1999b; Nautiyal et al. 2000). Since the field conditions are much more complex, further work is needed to clarify how these mechanisms could be used and managed in agricultural soils.

Mineralization of organic P compounds is linked with the action of the enzymes phosphatases and phosphohydrolases. They are classified as acid or alkaline depending on the optimal pH for catalytic activity, and specific or nonspecific according to the specificity for substrate. Some organic substances, such as nucleic acids, phospholipids, sugar phosphates are easily decomposable, while other, such as phytic acid, polyphosphate and phosphonate, are more stable compounds (Rodríguez and Fraga 1999; Gyaneshwar et al. 2002). A considerable part of soil P is used by soil microorganisms to build their cells and often they have a higher P uptake efficiency than plant roots. The P incorporated by soil microorganisms becomes rapidly available to plants after death and lysis of microbial cells (Deubel and Merbach 2005).

3.2.2 P-Solubilizing Microorganisms and Their Use to Improve Plant P Nutrition

P-solubilization has been reported for a great number of saprophytic bacteria and fungi. Bacterial species able to solubilize insoluble inorganic phosphates and/or to mobilize organic P-compounds belong to the genera *Achromobacter*, *Aerobacter*, *Agrobacterium*, *Arthrobacter*, *Bacillus*, *Burkholderia*, *Chyseobacterium*, *Citrobacter*, *Delftia*, *Enterobacter*, *Erwinia*, *Flavobacterium*, *Gordonia*, *Klebsiella*, *Micrococcus*, *Phyllobacterium*, *Proteus*, *Pseudomonas*, *Rhodococcus*, *Serratia* (Illmer and Schinner 1995; Rodríguez and Fraga 1999; Babana and Antoun 2005; Chung et al. 2005; El-Komy 2005; Chen et al. 2006; Pandey et al. 2006; Fernández et al. 2007; Megha et al. 2007; Ahmad et al. 2008), *Azotobacter* (Kumar et al. 2001; Reyes et al. 2006; Ahmad et al. 2008) and *Azospirillum* (Bashan et al. 2004).

Generally, fungi exhibit greater P-solubilizing ability than bacteria. Mobilization of soil P has been reported for fungal species of the genera *Aspergillus* (Illmer and Schinner 1995; Babana and Antoun 2005; Vassilev et al. 2006a), *Penicillium* (Wakelin et al. 2004; Babana and Antoun 2005; Reyes et al. 2006), and *Trichoderma* (Altomare et al. 1999; Rudresh et al. 2005a, b). The most powerful P mobilizers are mycorrhizal fungi, that as well as being able to solubilize and mineralize soil P can also absorb and transport P into the host root and are, therefore, directly involved in plant P uptake (Kothari et al. 1990; Koide 1991; Brundrett et al. 1996).

Increased P uptake by plants and plant growth promotion following the application of P-solubilizing soil microorganisms has been demonstrated under controlled and field conditions (Table 3), offering promising perspectives of reduction to some extent, or even replacement of synthetic P fertilizers (Igual and Rodríguez – Barrueco 2007).

Table 3 P-solubilizing microorganisms for which evidences exist that plant growth promotion and/or crop yield increase occur via P-solubilization and improvement of P-uptake

Microorganism	Crop	Conditions	Effect	Reference
Glomus etunicatum	Tomato	Pot experiment	Increased above-ground dry mass, root length, phosphorus content and yield under low phosphorus conditions	Bryla and Koide (1998)
			Improved vegetative (leaf area, days until first flower, leaf P concentration) and reproductive traits (total flower production, fruit mass, seed number, pollen production per plant, mean pollen production per flower)	Poulton et al. (2002)
Bacillus sp.	Barley	Pot experiment	Total biomass weight increased by 20.3–25.7% over the control as compared with 18.9% and 35.1% increases obtained with P and NP application, respectively	Canbolat et al. (2006)
Pseudomonas fluorescens	Black pepper	Pot experiment	Increased root biomass (30–135%), root length (12–127%), root area (43–200%), number of root tips (82–137%); significantly higher dry matter yields (23.64–40.83%) in treated plants compared to untreated control; enhanced nutrient uptake by the plants (P – 122%, N – 65%, and K – 3–25% over control)	Diby et al. (2005)
Bacillus megaterium, *Bacillus licheniformis*	Wheat and spinach	Pot experiment	Increased P concentration in plant tissues; enhancement of plant growth	Çakmakçi et al. (2007b)
Mixed inocula of *Glomus* spp.	Tomato	Pot experiment	Higher shoot length and biomass; increased host tissue N and P concentrations	Edathil et al. (1996)

(continued)

Table 3 (continued)

Microorganism	Crop	Conditions	Effect	Reference
Co-inoculation of *Azospirillum lipoferum* and *Bacillus megaterium*	Wheat	Pot experiment	High shoot dry weight, increased total N yield and the shoot P content by 37 and 53%	El-Komy (2005)
Single or combined inoculations with *P. fluorescens* and/or *Glomus mosseae*	Tomato	Pot experiment	More developed root systems (increased total root length, surface area, number of tips). The co-inoculation doubled or tripled the P content value compared to controls	Gamalero et al. (2004)
Bacillus pumilus	Wheat	Pot experiment	Increased plant biomass, root length, and total N and P contents in plants	Hafeez et al. (2006)
Combination of *Rhizobium* sp., *Pseudomonas striata* and *G. fasciculatum*	Chickpea	Pot experiment under controlled conditions	Increased grain, straw yields and P uptake; improved nodulation	Zaidi et al. (2003)
Combination of *Bacillus circulans, Cladosporium herbarum* and *Glomus fasciculatum*	Mungbean	Greenhouse experiment	Increased grain, straw yields; enhanced P-uptake; better N nutrition	Singh and Kapoor (1998)
Single, dual or triple combinations of *Glomus intraradices*, *Pseudomonas* sp., and *Aspergillus awamori*	Wheat	Field experiment	Enhanced plant height, increased grain yield and grain P-concentration	Babana and Antoun (2006)
Bacillus sp.	Raspberry	Field experiment	Increased yield (up to 74.9%), cane length (up to 15.0%), number of cluster per cane (up to 28.7%) and number of berries per cane (up 36.0%) compared with the control. Increased content of available P in soil	Orhan et al. (2006)

(continued)

Table 3 (continued)

Microorganism	Crop	Conditions	Effect	Reference
N_2-fixing and P-solubilizing strains of *Bacillus* spp. in triple combination	Sugar beet and barley	Field experiment	Increased sugar beet root (up to 12.7%) and barley yields (9.3%) over control, while yield increases from NP applications is up to 20.7–25.9%	Şahin et al. (2004)
Bacillus sp.	Hot pepper	Field experiment	Increased P availability from 12% to 21% and K availability from 13% to 15%; improved nutrient uptake; plant photosynthesis increased by 16%, leaf area increased by 35%; biomass and fruit yield increased by 23% and 30%, respectively	Supanjani et al. (2006)
Triple combinations of *Mesorhizobium ciceri*, *A. chroococcum*, *Pseudo-monas* sp., or *Bacillus* sp.	Chickpea	Field experiment	Tripled the seed yield and increased grain protein; 2.14-fold higher P-uptake; N-uptake 10–16% higher than the control	Wani et al. (2007)
P-solubilizing bacteria	Wheat	Pot and field experiments	Increased fresh and dry weights and total P content (%) in leaf sheath	Chen et al. (2008)
Co-inoculation of *B. megaterium* and *Trichoderma sp.*	Chickpea	Pot and field experiments	*Pot experiments*: Increased biomass: about 3 times over conventionally fertilized control and about 5 times over control fertilized with rock-phosphate (RP) *Field experiments*: Biomass about 2–3 times over conventionally fertilized control or RP control; grain yield about 3 times over conventionally fertilized control and about 7–8 times over RP control	Rudresh et al. (2005b)

Many reports suggest that inoculation with P-solubilizing microorganisms not always improve P uptake by plants and that P-solubilization is not always the only, or the main, mechanism responsible for the enhancement of plant growth. In fact, P-solubilizing microorganisms can promote plant growth also through mechanisms other than P-mobilization (Çakmakçi et al. 2006; Çakmakçi et al. 2007a; Ahmad et al. 2008). For instance, soybean plants, inoculated with P-solubilizing bacteria were taller and had a more balanced N/P ratio compared to the non-inoculated control. However, the P content of the plants did not increase (Fernández et al. 2007). Similar results were obtained with canola plants inoculated with P-solubilizers and grown in P-deficient soil amended with rock phosphate (de Freitas et al. 1997). Cress plants inoculated with the P-solubilizing fungi *Penicillium simplicissimum* and *Penicillium aurantiogriseum* had higher dry weights than uninoculated control plants, but only *P. simplicissimum* was able to increase the plant P-content (Illmer et al. 1995). *Pseudomonas jessenii* in combination with *Mesorhizobium ciceri* synergistically improved the growth of chickpea, but no changes in P and N content of shoots or grains were observed (Valverde et al. 2006). Although in none of the above instances the mechanism of plant growth promotion by P-solubilizing microorganisms was ascertained, in many cases production of phytohormones or hormone-like substances that stimulate the root system development was proposed as a likely explanation.

3.2.3 Strategies for Better Exploitation of P-Solubilizing Microorganisms

The P-solubilizing capabilities of soil microorganisms can be effectively exploited by combination with natural mineral P-fertilizers, such as rock phosphates. These rocks contain a high concentration of insoluble or sparingly soluble phosphates (mostly calcium phosphate) in nodular or compact masses. Rock phosphate is less expensive than chemical P-fertilizers and suitable for organic farming. Fungi are very efficient solubilizers of rock phosphates. For instance, as determined by total P_2O_5 extraction, an almost complete (94.80%) mobilization of P from rock phosphate in a liquid culture medium was achieved with *Aspergillus niger* (Bojinova et al. 2008). Barroso and Nahas (2007) investigated the ability of *A. niger* to solubilize hardly soluble phosphates and found an increase of the amount of soluble phosphate from 44% to 179% in inoculated soil, compared to the non-inoculated control. This particular fungus was active in dissolving aluminum phosphate more than other phosphates. The effect of solubilization of insoluble phosphates by *A. niger* depended on the addition of a carbon source (molasses) but decreased as soon as the carbon source was mineralized in the soil. The literature reports some examples of rock phosphate solubilization also by bacteria and Actinomycetes, such as *Streptomyces* sp. and *Micromonospora* sp. (Hamdali et al. 2008). A beneficial effect of the phosphate solubilizing bacterium *Bacillus megatherium* var. *phosphaticum* was observed on sugarcane growth, yield and quality. The application of P-solubilizing bacteria increased the plant available P in soil and reduced the need for P input by 25%. Furthermore it was found that rock phosphate could replace 50% of super phosphate when applied in combination with P-solubilizers (Sundara et al. 2002). Rock phosphates in combination with a P-solubilizing *Bacillus* strain were as effective as soluble fertilizers in hot pepper (Supanjani et al. 2006). Rock phosphate solubilization is a complex phenomenon and depends on many factors. Xiao et al. (2008) investigated rock phosphate solubilization by some fungi *in vitro* and reported the optimal temperature, pH, P-source, rock phosphate concentration and particle size, and C:N ratio necessary to maximize solubilization of phosphates. However, it should be stressed that although the capability to solubilize rock phosphate *in vitro* is regarded as an index of potential plant-growth promoting activity, there is no guarantee that one particular isolate which possesses this ability will promote plant growth in the field, and *vice versa* (Cattelan et al. 1999; Vessey 2003).

Another possible way to increase the efficiency of P-solubilizing microorganisms is to use mixed inocula. Numerous researchers have tried combined inocula of microbes and found some advantages over single-strain inoculation (see references in Table 3). Synergistic effects between soil bacteria and fungi have been demonstrated. P-solubilizing bacteria are known as mycorrhiza-helper-bacteria due to enhanced effect on plant growth and P nutrition when acting in combination with AM fungi (Barea et al. 2002). Some evidences indicate that bacteria (for instance *Bacillus polymyxa* and *Azospirillum brasilense*) solubilize sparingly soluble phosphates more easily than AM fungi (particularly *Glomus aggregatum*). Mycorrhizal hyphae absorb the released P ions and maintain low concentration of soluble P. As a consequence, bacteria are prompted to solubilize more phosphates and this results in continuous P mobilization (Ratti et al. 2001). It was shown that after inoculation of wheat seeds with P-solubilizing microorganisms in combination with the AM fungus *Glomus intraradices*, grain yield was comparable to that obtained from wheat fertilized with the expensive fertilizer diammonium phosphate (Babana and Antoun 2006). In the same experiment, either in the absence of any P fertilization or with rock phosphate as the sole P fertilizer, the co-inoculation of *G. intraradices* and the P-solubilizing microorganisms *Pseudomonas* sp. BR2 and *Aspergillus awamori* Nakazawa C1 significantly increased yield and P-content of grain compared to the non-inoculated controls.

Soil fungi and bacteria have the potential to increase the availability of the enormous P reserves of soil to plants. Numerous species of soil and rhizosphere microorganisms are directly or indirectly involved in the fluxes between the different P pools. They may solubilize insoluble mineral phosphate complexes, mineralize organic phosphates or influence diffusion of phosphates from adsorbing surfaces to soil solution. Mechanisms of acidification, chelating and enzymatic activities, as well as production of surface active agents are involved in microbial solubilization of insoluble phosphates. In addition to direct P-mobilization, P-solubilizing microorganisms may improve P-uptake and stimulate plant growth through production of hormones or hormone-like substances, which affect root system development and efficiency. Increased P uptake by plants and plant growth promotion following the application of P-solubilizing soil microorganisms has been demonstrated under controlled and field conditions, offering promising perspectives of reduction and rationalization of the use of synthetic P fertilizers. However, since the conditions in the soil environment are much more complex than those in controlled one, more reliable information concerning phosphate solubilization by soil microflora is obtained from field experiments. Under field conditions, particularly successful strategies seem to be the combined use of P-solubilizing microorganisms with mineral fertilizers such as rock-phosphate, and application of mixed inocula of different biofertilizers.

3.3 Other Macronutrients

Besides N and P, other essential elements used by plants in relatively large amounts for plant growth are potassium (K), calcium (Ca), magnesium (Mg), and sulfur (S). The main natural sources of these elements for plant nutrition are soil minerals and organic matter (Barker and Pilbeam 2007). Although the majority of investigations are focused on the contribution of soil microbes to N and P nutrition of plants, there is some evidence that soil microorganisms can influence K, Ca, Mg and S nutrition of higher plants as well. K, Ca and Mg cations form sparingly soluble complexes with phosphorus in alkaline soils. Therefore, the activity of microorganisms that dissolve insoluble phosphates also results in mobilization these cations. In particular, organic acids released by soil microorganisms are a powerful factor for mobilization of these metal nutrients.

K is involved in enzyme activation, protein synthesis, photosynthesis and respiration. Soil microorganisms belonging to the genus *Bacillus* sp. were found to solubilize K-containing rocks and increase K availability in soil, K uptake and growth of hot pepper plants (Supanjani et al. 2006). These K solubilizing bacteria have been proposed as a sustainable alternative to synthetic K-fertilizers for hot pepper. Likewise, K-bearing minerals were solubilized by *Aspergillus fumigatus* (Lian et al. 2008). However, improved K nutrition of plants is not always related to solubilization of sparingly soluble materials. Enhanced K uptake of bacterized black pepper plants in the absence of significant K mobilization was explained with the involvement of indirect mechanisms (Diby et al. 2005). This effect was attributed to the increased development and efficiency of the root system resulting from the inoculation of the PGPR.

Ca activates enzymes, is a structural component of cell membranes and walls, influences water movement in cells and is necessary for cell growth and division. In addition to dissolution of Ca-phosphates by P-solubilizing microorganisms, Ca availability can be increased by soil microorganisms through humification of soil organic matter. Humic and fulvic acids that are released during microbial humification can bind Ca in soluble humates and fulvates, making Ca available to plants (Ness and Vlek 2000).

AM fungi can contribute to K, Ca and Mg nutrition of maize grown in soil with limited availability of these nutrients (Liu et al. 2002). The concentrations of Ca, and S (beside P) in shoots of AM fungus-infected tomato plants grown at low level of added N were found to be significantly higher than in non-inoculated plants (Cavagnaro et al. 2006). This was probably due to the greater foraging of plants by the AM fungus hyphae and the consequent increased uptake of nutrients from the soil.

3.4 Micronutrients

Micronutrients are those elements that are required only in traces (less than $100\,mg\,kg^{-1}$ of plant dry matter) for balanced plant nutrition (Marschner 1995). Among the trace elements, iron (Fe), manganese (Mn), zinc (Zn), copper (Cu), molybdenum (Mo), and boron (B), are essential to the normal growth of plant. Sodium (Na), cobalt (Co) and selenium (Se) are beneficial, as they can compensate for toxic effects of other elements. They also may replace mineral nutrients in some other less specific functions such as the maintenance of osmotic pressure. Trace elements are involved in a number of physiological and metabolic processes. They are present as ions in cells, function as enzyme co-factors, regulate osmotic adjustment, signalling and are involved in redox reactions (electron transfer) (Marschner 1995).

The importance of trace minerals to the nutrition of plants in agriculture for crop productivity, produce quality and nutritional aspects of derived foods and feeds, is well-known. The increase in cultivation intensity with the increasing demand for higher yields has led to a faster depletion of micro-elements from the soil. As a result deficiencies occur in many intensively exploited soils. Moreover, in soil, micro-elements are not always present in the soil solution. Their availability to plants is strongly influenced by a number of soil factors, particularly pH, carbonate content and oxidation-reduction reactions (Viets 1961; He et al. 2005). The biological activities of microorganisms in the rhizosphere to a large degree mediate the solubility, and hence the availability, of metal ions at root surface (Marschner 1995). In fact, soil microorganisms can alter soil pH and modify the equilibrium of many chemical and biochemical reactions, such as precipitation/dissolution, adsorption/desorption, complexation/dissociation, and oxidation/reduction, of metal cations. Precipitation–dissolution is an important process that controls the solubility of trace elements in calcareous soils and soils with a pH above 7.0. Soil microorganisms may release extracellularly organic acids, such as citric acid, oxalic acid, malic acid gluconic acid, and others. Localized acidification by organic acids not only mobilizes phosphorus, but also Fe, Mn, and Zn in the rhizosphere and increases their uptake rates and contents in the plant (Evans 1991; Jones and Darrah 1994). In addition to acidification, organic acids are also strong chelators that can sequestrate metal ions (Cline et al. 1982). Chelation is crucial in controlling metal ion availability to plants. Chelation is a process during which trace elements form stable complexes with organic or inorganic ligands, which enhances dissolution and bioavailability of trace elements. Production of high-affinity Fe-chelating metabolites, named siderophores, by soil microorganisms has been reported to be associated to increased availability of Fe in the plant rhizosphere (Crowley et al. 1991). Beside Fe, siderophore can form complexes with other metal micronutrients (Shenker et al. 1996; Leach et al. 2007), although with lower specificity and stability, and thus contribute to their mobilization and uptake by plants. Microbial polymers, such as chitin and extracellularly released proteins, also have complexing properties (Kurita et al. 1979; Birch and Bachofen 1990; Chen et al. 1995) and can serve as ligands to trace elements in soil. In aerated soils, rhizosphere microorganisms may mobilize microelements by reduction, favoured by root excretion of protons. This mechanism of reductive dissolution of insoluble or sparingly soluble oxides is particularly important for mobilization of Mn and Fe, which in calcareous and alkaline soils are immobilized in insoluble oxidized forms.

Other indirect beneficial effects of soil microorganisms on micronutrient uptake by plants may come from changes of the root system architecture and extension, or in the composition of microbial community of the rhizosphere as a consequence of modified root exudation. For instance, in the plant/mycorrhiza association root colonization alters the morphology and physiology of the host root and quality and quantity of root exudates, resulting in a facilitated nutrient acquisition and transport, and a new microbial equilibrium in the rhizosphere (Linderman 1988; George et al. 1994). This change of the ecological conditions of the soil zone immediately surrounding the mycorrhizal root has been termed "mycorrhizosphere effect" (Linderman 1988; Duponnois et al. 2008).

While other elements are also involved, Fe, Mn and Zn deficiencies have the greatest impact on the yields and quality of agricultural produce worldwide. On global scale, Zn deficiency is more widespread than that of any other micronutrient. This particular deficiency also has a strong social impact, since Zn deficiency in food is a major cause of severe health disorders, especially in developing countries (Cakmak 2002, 2008). Herein we will review the effect of soil microorganisms on bioavailability of Fe, Mn and Zn in soil and their uptake by crop plants. Examples will be given of the potential for beneficial soil microorganisms to improve plant nutrition and growth under both deficient and toxic conditions.

3.4.1 Iron

In plants, iron (Fe) is critical for chlorophyll formation and photosynthesis and is important in enzyme systems and respiration. In fact, it is a component of the Fe-porphyrin complex (heme) of the cytochromes. Cytochromes are constituents of the redox system in chloroplasts and in mitochondria. Other heme enzymes are catalase and peroxidases. Iron is also part of several nonheme proteins, such as ferredoxin, superoxide dismutase, aconitase, and lipoxygenases (Marschner 1995).

Fe is found in the soil mainly as insoluble or sparingly soluble oxyhydroxide polymers of the general composition FeOOH or as Fe(III) chelates, especially at neutral or alkaline pH. Fe must be in the reduced form [Fe(II)] to be taken up by plant (Marschner 1995). Thus in order for root to take up Fe, Fe(III) oxides must be beforehand solubilised. All plants except the grasses have a mechanism to acquire Fe that involves the initial reduction of Fe(III) by a plasma membrane – bound Fe(III) chelate reductase, followed by transport of Fe(II) across the root epidermal cell membrane. This is called strategy I. Grasses have a different mechanism (strategy II). Strategy II plants release phytosiderophores (low-molecular-weight, nicotianamine-derived, high-affinity Fe(III)-chelating compounds). These compounds solubilize and bind ferric Fe in the rhizosphere. The phytosiderophores are recognized by specific membrane receptors and are taken up by the roots (Fox and Guerinot 1998). Biosynthesis and release of phytosiderophores is up-regulated by Fe- and Zn-deficiency. A similar chelating strategy is used by a wide variety of bacteria and fungi (Crowley et al. 1991; Renshaw et al. 2002). With the exception of strictly anaerobic bacteria and some yeasts, nearly all the bacterial and fungal species produce siderophores (Neilands 1995). Likewise phytosiderophores, the biosynthesis and release of microbial high-affinity Fe(III)-chelating compounds (siderophores) is enhanced by low iron availability. Under these conditions, the production of siderophores is used to gain an ecological advantage over other microorganisms that compete for the limited iron supply. Siderophore production by microorganisms can be beneficial to plants in two ways. Firstly, siderophores can solubilise Fe(III) by chelation and make it available to the plant (Crowley et al. 1991), either directly or through ligand exchange coupled with phytosiderophores. Secondly, siderophore production by non-pathogenic microorganisms can suppress growth of pathogenic microorganisms by depriving the pathogen of iron (Mercado-Blanco and Bakker 2007). However, siderophores are also produced by plant-pathogenic microorganisms and in some instances they are regarded as virulence factors (Haas et al. 2008). Lemanceau et al. (1985) found that non-pathogenic *Fusarium* species produced more siderophores than pathogenic species. In this way they were able to compete more effectively for iron and suppress growth of pathogenic *Fusaria*. The ability to utilize several types of siderophores could also be advantageous in competing with other organisms. It may allow a particular microbe to grow in a variety of environmental conditions and use siderophores produced by competing microorganisms (van der Helm and Winkelmann 1994). For example, ferrichrome, which is made by fungi but not by bacteria, is transported by many bacterial species (Jurkevitch et al. 1992). Hence, the effectiveness of a beneficial microorganism in suppressing a plant pathogen and promote plant growth depends both on its siderophore production efficiency and its capability to synthesize and transport into the cell different types of siderophores. Among the plant beneficial bacteria, fluorescent pseudomonads have long been known to produce and release siderophores in the rhizosphere. The role of these compounds in plant growth promotion and biological control of plant pathogens exhibited by *Pseudomonas* spp., viz. *P. fluorescens*, *P. putida*, and *P. aeruginosa*, has been extensively studied and reviewed (Leong 1986; Höfte et al. 1994; Compant et al. 2005; Mercado-Blanco and Bakker 2007). About the latter species, it should be noticed that *P. aeruginosa* is known to be an opportunistic human pathogen and this restricts its application in agriculture. However, biochemical and genetic analyses indicate that *P. aeruginosa* strains from soil are distinct from clinical isolates (Cornelis et al. 1989; Ferguson et al. 2001). This, along with more in-depth epidemiological and molecular studies might possibly lead to a revision of the taxonomy of the species in the future and to a reconsideration of this bacterium as bioinoculant.

Pyoverdines (also known as pseudobactins) are the prevalent class of siderophores produced by fluorescent *Pseudomonas* (Cornelis and Matthijs 2007). They are yellow-green water-soluble chromopeptides that fluoresce under UV light at 336 nm. They have both catechol and hydroxamate chelating groups, which results in a very high affinity for iron (Leong 1986). Besides pyoverdines, some beneficial *Pseudomonas* spp. strains may produce additionally one or more different types of siderophores, such as salicylic acid, dihydroaeruginoic acid, pyochelin (*P. aeruginosa*) (Buysens et al. 1996) or its stereoisomeric form enantio-pyochelin (Youard et al. 2007) and pseudomonine (*P. fluorescens*) (Anthoni et al. 1995; Mercado-Blanco et al. 2001).

The beneficial effect of fluorescent pseudomonads on plant-growth and the importance of siderophore production are widely acknowledged and reviewed (Kloepper et al. 1980; Compant et al. 2005; Mercado-Blanco and Bakker 2007). The most convincing evidence has been the fact that siderophore-deficient mutants obtained by chemical (Becker and Cook 1988) or transposon mutagenesis (Höfte et al. 1991) were unable to induce plant growth stimulation. Nevertheless, the actual effects of beneficial pseudomonads on iron plant nutrition may be influenced by several factors, including plant species, variety, the siderophore-producing *Pseudomonas* strain and the experimental conditions. For example, dicot and monocot plants vary in their iron uptake rates from *P. putida* Fe-pseudobactin (Barness et al. 1991). None or little influence on Fe acquisition by oat (*Avena sativa* L.) plants grown in a calcareous soil was observed when the roots were inoculated with *Pseudomonas* strains that produced high amounts of siderophores (Alexander and Zuberer 1993). In carnation (*Dianthus caryophyllus* L.), ferric pseudobactin-358 showed differential effects between two cultivars that differed in their ferric reducing activity. Chlorophyll synthesis was stimulated in the cultivar with the highest activity (Duijff et al. 1994). Environmental conditions and composition of root exudates may influence pyoverdine (as well as other siderophores) production (Duffy and Défago 1999). This can stimulate or reduce the effect of the bioinoculant on either plant-growth promotion or pathogen suppression.

Siderophores also vary in their chemical properties that influence the ability of plants to remove Fe from the chelator. For example, the Fe-chelates formed by pyoverdine-type siderophores are highly stable and have been found to be poorly utilized by maize and cucumber plants in short-term experiments (Walter et al. 1994). However, over a longer period of time, the Fe nutrition of plants may be increased, for example, through microbial degradation of siderophores that results in increased extracellular supply of Fe at the root surface (Becker et al. 1985; Crowley et al. 1988; Walter et al. 1994).

Iron plays a central role in nitrogen fixation as a constituent of the components I and II and of the iron-molybdenum cofactor (FeMo-co) of the nitrogenase enzyme complex (Kim and Rees 1992; Rubio and Ludden 2008). In addition, iron is required for several physiological processes that are important for successful infection of legume roots by *Rhizobia* (Barton et al. 1994). It is, therefore, no surprise that siderophores are also produced by both symbiotic and free-living (nonsymbiotic) nitrogen fixing bacteria, e.g. *Rhizobium meliloti* (Smith et al. 1985; Johnston 2004), *R. leguminosarum* (Storey et al. 2006), and *Azotobacter vinelandii* (Cornish and Page 1998, 2000), although production may vary among strains (Carson et al. 1992). Although there is no direct evidence that siderophores produced by nitrogen fixing bacteria have an effect on iron nutrition of plants, it is likely that they affect positively both nitrogen fixation and suppression of plant pathogens, thus exerting indirect beneficial effects on plant growth. Pacovsky (1988) inoculated sorghum plants with a strain of *Azospirillum brasilense*, while control (non-inoculated) plants were fertilized with mineral N. *Azospirillum*-inoculated sorghum plants were found to contain more Fe than N-amended plants. Nitrogen fixing bacteria may be adversely affected by competition and antagonism from other soil organisms in the rhizosphere, including siderophore-producing *Pseudomonas* spp. For instance, *Rhizobium* spp. showed reduced nodulation capacity when co-inoculated with different strains of root colonising *Pseudomonas* spp. (Burla et al. 1996; Fuhrmann and Wollum 1989). Other negative effects of *Pseudomonas* spp. on rhizobia are inhibition of nitrogenase activity (de Freitas et al. 1993) and reduced rhizobial growth *in vitro* (Berggren et al. 2001) and in the rhizosphere (Bolton et al. 1990). These detrimental effects are possibly associated to the production of interfering secondary metabolites (Keel et al. 1992; Shanahan et al. 1992), including the production of siderophores under iron limited conditions (Fuhrmann and Wollum 1989). This, of course, may have important agronomic implications with regard to the combined use of nitrogen fixing and other plant-growth promoting bioinoculants.

Nearly all fungi produce siderophores in response to iron-limited conditions (for reviews on this topic, see van der Helm and Winkelmann 1994; Renshaw et al. 2002; Winkelmann 2007; Haas et al. 2008). Siderophores have been isolated and characterized from cultures of several fungal species known for their biocontrol and plant-growth promoting activities (van der Helm and Winkelmann 1994). The role of these compounds in the mechanisms of disease suppression and plant-growth enhancement is still to be experimentally shown. Nevertheless, it has been shown that both Strategy I and Strategy II plants can take up iron from fungal siderophores (Shenker et al. 1992; Wang et al. 1993; Yehuda et al. 1996; Hördt et al. 2000) via ferric chelate reduction by the roots and subsequent transport throughout the plant (Miller et al. 1994; Johnson et al. 2002). Unlike bacteria, which produce siderophores containing a variety of functional groups, all of the fungal siderophores characterized so far are of the hydroxamate-type with the exception of the polycarboxylate rhizoferrin. Fungal hydroxamate siderophores can be divided into three structural families: fusarinins, coprogens and ferrichrome. Fungal siderophores, such as rhizoferrin, coprogen and rhodotorulic acid, have been studied for use as micronutrient biofertilizers for root (Shenker et al. 1992) and foliar applications (Fernández et al. 2005).

Strains of *Trichoderma* have been shown to produce coprogen, coprogen B, and ferricrocin as the major trihydroxamate siderophores (Anke et al. 1991). The plant-growth-promoting and biocontrol strain *T. harzianum* 1295-22 (T-22) was found to be able to solubilise the sparingly soluble ferric oxide (Fe_2O_3) *in vitro* by a multiple mechanism that involved both chelation and reduction of Fe(III) (Altomare et al. 1999). The cell-free culture filtrate of the strain T-22 was positive to the chrome azurol S assay, a chemical method for determination of general chelating activity, but the nature of the chelating substances was not determined. The production and release of chelating compounds by T-22 was a constitutive trait and did not require Fe deficiency for its expression (Altomare et al. 1999). Subsequently, similar results have been found using different *T. harzianum* strains intended for bioremediation (Adams et al. 2007b) or biocontrol (Küçük et al. 2008). Yedidia et al. (2001) reported an increase up to 30% of Fe concentration in cucumber plants grown in soil amended with *T. harzianum*. Similarly a significant increase in the concentration of Fe (as well as P, Cu, Zn, Mn and Na) was observed in inoculated roots in an axenic hydroponic system, in which higher uptake of Fe was noted in the roots as early as 48 h post-inoculation (Yedidia et al. 2001). The siderophores produced by *Gliocladium virens* (now *Trichoderma virens*), another biocontrol fungus, have also been isolated and characterised (Jalal et al. 1986, 1987).

Hydroxamate siderophores can be produced by ericoid mycorrhizal fungi (Haselwandter et al. 1992). In AM fungi, the only report so far suggesting siderophore production comes from an experiment with plants grown in nutrient solution (Cress et al. 1986). Not much information is available on the effect of AM fungi on Fe uptake by plants (George et al. 1994) and the results of the few studies carried out so far are often inconsistent. In soybean, Pacovsky and Fuller (1988) observed a decrease in Fe concentration in the presence of AM fungi, whereas for maize an increase of shoot Fe concentration has been described by Clark and Zeto (1996). In another experiment, total Fe uptake by soybean and maize was found to be higher in mycorrhizal plants (Lambert et al. 1979). By using a technique that allowed the spatial separation of the growing zones of roots and hyphae of *Glomus mosseae* in soil, and radiolabelled elements (^{32}P, ^{59}Fe), Caris et al. (1998) measured the contribution of external mycorrhizal hyphae to P and Fe uptake of peanut and sorghum plants. Mycorrhizal inoculation had no significant influence on the concentration of labelled Fe in shoots of peanut plants, but increased its concentration in the shoots of sorghum plants. The uptake of labelled Fe from soil by sorghum was particularly high under conditions producing a low Fe nutritional status of the plants.

Based on the results shown by Caris et al. (1998) it appears that hyphae of AM fungi are able to mobilise and/or take up Fe from soil and translocate it to the plant. AM fungi might affect Fe uptake of plants also indirectly, for example by modifying root morphology, composition of the microbial populations in the mycorrhizosphere (Nogueira et al. 2007), or solubilising insoluble iron phosphates (Bolan et al. 1987). In conclusion, to date no general assumption can be made about the effect of AM fungi on iron nutrition of plants, since Fe acquisition by mycorrhizal plants seems to be strongly dependent on experimental conditions, including type of plant and fungus, soil pH (Clark and Zeto 1996), soil supply of P, and on soil temperature (Raju et al. 1990).

Beside siderophores, other microbial products, including organic acids, peptides, proteins, phenolics, and chitin may exhibit chelating properties toward Fe and other metal ions (Birch and Bachofen 1990), although with lower affinity than siderophores. In plants, proton release is enhanced under Fe deficiency, which lowers rhizosphere pH and thereby increases the solubility of Fe(III). A number of soil-inhabiting fungi and bacteria have been shown to release organic acids, e.g. citric acid, malic acid and oxalic acid, which have been especially associated to solubilisation of phosphates (see for instance Khan et al. 2007 and references therein). Organic acids may also give a contribution to solubilisation of Fe (as well as to solubilisation of P, Mn, and Zn) through acidification of the microenvironment near roots and sequestration of Fe(III). Finally, solubilisation of Fe may also occur through reduction of Fe(III) to Fe(II). Altomare et al. (1999) showed that *T. harzianum* strain T-22 is able to produce diffusible metabolites that reduce Fe(III).

3.4.2 Manganese

In plants manganese (Mn) acts as a cofactor, activating about 35 different enzymes, most of which catalyze oxidation-reduction, decarboxylation, and hydrolytic reactions. In addition, Mn is a structural component of two enzymes, namely the Mn-protein in photosystem II and the Mn-containing superoxide dismutase (Burnell 1988). Thus, Mn has a profound influence on three essential metabolic functions: photosynthesis, N metabolism (especially the reduction of nitrate), and biosynthesis of aromatic ring compounds. These latter are precursors for aromatic amino acids, auxins, phenols and lignin, compounds which have a major effect on plant growth and resistance to diseases (Huber and Wilhelm 1988).

Manganese can exist in the oxidation states 0, II, III, IV, VI and VII. In natural systems, however, it usually exists in the II (Mn^{2+}) or IV (Mn^{4+}) oxidation state. The III oxidation state is also possible but it is generally unstable (Hughes and Williams 1988). The oxidized forms of Mn are insoluble, therefore it is available to plants only in the reduced form (Mn^{2+}). The oxidation state of soil Mn mainly depends on soil reaction (pH values below 6 favor reduction and values above 6.5 favor oxidation), but rhizosphere microorganisms may have a dramatic influence on Mn availability, due to their redox activity (Bromfield 1978).

Under Mn-deficient conditions, photosynthesis is reduced, nitrogen is inefficiently metabolized and defence reactions are severely limited. Yield of many cereals, including oats, wheat, barley, sorghum and rice, and legumes are particularly sensitive to low (as well as high) levels of Mn in soil (Campbell and Nable 1988). Manganese deficiency in wheat results in decreased levels of phenols, especially of alkali-labile phenols, in the shoots (Brown et al. 1984). The role of Mn in lignin/phenol synthesis is extremely important, as it relates to plant defence mechanisms in disease resistance. In fact, the correlation of several plant diseases (e.g., take-all disease of wheat, wilt of tomato, blast of rice, and common scab of potato) with low Mn availability in soil has long been known (Huber and Wilhelm 1988).

Take-all disease, root-, crown-, and foot-rot of cereals caused by the soil-borne fungus *Gaeumannomyces graminis* is a major disease in temperate climates throughout the world (Garrett 1981). Severity and grain yield losses are greatly influenced by Mn availability to plants. Take-all is prevalent on alkaline conditions, characteristic of immobilized Mn and consequent Mn-deficiency (Huber and McCay-Buis 1993). Virulence of *G. graminis* strains is positively correlated to the capability to oxidize Mn and induce Mn deficiency in wheat plants (Graham and Webb 1991). Conversely, a significant reduction of disease severity and yield loss was obtained in field trials by seed treatment with Mn-reducing fluorescent pseudomonad isolates (Weller and Cook 1983; Huber et al. 1989). Take-all was less severe in wheat grown from seed inoculated with *Bacillus cereus*, a manganese reducer common in wheat rhizosphere (Huber and McCay-Buis 1993). Also, reduced disease severity and increased manganese uptake and growth were found in wheat seedlings treated with unidentified manganese-reducing bacteria (Marschner et al. 1991). Actinomycetes, including members of the genus *Streptomyces*, and bacteria from the genera *Pseudomonas* and *Bacillus* have been reported to be Mn-reducers (Posta et al. 1994; Nogueira et al. 2007). However, some *Pseudomonas* and *Bacillus* species (e.g. *P. fluorescens* and *B. subtilis*) are able to oxidize as well as reduce soil Mn (Huber and McCay-Buis 1993; De Schamphelaire et al. 2007). The effect of these bacteria on Mn availability for plants is strongly influenced by nitrification processes that occur in soil. For instance, nitrification that follows the application of ammonium nitrogen is associated with an increase in the oxidation potential of soil bacteria and thereby an increase in the levels of insoluble forms of Mn (mostly MnO_2) (Huber and McCay-Buis 1993; Vandenabeele et al. 1995).

Altomare et al. (1999) showed that the biocontrol agent and plant growth promoting fungus *Trichoderma harzianum* was able to solubilise MnO_2, as well as P, insoluble Fe and Zn, in an *in vitro* system. Since solubilization of MnO_2 could not be attributed to either acidification of the substrate or chelation of Mn ions, a mechanism based on reduction of Mn^{4+} was hypothesised (Altomare et al. 1999). Yedidia et al. (2001) reported that *T. harzianum* increased Mn, Fe and Zn content of cucumber plants grown in either hydroponic culture or soil. In contrast, Öğüt and Er (2006) found that inoculation of bean plants with *T. harzianum* resulted in a reduced concentration of micronutrients (Fe, Mn, Zn, and Cu) in seeds and 45-day-old plants. Overall, these findings suggest that the *T. harzianum* strain and/or the experimental conditions used may be critical for Mn solubilisation and uptake by plants.

While it is known that AM colonization significantly decreases the Mn content of plants grown under excess Mn conditions (see below), it has also been reported that it may increase the plant Mn concentration in low Mn soils (Menge et al. 1982). Mycorrhizae are known to strongly affect root exudation and this may play a significant role in determining the relative levels of Mn-reducing and –oxidizing microorganisms (Kothari et al. 1991a). All the microorganisms that acidify environment in the rhizosphere (see solubilization of P) may also increase the availability of manganese to plants.

Globally, Mn toxicity is possibly an even more serious problem than Mn deficiency. Mn toxicity, as well as aluminium toxicity, is a major factor for reduced crop productivity in acid soils. Symptoms of Mn toxicity and the concentration of Mn that causes toxicity show a wide variability among plant species and also among varieties within species, because the phytotoxic mechanisms of Mn involve different biochemical pathways in different plant genotypes (El-Joual and Cox 1998).

The concentration of Mn in shoots and roots of mycorrhizal plants is often lower than in non-mycorrhizal plants. AM fungi are therefore able to decrease the uptake of Mn (Arines et al. 1989; Kothari et al. 1991a; Nogueira et al. 2004) and alleviate Mn toxicity (Bethlenfalvay and Franson 1989; Yano and Takaki 2005). A possible reason can be that mycorrhizae can affect the balance between Mn-reducing and Mn-oxidizing microorganisms in the mycorrhizosphere. Kothari et al. (1991a) and Posta et al. (1994) observed that mycorrhiza stimulated the Mn-oxidizing bacteria in alkaline soils. Nogueira et al. (2004), using a sandy and acidic soil (pH 5), also found that Mn-oxidizing bacteria were stimulated in the mycorrhizosphere. In the above situations, the increase of Mn-oxidizing bacteria in soil was associated with a reduced Mn availability or uptake by plants. On the other hand, Nogueira et al. (2007) reported that, in condition of adequate supply of P (30 mg kg^{-1} of soluble P), the Mn-reducing bacteria were stimulated in the rhizosphere of mycorrhizal soybean plants, while Mn-oxidizing bacteria were reduced. The increase of Mn-reducing and decrease of Mn-oxidizing bacteria coincided with more extractable Mn in the soil and Mn toxicity. However the mycorrhizal plants showed Mn toxicity symptoms only during the earlier growth stages, but these were alleviated later in the growth period. The authors hypothesized that the better P nutrition of mycorrhizal plants facilitated the cell detoxification by stimulating the ATP-dependent sequestration of Mn or Mn-chelates in the vacuoles or formation of low-solubility P–Mn complexes and their precipitation or compartmentation. Beneficial effects on plants exposed to Mn toxicity have been reported for several glomalean AM fungi, e.g. *Glomus mosseae, G. fasciculatum*, *G. tenue*, *G. epigaeum*, and *G. macrocarpum* (Arines et al. 1989; Kothari et al. 1991a). However, the level of protection against Mn toxicity may vary among species (Cardoso and Kuyper 2006).

3.4.3 Zinc

Zinc is involved in a wide variety of metabolic processes. The metabolic functions of zinc are based on its strong tendency to form tetrahedral complexes with N-, O-, and S-ligands and it thereby plays both a functional (catalytic or co-catalytic) and a structural role in plant enzyme reactions. Zn-binding motifs are present in proteins that are involved in all aspects of metabolism, including carbohydrate, lipid, protein and nucleic acid synthesis and degradation (Auld 2001). Several important plant enzymes, e.g. alcohol dehydrogenase Cu/Zn superoxide dismutase (SOD) and carbonic anhydrase, contain structurally bound Zn (Coleman 1992). Zinc is also a structural component of the ribosomes and it is essential for their structural integrity and function. In zinc deficient plants the protein synthesis and the protein content are drastically reduced whereas amino acids accumulate. Zinc is taken up predominantly as a divalent cation (Zn^{2+}), but at high pH it is presumably also taken up as a monovalent cation ($ZnOH^{+}$) (Marschner 1995).

Zn deficiency is a common problem in many important crops and is a cause of poor yields and low Zn content in agriculture commodities, with a major negative impact on dietary Zn intake and human health, especially in developing countries (Hotz and Brown 2004). Approximately 30% of the world's soils are Zn deficient, and nearly 50% of the cereal-grown areas in the world have soils with low levels of Zn available to plants (Graham and Welch 1996; Cakmak 2002). Low solubility rather than the total amount of Zn present in soils is the major reason for the widespread occurrence of Zn deficiency in crop plants. Zinc deficiency is common in plants grown in calcareous soils and in highly weathered acid soils with poor drainage. In the former case zinc deficiency is often associated with iron deficiency (lime chlorosis). The low availability of zinc in calcareous soils with a high pH results mainly from the adsorption of zinc to clay or $CaCO_3$ rather than the formation of $Zn(OH)_2$ or $ZnCO_3$ (Trehan and Sekhon 1977). Zinc deficiency may also raise after large applications of phosphate fertilizers. Several factors are involved in the P-induced Zn deficiency, including soil, plant and microbial factors. High phosphorus levels can decrease solubility of zinc in soils (Loneragan et al. 1979), due to fixation as highly insoluble metal phosphates. Increased P availability improves plant growth and as a result "dilutes" the total amount of zinc over a larger plant biomass. Finally, a high phosphate supply is often associated with reduced root growth, a lesser degree of mycorrhizal root infection and a reduced activity of the mycorrhizal fungi present. In dicotyledons, Zn deficiency results in shortening of the internodes ("rosetting") and a severe reduction of leaf size often combined with chlorosis which may be either highly contrasting or diffusive ('mottle leaf'). In cereals, chlorotic bands along the midrib and red, spot-like discoloration caused by anthocyanins on the leaves are common symptoms (Marschner 1995).

In deficient soils, adequate Zn uptake by plants and Zn content of plant-derived food might be achieved with the use of appropriate microorganisms. The capability of the biocontrol and plant growth promoting fungus *T. harzianum* strain T-22 to solubilise metallic Zn (oxidation state 0) *in vitro* was proven by Altomare et al. (1999). T-22 proved to have the ability in culture to accelerate the oxidative dissolution of metallic zinc, releasing Zn^{2+}. In this case, the effect of the fungus was thought to include the release of complexing ligands which sequestered Zn^{2+}, thereby increasing the dissolution of metallic zinc in the culture medium. Inoculation with *T. harzianum* resulted in increased Zn uptake by cucumber plants (Yedidia et al. 2001) and crack willow saplings (Adams et al. 2007a). In contrast, in field experiments conducted in Turkey, inoculation of bean plants with *T. harzianum* significantly reduced the concentration of Zn and other micronutrients in seeds and the cumulative Zn uptake by plants (Öğüt and Er 2006).

The ability to solubilize zinc fixed as zinc phosphate has been reported for the bacteria *Pseudomonas fluorescens* (Di Simine et al. 1998), *P. aeruginosa*, *Acinetobacter lwoffii*, and *Bacillus thuringiensis* (Fasim et al. 2002; Shahab and

Ahmed 2008). Desorption of zinc by culture filtrates of *T. harzianum*, *T. reesei* and the white-rot fungus *Coriolus versicolor* from wood charcoal was shown by Adams et al. (2007b). In this latter study, metal chelation via organic acids and proteins were the main mechanisms by which the fungi increased Zn desorption. The capability to desorb Zn from soil colloids has been reported also for other microorganisms, including *Bacillus* and *Pseudomonas* and fungi of the genera *Aspergillus* and *Penicillium* (Castro et al. 2000). Both solubilisation of zinc phosphates and desorption from adsorbing matrixes are mechanisms of mobilization of zinc that might be expected to increase the availability of this metal ion in the soil and prevent P-induced Zn deficiency in soils with a high organic matter or clay content, which have a high affinity for metals. However, while these microorganisms may have a potential as Zn nutrition enhancer, there is no direct experimental evidence that they actually increase Zn uptake by plants. Therefore their usefulness for improvement of plant Zn nutrition remains to be determined.

Root inoculation with mycorrhiza can increase the level of zinc in plant tissue and improve Zn nutrition of crop plants (Kothari et al. 1991b; Hamilton et al. 1993; Sharma et al. 1994; Cavagnaro 2008). This is particularly apparent if Zn content of AM plants is compared with that of P-fertilized, non-AM plants of similar weight, growth stage, and P status, as was shown for soybean (Pacovsky 1986). Application of P fertilizer reduces plant Zn uptake via the mycorrhizal symbiosis. In fact, high supply of P reduces mycorrhiza colonization and functioning, leading to a lower Zn uptake (Lambert et al. 1979).

Under Zn deficient conditions, mycorrhizal symbiosis improves Zn nutrition of plants by increasing its transport and subsequent uptake by roots (Sharma et al. 1994). Mycorrhizal plants often develop a more voluminous and exploratory root system. An increase of Zn transport in mycorrhizal plants results from an increased capability of the root system to intercept Zn ions from the soil solution. Also, Zn in soil solution is transported towards the roots by a flow of water that is generated by plant suction. Mycorrhizal plants have shown to absorb more water than non mycorrhizal plants (Safir et al. 1971). Therefore, AM fungi may facilitate Zn interception also by increasing the flow of soil solution towards the root (Sharma et al. 1994; Cavagnaro 2008). Zn uptake by mycorrhizal plants may be increased also via translocation of phosphorus. Mycorrhizal hyphae translocate phosphorus mainly as polyphosphates, which can bind zinc as well as other cations, viz. calcium, iron and manganese, and are then translocated into the plant roots as Zn-polyphosphate complex (Sharma et al. 1994). However, the co-regulation of P and Zn translocation in mycorrhizal plants is complex and far from being completely elucidated. Contradicting findings have been reported in this regard. Jansa et al. (2003) noted a correlation of P and Zn uptake rates by maize plants colonized by *G. intraradices*, while Mehravaran et al. (2000) found that different species of AM fungi differed in their ability to take up P and Zn and found no correlation between the uptake of these two elements in mycorrhizal plants.

Experiments in an axenic culture and soil with addition of different Zn quantities (ranging from 0 to 10 mM as $ZnSO_4$) showed that the Zn supply has significant effects on root colonization by AM fungi in both the extra-radical and the intra-radical phases of the process (Pawlowska and Charvat 2004). Increased Zn supply decreased spore germination, density of spores and hyphal extension of *Glomus etunicatum* and *G. Intraradices* significantly. The latter was more tolerant to high Zn supply. The addition of Zn to soil was reported to have either negative (Liu et al. 2000) or positive (Seres et al. 2006) effects on the extra radical growth of *G. intraradices* on maize roots. During the intra-radical phase, the colonisation of roots by AM fungi often decreases with increasing Zn addition to soil (Gildon and Tinker 1983; Bi et al. 2003; Chen et al. 2004), although also positive (Lee and George 2005) and neutral (Ortas et al. 2002) responses have been reported. Overall, literature data suggest that plant and fungal species or genotype are very important determinants of mycorrhizal response to soil Zn supply.

When zinc availability is high, zinc toxicity can readily be induced in non tolerant plants, resulting in root elongation inhibition and chlorosis of young leaves. The latter symptom may also be an indirect effect of Fe, Mn or Mg deficiencies which are often associated to high zinc supply (Marschner 1995). At high soil Zn concentrations, colonization of roots by AM fungi can result in lower levels of Zn in plant tissues and reduce toxicity damage. There is experimental evidence that AMF can function as a bioaccumulator for Zn and reduce the uptake by and the translocation to the plant under Zn toxicity conditions.

Hildebrandt et al. (1999) and Kaldorf et al. (1999) showed that the AM fungus *G. intraradices* was able to confer Zn tolerance to a variety of plants, including tomato and maize, grown in heavy metal polluted soils. Heggo et al. (1990) demonstrated that at high heavy metal concentrations in soil, arbuscular mycorrhizal infection reduce the concentrations of Zn, Cd and Mn in soybean leaves. Under conditions of surplus of available Zn in soil, improved P nutrition of mycorrhizal plants might help to tolerate Zn toxicity through the resulting increase in plant growth. Other possible mechanisms of attenuation of Zn toxicity of mycorrhizal plants may be immobilization of the metal in the mycorrhizosphere by binding to mycorrhizal structures, compartmentation in fungal cytoplasm or accumulation (partitioning) in restricted plant parts (tissues or organs) (Christie et al. 2004). The mycelium of mycorrhizal fungi has been reported to possess very strong metal binding capacity (Joner et al. 2000). Mycelium of *G. mosseae* was able to bind almost 3% of Zn on a dry matter basis, without reaching the adsorption saturation point (Christie et al. 2004). Recently glomalin, an insoluble glycoprotein produced by hyphae of AM fungi, was shown to bind heavy metals such as Cu, Cd, Pb (González-Chávez et al. 2004), and a similar effect on Zn is likely. Studies on element localization carried out with microbeam analysis (Kaldorf et al. 1999) and with electron-dispersive X-ray spectrometry (González-Guerrero et al. 2008) showed that Zn and other heavy metals are deposited mainly in the inner root parenchyma cells of mycorrhizal plants (where intraradical hyphae and most of the fungal structures are located), particularly in fungal cell walls and as electron-dense granules in fungal cytoplasm. The biochemical and molecular aspects related to plant and fungal genes involved in these mechanisms and their expression have been recently reviewed by Hildebrandt et al. (2007).

Given the bivalent function of mycorrhizae, which are able to enhance Zn uptake by plant at low soil concentrations and reduce uptake at high soil concentrations, there is a critical soil Zn concentration below which Zn uptake by the host plant is enhanced by AM fungi and above which Zn translocation, particularly to the aerial parts, is reduced (Chen et al. 2003). This critical value should be established for different soil types and plant species.

4 Conclusion

All data presented above clearly show that many soil fungi, including mycorrhizal fungi, and bacteria exhibit beneficial effects for higher plants. They have proven to be able to promote plant growth and increase crop productivity and can contribute significantly to the mineral nutrition of crop plants. These beneficial microorganisms influence directly plant nutrition and growth through various mechanisms of action. These include atmospheric nitrogen fixation, breakdown of organic matter, solubilization of sparingly soluble minerals, release of chelating compounds and biologically active substances such as phytohormones, vitamins and enzymes, and increase of the absorbing surface and soil-penetrating capability of root systems. In addition, beneficial soil microorganisms improve soil fertility and plant growth through other indirect mechanisms, such as amelioration of soil structure, enhancement of plant resistance to abiotic stresses, suppression of phytopathogenic fungi and bacteria or induction of plant resistance to pathogens. In many cases, the combination of different biofertilizers that operate through different modes of action has been reported to result in synergistic beneficial effects on plant nutrition and growth.

Conventional agriculture relies heavily on synthetic N fertilizers, but more than half of the applied N is lost. This leads to considerable environmental risks, including nitrate occurrence in groundwater and represents an economic loss for the farmers. Non symbiotic soil bacteria, able to fix atmospheric N_2 are widely distributed in bulk soil, root surface, root interior and plant aerial tissues. The application of non symbiotic N_2-fixing bacteria can improve N management in more "low-input" direction, and can constitute an integral part of sustainable agriculture.

Most soils contain enormous reserves of P that are unavailable to plants. Numerous species of soil and rhizosphere microorganisms may solubilize insoluble mineral phosphate complexes, mainly through acidification and production of organic acids. They also mineralize organic phosphates and influence diffusion of phosphates from adsorbing surfaces to soil solution. Mycorrhizal fungi possess high-affinity P uptake mechanisms and thus improve the P uptake of mycorrhizal plants. In addition, AMF determine modifications of the root architecture, such as increase of total root surface and volume, number of tips, degree of root branching and enhance the soil-exploration capability of roots. Under field conditions, the combined use of P-solubilizing microorganisms with mineral fertilizers such as rock-phosphate has often given successful results.

Trace minerals are of great importance to plant nutrition, crop productivity, produce quality and nutritional aspects of derived foods and feeds. The increase in cultivation intensity with the increasing demand for higher yields has led to a faster depletion of micro-elements from the soil. While other elements are also involved, Fe, Mn and Zn deficiencies have the greatest impact on the yields and quality of agricultural produce worldwide. Conversely, excess conditions of micro-elements may occur in acid or polluted soils and result in plant toxicity. In either cases, the biological activities of microorganisms in the rhizosphere can mediate the solubility, and hence the availability, of metal ions at root surface. Soil microorganisms have the capability to alter soil pH and modify the equilibrium of many chemical and biochemical reactions, such as precipitation/dissolution, adsorption/desorption, complexation/dissociation, and oxidation/reduction of metal cations and thus regulate the plant uptake. In excess conditions, AMF function as bioaccumulator and can reduce the uptake by and the translocation to the plant of metal ions and attenuate toxic effects.

In conclusion, the application of beneficial microorganisms as biofertilizers is a promising technology. It may have multiple beneficial effect on soil fertility, plant growth and health. Microbials, applied alone or in combination with mineral or organic fertilizers, have the potential to be part of a set of agricultural practices aimed to reduce environmental pollution and economic and social costs associated with the massive use of synthetic fertilizers. Beneficial soil microbes are low-input and multifunctional agents and a renewable resource. Therefore they should be extensively studied and exploited as an integral part of sustainable agriculture. However, maximization of the benefit from biofertilizers requires the design of an overall strategy for their production and effective use. This should be based on elaboration, promotion and dissemination of new agricultural practices which represent effective, though more ethical, alternatives to chemical fertilization.

5 Perspectives

The scientific interest in low-input and environmental-friendly agriculture has grown enormously in the last few decades. Despite the substantial research activity on beneficial soil microorganisms, there is still need of more in-depth understanding of the genetic bases of plant-microbe interaction. Furthermore little is known about the mechanisms driving biofertilizers and their molecular determinants. Scientific and technological developments are expected to improve survival, establishment and performance of microbial strains when applied to soil and to create synergistic mixtures and formulations. Further investigations are desirable with regard to soil microbial diversity and the exploitation of indigenous strains which possess better potential to promote plant growth when applied at local agriculture. Also, our current knowledge about composition of soil microbial communities, their interactions and effects on soil fertility and plant physiology is still largely incomplete. The relatively new research field of metagenomics enables studies of organisms that are not easily cultured in a laboratory as well as studies of organisms in their natural environment and opens new and exciting perspectives for study of soil biodiversity and microbial genes' functions in the complex soil environment.

Acknowledgements The authors would like to thank the three anonymous referees for critical reading of the manuscript and constructive suggestions. Dr. I. Tringovska acknowledges the financial support of Human Resources Development Operational Programme, co-financed by the European Union through European Social Fund under grant no. BGPO001/07/3.3-02/30.

References

Adams P, De-Leij FAAM, Lynch JM (2007a) *Trichoderma harzianum* Rifai 1295-22 mediates growth promotion of crack willow (*Salix fragilis*) saplings in both clean and metal-contaminated soil. Microb Ecol 54:306–313

Adams P, Lynch JM, De Leij FA (2007b) Desorption of zinc by extracellularly produced metabolites of *Trichoderma harzianum, Trichoderma reesei* and *Coriolus versicolor*. J Appl Microbiol 103:2240–2247

Adesemoye AO, Torbert HA, Kloepper JW (2009) Plant growth-promoting rhizobacteria allow reduced application rates of chemical fertilizers. Microb Ecol 58:921–929

Ahmad F, Ahmad I, Khan MS (2008) Screening of free-living rhizospheric bacteria for their multiple plant growth promoting activities. Microbiol Res 163:173–181

Aikio S, Ruotsalainen A (2002) The modelled growth of mycorrhizal and non-mycorrhizal plants under constant versus variable soil nutrient concentration. Plant Physiol 12:257–261

Alexander DB, Zuberer DA (1993) Responses by iron-efficient and inefficient oat cultivars to inoculation with siderophores-producing bacteria in a calcareous soil. Biol Fert Soils 16:118–124

Al-Karaki GN (2000) Growth of mycorrhizal tomato and mineral acquisition under salt stress. Mycorrhiza 10:51–54

Al-Karaki GN (2006) Nursery inoculation of tomato with arbuscular mycorrhizal fungi and subsequent performance under irrigation with saline water. Sci Hortic 109:1–7

Al-Karaki GN, Al-Raddad A (1997) Effects of arbuscular mycorrhizal fungi and drought stress on growth and nutrient uptake of two wheat genotypes differing in drought resistance. Mycorrhiza 7:83–88

Allen EB, Swenson W, Querejeta JI, Egerton-Waburton LM, Treseder KK (2003) Ecology of mycorrhizae: a conceptual framework for complex interactions among plants and fungi. Ann Rev Phytopathol 41:271–303

Altomare C, Norvell WA, Björkman T, Harman GE (1999) Solubilization of phosphates and micronutrients by the plant growth-promoting and biocontrol fungus *Trichoderma harzianum* Rifai 1295-22. Appl Environ Microbiol 65:2926–2933

Amijee F, Tinker PB, Stribley DP (1989) The development of endomycorrhizal root systems. VII. A detailed study of effects of soil phosphorus on colonization. New Phytol 111:435–446

Anke H, Kinn J, Bergquist K-E, Sterner O (1991) Production of siderophores by strains of the genus *Trichoderma*: Isolation and characterization of the new lipophilic coprogen derivative, palmitoylcoprogen. Biometals 4:176–180

Anthoni U, Christophersen C, Nielsen PH, Gram L, Petersen BO (1995) Pseudomonine, an isoxazolidone with siderophoric activity from *Pseudomonas fluorescens* AH2 isolated from Lake Victorian Nile perch. J Nat Prod 58:1786–1789

Arines J, Vilariño A, Sainz M (1989) Effect of different inocula of vesicular-arbuscular mycorrhizal fungi on manganese content and concentration in red clover (*Trifolium pratense* L.) plants. New Phytol 112:215–219

Arshad M, Frankenberger W (1991) Microbal production of plant hormones. Plant Soil 133:1–8

Artursson V, Finlay R, Jansson J (2006) Interactions between arbuscular mycorrhizal fungi and bacteria and their potential for stimulating plant growth. Environ Microbiol 8:1–10

Atkins CA (1984) Efficiencies and inefficiencies in the legume/Rhizobium symbiosis – a review. Plant Soil 82:273–284

Auld DS (2001) Zinc coordination sphere in biochemical zinc sites. Biometals 14:271–313

Azcon R, Ruiz-Lozano JM, Rodriguez R (2001) Differential contribution of arbuscular mycorrhizal fungi to the plant uptake (^{15}N) under increasing N supply to the soil. Can J Bot 79:1175–1180

Babana AH, Antoun H (2005) Biological system for improving the availability of Tilemsi phosphate rock for wheat (*Triticum aestivum* L.) cultivated in Mali. Nut Cycl Agroecosyst 72:147–157

Babana AH, Antoun H (2006) Effect of Tilemsi phosphate rock-solubilizing microorganisms on phosphorus uptake and yield of field-grown wheat (*Triticum aestivum* L.) in Mali. Plant Soil 287:51–58

Baldani JI, Caruso L, Baldani VLD, Goi SR, Döbereiner J (1997) Recent advances in BNF with non-legume plants. Soil Biol Biochem 29:911–922

Barakrt MAS, Gabr SM (1998) Effect of different biofertilizer types and nitrogen fertilizer levels on tomato plants. Alex J Agric research 43:149–160

Barea JM, Azcón R, Azcón–Aguilar C (2002) Mycorrhizosphere interactions to improve plant fitness and soil quality. Antonie Leeuwenhoek 81:343–351

Barker A, Bryson G (2007) Nitrogen. In: Barker AV, Pilbeam DJ (eds) Handbook of plant nutrition. Taylor & Francis, Boca Raton, pp 21–50

Barker AV, Pilbeam DJ (eds) (2007) Handbook of plant nutrition. Taylor & Francis, Boca Raton

Barness E, Chen Y, Hadar Y, Marschner H, Romheld V (1991) Siderophores of *Pseudomonas putida* as an iron source for dicot and monocot plants. Plant Soil 130:231–241

Barroso CB, Nahas E (2007) Solubilization of hardly soluble iron and aluminum phosphates by the fungus *Aspergillus niger* in the soil. In: Vélazquez E, Rodríguez-Barrueco C (eds) First international meeting on microbial phosphate solubilization. Developments in plant and soil sciences, vol 102. Springer, Dordrecht, pp 193–198

Barton LL, Vester CR, Gill PR, Jr Neilands JB (1994) The role of siderophores in symbiotic nitrogen fixation by legume root nodule bacteria. In: Manthey JA, Crowley DE, Luster DG (eds) Biochemistry of metal micronutrients in the rhizosphere. CRC Press, Boca Raton, pp 55–66

Bashan Y (1999) Interactions of *Azospirillum* spp. in soils: a review. Biol Fertil Soils 29:246–256

Bashan Y, Holguin G, de Bashan LE (2004) *Azospirillum*-plant relationships: physiological, molecular, agricultural, and environmental advances (1997–2003). Can J Microbiol 50:521–577

Becker JO, Cook RJ (1988) Role of siderophores in suppression of *Pythium* species and production of increased-growth response of wheat by fluorescent pseudomonads. Phytopathology 78: 778–782

Becker JO, Messens E, Hedges RW (1985) The influence of agrobactin on the uptake of ferric iron by plants. FEMS Microbiol Ecol 31:171–175

Behl RK, Sharma H, Kumar V, Narula N (2003) Interactions amongst mycorrhiza, *Azotobacter chroococcum* and root characteristics of wheat varieties. J Agron Crop Sci 189:151–155

Beneduzi A, Peres D, Vargas LK, Bodanese-Zanettini M, Passaglia L (2008) Evaluation of genetic diversity and plant growth promoting activities of nitrogen-fixing bacilli isolated from rice fields in South Brazil. Appl Soil Ecol 39:311–320

Berg G (2009) Plant–microbe interactions promoting plant growth and health: perspectives for controlled use of microorganisms in agriculture. Appl Microbiol Biotechnol 84:11–18

Berggren I, van Vuurde JWL, Mårtensson AM (2001) Factors influencing the effect of deleterious *Pseudomonas putida* rhizobacteria on initial infection of pea roots by *Rhizobium leguminosarum* bv. Viceae. Appl Soil Ecol 17:97–105

Bethlenfalvay GJ, Franson RL (1989) Manganese toxicity alleviated by mycorrhizae in soybean. J Plant Nutr 12:953–970

Bi YL, Li XL, Christie P (2003) Influence of early stages of arbuscular mycorrhiza on uptake of zinc and phosphorus by red clover from a low-phosphorus soil amended with zinc and phosphorus. Chemosphere 50:831–837

Birch L, Bachofen R (1990) Complexing agents from microorganisms. Experientia 46:827–834

Bojinova D, Velkova R, Ivanova R (2008) Solubilization of Morocco phosphorite by *Aspergillus niger.* Bioresour Technol 99:7348–7353

Bolan NS, Robson AD, Barrow NJ (1987) Effects of vesicular-arbuscular mycorrhiza on the availability of iron phosphates to plants. Plant Soil 99:401–410

Bolton H, Elliott LF, Turco RF, Kennedy AC (1990) Rhizoplane colonization of pea seedlings by *Rhizobium leguminosarum* and a deleterious root colonizing *Pseudomonas* sp. and effect on plant growth. Plant Soil 123:121–124

Bromfield SM (1978) The effect of manganese-oxidizing bacteria and pH on the availability of manganous ions and manganese oxides to oats in nutrient solutions. Plant Soil 49:23–39

Brown PH, Graham RD, Nicholas DJD (1984) The effects of managanese and nitrate supply on the levels of phenolics and lignin in young wheat plants. Plant Soil 81:437–440

Brundrett M, Bougher N, Dell N, Grove T, Malajczuk N (1996) Working with mycorrhizas in forestry and agriculture. Australian Centre for International Agricultural Research, Canberra, Monograph 32, 374 p

Bryla D, Koide R, 1998. Mycorrhizal response of two tomato genotypes relates to their ability to acquire and utilize phosphorus. Ann Bot 82, 849–857

Burdman S, Jurkevitch E, Okon Y (2000) Recent advances in the use of plant growth promoting rhizobacteria (PGPR) in agriculture. In: Subba RN, Dommergues YR (eds) Microbial interactions in agriculture and forestry, vol II. Science Publishers, Plymouth, pp 29–250

Bürgmann H, Meier S, Bunge M, Widmer F, Zeyer J (2005) Effects of model root exudates on structure and activity of a soil diazotroph community. Environ Microbiol 7:1711–1724

Burla M, Goverde M, Schwinn FJ, Wiemken A (1996) Influence of biocontrol organisms on root pathogenic fungi and on the plant symbiotic micro-organisms *Rhizobium phaseoli* and *Glomus mosseae.* J Plant Dis Protect 103:156–163

Burnell JN (1988) The biochemistry of manganese in plants. In: Graham RD, Hannam J, Uren NC (eds) Manganese in soils and plants. Kluwer, Dordrecht, pp 125–137

Buysens S, Heungens K, Poppe J, Höfte M (1996) Involvement of pyochelin and pyoverdin in suppression of *Pythium*-induced damping-off of tomato by *Pseudomonas aeruginosa* 7NSK2. Appl Environ Microbiol 62:865–871

Cakmak I (2002) Plant nutrition research: priorities to meet human needs for food in sustainable ways. Plant Soil 247:3–24

Cakmak I (2008) Enrichment of cereal grains with zinc: agronomic or genetic biofortification? Plant Soil 302:1–17

Çakmakçi R, Dönmez F, Aydin A, Şahin F (2006) Growth promotion of plants by plant growth-promoting rhizobacteria under greenhouse and two different field soil conditions. Soil Biol Biochem 38:1482–1487

Çakmakçi R, Dönmez F, Erdoğan Ü (2007a) The effect of plant growth promoting rhizobacteria on barley seedling growth, nutrient uptake, some soil properties, and bacterial counts. Turk J Agric For 31:189–199

Çakmakçi R, Erat M, Erdoğan Ü, Dönmez F (2007b) The influence of plant growth-promoting rhizobacteria on growth and enzyme activities in wheat and spinach plants. J Plant Nutr Soil Sci 170:288–295

Campbell LC, Nable RO (1988) Physiological functions of manganese in plants. In: Graham RD, Hannam J, Uren NC (eds) Manganese in soils and plants. Kluwer, Dordrecht, pp 139–154

Canbolat MY, Bilen S, Çakmakçi R, Şahin F, Aydin A (2006) Effect of plant growth-promoting bacteria and soil compaction on barley seedling growth, nutrient uptake, soil properties and rhizosphere microflora. Biol Fertil Soils 42:350–357

Cardoso IM, Kuyper TW (2006) Mycorrhizas and tropical soil fertility. Agric Ecosyst Environ 116:72–84

Caris C, Hördt W, Hawkins H-J, Römheld V, George E (1998) Studies of iron transport by arbuscular mycorrhizal hyphae from soil to peanut and sorghum plants. Mycorrhiza 8:35–39

Carson KC, Holliday S, Glenn AR, Dilworth MJ (1992) Siderophore and organic acid production in root nodule bacteria. Arch Microbiol 157:264–271

Castro IM, Fietto JLR, Vieira RX, Tropia MJM, Campos LMM, Paniago EB, Brandao RL (2000) Bioleaching of zinc and nickel from silicates using *Aspergillus niger* cultures. Hydrometallurgy 57:39–49

Cattelan AJ, Hartel PG, Fuhrmann JJ (1999) Screening for plant growth promoting rhizobacteria to promote early soybean growth. Soil Sci Soc Am J 63:1670–1680

Cavagnaro TR (2008) The role of arbuscular mycorrhizas in improving plant zinc nutrition under low soil zinc concentrations: a review. Plant Soil 304:315–325

Cavagnaro TR, Jackson LE, Six J, Ferris H, Goyal S, Asami D, Scow KM (2006) Arbuscular mycorrhizas, microbial communities, nutrient availability, and soil aggregates in organic tomato production. Plant Soil 282:209–225

Chan YK, Barraquio WL, Knowles R (1994) N_2-fixing pseudomonads and related soil bacteria. FEMS Microbiol Rev 13:95–118

Chen J-H, Czajka DR, Lion LW, Shuler ML, Ghiorse WC (1995) Trace metal mobilization in soil by bacterial polymers. Environ Health Perspect 103(suppl 1):53–58

Chen BD, Li XL, Tao HQ, Christie P, Wong MH (2003) The role of arbuscular mycorrhiza in zinc uptake by red clover growing in a calcareous soil spiked with various quantities of zinc. Chemosphere 50:839–846

Chen BD, Shen H, Li XL, Feng G, Christie P (2004) Effects of EDTA application and arbuscular mycorrhizal colonization on growth and zinc uptake by maize (*Zea mays* L.) in soil experimentally contaminated with zinc. Plant Soil 261:219–229

Chen YP, Rekha PD, Arun AB, Shen FT, Lai WA, Young CC (2006) Phosphate solubilizing bacteria from subtropical soil and their tricalcium phosphate solubilizing activity. Appl Soil Ecol 34:33–41

Chen Z, Ma S, Liu L (2008) Studies on phosphorus solubilizing activity of a strain of phosphobacteria isolated from chestnut type soil in China. Bioresour Technol 99:6702–6707

Choudhury ATMA, Kennedy IR (2004) Prospects and potentials for systems of biological nitrogen fixation in sustainable rice production. Biol Fertil Soils 39:219–227

Christie P, Li XL, Chen BD (2004) Arbuscular mycorrhiza can depress translocation of zinc to shoots of host plants in soils moderately polluted with zinc. Plant Soil 261:209–217

Chung H, Parka M, Madhaiyana M, Seshadri S, Songb J, Chob H, Sa T (2005) Isolation and characterization of phosphate solubilizing bacteria from the rhizosphere of crop plants of Korea. Soil Biol Biochem 37:1970–1974

Clark RB, Zeto SK (1996) Mineral acquisition by mycorrhizal maize grown on acid and alkaline soil. Soil Biol Biochem 28:1495–1503

Cleveland CC, Townsend AR, Schimel DS, Fisher H, Howarth RW, Hedein LO, Perakis SS, Latty EF, Von Fischer JC, Elseroad A, Watson MF (1999) Global patterns of terrestrial biological nitrogen (N2) fixation in natural ecosystems. Glob Biogeochem Cycles 13:623–645

Cline GR, Powell PE, Szaniszlo PJ, Reid CPP (1982) Comparison of the abilities of hydroxamic, synthetic, and other natural organic acids to chelate iron and other ions in nutrient solution. Soil Sci Soc Am J 46:1158–1164

Cocking E (2003) Endophytic colonization of plant roots by nitrogen-fixing bacteria. Plant Soil 252:169–175

Coleman JE (1992) Zinc proteins: enzymes, storage proteins, transcription factors, and replication proteins. Annu Rev Biochem 61:897–946

Colla G, Rouphael Y, Cardarelli M, Tullio M, Rivera CM, Rea E (2008) Alleviation of salt stress by arbuscular mycorrhizal in zucchini plants grown at low and high phosphorus concentration. Biol Fertil Soils 44:501–509

Compant S, Duffy B, Nowak J, Clément C, Barka EA (2005) Use of plant growth-promoting bacteria for biocontrol of plant diseases: principles, mechanisms of action, and future prospects. Appl Environ Microbiol 71:4951–4959

Cornelis P, Matthijs S (2007) *Pseudomonas* siderophores and their biological significance. In: Varma A, Chincholkar S (eds) Microbial siderophores. Springer, Berlin/Heidelberg, pp 193–203

Cornelis P, Hohnadel D, Meyer J-M (1989) Evidence for different pyoverdine-mediated iron uptake systems among *Pseudomonas aeruginosa* strains. Infect Immun 57:3491–3497

Cornish AS, Page WJ (1998) The catecholate siderophores of *Azotobacter vinelandii*: their affinity for iron and role in oxygen stress management. Microbiology 144:1747–1754

Cornish AS, Page WJ (2000) Role of molybdate and other transition metals in the accumulation of protochelin by *Azotobacter vinelandii.* Appl Environ Microbiol 66:1580–1586

Cress WA, Johnson GV, Barton LL (1986) The role of endomycorrhizal fungi in iron uptake by *Hillaria jamesii.* J Plant Nutr 9:547–556

Crowley DE, Reid CPP, Szaniszlo PJ (1988) Utilization of microbial siderophores in iron acquisition by oat. Plant Physiol 87:680–685

Crowley DE, Wang YC, Reid CPP, Szaniszlo PJ (1991) Mechanisms of iron acquisition from siderophores by microorganisms and plants. Plant Soil 130:179–198

Das A, Prasad M, Shivay YS, Subha KM (2004) Productivity and sustainability of cotton (*Gossypium hirsutum* L.) – wheat (*Triticum aestivum* L.) cropping system as influenced by prilled urea, farmyard manure and *Azotobacter.* J Agron Crop Sci 190:298–304

de Freitas JR (2000) Yield and N assimilation of winter wheat (*Triticum aestivum* L., var. Norstar) inoculated with rhizobacteria. Pedobiologia 44:97–104

de Freitas JR, Gupta VVSR, Germida JJ (1993) Influence of *Pseudomonas syringae* R25 and *P. putida* R105 on the growth and N_2 fixation (acetylene reduction activity) of pea (*Pisum sativum L.*) and field bean (*Phaseolus vulgaris L.*). Biol Fertil Soils 16:215–220

de Freitas JR, Banerjee MR, Germida JJ (1997) Phosphate-solubilizing rhizobacteria enhance the growth and yield but not phosphorus uptake of canola (*Brassica napus L.*). Biol Fertil Soils 24:358–364

de Miranda JCC, Harris PJ, Wild A (1989) Effects of soil and plant phosphorus concentrations on vesicular-arbuscular mycorrhizae in sorghum plants. New Phytol 112:405–410

de Oliveira Pinheiro R, Boddey L, James EK, Sprent JI, Boddey RM (2002) Adsorption and anchoring of *Azospirillum* strains to roots of wheat seedlings. Plant Soil 246:151–166

De Schamphelaire L, Rabaey K, Boon N, Verstraete W (2007) Minireview: the potential of enhanced manganese redox cycling for sediment oxidation. Geomicrobiol J 24:547–558

Deubel A, Merbach W (2005) Influence of microorganisms on phosphorus bioavailability in soils. In: Buscot F, Varma A (eds) Microorganisms in soils: roles in genesis and functions. Springer, Berlin, pp 177–191

Di Simine CD, Sayer JA, Gadd GM (1998) Solubilization of zinc phosphate by a strain of *Pseudomonas fluorescens* isolated from forest soil. Biol Fertil Soils 28:87–94

Diby P, Sarma YR, Srinivasan V, Anandaraj M (2005) *Pseudomonas fluorescens* mediated vigour in black pepper (*Piper nigrum* L.) under green house cultivation. Ann Microbiol 55:171–174

Dobbelaere S, Vanderleyden J, Okon Y (2003) Plant growth-promoting effects of diazotrophs in the rhizosphere. Crit Rev Plant Sci 22:107–149

Döbereiner J (1997) Biological nitrogen fixation in the tropics: social and economical contributions. Soil Biol Biochem 29:771–774

Dossier Third World Network (1990) Return to the good earth: damaging effects of modern agriculture and the case for ecological farming. Third World Network, Penang, 570 p

Duffy BK, Défago G (1999) Environmental factors modulating antibiotic and siderophore biosynthesis by *Pseudomonas fluorescens* biocontrol strains. Appl Environ Microbiol 65:2429–2438

Duffy B, Keel C, Défago G (2004) Potential role of pathogen signaling multitrophic plant-microbe interactions involved in disease protection. Appl Environ Microbiol 70:1836–1844

Duijff BJ, Bakker PAHM, Schippers B (1994) Ferric pseudobactin 358 as an iron source for carnation. J Plant Nutr 17:2069–2078

Duponnois R, Galiana A, Prin Y (2008) The mycorrhizosphere effect: a multitrophic interaction complex improves mycorrhizal symbiosis and plant growth. In: Siddiqui ZA, Akhtar MS, Futai K (eds) Mycorrhizae: sustainable agriculture and forestry. Springer, Dordrecht, pp 227–240

Edathil TT, Manian S, Udaiyan K (1996) Interaction of multiple VAM fungal species on root colonization, plant growth and nutrient status of tomato seedlings (*Lycopersicon esculentum* Mill.). Agric Ecosyst Environ 59:63–68

Egamberdiyeva D (2007) The effect of plant growth promoting bacteria on growth and nutrient uptake of maize in two different soils. Appl Soil Ecol 36:184–189

Egorenkova IV, Konnova SA, Skvortsov IM, Ignatov VV (2000) Investigation of the initial stages of interaction of the bacterium *Azospirillum brasilense* with wheat seedling roots: adsorption and root hair deformation. Microbiol Moscow Russ Ed 69:103–108

El-Joual T, Cox DA (1998) Manganese toxicity in plants. J Plant Nutr 21:353–356

Elkan GH (1992) Biological nitrogen fixation systems in tropical ecosystems: an overview. In: Mulongoy K, Gueye M, Spencer DSC (eds) Biological nitrogen fixation and sustainability of tropical agriculture. Wiley, Chichester, pp 27–40

El-Komy HMA (2005) Coimmobilization of *A. lipoferum* and *B. megaterium* for successful phosphorus and nitrogen nutrition of wheat plants. Food Technol Biotechnol 43:19–27

El-Zemrany H, Czarnes S, Hallett PD, Alamercery S, Bally R, Monrozier LJ (2007) Early changes in root characteristics of maize (*Zea mays*) following seed inoculation with the PGPR *Azospirillum lipoferum* CRT1. Plant Soil 291:109–118

Evans A (1991) Influence of low molecular weight organic acids on zinc distribution within micronutrient pools and zinc uptake by wheat. J Plant Nutr 14:1307–1318

FAO (2002) World agriculture: towards 2015/2030. Summary report. Food and Agriculture Organization of the United Nations, Rome

Fasim F, Ahmed N, Parsons R, Gadd GM (2002) Solubilization of zinc salts by bacterium isolated by the air environment of tannery. FEMS Microb Lett 213:1–6

Feng G, Zhang F, Li X, Tian C, Tang C, Rengel Z (2002) Improved tolerance of maize plants to salt stress by arbuscular mycorrhiza is related to higher accumulation of soluble sugars in roots. Mycorrhiza 12:185–190

Ferguson MW, Maxwell JA, Vincent TS, da Silva J, Olson JC (2001) Comparison of the *exoS* gene and protein expression in soil and clinical isolates of *Pseudomonas aeruginosa*. Infect Immun 69:2198–2210

Fernández V, Ebert G, Winkelmann G (2005) The use of microbial siderophores for foliar iron application studies. Plant Soil 272:245–252

Fernández L, Zalba P, Gómez M, Sagardoy M (2007) Phosphate-solubilization activity of bacterial strains in soil and their effect on soybean growth under greenhouse conditions. Biol Fertil Soils 43:805–809

Fischer G, Shah M, van Velthuizen H (2002) Climate change and agricultural vulnerability. International Institute for Applied Systems Analysis, Vienna

Fox TC, Guerinot ML (1998) Molecular biology of cation transport in plants. Annu Rev Plant Physiol Plant Mol Biol 49:669–696

Fuhrmann J, Wollum AG (1989) Nodulation competition among *Bradyrhizobium japonicum* strains as influenced by rhizosphere bacteria and iron availability. Biol Fertil Soils 7:108–112

Furina EK, Bonartseva GA (2007) The effect of combined and separate inoculation of alfalfa plants with *Azospirillum lipoferum* and *Sinorhizobium meliloti* on denitrification and nitrogen-fixing activities. Appl Biochem Microbiol 43:286–291

Galloway JN, Schlesinger WH, Levy IH, Michaels A, Schnoor JL (1995) Nitrogen fixation: anthropogenic enhancement – environmental response. Glob Biochem Cycles 9:235–252

Gamalero E, Martinotti MG, Trotta A, Lemanceau P, Berta G (2002) Morphogenetic modifications induced by *Pseudomonas fluorescens* A6RI and *Glomus mosseae* BEG12 in the root system of tomato differ according to plant growth conditions. New Phytol 155:293–301

Gamalero E, Trotta A, Massa N, Copetta A, Martinotti MG, Berta G (2004) Impact of two fluorescent pseudomonads and an arbuscular mycorrhizal fungus on tomato plant growth, root architecture and P acquisition. Mycorrhiza 14:185–192

Garrett SD (1981) Introduction. In: Asher MJC, Shipton PJ (eds) Biology and control of take-all. Academic, New York, pp 1–14

George E, Römheld V, Marschner H (1994) Contribution of mycorrhizal fungi to micronutrient uptake by plants. In: Manthey JA, Crowley DE, Luster DG (eds) Biochemistry of metal micronutrients in the rhizosphere. CRC Press, Boca Raton, pp 93–109

Gildon A, Tinker PB (1983) Interactions of vesicular-arbuscular mycorrhizal infection and heavy metals in plants. I. The effects of heavy metals on the developmnet of vesicular-arbuscular mycorrhizas. New Phytol 95:247–261

González-Chávez MC, Carrillo-González R, Wright SF, Nichols KA (2004) The role of glomalin, a protein produced by arbuscular mycorrhizal fungi, in sequestering potentially toxic elements. Environ Pollut 130:317–323

González-Guerrero M, Melville L-H, Ferrol N, Azcón-Aguilar C, Peterson R-L (2008) Ultrastructural localization of heavy metals in the extraradical mycelium and spores of the arbuscular mycorrhizal fungus *Glomus intraradices*. Can J Microbiol 54:103–110

González-López J, Rodelas B, Pozo C, Salmerón-López V, Martínez-Toledo MV, Salmerón V (2005) Liberation of amino acids by heterotrophic nitrogen fixing bacteria. Amino Acids 28:363–367

Govindarajan M, Balandreau J, Kwon S, Weon H, Cunthipuram L (2008) Effects of the inoculation of *Burkholderia vietnamensis* and related endophytic diazotrophic bacteria on grain yield of rice. Microb Ecol 55:21–37

Graham RD, Webb MJ (1991) Micronutrients and disease resistance and tolerance in plants. In: Mortvedt JJ, Cox FR, Shuman LM, Welch RM (eds) Micronutrients in agriculture, 2nd edn. Soil Science Society of America, Madison, pp 329–370

Graham RD, Welch RM (1996) Breeding for staple-food crops with high micronutrient density: working papers on agricultural strategies for micronutrients, vol 3. International Food Policy Institute, Washington

Grifoni A, Bazzicalupo M, Di Serio C, Farcelli S, Fani R (1995) Identification of *Azospirillum* strains by restriction fragment length polymorphism of 16S rDNA and the histidine operon. FEMS Microbiol Lett 127:85–91

Gryndler M, Larsen J, Hršelová H, Řezáčová V, Gryndlerová H, Kubát J (2006) Organic and mineral fertilization, respectively, increase and decrease the development of external mycelium of arbuscular mycorrhizal fungi in a long-term field experiment. Mycorrhiza 16:159–166

Gyaneshwar P, Naresh-Kumar G, Parekh LJ, Poole PS (2002) Role of soil microorganisms in improving P nutrition of plants. Plant Soil 245:83–93

Haas D, Défago G (2005) Biological control of soil-borne pathogens by fluorescent pseudomonads. Nat Rev Microbiol 3:307–319

Haas H, Eisendle M, Turgeon BG (2008) Siderophores in fungal physiology and virulence. Annu Rev Phytopathol 46:149–187

Hafeez FY, Yasmin S, Ariani D, Rahman M, Zafar Y, Malik KA (2006) Plant growth-promoting bacteria as biofertilizer. Agron Sustain Dev 26:143–150

Hamdali H, Bouizgarne B, Hafidi M, Lebrihi A, Virolle MJ, Ouhdouch Y (2008) Screening for rock phosphate solubilizing Actinomycetes from Moroccan phosphate mines. Appl Soil Ecol 38:12–19

Hamilton MA, Westermann DT, James DW (1993) Factors affecting zinc uptake in cropping systems. Soil Sci Soc Am J 57:1310–1315

Han SO, New PB (1998) Variation in nitrogen fixing ability among natural isolates of *Azospirillum*. Microb Ecol 36:193–201

Harman GE, Björkman T (1998) Potential and existing uses of *Trichoderma* and *Gliocladium* for plant disease control and plant growth enhancement. In: Harman GE, Kubicek CP (eds) Trichoderma and gliocladium, vol 2. Taylor & Francis, London, pp 229–265

Harman GE, Howell CR, Viterbo A, Chet I, Lorito M (2004) *Trichoderma* species: opportunistic, avirulent plant symbionts. Nat Rev Microbiol 2:43–56

Haselwandter K, Dobernigg B, Beck W, Jung G, Cansier A, Winkelmann G (1992) Isolation and identification of hydroxamate siderophores of ericoid mycorrhizal fungi. Biometals 5:51–56

He ZL, Yang XE, Stoffella PJ (2005) Trace elements in agroecosystems and impacts on the environment. J Trace Elem Med Biol 19:125–140

Heggo A, Angle JS, Chaney RL (1990) Effects of vesicular-arbuscular mycorrhizal fungi on heavy metal uptake by soybeans. Soil Biol Biochem 22:865–869

Hildebrandt U, Kaldorf M, Bothe H (1999) The zinc violet and its colonization by arbuscular mycorrhizal fungi. J Plant Physiol 154:709–711

Hildebrandt U, Regvar M, Bothe H (2007) Arbuscular mycorrhiza and heavy metal tolerance. Phytochemistry 68:139–146

Hodge A (2001) Arbuscular mycorrhizal fungi influence decomposition of, but not plant nutrient capture from, glycine patches in soil. New Phytol 151:725–734

Hodge A (2003) Plant nitrogen capture from organic matter as affected by spatial dispersion, interspecific competition and mycorrhizal colonization. New Phytol 157:303–314

Hodge A, Campbell CD, Fitter AH (2001) An arbuscular mycorrhizal fungus accelerates decomposition and acquires nitrogen directly from organic material. Nature 413:297–299

Hoefsloot G, Termorshuizen AJ, Watt DA, Cramer MD (2005) Biological nitrogen fixation is not a major contributor to the nitrogen demand of a commercially grown South African sugarcane cultivar. Plant Soil 277:85–96

Höfte M, Seong KY, Jurkevitch E, Verstraete W (1991) Pyoverdin production by the plant growth beneficial *Pseudomonas* strain 7NSK$_2$: ecological significance in soil. Plant Soil 130:249–257

Höfte M, Vande WM, Verstraete W (1994) Role of siderophores in plant growth promotion and plant protection by fluorescent pseudomonads. In: Manthey JA, Crowley DE, Luster DG (eds) Biochemistry of metal micronutrients in the rhizosphere. CRC Press, Boca Raton, pp 81–92

Hördt W, Römheld V, Winkelmann G (2000) Fusarinines and dimerum acid, mono- and dihydrate siderophores from *Penicillium chrysogenum*, improve iron utilisation by strategy I and strategy II plants. Biometals 13:37–46

Hotz C, Brown KH (2004) Assessment of the risk of zinc deficiency in populations and options for its control. Food Nutr Bull 25:94–204

Howard JB, Rees DC (1996) Structural basis of biological nitrogen fixation. Chem Rev 96: 2965–2982

Huber DM, McCay-Buis TS (1993) A multiple component analysis of the take-all disease of cereals. Plant Dis 77:437–447

Huber DM, Wilhelm NS (1988) The role of manganese in resistance to plant diseases. In: Graham RD, Hannam RJ, Uren NC (eds) Manganese in soils and plants. Kluwer, Dordrecht, pp 155–173

Huber DM, El-Nasshar H, Moore LW, Mathre DE, Wagner JE (1989) Interaction between a peat carrier and bacterial seed treatments evaluated for biological control of the take-all diseases of wheat (*Triticum aestivum* L.). Biol Fertil Soils 8:166–171

Hughes NP, Williams RJP (1988) An introduction to manganese biological chemistry. In: Graham RD, Hannam RJ, Uren NC (eds) Manganese in soils and plants. Kluwer, Dordrecht, pp 7–19

IEA (International Energy Agency) (2004) Biofuels for transport: an international perspective. OECD, Paris

Igual JM, Rodríguez-Barrueco C (2007) Fertilizers, food and environment. In: Vélazquez E, Rodríguez-Barrueco C (eds) First international meeting on microbial phosphate solubilization, vol 102, Developments in Plant and Soil Sciences. Springer, Dordrecht, pp 199–202

Illmer P, Schinner F (1995) Solubilization of inorganic calcium phosphates-solubilization mechanisms. Soil Biol Biochem 27:257–263

Illmer P, Barbato A, Schinner F (1995) Solubilization of hardly-soluble $AlPO_4$ with P-solubilizig microorganisms. Soil Biol Biochem 27:265–270

Ingham E (1998) Fungi, glue and risky soils. BioCycle 39:86–87

Jalal MA, Love SK, van der Helm D (1986) Siderophore mediated iron(III) uptake in *Gliocladium virens*. 1. Properties of cis-fusarinine, trans-fusarinine, dimerum acid, and their ferric complexes. J Inorg Biochem 28:417–430

Jalal MA, Love SK, van der Helm D (1987) Siderophore mediated iron(III) uptake in *Gliocladium virens*. 2. Role of ferric mono- and dihydroxamates as iron transport agents. J Inorg Biochem 29:259–267

Jansa J, Mozafar A, Frossard E (2003) Long-distance transport of P and Zn through the hyphae of an arbuscular mycorrhizal fungus in symbiosis with maize. Agronomie 23:481–488

Javaid A (2009) Arbuscular mycorrhizal mediated nutrition in plants. J Plant Nutr 32:1595–1618
Jenkinson DA (2001) The impact of humans on the nitrogen cycle, with focus on temperate arable agriculture. Plant Soil 228:3–15
Johnson GV, Lopez A, La Valle FN (2002) Reduction and transport of Fe from siderophores: reduction of siderophores and chelates and uptake and transport of iron by cucumber seedlings. Plant Soil 241:27–33
Johnston AWB (2004) Mechanisms and regulation of iron uptake in the Rhizobia. In: Crosa JH, Mey AR, Payne SM (eds) Iron transport in bacteria. ASM Press, Washington, DC, pp 469–488
Joner EJ (2000) The effect of long-term fertilization with organic or inorganic fertilizer on mycorrhiza mediated phosphorus uptake in subterranean clover. Biol Fertil Soils 32:435–440
Joner EJ, Briones R, Leyval C (2000) Metal-binding capacity of arbuscular mycorrhizal mycelium. Plant Soil 226:227–234
Jones DL, Darrah PR (1994) Role of root derived organic acids in the mobilization of nutrients from the rhizosphere. Plant Soil 166:247–257
Jones KM, Kobayashi H, Davies BW, Taga ME, Walker GC (2007) How rhizobial symbionts invade plants: the *Sinorhizobium-Medicago* model. Nat Rev Microbiol 5:619–633
Jurkevitch E, Hadar Y, Chen Y, Libman J, Shanzer A (1992) Iron uptake and molecular recognition in *Pseudomonas putida*: receptor mapping with ferrichrome and its biomimetic analogs. J Bacteriol 174:78–83
Kahinidi JHP, Woomer P, George T, de Souza MFM, Karanja NK, Giller KE (1997) Agricultural intensification, soil biodiversity and ecosystem function in the tropics: the role of nitrogen-fixing bacteria. Appl Soil Ecol 6:55–76
Kaldorf M, Kuhn AJ, Schröder WH, Hildebrandt U, Bothe H (1999) Selective element deposits in maize colonized by a heavy metal tolerance conferring arbuscular mycorrhizal fungus. J Plant Physiol 154:718–728
Kaya C, Higgs D, Kirnak H, Tas I (2003) Mycorrhizal colonization improves fruit yield and water use efficiency in water melon (*Citrullus lanatus* Thunb) grown under well-watered and water-stressed conditions. Plant Soil 253:287–292
Keel C, Schnider U, Maurhofer M, Voisard C, Laville J, Burger U, Wirthner P, Haas D, Défago G (1992) Suppression of root diseases by *Pseudomonas fluorescens* CHAO10: importance of bacterial secondary metabolite 2,4-diacetylphloroglucinol. Mol Plant Microbe Interact 5:4–13
Kennedy AC (1999) Microbial diversity in agroecosystem quality. In: Collins WW, Qualset CO (eds) Biodiversity in agroecosystems. CRC Press, Boca Raton, pp 1–17
Kennedy I, Choudhury A, Kecskés M (2004) Non-symbiotic bacterial diazotrophs in crop-farming systems: can their potential for plant growth promotion be better exploited? Soil Biol Biochem 36:1229–1244
Khan MS, Zaidi A, Wani PA (2007) Role of phosphate-solubilizing microorganisms in sustainable agriculture – A review. Agron Sustain Dev 27:29–43
Khush GS (1999) Green revolution: preparing for the 21st century. Genome 42:646–655
Kim J, Rees DC (1992) Crystallographic structure and functional implications of the nitrogenase molybdenum-iron protein from *Azotobacter vinelandii*. Nature 360:553–560
Kim KY, Jordan D, McDonald GA (1998) *Enterobacter agglomerans*, phosphate solubilizing bacteria, and microbial activity in soil: effect of carbon sources. Soil Biol Biochem 30: 995–1003
Kloepper JW (1993) Plant growth-promoting rhizobacteria as biological control agents. In: Jr Metting FB (ed) Soil microbial ecology: applications in agricultural and environmental management. Marcel Dekker, New York, pp 255–274
Kloepper JW, Schroth MN (1981a) Plant growth-promoting rhizobacteria and plant growth under gnotobiotic conditions. Phytopathology 71:642–644
Kloepper JW, Schroth MN (1981b) Relationship of in vitro antibiosis of plant growth-promoting rhizobacteria to plant growth and the displacement of root microflora. Phytopathology 71:1020–1024
Kloepper JW, Leong J, Teintze M, Schroth MN (1980) Enhanced plant growth by siderophores produced by plant growth-promoting rhizobacteria. Nature 286:883–884

Koide RT (1991) Nutrient supply, nutrient demand and plant response to mycorrhizae infection. New Phytol 117:365–386

Kothari SK, Marschner H, Römheld V (1990) Direct and indirect effects of VA mycorrhizal fungi and rhizosphere microorganisms on acquisition of mineral nutrients by maize (*Zea mays* L.) in a calcareous soil. New Phytol 116:637–646

Kothari SK, Marschner H, Römheld V (1991a) Effect of a vesicular-arbuscular mycorrhizal fungus and rhizosphere micro-organisms on manganese reduction in the rhizosphere and manganese concentrations in maize (*Zea mays* L.). New Phytol 117:649–655

Kothari SK, Marschner H, Römheld V (1991b) Contribution of the VA mycorrhizal hyphae in acquisition of phosphorus and zinc by maize grown in a calcareous soil. Plant Soil 131:177–185

Küçük Ç, Kivanç M, Kinaci E, Kinaci G (2008) Determination of the growth and solubilisation capabilities of *Trichoderma harzianum* T1. Biologia 63:167–170

Kuiper I, Kravchenko LV, Bloemberg GV, Lugtenberg BJJ (2002) *Pseudomonas putida* strain PCL1444, selected for efficient root colonization and naphtalene degradation, effectively utilizes root exudate components. Mol Plant Microbe Interact 15:734–741

Kumar V, Singh K (2001) Enriching vermicompost by nitrogen fixing and phosphate solubilizing bacteria. Bioresour Technol 76:173–175

Kumar V, Behl KR, Narula N (2001) Establishment of phosphate-solubilizing strains of *Azotobacter chroococcum* in the rhizosphere and their effect on wheat cultivars under green house conditions. Microbiol Res 156:87–93

Kurita K, Sannan T, Iwakura Y (1979) Studies on chitin. VI. Binding of metal cations. J Appl Polym Sci 23:511–515

Lambert DH, Baker DE, Cole HJR (1979) The role of mycorrhizae in the interactions of phosphorus with zinc, copper, and other elements. Soil Sci Soc Am J 43:976–980

Leach LH, Morris JC, Lewis TA (2007) The role of the siderophore pyridine-2,6-bis (thiocarboxylic acid) (PDTC) in zinc utilization by *Pseudomonas putida* DSM 3601. Biometals 20:717–726

Lee YJ, George E (2005) Contribution of mycorrhizal hyphae to the uptake of metal cations by cucumber plants at two levels of phosphorus supply. Plant Soil 278:361–370

Lemanceau P, Alabouvette C, Meyer JM (1985) Production of fusarinine and iron assimilation by pathogenic and non-pathogenic *Fusarium*. In: Swinburne TR (ed) Iron, siderophores and plant diseases. Plenum Press, London, pp 251–259

Leong J (1986) Siderophores: their biochemistry and possible role in the biocontrol of plant pathogens. Annu Rev Phytopathol 24:187–209

Lian B, Wang B, Pan M, Liu C, Teng HH (2008) Microbial release of potassium from K-bearing minerals by thermophilic fungus *Aspergillus fumigatus*. Geochim Cosmochim Acta 72:87–98

Lin Q, Zhao X, Zhao Z, Li B (2002) Rock phosphate solubilization mechanisms of one fungus and one bacterium. Agric Sci China 1:1023–1028

Lin T, Huang H, Shen F, Young C (2006) The protons of gluconic acid are the major factor responsible for the dissolution of tricalcium phosphate by *Burkholderia cepacia* CC-A174. Bioresour Technol 97:957–960

Linderman RG (1988) Mycorrhizal interactions with the rhizosphere microflora: the mycorrhizosphere effect. Phytopathology 78:366–371

Liu A, Hamel C, Hamilton RI, Ma BL, Smith DL (2000) Acquisition of Cu, Zn, Mn and Fe by mycorrhizal maize (*Zea mays* L.) growth in soil at different P and micronutrient levels. Mycorrhiza 9:331–336

Liu A, Hamel C, Elmi A, Costa C, Ma B, Smith DL (2002) Concentrations of K, Ca and Mg in maize colonized by arbuscular mycorrhizal fungi under field conditions. Can J Soil Sci 82: 271–278

Loneragan JF, Grove TS, Robson AD, Snowball K (1979) Phosphorus toxicity as a factor in zinc-phosphorus interactions in plants. Soil Sci Soc Am J 43:966–972

Lucy M, Reed E, Glick BR (2004) Applications of free living plant growth-promoting rhizobacteria. Antonie Leeuwenhoek 86:1–25

Lynch JM, Whipps JM (1990) Substrate flow in the Rhizosphere. Plant Soil 129:1–10

Mahendran PP, Chandramani P (1998) NPK uptake, yield and starch content of potato cv. Kufri Jyotis influenced by certain bio-fertilizers. J Indian Potato Assoc 25:50–52

Mahendran PP, Kumar N (1998) Effect of biofertilizers on tuber yield and certain quality parameters of potato CV. Kufri Jyoti. S Indian Hortic 46:97–98

Mahendran PP, Kumar N, Sraswathy S (1996) Studies of the effect of bio-fertilizers on potato (*Solnum tuberosum* L.). S Indian Hortic 44:79–82

Malik KA, Bilal R, Mehnaz S, Rasul G, Mirza MS, Ali S (1997) Association of nitrogen-fixing, plant-growth-promoting rhizobacteria (PGPR) with kallar grass and rice. Plant Soil 194:37–44

Marschner H (1995) Mineral nutrition of higher plants, 2nd edn. Academic, London

Marschner P, Ascher JS, Graham RD (1991) Effect of manganese-reducing rhizosphere bacteria on the growth of *Gaeumannomyces graminis* var. *tritici* and on manganese uptake by wheat (*Triticum aestivum* L.). Biol Fertil Soils 12:33–38

Martin C, Stutz J (2004) Interactive effects of temperature and arbuscular mycorrhizal fungi on growth, P uptake and root respiration of *Capsicum annuum* L. Mycorrhiza 14:241–244

Mayak S, Tirosh T, Glick BR (2004) Plant growth-promoting bacteria confer resistance in tomato plants to salt stress. Plant Physiol Biochem 42:565–573

Medina A, Probanza A, Gutierrez Mañero FJ, Azcón R (2003) Interactions of arbuscular-mycorrhizal fungi and *Bacillus* strains and their effects on plant growth, microbial rhizosphere activity (thymidine and leucine incorporation) and fungal biomass (ergosterol and chitin). Appl Soil Ecol 22:15–28

Megha YJ, Alagawadi AR, Krishnaraj PU (2007) Multiple beneficial functions of fluorescent pseudomonads of Western Ghats of Uttara Kannada District. Karnataka. J Agric Sci 20: 305–309

Mehravaran H, Mozafar A, Frossard E (2000) Uptake and partitioning of P-32 and Zn-65 by white clover as affected by eleven isolates of mycorrhizal fungi. J Plant Nutr 23:1385–1395

Mena-Violante HG, Ocampo-Jimenez O, Dendooven L, Martinez-Soto G, Gonzalez-Castaneda J, Davies FT, Olalde-Portugal V (2006) Arbuscular mycorrhizal fungi enhance fruit growth and quality of chile ancho (*Capsicum annuum* L. cv San Luis) plants exposed to drought. Mycorrhiza 16:261–267

Menge JA, Jarrell WM, Labanauskas CK, Ojala JC, Huszar C, Johnson ELV, Sibert D (1982) Predicting mycorrhizal dependency of troyer citrange on *Glomus fasciculatus* in California citrus soils and nursery mixes. Soil Sci Soc Am J 46:762–768

Mengel K (2008) Nutrient potentials. In: Chesworth W (ed) Encyclopedia of soil science. Springer, Dordrecht, pp 494–500

Mercado-Blanco J, Bakker PAHM (2007) Interactions between plants and beneficial *Pseudomonas* spp.: exploiting bacterial traits for crop protection. Antonie Leeuwenhoek 92:367–389

Mercado-Blanco J, van der Drift KMGM, Olsson PE, Thomas-Oates JE, van Loon LC, Bakker PAHM (2001) Analysis of the pmsCEAB gene cluster involved in biosynthesis of salicylic acid and the siderophore pseudomonine in the biocontrol strain *Pseudomonas fluorescens* WCS374. J Bacteriol 183:1909–1920

Meunchang S, Panichsakpatana S, Weaver RW (2006) Tomato growth in soil amended with sugar mill by-products compost. Plant Soil 280:171–176

Miller GW, Hasegawa S, Shigematsu A, Welkie GW (1994) Mechanisms of iron uptake from rhodotorulate-iron by tomato. In: Manthey JA, Crowley DE, Luster DG (eds) Biochemistry of metal micronutrients in the rhizosphere. CRC Press, Boca Raton, pp 267–284

Mrkovački N, Milic Y (2001) Use of *Azotobacter chroococcum* as potentially useful in agricultural application. Ann Microbiol 51:145–158

Mrkovački N, Mezei S, Verešbaranji I, Popović M, Sarić Kovačev L (1997) Associations of sugar beet and nitrogen-fixing bacteria *in vitro*. Biol Plant 39:419–425

Muthukumarasamy R, Govindarajan M, Vadivelu M, Revathi G (2006) N-fertilizer saving by the inoculation of *Gluconacetobacter diazotrophicus* and *Herbaspirillum* sp. in micropropagated sugarcane plants. Microbiol Res 161:238–245

Nagesh M, Parvatha-Reddy P (2004) Biochemical changes in *Glomus fasciculatum* colonized roots of *Lycopersicon esculentum* in presence of *Meloidogyne incognita*. Indian J Exp Biol 42:721–727

Narula N, Kumar V, Singh B, Bhatia R, Lakshminarayana K (2005) Impact of biofertilizers on grain yield in spring wheat under varying fertility conditions and wheat-cotton rotation. Arch Agr Soil Sci 51:79–89

Nautiyal CS, Bhadauria S, Kumar P, Lal H, Mondal R, Verma D (2000) Stress induced phosphate solubilization in bacteria isolated from alkaline soils. FEMS Microbiol Lett 182:291–296

Neilands JB (1995) Siderophores: structure and function of microbial iron transport compounds. J Biol Chem 270:26723–26726

Ness R, Vlek P (2000) Mechanism of calcium and phosphate release from hydroxy- apatite by mycorrhizal hyphae. Soil Sci Soc Am J 64:949–955

Nogueira MA, Magalhães GC, Cardoso EJBN (2004) Manganese toxicity in mycorrhizal and phosphorus-fertilized soybean plants. J Plant Nutr 27:141–156

Nogueira MA, Nehls U, Hampp R, Poralla K, Cardoso EJBN (2007) Mycorrhiza and soil bacteria influence extractable iron and manganese in soil and uptake by soybean. Plant Soil 298: 273–284

Nuruzzaman M, Ashrafuzzaman M, Zahurul IM, Rafiqul IM (2003) Field efficiency of biofertilizers on the growth of okra (*Abelmoschus esculentus* [(L.)Moench]). J Plant Nutr Soil Sci 166:764–770

OECD-FAO (2007) Agricultural outlook 2007–2016. OECD Publishing, Paris

Öğüt M, Er F (2006) Micronutrient composition of field-grown dry bean and wheat inoculated with *Azospirillum* and *Trichoderma*. J Plant Nutr Soil Sci 169:699–703

Öğüt M, Akdağ C, Düzdemir O, Ali Sakin M (2005) Single and double inoculation with *Azospirillum/Trichoderma*: the effects on dry bean and wheat. Biol Fertil Soils 41:262–272

Okon Y, Itzigsohn R (1995) The development of *Azospirillum* as a commercial inoculant for improving crop yields. Biotechnol Adv 13:415–424

Oliveira ALM, Urquiaga S, Döbereiner J, Baldani JI (2002) The effect of inoculating endophytic N_2-fixing bacteria on micropropagated sugarcane plants. Plant Soil 242:205–215

Olsson P, van Aarle I, Allaway W, Ashford A, Rouhier H (2002) Phosphorus effects on metabolic processes in monoxenic arbuscular mycorrhiza cultures. Plant Physiol 130:1162–1171

Orhan E, Esitken A, Ercisli S, Turan M, Sahin F (2006) Effects of plant growth promoting rhizobacteria (PGPR) on yield, growth and nutrient contents in organically growing raspberry. Sci Hortic 111:38–43

Ortas I, Ortakei D, Kaya Z, Çinar A, Önelge N (2002) Mycorrhizal dependency of sour orange in relation to phosphorus and zinc nutrition. J Plant Nutr 26:1263–1279

Ozturk A, Caglar O, Sahin F (2003) Yield response of wheat and barley to inoculation of plant growth promoting bacteria at various levels of nitrogen fertilization. J Plant Nutr Soil Sci 166:262–266

Pacovsky RS (1986) Micronutrient uptake and distribution in mycorrhizal or phosphorus-fertilized soybeans. Plant Soil 95:379–388

Pacovsky RS (1988) Influence of inoculation with *Azospirillum brasilense* and *Glomus fasciculatum* on sorghum nutrition. Plant Soil 110:283–287

Pacovsky RS (1990) Development and growth effects in the *Sorghum-Azospirillum* association. J Appl Bacteriol 68:555–563

Pacovsky RS, Fuller G (1988) Mineral and lipid composition of *Glycine-Glomus-Bradyrhizobium* symbioses. Physiol Plant 72:733–746

Pandey A, Trivedi P, Kumar B, Palni LMS (2006) Characterization of a phosphate solubilizing and antagonistic strain of *Pseudomonas putida* (b0) isolated from a sub-alpine location in the Indian Central Himalaya. Curr Microbiol 53:102–107

Pawlowska TE, Charvat I (2004) Heavy-metal stress developmental patterns of arbuscular mycorrhizal fungi. Appl Environ Microbiol 70:6643–6649

Pedraza R (2008) Recent advances in nitrogen-fixing acetic acid bacteria. Int J Food Microbiol 125:25–35

Pepper D (2008) The toxic consequences of the Green Revolution. US News & World Report, July 7 (http://www.usnews.com/articles/news/world/2008/07/07/the-toxic-consequences-of-the-green-revolution.html)

Peters S, Habte M (2001) Optimizing solution P concentration in a peat-based medium for producing mycorrhizal seedlings in containers. Arid Land Res Manage 15:359–371

Piotrowski JS, Denich T, Klironomos JN, Graham JM, Rillig MC (2004) The effects of arbuscular mycorrhizas on soil aggregation depend on the interaction between plant and fungal species. New Phytol 164:365–373

Plante AF (2007) Soil biochemical cycling of inorganic nutrients and metals. In: Paul EA (ed) Soil microbiology, ecology, and biochemistry, 3rd edn. Academic, Oxford, pp 389–432

Porcel R, Ruiz-Lozano JM (2004) Arbuscular mycorrhizal influence on leaf water potential, solute accumulation, and oxidative stress in soybean plants subjected to drought stress. J Exp Bot 55:1743–1750

Posta K, Marschner H, Römheld V (1994) Manganese reduction in the rhizosphere of mycorrhizal and nonmycorrhizal maize. Mycorrhiza 5:119–124

Poulton JL, Bryla D, Koide RT, Stephenson AG (2002) Mycorrhizal infection and high soil phosphorus improve vegetative growth and the female and male functions in tomato. New Phytol 154:255–264

Powlson D, Hirsch P, Brookes P (2001) The role of soil microorganisms in soil organic matter conservation in the tropics. Nut Cycl Agroecosyst 61:41–51

Preston GM (2004) Plant perceptions of plant growth-promoting *Pseudomonas*. Phil Trans R Soc Lond B 359:907–918

Raju PS, Clark RB, Ellis JR, Maranville JW (1990) Effects of species of VA-mycorrhizal fungi on growth and mineral uptake of sorghum at different temperatures. Plant Soil 121:165–170

Ratti N, Kumar S, Verma HN, Gautam SP (2001) Improvement in bioavailability of tricalcium phosphate to *Cymbopogon martinii* var. *motia* by rhizobacteria, AMF and *Azospirillum* inoculation. Microbiol Res 156:145–149

Renshaw JC, Robson GD, Trinci APJ, Wiebe MG, Livens FR, Collison D, Taylor RJ (2002) Fungal siderophores: structures, functions and applications. Mycol Res 106:1123–1142

Reyes I, Bernier L, Simard RR, Antoun H (1999a) Effect of nitrogen source on the solubilization of different inorganic phosphates by an isolate of *Penicillium rugulosum* and two UV-induced mutants. FEMS Microbiol Ecol 28:281–290

Reyes I, Bernier L, Simard RR, Tanguay P, Antoun H (1999b) Characteristics of phosphate solubilization by an isolate of a tropical *Penicillium rugulosum* and two UV-induced mutants. FEMS Microbiol Ecol 28:291–295

Reyes I, Baziramakenda R, Bernier L, Antoun H (2001) Solubilization of phosphate rocks and minerals by wild-type strain and two UV-induced mutants of *Penicillium rugulosum*. Soil Biol Biochem 33:1741–1747

Reyes I, Valery A, Valduz Z (2006) Phosphate-solubilizing microorganisms isolated from rhizospheric and bulk soils of colonizer plants at an abandoned rock phosphate mine. Plant Soil 287:69–75

Ribaudo CM, Rondanini DP, Curá JA, Fraschina AA (2001) Response of *Zea mays* to the inoculation with *Azospirillum* on nitrogen metabolism under greenhouse conditions. Biol Plant 44:631–634

Rillig MC, Mummey DL (2006) Mycorrhizas and soil structure. New Phytol 171:41–53

Rodrigues EP, Rodrigues LS, Martinez de Oliveira AL, Baldani VLD, dos Santos TKR, Urquiaga S, Reis VM (2008) *Azospirillum amazonense* inoculation: effects on growth, yield and N_2 fixation of rice (*Oryza sativa* L.). Plant Soil 302:249–261

Rodríguez H, Fraga R (1999) Phosphate solubilizing bacteria and their role in plant growth promotion. Biotechnol Adv 17:319–339

Rubio LM, Ludden PW (2008) Biosynthesis of the iron-molybdenum cofactor of nitrogenase. Annu Rev Microbiol 62:93–111

Rudresh DL, Shivaprakash MK, Prasad RD (2005a) Effect of combined application of *Rhizobium*, phosphate solubilizing bacterium and *Trichoderma* spp. on growth, nutrient uptake and yield of chickpea (*Cicer aritenium* L.). Appl Soil Ecol 28:139–146
Rudresh DL, Shivaprakash MK, Prasad RD (2005b) Tricalcium phosphate solubilizing abilities of *Trichoderma* spp. in relation to P uptake and growth and yield parameters of chickpea (*Cicer arietinum* L.). Can J Microbiol 51:217–222
Ruiz-Lozano JM, Azon R, Gomez M (1995) Effects of arbuscular-mycorrhizal *Glomus* species on drought tolerance: physiological and nutritional plant responses. Appl Environ Microbiol 61:456–460
Ruiz-Lozano JM, Azcon R, Gomez M (1996) Alleviation of salt stress by arbuscular mycorrhizal *Glomus* species in *Lactuca sativa* plants. Physiol Plant 98:767–772
Saad MS, Sabuddin ASA, Yunus AG, Shamsuddin ZH (1999) Effects of *Azospirillum* inoculation on sweet potato on sandy tin-tailing soil. Commu Soil Sci Plant Anal 30:1583–1592
Safir GR, Boyer JS, Gerdemann JW (1971) Mycorrhizal enhancement of water transport in soybean. Science 172:581–583
Şahin F, Çakmakçi R, Kantar F (2004) Sugar beet and barley yields in relation to inoculation with N_2-fixing and phosphate solubilizing bacteria. Plant Soil 265:123–129
Sanchez CA (2007) Phosphorus. In: Barker AV, Pilbeam DJ (eds) Handbook of plant nutrition. Taylor & Francis, Boca Raton, pp 51–91
Saubidet M, Fatta N, Barneixet A (2002) The effect of inoculation with *Azospirillum brasilense* on growth and nitrogen utilization by wheat plants. Plant Soil 245:215–222
Schnepf R (2006) European union biofuels policy and agriculture: an overview. The Library of Congres, Washington, DC
Schroeder MS, Janos DP (2004) Phosphorus and intraspecific density alter plant responses to arbuscular mycorrhizas. Plant Soil 264:335–348
Seres A, Bakonyi G, Posta K (2006) Zn uptake by maize under the influence of AM-fungi and Collembola *Folsomia candida*. Ecol Res 21:692–697
Shahab S, Ahmed N (2008) Effect of various parameters on the efficiency of zinc phosphate solubilization by indigenous bacterial isolates. Afr J Biotechnol 7:1543–1549
Shaharoona B, Arshad M, Zahir AZ, Khalid A (2006) Performance of *Pseudomonas* spp. containing ACC-deaminase for improving growth and yield of maize (*Zea mays* L.) in the presence of nitrogenous fertilizer. Soil Biol Biochem 38:2971–2975
Shaharoona B, Naveed M, Arshad M, Zahir ZA (2008) Fertilizer-dependent efficiency of Pseudomonads for improving growth, yield, and nutrient use efficiency of wheat (*Triticum aestivum* L.). Appl Microbiol Biotechnol 79:147–155
Shanahan P, O'Sullivan DJ, Simpson P, Glennon JD, O'Gara F (1992) Isolation of 2,4-diacetylphloroglucinol from a fluorescent pseudomonad and investigation of physiological parameters influencing its production. Appl Environ Microbiol 58:353–358
Shantharam S, Mattoo AK (1997) Enhancing biological nitrogen fixation: an appraisal of current and alternative technologies for N input into plants. Plant Soil 194:205–216
Sharma AK, Srivastiva PC, Johri BN (1994) Contribution of the VA mycorrhiza to zinc uptake in plants. In: Manthey JA, Crowley DE, Luster DG (eds) The biochemistry of micronutrients in the rhizosphere. CRC Press, Boca Raton, pp 111–123
Shekhar NC, Bhadauria S, Kumar P, Lal H, Mondal R, Verma D (2000) Stress induced phosphate solubilization in bacteria isolated from alkaline soils. FEMS Microbiol Lett 182:291–296
Shenker M, Oliver I, Helmann M, Hadar Y (1992) Utilization by tomatoes of iron mediated by a siderophore produced by *Rhizopus arrhizus*. J Plant Nutr 15:2173–2182
Shenker M, Hadar Y, Chen Y (1996) Stability constants of the fungal siderophore rhizoferrin with various microelements and calcium. Soil Sci Soc Am J 60:1140–1144
Shiva V (1991) The Green Revolution in the Punjab. Ecologist 21:57–60
Shokri S, Maadi B (2009) Effects of arbuscular mycorrhizal fungus on the mineral nutrition and yield of *Trifolium alexandrinum* plants under salinity stress. J Agron 8:79–83

Siddiqui IA, Shaukat SS (2004) *Trichoderma harzianum* enchances the production of nematicidal compounds in vitro and improves biocontrol of *Meloidogyne javanica by Pseudomonas fluoresces* in tomato. Lett Appl Microbiol 38:169–176

Singh S, Kapoor KK (1998) Effects of inoculation of phosphate-solubilizing microorganisms and an arbuscular mycorrhizal fungus on mungbean grown under natural soil conditions. Mycorrhiza 7:249–253

Smith MJ, Shoolery JN, Schwyn B, Holden I, Neilands JB (1985) Rhizobactin, a structurally novel siderophore from *Rhizobium meliloti*. J Am Chem Soc 107:1739–1743

Soon YK (2008) Phosphorus cycle. In: Chesworth W (ed) Encyclopedia of soil science. Springer, Dordrecht, pp 547–555

Spaink HP (2000) Root nodulation and infection factors produced by rhizobial bacteria. Annu Rev Microbiol 54:257–288

Steenhoudt O, Vanderleyden J (2000) *Azospirillum*, a free-living nitrogen-fixing bacterium closely associated with grasses: genetic, biochemical and ecological aspects. FEMS Microbiol Rev 24:487–506

Stein T, Hayen-Schneg N, Frendrik I (1997) Contribution of BNF by *Azoarcus* sp. BH72 in *Sorghum vulgare*. Soil Biol Biochem 29:969–971

Storey EP, Boghozian R, Little JL, Lowman DW, Chakraborty R (2006) Characterization of 'Schizokinen'; a dihydroxamate-type siderophore produced by *Rhizobium leguminosarum* IARI 917. Biometals 19:637–649

Subramanian KS, Charest C (1997) Nutritional, growth, and reproductive responses of maize (*Zea mays* L.) to arbuscular mycorrhizal inoculation during and after drought stress at tasselling. Mycorrhiza 7:25–32

Subramanian KS, Charest C (1999) Acquisition of N by external hyphae of an arbuscular mycorrhizal fungus and its impact on physiological responses in maize under drought-stressed and well-watered conditions. Mycorrhiza 9:69–75

Subramanian KS, Santhanakrishnan P, Balasubramanian P (2006) Responses of field grown tomato plants to arbuscular mycorrhizal fungal colonization under varying intensities of drought stress. Sci Hortic 107:245–253

Sundara B, Natarajan V, Hari K (2002) Influence of phosphorus solubilizing bacteria on the changes in soil available phosphorus and sugarcane and sugar yields. Field Crops Res 77:43–49

Supanjani HSH, Jung JS, Lee KD (2006) Rock phosphate-potassium and rock-solubilising bacteria as alternative, sustainable fertilizers. Agron Sustain Dev 26:233–240

Tawaraya K, Watanabe S, Yoshida E, Wagatsuma T (1996) Effect of onion (*Allium cepa*) root exudates on the hyphal growth of *Gigaspora margarita*. Mycorrhiza 6:57–59

Tejera N, Lluch C, Martínez-Toledo MV, González-López J (2005) Isolation and characterization of *Azotobacter* and *Azospirillum* strains from the sugarcane rhizosphere. Plant Soil 270:223–232

Tian CY, Feng G, Li XL, Zhang FS (2004) Different effects of arbuscular mycorrhizal fungal isolates from saline or non-saline soil on salinity tolerance of plants. Appl Soil Ecol 26:143–148

Toussaint JP, Smith FA, Smith SE (2007) Arbuscular mycorrhizal fungi can induce the production of phytochemicals in sweet basil irrespective of phosphorus nutrition. Mycorrhiza 17:291–297

Trehan SP, Sekhon GS (1977) Effect of clay, organic matter and $CaCO_3$ content on zinc adsorption by soils. Plant Soil 46:329–336

UN (2004) World population to 2300. United Nations, New York

Valverde A, Burgos A, Fiscella T, Rivas R, Velázquez E, Rodríguez-Barrueco C, Cervantes E, Chamber M, Igual JM (2006) Differential effects of coinoculations with *Pseudomonas jessenii* PS06 (a phosphate-solubilizing bacterium) and *Mesorhizobium ciceri* C-2/2 strains on the growth and seed yield of chickpea under greenhouse and field conditions. Plant Soil 287:43–50

van der Helm D, Winkelmann G (1994) Hydroxamates and polycarboxylates as iron transport agents (siderophores) in fungi. In: Winkelmann G, Winge D (eds) Metal ions in fungi. Marcel Dekker, New York, pp 39–98

Vandenabeele J, De Beer D, Germonpré R, Van De Sande R, Verstraete W (1995) Influence of nitrate on manganese removing microbial consortia from sand filters. Water Res 29:579–587

Vassilev N, Medina A, Azcon R, Vassileva M (2006a) Microbial solubilization of rock phosphate on media containing agro-industrial wastes and effect of the resulting products on plant growth and P uptake. Plant Soil 287:77–84

Vassilev N, Vassileva M, Nikolaeva I (2006b) Simultaneous P-solubilizing and biocontrol activity of microorganisms: potentials and future trends. Appl Microbiol Biotechnol 71:137–144

Verma TS, Thakur PC, Ajeet S (1997) Effect of biofertilizers on vegetable and seed yield of cabbage. Vegetable Sci 24:1–3

Vessey K (2003) Plant growth promoting rhizobacteria as biofertilizers. Plant Soil 255:571–586

Viets FG (1961) Chemistry and availability of micronutrients in soils. J Agric Food Chem 10:174–178

Villegas J, Fortin JA (2001) Phosphorus solubilization and pH changes as a result of the interactions between soil bacteria and arbuscular mycorrhizal fungi on a medium containing NH_4+ as nitrogen source. Can J Bot 79:865–870

Vivas A, Marulanda A, Ruiz-Lozano J, Barea J, Azcón R (2003) Influence of a *Bacillus* sp. on physiological activities of two arbuscular mycorrhizal fungi and on plant responses to PEG-induced drought stress. Mycorrhiza 13:249–256

Vosátka M, Albrechtová J (2009) Benefits of arbuscular mycorrhizal fungi to sustainable crop production. In: Khan MS, Zaidi A, Musarrat J (eds) Microbial strategies for crop improvement. Springer, Berlin/Heidelberg, pp 205–225

Wakelin SA, Warren RA, Harvey PR, Ryder MH (2004) Phosphate solubilization by *Penicillium* spp. closely associated with wheat roots. Biol Fertil Soils 40:36–43

Walter A, Römheld V, Marschner H, Crowley DE (1994) Iron nutrition of cucumber and maize: effect of *Pseudomonas putida* YC3 and its siderophore. Soil Biol Biochem 26:1023–1031

Wang Y, Brown HN, Crowley DE, Szaniszlo PJ (1993) Evidence for direct utilization of a siderophore, ferrioxamine B, in axenically grown cucumber. Plant Cell Environ 16:579–585

Wange S (1996) Effect of biofertilizers under graded nitrogen levels on carrot. Ann Plant Physiol 10:96–98

Wani P, Khan MS, Zaidi A (2007) Synergistic effects of the inoculation with nitrogen-fixing and phosphate-solubilizing rhizobacteria on the performance of field-grown chichpea. J Plant Nutr Soil Sci 170:283–287

Welch SA, Taunton AE, Banfield JF (2002) Effect of microorganisms and microbial metabolites on apatite dissolution. Geomicrobiol J 19:343–367

Weller DM, Cook RJ (1983) Suppression of take-all of wheat by seed treatments with fluorescent pseudomonads. Phytopathology 73:463–469

White PJ, Hammond JP (2008) Phosphorus nutrition of terrestrial plants. In: White PJ, Hammond JP (eds) The ecophysiology of plant-phosphorus interactions. Springer, Dordrecht, pp 51–82

Windham MT, Elad Y, Baker R (1986) A mechanism for increased plant growth induced by *Trichoderma* spp. Phytopathology 76:518–521

Winkelmann G (2007) Ecology of siderophores with special reference to the fungi. Biometals 20:379–392

Wu SC, Caob ZH, Li ZG, Cheunga KC, Wong MH (2005) Effects of biofertilizer containing N-fixer, P and K solubilizers and AM fungi on maize growth: a greenhouse trial. Geoderma 125:155–166

Xiao CQ, Chi RA, Huang XH, Zhang WX, Qiu GZ, Wang DZ (2008) Optimization of rock phosphate solubilization by phosphate-solubilizing fungi isolated from phosphate mines. Ecol Eng 33:187–193

Yano K, Takaki M (2005) Mycorrhizal alleviation of acid soil stress in sweet potato (*Ipomoea batatas*). Soil Biol Biochem 37:1569–1572

Yasmin F, Othman R, Sijam K, Saad MS (2007) Effect of PGPR inoculation on growth and yield of sweetpotato. J Biol Sci 7:421–424

Yedidia I, Srivastva AK, Kapulnik Y, Chet I (2001) Effect of *Trichoderma harzianum* on microelement concentrations and increased growth of cucumber plants. Plant Soil 235:235–242

Yehuda Z, Shenker M, Römheld V, Marschner H, Hadar Y, Chen Y (1996) The role of ligand exchange in the uptake of iron from microbial siderophores by gramineous plants. Plant Physiol 112:1273–1280

Youard ZA, Mislin GLA, Majcherczyk PA, Schalk IJ, Reimmann C (2007) *Pseudomonas fluorescens* CHA0 produces enantio-pyochelin, the optical antipode of the *Pseudomonas aeruginosa* siderophore pyochelin. J Biol Chem 282:35546–35553

Zaidi A, Khan MS, Amil M (2003) Interactive effect of rhizotrophic microorganisms on yield and nutrient uptake of chickpea (*Cicer arietinum* L.). Eur J Agron 19:15–21

Human Labour and Green Manure, Two Overlooked Factors for Energy Analysis in Agriculture

J.Y. Wu, M. Martinov, and Vito I. Sardo

Abstract Agriculture is both a "producer" and a "consumer" of energy. Since useful energy is in ever shorter supply and energy use is a pollution factor solutions are needed for enhancing production while reducing inputs. To this end it is necessary to get accurate data and adopt unequivocal analytic procedures permitting to compare and rank contrasting agricultural systems. In this report we show that two basic points are frequently mishandled in elaborating energy balances in agriculture: (1) the energy input of human labour and (2) the energy requirement for producing and applying organic manure. We show that the energy in human labour is often underestimated, probably at about 1/100 of its real value. It is also demonstrated that obtaining nitrogen from green manure is not less energy demanding than from synthetic fertilizers, opposite to the general belief. The correct evaluation of such two items would radically change the results usually obtained when working out energy balances, particularly when comparing organic and non-organic systems. Some scientific papers flawed with wrong assumptions and consequently conveying wrong information are listed. Spreading such misinformation seriously threatens the progress towards the sustainability of agricultural systems, since human labour and green manure are heavily conditioning energy balances, that in turn impact social, economical and environmental sustainability.

J.Y. Wu (✉)
Research and Development Center for Grasses and Environment, Beijing Academy of Agriculture and Forestry Sciences, Beijing 100097, P.R. China

M. Martinov
Department of Environmental Engineering, University of Novi Sad, Trg Dositeja Obradovića 6, 21000 Novi Sad, Serbia

V.I. Sardo
Department of Agricultural Engineering, University of Catania, via Santa Sofia, 100, 95123 Catania, Italy
e-mail: sardov@unict.it

E. Lichtfouse (ed.), *Genetics, Biofuels and Local Farming System*, Sustainable Agriculture Reviews 7, DOI 10.1007/978-94-007-1521-9_7,

Keywords Organic manure • Organic agriculture • Energy balance • Sustainable agriculture

1 Introduction

Agriculture is at the same time a "producer" and a "consumer" of energy: since useful energy is in ever shorter supply it is of interest to pursue those solutions conducive to its most rational use. At the same time, acknowledging that energy input is a factor of pollution to the extent that the energy intensity of a system has been suggested as an indicator of its sustainability (e.g. Hülsbergen et al. 2002; Bailey et al. 2003), those solutions must be favoured permitting to reduce it to the possible extent.

To advance towards such two objectives, accurate basic data should be assumed as a starting point in the elaboration of energy analysis and rigorously standardized procedures adopted, enabling to assess unambiguously the value of contrasting solutions and systems. Assessing energy inputs with the possible accuracy is also important since they are tightly linked to economic and environmental issues: for instance the mechanical energy developed by humans is much more costly and polluting than electric energy, and is of a lower effectiveness, hence the need to confine it to the lowest possible level.

The uncertainty about the modes to reach sustainability in agriculture – actually even in defining it (e.g. DFID 2004; Pretty et al. 2006; UNEP 2010) – prompted the mushrooming of a number of more or less serious "schools of thought", each more or less explicitly claiming to possess the recipe or even the panacea for securing sustainability.

In principle, the higher the claimed efficacy of the proposed "system" or "philosophy", the lower the reliability of the claimant: sustainability in agriculture is in fact such a complex, elusive issue that no pre-concocted rules can be expected to apply to all agronomic and social conditions. Every farm has its own specificities and all the resources of science and technology should be applied in a flexible approach to achieve the best results in every specific case rather than attempting to fit farms to the Procrustean bed of a particular ideology. "*The idea of agricultural sustainability, though, does not mean ruling out any technologies or practices on ideological grounds. If a technology works to improve productivity for farmers and does not cause undue harm to the environment, then it is likely to have some sustainability benefits.*" (Pretty 2008). A similarly well balanced approach, refraining from extreme positions, is adopted by the UNEP (2010): "*Organic and conventional agriculture should not be seen as contradictions, or intrinsically "good" and "bad" for agricultural sustainability, but as complementary sources from which the best elements should be borrowed and applied in appropriate contexts*".

Elaborating energy balances is a challenging but necessary step in the comparison and ranking of agricultural systems and ultimately in the selection of the most sustainable solutions, due to the impact of energy input on sustainability. Reliable energy analyses permit in fact to pinpoint possible savings and evidence sources of wastes and pollution.

A trend can be easily perceived in organic farming supporters to minimize energy requirement of human labour and green manure – typical components of organic systems "philosophy" – which has brought about diffuse misconceptions and the spreading of erroneous data, in turn affecting sometimes the approaches and conclusions even of serious researchers. Since at the source of the confusion in such approaches and conclusions there is the endless repetition of codswallops that has given them the dignity of truths, it can be perhaps useful to try to examine in some detail these two key factors, human labour and green manure, crucial in the elaboration and comparison of energy balances.

2 Human Labour

Estimating the energy input of human labour is not easy, because defining the boundaries of the system is a matter for discussion, as well illustrated by Giampietro et al. (2006). Proceeding by ever widening system boundaries the steps are as follows.

1. In the "Encyclopedia of Earth" Cleveland (2010) states: "*The biological energy equivalent of labor is the fuel burned when human labor does mechanical work. This quantity of fuel can be measured directly by a respirometer*" and for a person doing a desk work results about 0.3 MJ/h while for a manual labourer is about 0.6 MJ/h
2. According to Dawson (2008) "*the average daily energy content (calorific value) of the food consumed by the UK public is 9.5 MJ per person*". This means almost 0.4 MJ/h, neglecting the energy needed for *producing, processing and distributing* the food: such a figure matches closely enough that reported in step 1
3. The quantity of fuel used *to produce, process and distribute* the amount of food consumed by each person is equivalent to about 5.7 MJ/h (Cleveland 2010): therefore for a manual labourer the total of $0.6 + 5.7 = 6.3$ MJ/h is reached
4. Ortiz-Cañavate and Hernanz (1999) reviewing the previous researches wrote: "*Fluck [...] summarizes nine methods. The first five methods include muscular energy and energy sequestered in food, ranging from 1 to 93.2 MJ/day, and the last four methods consider indirect energy... and the style of life. Values range between 510 MJ/day and 1,450 MJ/day, although the latter seems too high to be taken into account.*"
5. According to Lawyer (2008) in the US "*the energy to support a worker is about 100 times his food caloric energy (metabolic energy)*": for a manual worker this leads to 60 MJ/h (0.6 MJ/h × 100)

6. Bankier and Gale (2006) estimated a worker's energy 967 MJ/day, namely 138 MJ/h for a 7 h workday, considering incorrect to base estimates on workers' metabolic energy only, "*usually giving insignificantly small figures*"
7. In several papers Pimentel estimated energy input for labour 40.000 kcal/h, based on a yearly consumption of 8.000 l oil equivalent per person and 2,000 working hours (e.g. Pimentel 2003; Pimentel and Patzek 2005; Pimentel 2006). In his words "*This energy represents inputs for roads, schools, police, fire protection and others. It does not include any energy inputs for the family*" (Pimentel 2006). This energy is equivalent to 167.6 MJ/h
8. Furthermore it is claimed that the energy cost for producing and maintaining labour in the household sector about doubles that for a single person, since a sustainable society needs families with children to replace ageing workers. Additionally, aged and disabled workers must be cared for (Lawyer 2008).
9. According to Giampietro et al. (2006) the energy input of labour ranges from a minimum of 0.4 MJ/h if only "extra metabolic energy" due to the work (i.e. the total energy consumption minus metabolic rate) is considered, to a maximum of 400 MJ/h when estimating "exosomatic energy input" at society level
10. Odum (1996) wrote: "*Directly and indirectly it takes about 1,000 kcal of sunlight to make a kilocalorie of organic matter, about 40,000 to make a kilocalorie of coal, about 170,000 kcal to make a kilocalorie of electrical power, and 10 million or more to support a typical kilocalorie of human service*" – in other terms producing "*one typical kilocalorie of human service*" is as demanding as about 60 "*kilocalories of electrical power*"

It is evident how the issue lends itself to endless debate and finding a way out is not simple, however it seems obvious that labour *should* be assigned a reasonable energy cost, in terms of direct and indirect energy, like all the other goods and services. Considering it at zero or negligible levels as many researchers do is incorrect and misleading, since it would contribute to convey biased indications, particularly whenever different farming systems with different labour requirements are compared. The same way that the indirect energy input to manufacture a tractor is factored in even if it lies idle 300 days per year, indirect energy to sustain labourers should be considered. Furthermore, it is evidently not correct to neglect the direct and indirect energy – and the related pollution – spent in transporting the labourers to their workplace and back.

Cleveland (2010) comments: "*labor also has an energy cost associated with its use. These energy costs can be separated into three components: (1) the caloric value of the food the worker consumes; (2) the embodied energy of that food (i.e., the direct plus indirect fuel used to produce food); and (3) the fuel purchased with the wages and salaries of labor. Obviously, there are important differences between human labor and other factors, but this does not alter the fact that labor requires a continuous input of energy to sustain itself.*"

In the light of the above considerations the values of 0.62 MJ/h assumed by Lillywhite et al. (2007), 0.8 MJ/h by Reganold et al. (2001), 1.96 MJ/h by Gündoğmus (2006), 2.2 MJ/h by Strapatsa et al. (2006), Kaltsas et al. (2007),

Guzmàn and Alonso (2008), 2.3 MJ/h by Hatirli et al. (2006) appear all a gross underestimation because they hardly consider energy in food and even neglect the embodied energy of food itself. Furthermore none of these authors conveys a convincing explanation for such low values.

The range of estimates of Bankier and Gale (2006) and Pimentel (2006) seems more reasonable, since unlike the others they are supported by a logical explanation.

Of course since such estimates refer to developed countries – specifically those of Bankier and Gale (2006) refer to the US and Australia – with a high unit fuel consumption, 8,000 l oil-equivalent per person-year, they should be proportionally reduced when referring to developing countries.

Odum's emergy-based estimates are referring to still uncertain and not universally accepted values and cannot be used to elaborate reasonably accurate energy balances, also because labour energy would dwarf and make meaningless all the other inputs.

It must be reminded that a higher energy input inescapably entails higher levels of pollution and consequently human labour is by far more polluting than machine work, which makes it less sustainable under the environmental point of view.

Parallel to some extent to energy aspects, also the economic aspects of human labour are quite peculiar: compared to the cost of electric energy currently at about euro 0.2/kWh, the mechanical energy released by humans – although of a lower quality – costs about 6–300 times as much in developing and developed countries, respectively. It is therefore surprising that the higher labour amount typically required by organic agriculture compared to non-organic (averaging 32% in the UK according to DEFRA (2008); the same percentage of 32% is given by Pimentel (2006), for corn and soybean in the USA, while about 33% is the figure suggested by Ziesemer (2007)) be frequently applauded as a major achievement. Only persons unfamiliar with farms and their tight balances can consider a higher labour requirement in a positive light.

On the opposite, consideration should be given to the frequent difficulty in the developed countries to find skilled or even unskilled labourers at the appropriate time for the required cultural practices – for instance Lillywhite et al. (2007) wrote: "*The difficulties that can arise are illustrated by the case of the German asparagus harvest in the spring of 2007. Difficulties with finding sufficient casual workers to pick the crop meant that as much as 15% was left in the ground on some farms. Part of the problem was a lesser availability of Polish workers because of better opportunities at home or elsewhere.*" The "philosophy" of organic farming supports an ever growing use of human labour. In her celebrated "*classic text in the organic movement*" lady Eve Balfour, that icon of organic movement, proclaimed: " *Labour-intensive small units will always be able to produce spectacularly more per acre than the large mechanised farms*" (Balfour 1977). Later, Altieri echoed: "*in large holdings, when capital (mechanization) is substituted for labor, this lowers yields more often than it raises them*" (Altieri 2002). It would be interesting to explore if the supporters of manual working would not change their mind after experiencing one single day's backbreaking hand weeding or "bucket irrigating" in the field (Fig. 1).

Fig. 1 "Bucket irrigation" in a drought-stricken crop in China

For a matter of respect to human dignity, man must be considered a *governor* of energy rather than a *supplier* doing the same activity that can be done much more cheaply and with less pollution by animals or machines; future improvement of organic production should be focused toward human labour reduction, i.e. development of mechanisation apt to replace it. Most principles of organic production deserve to be supported, but further research and development activities must enable reduction of human labour to the possible extent.

In conclusion, since notoriously sustainability is based on the three pillars of environmental, economic and social aspects, it can be safely stated that under the standpoint of human labour, organic farming philosophy is conflicting against all such three pillars and is therefore unsustainable.

3 About Organic Manure

An analysis of energy balance in green manures can supply interesting information evidencing the fallacy of the widely accepted assumption that they permit considerable energy savings compared to synthetic nitrogen fertilizer. This assumption is practically constant, monotonously repeated with almost identical wording and acritically accepted in all the papers dealing with this issue.

It has been even maintained that the declining oil reserves make conventional food production unsustainable, due to the high energy required for manufacturing N, while conversely biological N fixation is claimed to be an "energy-free" process. For instance in Crews and People (2004) alarmed words "*Given that ultimate reserves (known and predicted resources) of fossil fuels will be largely exhausted in the later half of the 21*st *century [Rogner 2000], is it wise to put faith in the development of an unproved or unknown energy source to maintain our species*

Table 1 Comparison of energy input in clover and faba beans

	Clover			Faba beans		
Item	Quantity	Energy rate	MJ/ha	Quantity	Energy rate	MJ/ha
Seeds	10 kg/ha[a]	166 MJ/kg[a]	1,660	200 kg/ha	17 MJ/kg	3,400
Harrowing and drilling	1	1,200 MJ/ha	1,200	1	1,200 MJ/ha	1,200
Rolling	1	250 MJ/ha	250	1	250 MJ/ha	250
Mowing-chopping	–	–	–	1	200 MJ/ha	200
Ploughing (25 cm)	1	1,700 MJ/ha	1,700	2	1,700 MJ/ha	3,400
Labour	6 h	167 MJ/h	1,000	10 h	167 MJ/h	1,670
		Total MJ/ha	**5,810**		**Total MJ/ha**	**10,120**

[a] Averaged values

beyond the next 100 years?". These considerations derive from their underlying assumption that "*Nitrogen biologically fixed by legumes is ultimately derived from solar energy, while [mineral] fertilizer N requires significant amounts of non-renewable fossil fuels or other commercial energy sources to produce*" (Crews and Peoples, ibidem). Bertilsson et al. (2008) touched on the subject writing "*There is a common belief that the high fossil fuel requirement in N fertilizer production is non-sustainable. This is a misconception, since the use of N fertilizers provides a net energy gain. If N fertilizers were to be completely replaced by biological N_2 fixation, net energy production would be significantly lower*". Unfortunately they did not further elaborate, undertaking an estimation of energy inputs to biological N.

In order to estimate the energy inputs to green manure some assumptions are necessary owing to the rather wide range of possible conditions. The seeds of legumes usually used for manuring can vary in size from small clover to big faba beans, which in turn implies different energy requirement for production, different amounts of seeds per hectare and different cultural practices.

Many data here are taken from the University of California's UC SAREP Online Cover Crop Database (downloaded March 2010). According to their indications seeding rates vary from a low of 10 kg/ha for clover (averaged values) to a high of 200 kg/ha for faba beans, or bell beans.

Data on energy input for seed production are taken from Heichel (1980), those for human labour from Pimentel (2006) and those for field operation and synthetic nitrogen are from Williams et al. (2006).

In Table 1 details are given for the energy budget of green manuring in the cases of clover and beans. For each of them the hypotheses of a low, intermediate and high N production are examined (Table 2) in order to capture the wide variety of data reported in the literature, omitting only extreme cases. Field operations are basically those suggested in UC SAREP Online Cover Crop Database, integrated through our farming experience. One ploughing to 25 cm is necessary to prepare the seedbed for faba beans before harrowing and drilling, and one more is needed to fully incorporate the biomass, clover or faba, into the soil since harrowing would leave a considerable part of the biomass exposed to the atmosphere (Fig. 2).

Table 2 Comparison of energy requirements of organic and mineral nitrogen

	Clover			Faba beans		
Yield level	Low	Intermediate	High	Low	Intermediate	High
N production from green manure (kg/ha)	100	125	150	200	250	300
Energy to produce and apply the same amount of synthetic N at 41 MJ/kg N[a]	4,100	5,125	6,150	8,200	10,250	12,300
Energy saving with green manure (difference with the total in Table 1, MJ/ha)	**− 1,710**	**−685**	**340**	**−1,920**	**130**	**2,180**

[a]From Williams et al. (2006)

Fig. 2 A two step operation: chopping (*right*) and then incorporating (*centre*) the green manure (Photo credit: Washington State University Extension)

In lands with semiarid or Mediterranean climate it is frequently necessary to apply a pre-irrigation before seeding crops for green manuring, which adds to energy requirements: here this is not considered.

Table 2 shows that the energy input to green manures is on the average not less than that to mineral fertilizers; only in the hypothesis of high-yielding crops can some limited savings in energy be *theoretically* achieved. But one major shortcoming of organic fertilizers depends on the well known lack of synchronization between nitrogen release in the soil and crop requirements, which can lead to an insufficient availability to plants in critical phenophases and a consequent leaching of unused nitrogen (e.g. Cameron and Wild 1984; Kirchmann and Thorvaldson 2000; Stopes et al. 2002; Aronsson et al. 2007; Russo et al. 2010). This implies that only a part of N produced by legumes is actually available to the succeeding crops (e.g. Berry et al. 2002; Sullivan 2003; Robertson and Vitousek 2009) and the remaining is lost. This in turn reduces its efficiency, and therefore in the comparison with mineral N the final outcome is much worse than that resulting from Table 2: *"even with best management practices there are new challenges associated with organic fertilizers, because the risk of losses to water and air is potentially larger, and the use efficiency is lower, compared with the use of equivalent quantities of inorganic fertilizers"* (Mc Neill et al. 2005). The adoption of organic nutrients is at the opposite end of precision-applied, high-frequency, high-efficiency fertigation: no precision, ultra-low frequency and low efficiency of application and use.

Even in the most favourable cases N from green manure is more inefficient and considerably more polluting than the vilified synthetic N. In Torstensson et al. (2005) words: "*What these results clearly suggest is that N use efficiency is less if green manures are used rather than inorganic N fertilizers*". And of course a lower efficiency automatically implies a higher pollution.

A special case refers to animal manure, often considered a farm waste requiring a negligible energy input: such is for instance the case of Reganold et al. (2001) and Tzilivakis et al. (2005): commenting Reganold's paper Trewavas (2004) argued that anyway the related energy inputs *are a cost on the world's energy use; quite simply there is no "free lunch"* (p. 773). In any case the considerable amount of energy for curing, transporting, distributing and burying them, not to mention GHG losses, cannot be overlooked.

A final comment: as Undersander and Laboski (2008) quite appropriately state "*it is important to note that [cover] crops will perform differently from year to year depending on rainfall patterns, temperature and other considerations*" (we specify: including variations in microbiological activities and redox potential) which adds to the difficulties for an appropriate and timely management of fertilization. But farmers need to know N availability with the possible precision: *"The variability in N fixation is of crucial concern to farmers since it is the key factor in organic rotations that determines final crop production and income*" (Goulding and Trewavas 2009).

In conclusion probably the best solution with regard to N-related agronomic and environmental problems is the time-honoured combination of *well managed* organic and synthetic fertilizers, permitting to take the most advantage of the two forms while reducing their drawbacks. This is supported by the findings in the long-term research conducted in Rothamsted (Goulding et al. 2008), demonstrating also that the application of synthetic nitrogen enhances soil organic matter content by stimulating crop growth and consequently the amount of organic wastes (Powlson et al. 2010).

In the short term, a research on soybean production recently conducted by Mandal et al. (2009) led to the same conclusions: "*a combination of NPK-fertilizer and organic manure (FYM) could be the viable nutrient management option for soybean production*".

The economic aspects, needless to say, are quite sharply in favour of mineral N, costing about one order of magnitude less than the organic, but this simplistic approach would overlook the multifarious, well known advantages of organic matter addition, including fertility enhancement, soil protection against erosion, overland flow reduction etc.

It is evident that rather than to defend organic or mineral nutrients, efforts should be addressed to improve their combined management, which is the key to nutrient efficiency, keeping in mind that the application *in either form* is however a threat to the environment (Robertson and Vitousek 2009).

4 Considerations on Crop Energy Analysis

From the considerations about energy in human labour and N fertilizers we can conclude that many comparative energy analyses for agriculture are flawed and misleading owing to erroneous key assumptions.

In the energy input comparisons between organic and non-organic systems the results frequently appear more or less biased in favour of organic systems. This is for instance the case of the papers reported in Table 3, not to mention all those issued by organic farming supporters, generally of a poor scientific value and/or not peer reviewed and therefore not considered here, like all those listed in Table 4.19 of Stolze et al. (2000), later expanded in Chap. 2 of a FAO report (Scialabba and Hattam 2002) and in turn copied by Gomiero et al. (2008).

It easy to find out that if the authors listed in Table 3 had assumed more correct inputs the indications resulting from their works would be different, with organic agriculture quite often requiring higher energy inputs, or emitting more CO_2, than non-organic, both in terms of unit yield and land surface, and ultimately resulting in a higher pollution. Details are shown in Table 3 below.

The result of such confounding conditions is that whenever compared to the so called conventional systems organic agriculture results almost constantly more "sustainable", with a better "energy efficiency" -whatever this expression can mean-, a better "environmental footprint" and a lower "environmental impact", be it true or not. It is time that a consistent, unequivocal and reliable procedure based on objective standards and a properly designed energy accounting protocol be defined in this domain of research for limiting the possibility of false and misleading statements.

5 Conclusion

Agricultural sustainability is such an important issue that all efforts should be undertaken to reduce the room for uncertainties and misrepresentations, presently far too large, and advance towards a sounder knowledge permitting to take ever more rationally-based decisions.

Table 3 Remarks on some authors' comparative analyses

Author	Human labour	Energy for organic manure	Comments
Robertson et al. (2000)	Not considered	Not considered	CO_2 emissions estimated zero in organic systems with legume covers
Reganold et al. (2001)	Estimated at 0.8 MJ/h	Not considered	Only energy required to distribute organic manure was considered
MAFF (2000)	Undisclosed (probably not considered)	Not considered	None
University of Manitoba (2005)	Undisclosed (probably not considered)	Not considered	None
Tzilivakis et al. (2005)	Not considered	Not considered	Only energy required to distribute organic manure was considered
Gündoğmus (2006)	Estimated at 1.96 MJ/h	Not considered	Only energy required to distribute organic manure was considered
Pimentel (2006)	Estimated at 167.6 MJ/h	Underestimated	Only embodied energy in vetch seeds was considered
Kaltsas et al. (2007)	Estimated at 2.2 MJ/h	Unspecified but underestimated	Less than 1/3 real energy consumption was allocated to organic fertilizers
Lillywhite et al. (2007)	Estimated at 0.62 MJ/h	Underestimated	Total energy for clover estimated about half the correct amount
Guzmàn and Alonso (2008)	Estimated at 2.2 MJ/h	Considered	None
Gelfand et al. (2010)	Not considered	Considered	Energy cost to obtain organic N not fully unveiled

Comparison between systems should be based on logical, unambiguous and homogeneous metrics, leaving no possibility to intended or unintended distortions of truth: while it can be understood that groups of more or less responsible fans continue to cultivate their "philosophies" and their a-scientific principles, the scientific community has the responsibility of focusing on the progress towards sustainable agriculture. Unfortunately it seems presently that objecting to organic principles, even when they are blatantly wrong, is often perceived as politically incorrect even by high-profile researchers and reputable institutions, and such perception brings about a sort of apprehension or timidity in affirming evident truths conflicting against them.

Serious research should break such fetters and freely progress in the sole interest of science, mankind and environment.

References

Altieri M (2002) Lectures at the course on sustainable agriculture. Mediterranean Agronomic Institute of Bari, Bari

Aronsson H, Torstensson G, Bergström L (2007) Leaching and crop uptake of N, P and K from organic and conventional cropping systems on a clay soil. Soil Use Manage 23:71–81. doi:10.1111/j.1475–2743.2006.00067.x

Bailey AP, Basford WD, Penlington N, Park JR, Keatinge JDH, Rehman T, Tranter RB, Yates CM (2003) A comparison of energy use in conventional and integrated arable farming systems in the UK. Agric Ecosyst Environ 97:241–253. doi:10.1016/S0167-8809(03)00115-4

Balfour E (1977) Towards a sustainable agriculture – the living soil. Available from soil and health library. http://www.soilandhealth.org/01aglibrary/01principles.html

Bankier C, Gale S (2006) Energy payback of roof mounted photovoltaic cells. The Environ Eng 7(4):11–14

Berry PM, Sylvester-Bradley R, Philipps L, Hatch DJ, Cuttle SP, Rayns FW, Gosling P (2002) Is the productivity of organic farms restricted by the supply of available nitrogen? Soil Use Manage 18:248–255. doi:10.1079/SUM2002129

Bertilsson G, Kirschmann H, Bergström L (2008) Energy analysis of organic and conventional agricultural systems. In: Kirchmann H, Bergström L (eds) Organic crop production – ambitions and limitations. Springer, Dordrecht, pp 173–188, doi: 10.1007/978-1-40209316-6_9

Cameron KC, Wild A (1984) Potential aquifer pollution from nitrate leaching following the plowing of temporary grasslands. J Environ Qual 13:274–278

Cleveland C (Lead Author), Costanza R (Topic Editor) (2010) Net energy analysis. In: Cleveland CJ (ed) Encyclopedia of Earth. Environmental information coalition, national council for science and the environment, Washington, DC. First published in the Encyclopedia of Earth 14 Sep 2006, Last revised 23 Jul2010. Available at http://www.eoearth.org/article/Net_energy_analysis

Crews TE, People MB (2004) Legume versus fertilizer sources of nitrogen: ecological tradeoffs and human needs. Agric Ecosyst Environ 102:279–297. doi:10.1016/j.agee.2003.09.018

Dawson CJ (2008) Food, fertilisers and footprints – an environmental essay. Paper presented to The International Fertiliser Society at a meeting in London, on 17th Apr 2008. ISBN 978-0-85310-260-1

DEFRA, Department for Environment, Food and Rural Affairs (2008) The contribution that organic farming makes in supplying public goods. Available at http://randd.defra.gov.uk/Document.aspx?Document=EV02007_4601_FRP.pdf

DFID (UK Department For International Development) (2004) Agricultural sustainability. Working paper no. 12. DFID, London

Gelfand I, Snapp SS, Robertson GP (2010) Energy efficiency of conventional, organic, and alternative cropping systems for food and fuel at a site in the U.S. Midwest. Environ Sci Technol 44(10):4006–4011

Giampietro M, Mayumi K, Munda G (2006) Integrated assessment and energy analysis: quality assurance in multi-criteria analysis of sustainability. Energy 31(1):59–86. doi:10.1016/j.energy.2005.03.005

Gomiero T, Paoletti MG, Pimentel D (2008) Energy and environmental issues in organic and conventional agriculture. Crit Rev Plant Sci 27(4):239–254. doi:10.1080/07352680802225456

Goulding KWT, Trewavas AJ (2009) Can organic agriculture feed the world? AgBioView, 23 Jun. Available at http://www.agbioworld.org/

Goulding K, Jarvis S, Whitmore A (2008) Optimizing nutrient management for farm systems. Philos Trans R Soc B 363:667–680. doi:10.1098/rstb.2007.2177. Available at http://rstb.royalsocietypublishing.org/

Gündoğmus E (2006) Energy use on organic farming: a comparative analysis on organic versus conventional apricot production on small holdings in Turkey. Energy Convers Manage 47:3351–3359

Guzmàn GI, Alonso AM (2008) A comparison of energy use in conventional and organic olive oil production in Spain. Agric Syst 98:167–176

Hatirli SA, Ozkan B, Fert C (2006) Energy inputs and crop yield relationship in greenhouse tomato production. Renewable Energy 31(4):427–438

Heichel GH (1980) Assessing the fossil energy costs of propagating agricultural crops. In: Pimentel D (ed) Handbook of energy utilization in agriculture. CRC Press, Boca Raton

Hülsbergen K-J, Feil B, Diepenbrock W (2002) Rates of nitrogen application required to achieve maximum energy efficiency for various crops: results of a long-term experiment. Field Crops Res 77(1):61–76

Kaltsas AM, Mamolos AP, Tsatsarelis CA, Nanos GD, Kalburtji KL (2007) Energy budget in organic and conventional olive groves. Agric Ecosyst Environ 122:243–251

Kirchmann H, Thorvaldson G (2000) Challenging targets for future agriculture. Eur J Agron 12:145–161

Lawyer DS (2008) Human energy accounting. In: Encyclopedia of Earth. Available at www.lafn.org/~dave/

Lillywhite R, Chandler D, Grant W, Lewis K, Firth C, Schmutz U, Halpin D (2007) Environmental footprint and sustainability of horticulture (including potatoes) – a comparison with other agricultural sectors. Final report produced for DEFRA

MAFF, Ministry of Agriculture, Fisheries and Food (2000) Study OF0182 – project "Energy use in organic farming systems" final project report. Archived at http://orgprints.org/8169

MAFF, Ministry of Agriculture, Fisheries and Food (2006) Energy use in organic farming systems (OF0182). Available at http://www2.defra.gov.uk/research/project_data/More.asp?I=OF0182. Last modified 20 Aug 2009

Mandal KG, Hati KM, Misra AK (2009) Biomass yield and energy analysis of soybean production in relation to fertilizer-NPK and organic manure. Biomass Bioenergy 3(3):1670–1679. doi:10.1016/j.biombioe.2009.08.010

McNeill AM, Eriksen J, Bergstrom L, Smith KA, Marstorp H, Kirchmann H, Nilsson I (2005) Nitrogen and sulphur management: challenges for organic sources in temperate agricultural systems. Soil Use Manage 21:82–93. doi:10.1079/SUM2005303

Odum HT (1996) Environmental accounting: emergy and environmental decision making. Wiley, New York

Ortiz-Cañavate J, Hernanz JL (1999) Energy for biological systems. In: International Commission of Agricultural Engineering (ed) CIGR handbook of agricultural engineering, vol 5. American Society of Agricultural Engineers, St Joseph

Pimentel D (2003) Ethanol fuels: energy balance, economics, and environmental impacts are negative. Nat Resour Res 12(2):127–134, Jun 2003

Pimentel D (2006) Impacts of organic farming on the efficiency of energy use in agriculture. The Organic Centre, Cornell University, Ithaca

Pimentel D, Patzek TW (2005) Ethanol production using corn, switchgrass, and wood;biodiesel production using soybean and sunflower. Nat Resour Res 14(1):65–76. doi:10.1007/s11053-005-4679-8, Mar 2005

Powlson DS, Jenkinson DS, Johnston AE, Poulton PR, Glendining MJ, Goulding KWT (2010) Comments on "synthetic nitrogen fertilizers deplete soil nitrogen: a global dilemma for sustainable cereal production," by R.L. Mulvaney, S.A. Khan, and T.R. Ellsworth in the Journal of Environmental Quality 2009 38:2295–2314. J Environ Qual 39:749–752. doi:10.2134/jeq2010.0001le

Pretty J (2008) Agricultural sustainability: concepts, principles and evidence. Philos Trans R Soc B 363:447–465. doi:10.1098/rstb.2007.2163

Pretty JN, Noble AD, Bossio D, Dixon J, Hine RE, Penning De Vries FWT, Morison JIL (2006) Resource-conserving agriculture increases yields in developing countries. Environ Sci Technol 40(4):1114–1119

Reganold JP, Jerry D, Glover JP, Andrews PK, Hinman HR (2001) Sustainability of three apple production systems. Nature 410:926–930, 19 Apr 2001

Robertson GP, Vitousek PM (2009) Nitrogen in agriculture: balancing the cost of an essential resource. Annu Rev Environ Resour 34:97–125. doi:10.1146/annurev.environ.032108.105046

Robertson GP, Paul EA, Harwood RR (2000) Greenhouse gases in intensive agriculture: contributions of individual gases to the radiative forcing of the atmosphere. Science 289:1922–1925, Sep 2000

Rogner H (2000) Energy resources. In: Goldemberg J (ed) World Energy Assessment. United Nations Development Programme, New York, pp 135–171

Russo MA, Belligno A, Wu JY, Sardo VI (2010) Comparing mineral and organic nitrogen fertilizer impact on the soil-plant-water system in a succession of Three crops. Recent Res in Sci Technol 2(1):014–022, ISSN: 2076-5061

Scialabba NE, Hattam C (2002) Organic agriculture, environment and food security, vol 4, Environment and natural resources management series. FAO, Rome. ISBN 9251048193

Stolze M, Piorr A, Häring A, Dabbert S (2000) The environmental impacts of organic farming in Europe – organic farming in Europe: economics and policy. Universität Hohenheim, Stuttgart-Hohenheim. ISBN 3-933403-05-7

Stopes C, Lord EI, Philipps L, Woodward L (2002) Nitrate leaching from organic farms and conventional farms following best practice. Soil Use Manage 18:256–263. doi:10.1079/SUM2002128

Strapatsa AV, Nanos GD, Tsatsarelis CA (2006) Energy flow for integrated apple production in Greece. Agric Ecosyst Environ 116:176–180

Sullivan P (2003) Overview of cover crops and green manures – fundamentals of sustainable agriculture. ATTRA publication. IP024. http://attra.ncat.org/attra-pub/PDF/covercrop.pdf

Torstensson G, Aronsson H, Bergström L (2005) Nutrient use efficiencies and leaching of organic and conventional cropping systems in Sweden. Agron J 98:603–615. doi:10.2134/agronj2005.0224

Trewavas A (2004) A critical assessment of organic farming-and-food assertions with particular respect to the UK and the potential environmental benefits of no-till agriculture. Crop Prot 23:757–781

Tzilivakis J, Warner DJ, May M, Lewis KA, Jaggard K (2005) An assessment of the energy inputs and greenhouse gas emissions in sugar beet (Beta vulgaris) production in the UK. Agric Syst 85:101–119. doi:10.1016/j.agsy.2004.07.015

Undersander D, Laboski C (2008) Late summer planting legumes to produce nitrogen credits for next year. University of Wisconsin. http://www.uwex.edu/ces/crops/teamforagehttp://www.uwex.edu/ces/crops/teamforage/TF_2009_POW_and_EVAL.doc

UNEP/CBD/SBSTTA/14/INF/34 (2010) Sustainable agriculture and the sustainable use of agricultural biodiversity: concepts, trends and challenges. Available at http://www.cbd.int/sbstta14/meeting/documents.shtml

University of California Sustainable Agriculture Research and Education Program (2010) Online covercrop database. Available at www.sarep.ucdavis.edu/cgi-bin/ccrop.ex

University of Manitoba (2005) Welcome to the Glenlea study – Canadas oldest organic-conventional crop comparison study. Available at http://www.umanitoba.ca/faculties/afs/plant_science/glenlea/glenleacurrent.html

USDA, United States Department of Agriculture, Natural Resources Conservation Service (2005) Use of manures or legumes for nutrients – conservation activity job sheet, Dec 2005

Williams AG, Audsley E, Sandars DL (2006) Determining the environmental burdens and resource use in the production of agricultural and horticultural commodities. Main report. Defra research project IS0205. Cranfield University and Defra, Bedford. Available on www.silsoe.cranfield.ac.uk, and www.defra.gov.uk

Wu JY, Sardo V (2009) Sustainable vs organic agriculture – a review In: Lichtfouse E (ed) Sociology, organic farming, climate change and soil science, vol 3, Sustainable agriculture review. Springer, Dordrecht. doi: 10.1007/978-90-481-3333-8_3

Ziesemer J (2007) Energy use in organic food system. Food and Agriculture Organization of the United States, Rome

Sustaining Indian Agriculture in the Era of Climate Change

Manoranjan Mishra, Gyan Ranjan Panda, and B.K. Padhi

Abstract Climate change has moved to the top of the world's environmental agenda. Agriculture is an economic activity that is highly dependent on climate in order to produce the food and fiber necessary to sustain human life. The agriculture in India is highly sensitive to variations of weather and climate. The history of Indian agriculture reflect a series of adaptation. Soil, water, terrain, climate, technological developments and market factors provide constraints and opportunity for agricultural productions. Climate change has already been a reality for Indian farmers. These changes over time represent stimuli that affect the success of farming activities negatively and that prompt adjustments to altered circumstances. Since the inceptions of 'sustainable agriculture' in Rio Summit in 1992, a diverse range of approaches are put forward by researchers, state organization, non governmental development bodies, private organization in India but guided by the basic principle of sustainability. Sustainable agriculture must be economically viable, socially responsible and conserve land, water, genetic and other resources for future generation. The aim of this review is to highlight the potential impacts of climate change on agriculture in India and the status of mitigation and adaptation strategies to sustain the Indian agriculture.

Keywords Climate change • Sustainable agriculture • India • Grain • Food production • CO_2 • Wheat • Rice • Irrigation • Surface temperature

M. Mishra
Department of Geography and NRM, Sikkim University India, Gangtok, Sikkim, India
e-mail: geo.manu05@gmail.com

G.R. Panda
Research Officer in Centre for Budget and Governance Accountability, New Delhi, India
e-mail: panda.gyana77@gmail.com

B.K. Padhi (✉)
School of Environmental Sciences, Jawaharlal Nehru University, New Delhi 110067, India
e-mail: bkpadhi@gmail.com

E. Lichtfouse (ed.), *Genetics, Biofuels and Local Farming System*, Sustainable Agriculture Reviews 7, DOI 10.1007/978-94-007-1521-9_8,

• Precipitation • Rainfall • Flood • Sealevel rise • Cover crops • Green manure • Crop rotation • Intercropping • Pest control • Watershed • No tillage • Biofertilizer • Vermicompost • Raised bed • Furrow planting

1 Introduction

Climate induced natural disasters has a growing concern for their effects on agriculture and have stimulated academic, public and policy-level interests on the analysis of the impacts of climate variability on agricultural production systems (Lavalle et al. 2009). The spectrum of climate change will exacerbate its impact on agricultural production in most part of the world. The global environmental change juxtaposed with frequency and intensity of climate induced natural disasters like storms, droughts, floods alter the hydrological cycle and precipitation variability having implication in food availability in developing world (Dash et al. 2009; Dash et al. 2007; Rao et al. 2008). Indian agriculture represents a core part of the Indian economy and provides food and livelihood activities to more than 700 million people (Ministry of Agriculture 2001b). India makes up roughly 20% of gross domestic product (GDP) and provides nearly 52% of employment (as compared to 1% of GDP and 2% of employment for the US), with the majority of agricultural workers drawn from poorer segments of the population (FAO 2006; Kumar 2007). The share of agricultural production in exports is substantial (15% of export earning) and 57% weight in consumer price index. The growth in agricultural production also has direct impact on poverty eradication and also important factor for employment generation in rural areas (Planning Commission 1997, 2002). India's agriculture is highly sensitive to the vagaries of weather, particularly variability in rainfall. For instance, it is reported that about two-thirds of the sown area in the country is drought-prone and around 40 million hectares is flood-prone. Eighty percent of the total rainfall over the Indian sub-continent occurs during only 3 months (June-September), as a result of the South-West monsoon (Ministry of Finance 2001). Indian agriculture continues to be fundamentally climatic sensitive, with much of recent growth rates are being result of number of successive good monsoons (Kavikumar 2007).

Indian agriculture faces the dual challenge of feeding a billion people in a changing climatic and economic scenario. India accounts for some 200 million undernourished and an equivalent food gap of 600 million will lose 30 million tons cereal production due to climate change. According to World Bank, 2005 estimates 35.3%, 42% and 81% of rural population of India fall below poverty line of $1, $1.25 and $2 a day respectively. According to officially declared poverty line (NSS 2004–2005) 28% of total population in rural India are below poverty line. Among other vulnerable communities are children lees than 5 years of age are undernourished (48%) and also under weight (42.5%). The net impacts of climate change on agricultural output in India are uncertain, yet specific regions and certain groups of farmers, particularly those farming on marginal, rainfed lands, are likely

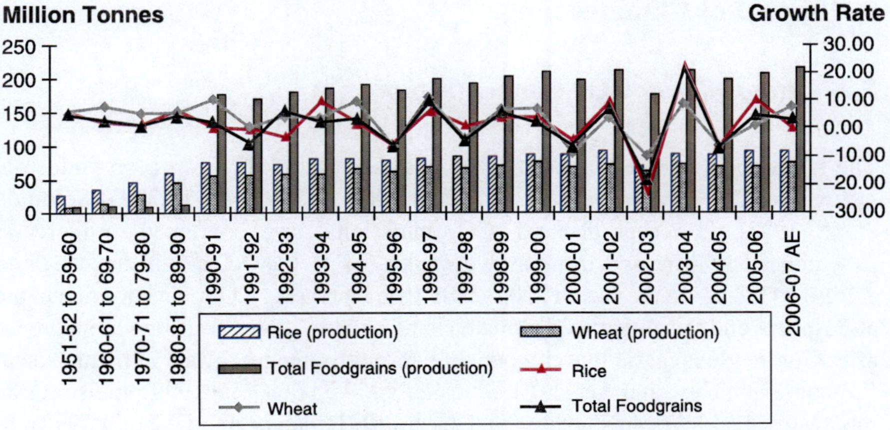

Fig. 1 Changes of total foodgrain productions, rice and wheat since 1950s

to suffer significant damages as the result of climate change (Dinar et al. 1998, cited in Selvaraj et al. 2002). The poorest people are likely to be hardest hit by the impacts of climate variability and change because they rely heavily on climate-sensitive sectors such as rainfed agriculture and fisheries. They also tend to be located geographically in more exposed or marginal areas, such as flood plains or nutrient-poor soils. The poor also are less able to respond due to limited human, institutional and financial capacity and have very limited ability to cope with climate impacts and to adapt to a changing hazard burden. India agriculture is vulnerable to the impact of climate change due to predominantly rain-fed agriculture, Himalayan glacier fed river basins, a large livestock population and a 7,500 km long coastal agriculture system.

Indian agriculture is also under performing due to less critical inputs – low level of public and private investments, lack of irrigation, lack of mechanization, heavy dependence on erratic monsoon and other climatic indicators. The productivity of total food grains have increased many fold due to first green revolution. However the production of rice and wheat have decreased since 1990 due to external factors like break in south west monsoon, changing pattern in rainfall, natural calamities and drought like conditions (shown in Fig. 1). The other factors like soil degradation, sinking water table and overuse of fertilizers also plays important role in productivity of crops. The policy implication relating to climate change impact in Indian agriculture are wide reaching as it will directly or indirectly affect larger population in India. Climate change being a complex problem needs proper scientific assessments urgent programmatic intervention by different ministry, farmers' institutions, their resources, knowledge and innovations.

2 Impact of Climatechange

2.1 Atmospheric CO_2 Concentration

The atmospheric concentration of CO_2 has increased 35% from a pre-industrial value, from 280 ppm by volume (ppmv) in 1,880 to 379 ppmv in 2005 (Solomon et al. 2007). In recent decades, CO_2 emissions have continued to increase: CO_2 emissions increased from an average of 6.4 ± 0.4 GtC a^{-1} in the 1990s to 7.2 ± 0.3 GtC a^{-1} in the period 2000–2005. The increasing CO_2 concentration in the atmosphere and the anticipated climate change due to global warming are likely to affect future global agricultural production through changes in rate of plant growth (Lemon 1983; Cure and Acock 1986; Rotter and Van de Geijn 1999), transpiration rate (Morison 1987; McNaughton and Jarvis 1991; Jacobs and DeBruin 1992). In principle, higher levels of CO_2 should stimulate photosynthesis in certain plants as they tend to suppress their photo-respiration. This should be true for the majority of species globally and especially in cooler and wetter habitats, including wheat, rice, barley, cassava and potato. Positive, but smaller effects on yields should be observed for tropical crops as maize, sugar cane, sorghum and millet, which are important for the food security of many developing countries, as well as pasture and forage grasses. In the Indian context Kumar and Parikh (2001a) showed that under doubled carbon dioxide concentration levels in the latter half of twenty first century the gross domestic product would decline by 1.4–3% points under various climate change scenarios.

2.2 Temperature

Between 1906 and 2005, the global average surface temperature has increased by 0.74°C(±0.18°C) (Solomon et al. 2007). The linear warming trend of the last 50 years (0.13°C per decade) is nearly twice that for the last 100 years. This rise in globally averaged temperatures since the mid-twentieth century is considered to be very likely due to the observed increase in anthropogenic greenhouse gas atmospheric concentrations (Solomon et al. 2007). The range in projections for the rise in global averaged surface temperatures from 1980 to 1999 to the end of the twenty-first century (2090–2099) is 1.1–6.4°C (Solomon et al. 2007). Increased surface temperature is expected to affect agricultural production worldwide (Table 1). According to Table 1 following generalization has been generalized that the aggregated impact are small (−1.5 to + 2.6%) in terms of changes in GDP from agriculture (similar to Tol et al. 2002). The developed countries are going get benefit in agricultural production due to climate change. For India, Kumar and Parikh (2001b) and Sanghi and Mendelsohn (2008) have used a variant of this approach and showed that a 2°C temperature rise and 7% increase in rainfall would lead to almost 10% loss in farm level net revenue (1990 net revenue). Higher

Table 1 Global impacts on agriculture[a] for a 2.5°C increase in global mean temperature (Source: Tol (2002)). SD: standard deviation. GDP: gross domestic product

Regions	Without adaptation		With adaptation		
	Best guess	SD	Best guess	SD	Mean
Latin America	−0.8	0.6	0.6	0.7	−0.1
South and South-east Asia	−0.7	0.3	0.6	0.3	0
Middle East	−0.4	0.4	0.6	0.5	0.1
Africa	−0.2	0.2	0.5	0.3	0.1
OECD-P	−0.2	1.6	0.8	1.6	0.3
OECD-A	−0.3	1.3	1	1.3	0.4
OECD-E	0.6	1	2.1	1.1	1.3
Eastern Europe and Former SU	0.9	1.2	2.7	1.1	1.8
Centrally planned Asia	1.7	1	3.1	1	2.4

[a] Expressed in percent change from reference Projection of GDP

temperatures will influence production patterns. Directly, as some plant growth and health may benefit from fewer freezes and chills, while some other crops may be damaged by higher temperatures; or indirectly through the temperature effect on water demand and supply, on the expansion of insects and plant diseases, on weeds expansion into different-latitude habitats.

2.3 *Precipitation*

Globally, rainfall is predicted to increase by about 25% by 2050 in response to climate change. However, the regional distribution of rainfall will be uneven (Houghton et al. 2001). Increased precipitation is very likely in high-latitudes, and decreased precipitation is likely in most subtropical regions, especially at the pole ward margins of the subtropics (Solomon et al. 2007).

In the most recent assessment, the Intergovernmental Panel on Climate Change reported significant increases in precipitation in eastern parts of North and South America, northern Europe and northern and central Asia, with drying in the Sahel, the Mediterranean, southern Africa and parts of southern Asia (Solomon et al. 2007). Long-term trends had not been observed for other regions. Changes in precipitation patterns are expected to affect agricultural productivity.

2.4 *Storms*

During the twenty-first century the Intergovernmental Panel on Climate Change projects that there is likely to be an increase in tropical cyclone peak wind intensities and increase in tropical cyclone mean and peak precipitation intensities in some areas as a result of global climate change (Houghton et al. 2001; Solomon et al. 2007). Storm surge heights are also predicted to increase if the frequency of strong winds and low pressures increase. This may occur if storms become more frequent or severe as a result of climate change (Church et al. 2001; Houghton et al. 2001; Solomon et al. 2007).

2.5 Flood

The impact of climate change on flood frequency is often defined by the percentage change in a flood peak of a given period. Floods can result in huge economic losses due to damage to infrastructure, property and agricultural land, and indirect losses in or beyond the flooded areas, such as production losses caused by damaged stock or roads, or the interruption of power generation and navigation (Douglas 2009; World Bank 2008). In India climate change appears to be influencing both the monsoon and tropical cyclones, the two prime drivers of flood events which will affected by changes in temperature and in rainfall. Rising temperatures will accelerate the rate of melting of snow and glacier ice, increasing seasonal peak flows of Himalayan headwaters. This in turn may lead to an increased frequency of flooding particularly along rivers whose channel capacity had been reduced by sedimentation (Aggarwal et al. 2004).

2.6 Sea-Level Rise

Global sea-level rise is one of the more certain outcomes of global warming, it is already likely taking place (12–22 cm occurred during the twentieth century), and several climate models project an accelerated rate of rise over coming decades (Cazenave and Nerem 2004; Church et al. 2001, 2004a; Holgate and Woodworth 2004; Thomas et al. 2004; Church and White 2006; Solomon et al. 2007). The range of projections for global sea-level rise from 1980 to 1999 to the end of the twenty-first century (2090–2099) is 0.18–0.59 m (Solomon et al. 2007). In India a possible sea-level rise of 15–38 cm by the 2050s would cause saline water to penetrate further inland (Ravi 2008) and ultimately damages the coastal agriculture.

3 Impact of Climate on Indian Agriculture

Agriculture has been the backbone of the Indian economy and it will continue to remain so for a long time. It has to support almost 17% of world population from 2.3% of world geographical area. However, climate change is likely to bring greater variability (Whetton 1994) and pose problems for national food supplies because agriculture is so dependent on the monsoon (Shukla 2003). It could also raise temperatures by 0.1–0.3°C by 2010, 0.4–2.0°C by 2070 (Whetton 1994), and 3.5–5°C by 2100 (Ravi 2008). These higher temperatures will affect water availability for food crops by their impacts on evapo-transpiration. Monsoon season (kharif) crops may be little affected by moderate temperature and CO_2 rises, but winter (rabi) crops could suffer decreased yields, due to large temperature increases and greater uncertainty of rainfall (Mall et al. 2006). Many of the gains in yield that would be caused by a doubling of atmospheric CO_2 would be more than offset by reduction in yield resulting from temperature and moisture changes (Vasudeva 2002; Attri and Rathore 2003).

Increase in storm precipitation will mean that severe flooding will happen more often, as it did in Gujarat in 2005, 2006 and 2007. Productive deltaic agricultural land will become more vulnerable to floods, to the impacts of possibly more severe tropical cyclones and to rising sea levels. Global warming may also threaten India food security if there is a negative effect on agriculture. Although, the effect of increasing CO_2 concentrations will increase the net primary productivity of plants, but climate changes, and the changes in disturbance regimes associated with them, may lead either to increased or decreased net ecosystem productivity. In many tropical and subtropical regions, potential yields are projected to decrease for most projected increases in temperature. The impacts of elevated CO_2 should be considered among others, in the context of, (A) changes in air temperature, particularly nocturnal temperature due to increase in CO_2 and other trace gases and changes in moisture availability and their effect on vegetative versus reproductive growth; (B) need for more farm resources, e.g. fertilizers; and (C) survival and distribution of pest populations, thus developing a new equilibrium between crops and pests (Krupa 2003).

The radical impact of climate change may disturb the equilibrium and tilt toward temperate countries than tropical countries. For example India is going to witness not only reduction in crop duration but also frequent climatic events like floods and droughts due to increased variability of annual monsoon rains and fast depletion of Himalayan glaciers (Chaudhary and Aggarwal 2009). In developing countries, 11% of arable land could be affected by climate change, including a reduction of cereal production in up to 65 countries, about 16% of agricultural GDP (FAO Committee on Food Security, Report of 31st Session 2005). The developing world's are already contend with chronic food problems. Agriculture typically plays a larger role and significant challenge to meet this need. For example, agriculture in India makes up roughly 20% of GDP and provides nearly 52% of employment (as compared to 1% of GDP and 2% of employment for the US), with the majority of agricultural workers drawn from poorer segments of the population (FAO 2006). There is a projection that overall food production may not be reduced but those least able to cope will likely bear the adverse impact (WRI 2005). The socio-economic systems of the developing countries are more vulnerable to climate change because of high reliance in natural resources, lower per capita GDP, high poverty and lack of safety nets. This negative impact of climate change on agriculture is likely to have a serious impact on poverty: recent estimates from across developing countries suggest that one percentage point of agricultural GDP growth increases the consumption of the three poorest deciles by 4–6% points (Ligon and Sadoulet 2007).

4 Sustainable Agriculture: A Shifting Interpretation in Indian Context

The history of Indian agriculture reflect a series of adaption like soil, water, terrain, climate, technological developments and market factors provide constraints and opportunity for agricultural productions. For instance India introduced 'Green Revolution' in early 1960s – technologies, high-yielding grain varieties, fertilizers, pesticides and irrigation and by early 1990, India was self sufficient in food production (Gadgil 1995; Swaminathan 2002). None of these factors playing major role in agricultural productions remain constant and their effects are interdependent (Fig. 1).

Climate change has already been a reality for Indian framers. These changes over time represent stimuli that affect the success of farming activities negatively and that prompt adjustments to altered circumstances. The capacity of a farming system to adapt to changing climate and weather conditions is based on its natural resource endowment and associated economic, social, cultural and political conditions (Wall and Smit 2005). The feasibility of these elements also constitutes the basis for sustainable agriculture, basically understood as agricultural production that: ensures ecologically sustainable, economically viable and socially acceptable (Smit and Smithers 1994). Friedrich and Gustafson (2007) described that combined resource-conserving technologies applied in conservation agriculture produce benefits for the farming sector, the environment and the general public. Since the inceptions of 'sustainable agriculture' in Rio Summit in 1992, a diverse range of approaches are put forward by researchers, state organization, non governmental development bodies, private organization in India but guided by the basic principle of sustainability. There is a general consensus among above stake holders in India with the line of FAO definition that "Sustainable agriculture must be economically viable, socially responsible and conserve land, water, genetic and other resources for future generation" (as cited by Sinha 1997). The approaches originally developed were counter movement against the conventional agricultural practices i.e. Green Revolution. Then the ecological dimension was placed in central place of conception. In recent time the pragmatic approach like social dimension especially were related to sustainable agriculture. Hence the overall goal of sustainable agriculture is to optimize yields (by making optimum use of land and water resources) without causing adverse short-term or long term impacts on nature, environment or society.

5 Traditional Approaches of Sustainable Agriculture Used in India

Since the beginning of an agricultural activity (traditionally placed after the last ice age 10,000 years ago), the role of environmental conditions in influencing soil properties, crops' growth and then land productivity and production has always been a paramount interest to farmers and then, much later, to agricultural scientists. In India traditional agriculture practices dates back to the Neolithic Age of 7,500–6,500 BC. The ancient Indian farmers tried to develop nature friendly farming systems by the cosmic forces with health and fertility of the land available. The first scientific approach of sustainable agriculture can be seen in later Vedic periods, 1,000–600 BC (Randhawa 1986; Pereira 1993). The essence was to live partnership rather than exploit the nature. According to Mahale and Sorée (1999), the classical Indian plant sciences (Vrikshayurveda) had tried to develop sustainable agricultural technologies and agronomic practices like genetic diversity, crop rotations and mixed cropping systems. Historical evidence shows that the yield per hectares was very high during eighteen centaury as compared to today's highest level. In eighteen centaury, Thomas Barnard, a British engineer conducted a survey of 800 villages near Madars and found that the average yield of wetland rice of 130 villages were 8.2 t/ha while present day average is 3.1 t/ha with extensive use of chemical and fertilizers. The wisdom gained and sustainable practices adopted by Indian farmers were passed down through generations and become ingrained in cultural outlook of the society. The practices of collection, selection and storage of seeds, genetic diversity and fertility of soil are interwoven in the belief systems, the rituals and dietary system of the communities. In past four decades, the traditional sustainable farming practices were eroded because of influx of modern conventional agriculture. The tradition agriculture practice sustained by Indian communities has gained renewed importance recently in era of climate change for sustaining Indian agriculture in light of climate change sustainable agriculture defined as an integrated farming system (with crops, trees, livestock etc.) which is based on locally adapted agro-diverse cropping patterns and use of local resources (natural resources and natural processes), based on local knowledge, skills and innovations. Many traditional agricultural practices around the India refrain from using chemical fertilizers and pesticides. They do this for various reasons: by tradition, because farmers cannot afford agrochemicals, they cannot buy them locally, or they do not know how to use them. Hence basic philosophy of sustainable agriculture should think earth as a living being and not try to exploit the nature for short term gain. The different traditional method of sustainable agriculture area describe below.

5.1 Bio-Dynamics

It is the science of life processes of the role of nature and cosmic forces in agriculture. This method which was practiced in different part of India considering

Earth as living being in living universe and part of spiritual matrix (Mahale and Sorée 1999). In this method the farming units are basically managed by principle and practices. It follows two steps for sustainable managing agricultural land. First step, farming operation on basis of astronomical calendar and other step as special preparation of agricultural land using indigenous spray and compost heap (Government of India 2001). This type of farming addresses soil and plant health, pest control, composting, the relationship of the soil and the plant, food quality, animal husbandry and animal welfare measures. It's basic objective is to address the problem of the agricultural land and managed it sustainable by using natural forces. Hence it is today perhaps the largest "specialized" organic farming system in the world.

5.2 *Green Manuring, Mulching, and Cover Cropping*

It is a tradition method to improve soil fertility and also supply sufficient nutrient to growing crops. A green manure is nothing more than a crop (nitrogen fixing legume) that is grown in field (45–50 days), then cut and incorporated to soil to provide sufficient amount (80–100 kg of N/ha^2) nitrogen fertilizers (Mishra and Nayak 2004). Potential green manures include sesbania (*Sesbania aculeata,dhaincha,dhunchi*), sun hemp (*Crotalaria juncea*), cowpea (*Vigna unguiculata*), mungbean (*Vigna radiata*), cluster bean (*Cyamopsistetragonoloba, guar*), berseem clover (*Trifolium alexandrinum*).

Surface mulches are organic materials used to coverthe soil especially around the plant to prevent soil from blowing, being washed, to reduce evaporation, to increase infiltration, to keep down weeds, to improve the structure and eventually to increase crop yields (Lomte and Pendke 2004). Cover crops are nitrogen fixing crops mostly used for covering soil in the fallow months, adding nitrogen to soil, suppressing weeds, preventing soil soil erosion and latter used as bio-mass or fodder.

5.3 *Crop Rotations, Multi-Cropping and Inter-Cropping*

Traditional Indian farmers till date follow systems of crop rotations, muti-crpping, inter cropping and poly culture to make maximum use of all inputs like soil, water, light at the minimum cost of the environment. Crop rotation may be defined as a process of growing different crops in succession on a piece of land in specific period of time without decreasing soil fertility (Sahai 1982). The combinations of crops are endless and also depend on great deal on local situations. Multi-cropping is traditional method of cultivating two or more crops. Traditional Indian farmers are known to have sown as many 15 crops at time (Sarma 2008). 'Baranaaja' is a example of traditional multi-cropping system practiced in Garhwal region. It is a simultaneous system where 12 or more food crops are grown in synergetic combination (Mishra et al. 2008 and Zardhari 2000).

Inter-cropping is an ancient traditional agronomic practices where cultivation where cultivation of another crop in spaces available between main crops (Awal et al. 2006). This method can increase yields, by using available resources, reduce weeds, insects and disease pressure and provide greater biological and economic stability (Vandermeer 1989).

5.4 Ecological Farming or Site-Appropriate Agriculture

This tries to cut down on costly inputs and minimize negative environmental impacts by making intelligent use of existing ecological factors (Reddy and Reddy 1991). It developed as an alternative to the increasingly intensive use of irrigation and fertilizers, and tries to free farmers from constraining factors in the local natural environment. The element of eco-farming are part of traditional farming system can be promoted among the poor farmers living in rain fed conditions. An important key element for promoting theses of agricultural practices is to reduce poverty in India. These can only be possible by catalytic development of people's association or 'sangams' that ensure people's participations. Ecological farming can provide large-scale carbon sinks and offer many other options for mitigation of climate change. In addition, farming with biodiversity is the most effective strategy to adapt agriculture to future climatic conditions. A mix of different crops and varieties in one field is a proven and highly reliable farming method to increase resilience to erratic weather changes. This farming are better than convectional agriculture because it protects nature by taking advantage of natural goods and services, such as biodiversity, nutrient cycling, soil regeneration and natural enemies of pests, and integrating these natural goods into agroecological systems that ensure food for all today and tomorrow.

5.5 Indigenous Pest and Disease Management

Indigenous pest management knowledge derived through long experience and perceptions accumulated by traditional farmers reducing the use of synthetic pesticides by integrating a range of ways to control pests and disease pathogens, from crop rotations to determining damage thresholds before applying plant protection products (Chhetry and Belbahri 2009). Against insects and disease, the main weapon of these farmers has been genetic diversity (Sofia et al. 2006). India once had 30,000 varieties of rice. These varieties were not used at random, but were delicately fitted into their appropriate ecological niches. The indigenous pest and disease management, age old knowledge fitted well ancient land use systems, yet need through validation in view changing agricultural scenario in India.

5.6 *Watershed Management*

The rehabilitation of degraded watershed areas has become a high priority. Watershed management aims to adapt land management practices in ecologically vulnerable hill and mountain regions to the natural carrying capacity by means of systematic management (Vani 1992, 2002). Unlike the other methods mentioned above, it is not an agricultural production system. Rather, it is a process that plans and regulates the use of land, water and other resources within a watershed area, in ways that sustain these resources. It involves not just technologies, but also devising policies and usage guidelines. It emphasizes adapting technical solution to socio-economic circumstances of users respecting the (often conflicting) needs of different resource users and attempting to reconcile their interests.

5.7 *Minimum Tillage System*

One of the most important principles of Conservation Agriculture (CA) is minimal soil disturbance. In minimum till system, the seed is placed into the soil by a seed drill without prior land preparation. This aims to conserve the soil structure and improve the water storage capacity of the soil. Introduced on a large farm level it is often combined with weed management through pesticides. Because it eliminates ploughing, agriculture needs less labour, so is a viable option for areas with labour shortages. By using crop rotations and intercropping, it reduces the risk through diversification.

6 Potential of Updated Modern Sustainable Agriculture Approaches Used in India

India had developed a vast and rich traditional agricultural knowledge since ancient times and presently finding solutions to problems created by over use of agrochemicals. Present days' modern conventional farming is not sustainable in consonance with economics, ecology, equity, energy and socio-cultural dimensions. The Green Revolution model of agriculture focus on high yielding seed varities and high external inputs resulted in monocrops and chemicalization of agriculture. Hence the native agricultural biodiversity destroyed. Introducing the concept of Conservation Agriculture (Resource Conserving Technologies) was an important breakthrough for sustaining productivity, natural resource base and economic growth of the farmers. Kansara (1995) and Deshpande (1996) had tried to update the vedic agricultural practices and promote these updated sustainable practices among the farmers of Maharastra. In recent case studies done in Tamil Nadu and Andra Pradesh by Winin Pereira (1990, 1993) and criticized the conventional western agricultural practices and stressed on importance updated traditional agricultural systems in India. Some of update methods of sustainable agriculture practiced in India are described below.

6.1 Bio-Fertilizers

Bio-fertilizers are microbial inoculants that artificially multiplied cultures of certain organism that fix nitrogen from air and improve soil fertility and crop productivity. This technique is known in India since ancient times but the role of biological nitrogen fixation was discovered more than century ago and commercial production of such biological process is recent interest and practices. Commonly discovered bio-fertilizers used in India are *Rhyzobium, Azotobacter, Azospirillum, Blue green Algae, Azolla and Phosphate solubilizing (Ghosh 2007). These microorganisms can only survive and play a commercially viable role in healthy living soil condition without ploughing. Bio-fertilizers* offer a new scientific approach to Indian agriculture holding a promise to balance many of the shortcomings of the conventional chemical based technology and certainly will help in sustain India agriculture in era of climate change. It is a product that is likely to be commercially promising and pragmatic impact in the long run once information becomes circulated adequately to producers and farmers through experience and communication.

6.2 Vermicompost

Vermitecnology is an important part of biotechnology where earthworms are used various type of organic wastes into valuable resources (Padma et al. 2002; Prabha et al. 2005). It is very much rich in nutrients like 1.5% nitrogen, 0.5% phosphorous and 0.8% potassium as well as other micro-nutrients. Earthworms are considered as the key-stone organism in regulating nutrient cycling processes in many ecosystems (Tripathi and Bhardwaj 2004). These worm eat almost any type of organic matter including bones, egg shells and consume their own weight of residue everyday converting into nutrient casts (Rangasamy and Jayanthi 2001). It is successful used in India to clean the environment as it is used a waste as raw material and covert the polluted costly conventional chemical farming to sustainable agriculture (Lal et al. 2003).

6.3 Zero Tillage System

This is one of the most important principle of sustainable agriculture, where the seed is placed into soil by a seed drill without prior land preparations. This approach has been tested and being practiced over 2 million hectares in India (RWC-CIMMYT 2005). This technology has high relevance in higher yielding and more mechanized area like northern western India. It has been by several scientist in India that no till system is better than conventional system in agriculture because it saves fuel, labour, irrigation, production cost, cut in emission of carbon dioxide along with positive health in soil health and environmental quality (Gupta et al. 2002, 2005; Ladha et al. 2003, Malik et al. 2005). It is proved to be an important step in conserving and sustaining Indian agriculture in scenario of global environmental change.

6.4 *Laser Land Leveling*

Unevenness of soil surface has major impact on farming operations, energy use cropping stand due to inhomogeneous water distribution and soil moisture. Therefore the land leveling is precursor to good agronomic, soil and crop management practices. The general, practices of land leveling used by the farmers in India is either through use of plankers drawn by draft animals or by small tractors. Farmers in Indo–Gangatic Plains especially Punjab, Haryana and Uttar Pradesh are using iron scrappers/ leveling boards drawn by 4-wheel tractors (Jat et al. 2002). Hence the tradition methods of land leveling are not only cumbersome and time consuming but also more expensive. Laser land leveling are few recent techniques used as inputs in intensively cultivated irrigated farming that meets the twin objective of achieving a better crop stand, save irrigation water and improves the input use efficiency (Jat et al. 2005).

6.5 *Raised Bed and Furrow Planting System*

This planting technology permits growing crops on beds with less water. During last decades a raised bed planting technique has been emerge in greater pace in Indo-gangatic plain (Dhillon et al. 2000). The major benefit from this system is to enhance the productivity and save the irrigation water. Typical irrigation by using this system are from 18% to 35% (RWC-CIMMYT 2003a, b; Hobbs and Gupta 2003; Jat et al. 2005). Farmers in India are now days are much interested in these techniques because of reducing the mechanical cultivation of weeds, reduced irrigation time and improve yields on fine textured soil prone to water logging.

6.6 *Crop Diversification*

It is like inclusion of certain crops in sequential and intercropping system help in mitigating the environmental problems arising on account of monoculture and obnoxious weeds. Nitrate leaching are threats inevitable under these agricultural production systems. Choices of appropriate cropping systems and management help in minimizing nitrate leaching. Crop diversification in India is slowly picking up momentum in favour of legume intercropping in cereals. The nature of diversification differ across the region due to existence of wide hetrogeneity in agro-climatic and socio-economic environments. This method will lead to sustainable farming and will be thrust for future farming in India.

6.7 *Crop Residue Management*

Crop residues are good source of plant nutrients (25% of nitrogen (N) and phosphate (P), 50% of sulphur (S) and 75% of potassium) and also important component for stability of agricultural systems (Jat et al. 2004). Burning of crop residue leads to loss of considerable amount of N, P, K and S (Sidhu and Beri 1989) but also contribute to global NO_2 and CO_2 budget (Grace et al. 2002) and also destruction of beneficial micro-flora of the soil (Jat and Pal 2000; Timsina and Connor 2001). Residue characteristics and soil management factors affect residue decomposition in the soil. Sufficient efforts are being made during recent past to develop efficient technologies but the development of new generation drills has been proved to be significant achievement in direction of conservation of agriculture.

7 Conclusion

Concerns about climate change are global and real. As all communities try to get adapted to the challenges of their local climate, they are today sensitive to its variations. Indian agricultural systems are threatened by the predicted effects of climate change because of their economic dependence on climate for development whose backbone is agriculture. The task of bringing in a paradigm change in India's agricultural system is by no means going to be a straight forward task. Fortunately, the Government of India has been constantly evolving in response to emerging challenges and is well placed for an accelerated evolution, consistent with new challenges facing the system. In the last couple of decades, the government has been implementing many plan and programmes for the development of the agricultural lands and non-agricultural lands whose implementations have added significant values to the sustaining food and land productivity in the case of evolving climatic vulnerability.

The vulnerability and risk of Indian agriculture due to weather fluctuations and climate variability can be minimized if future weather variation can be adequately predicted and a suitable process-based eco-physiological crop yield forecasting model can be identified to produce real-time yield forecasts. Scientists and farmers must join efforts to further understand crop–climate relationships and formulate viable, locally adapted production technologies that will address critical issues such as climate variability. In this context, conservation agriculture has been identified as the immediate as well as long-term solution. Conservation agriculture systems with suitable crop management packages will be the future technique for addressing the sustainability of agriculture in India. The major components of sustainability of Indian described above includes minimal soil disturbance, maintaining soil cover and crop diversification. This can only address the increasing population and shrinking land resources and impending food crisis in India. This methods of farming can only supported by interventionist action precludes traditional models of research and support systems and requires alternative but urgent programmatic interventions, led by farmers' institutions and their local resources, knowledge and innovations. Finally we proposed an integrated assessment exercise to evaluate climate change impact on agriculture.

References

Aggarwal PK (2007) Climate change: implications for Indian agriculture. Hydrol Rev 22:37–46

Aggarwal PK, Joshi PK, Ingram JSI, Gupta RK (2004) Adapting food systems of the Indo-Gangetic plains to global environmental change: key information needs to improve policy formulation. Environ Sci Policy 7:487–498

Attri SD, Rathore LS (2003) Simulation of impact of projected climate change on wheat in India. Int J Climatol 23:693–705

Awal et al. (2006) Radiation interception and use by maize/peanut intercrop canopy. Agric For Meteorol 139:74–83

Cazenave A, Nerem R (2004) Present-day sea level change: observations and causes. Rev Geophys 42:1–20

Chaudhary A, Aggarwal PK (2009) Impact of climate change on food security in India. In: Proceedings of International Conference on Energy and Environment, 19–21 March 2009 ISSN 2070-3740

Chhetry GKN, Belbahri L (2009) Indigenous pest and disease management practices in traditional farming systems in north east India. A review, J Plant Bree Crop Sci 1(3): 028–038

Church J, Gregory J, Huybrechts P, Kuhn M, Lambeck K, Nhuan M, Qin D, Woodworth P, 2001. Changes in sea level. Houghton J, Ding Y, Griggs D, Noguer M, van der Linden P, Dai X, Maskell K, Johnson C, Eds. In: Climate change 2001. The scientific basis (published for the intergovernmental panel on climate change). Cambridge University Press, Cambridge, New York, pp. 639–693, (Chapter 11)

Church JA, White NJ (2006) Sea-level rise at tropical Pacific and Indian Ocean islands. Global & Planetary Change, 53:155–168

Church J, White N, Coleman R, Lambeck K, Mitrovica J (2004) Estimates of the regional distribution of sea-level rise over the 1950 to 2000 period. J Climate 17:2609–2625

Commission P (1997) Ninth five year plan: 1997–2002 vol 2: thematic issues and sectoral programmes. Planning Commission, New Delhi

Cure JD, Acock B (1986) 'Crop responses to carbon dioxide doubling – a literature survey'. Agric For Meteorol 38:127–145

Dash SK, Jenamani RK, Kalsi SR, Panda SK (2007) Some evidence of climate change in twentieth-century India. Clim Change 85:299–321

Dash SK, Kulkarni MA, Mohanty UC, Prasad K (2009) Changes in the characteristics of rain events in India. J Geophys Res 114:D10109

Deshpande MS (1996) Rishi-Krishi. Shri Samarth Farm Sanshodhan Kendra, Khede-Ajra, Kolhapur

Dhillon SS, Hobbs PR Samra JS (2000) Investigations on bed planting system as an alternative tillage and crop establishment practice for improving wheat yields sustainably In: Proceedings 15th conference. International Soil Tillage Research Organisation. Fort Worth, 2-7 July 2000

Dinar et al. (1998) Measuring the impact of climate change on Indian agriculture. Washington, DC: World Bank Technical Paper 402

Douglas I (2009). Climate change, flooding and food security in south Asia. Food Secur 1, 127–136

FAO (2005) Impact of climate change, pests and diseases on food security and poverty reduction. Special event background document for the 31st Session of the Committee on World Food Security, Rome, 23–26 May 2005

FAO (2006) FAO Statistical yearbook (2005–2006). Food and Agricultural Organization, 2006. URL http://www.fao.org/statistics/yearbook/.

Friedrich T, Gustafson D (2007) Conservation agriculture: synergies of resource-conserving technologies in rice-based systems. Int Rice Comm Newsl 56:25–31

Gadgil S (1995) Climate change and agriculture – an Indian perspective. Curr Sci 9(8):649–659

Ghosh SN, Tarai RK (2007) Indian Journal of Fertilisers, 3:47–49

Government of India (2001) The report of the working group on organic and biodynamic farming, Planning Commission, Government of India, pp 1–25

Grace PR et al. (2002) Environmental concerns in rice-wheat system. In: Proceedings of the International Workshop on Developing Action Programme for Farm level Impact in Rice-Wheat system of the Indo-Gangetic Plains held during 25–27 September 2000 at New Delhi, India. Rice-Wheat Consortium paper series 14, New Delhi, India. pp 99–111

Gupta et al. (2002) Adopting conservation agriculture in rice-wheat systems of the Indo-Gangetic plains–new opportunities for saving on water. Paper presented at the "Water wise rice production workshop", 5–10 April 2002, IRRI, Philippines

Gupta RK, Jat ML, Sharma SK (2005) Resource conserving technologies for water savings and enhancing productivity. In: Proceedings of the National Symposium on efficient water management for eco-friendly sustainable and profitable agriculture, New Delhi, 1–3 Dec 2005, pp 181–182

Hobbs PR, Gupta RK (2003) Resource conserving technologies for wheat in rice-wheat systems. In: Ladha JK, Hill J, Gupta RK, Duxbury J, Buresh RJ (eds) Improving the productivity and sustainability of rice-wheat systems: issues and impact, vol 65, ASA Special Publication. ASA, Madison, pp 149–171, Chapter 7

Holgate SJ, Woodworth PL (2004) Evidence for enhanced coastal sea level rise during the 1990s. Geophys Res Lett 31:L07305

Houghton J, Ding Y, Griggs D, Noguer M, van der Linden P, Dai X, Maskell K, Johnson C (eds) (2001) Climate change 2001, The scientific basis (published for the intergovernmental panel on climate change). Cambridge University Press, Cambridge/ New York

Jacobs CMJ, DeBruin HAR (1992). 'The sensitivity of regional transpiration to land surface characteristics: significance of feedback'. J Climate 5, 683–698

Jat ML, Pal SS (2000) In-situ decomposition of rice straw using cellulose-degrading fungi in a typic Ustochrepts of Indo-Gangetic plain under rice-wheat system. In: Proceedings of national symposium on agronomy: challenges and strategies for the new millennium held during at Gujrat Agricultural University, Junagadh, 15–18 Nov 2000

Jat ML, Pal SS, Singh R (2002) Crop residue treatment with mineral salts and its recycling under different tillage management practices and their impact on crop yields and soil health in a rice-wheat cropping system in Indo-Gangetic plains. In: 17th world congress of soil science, Bangkok, 14–21 Aug 2002

Jat ML, Pal SS, Shukla L, Mathur JMS, Singh M (2004) Rice residue management using cellulolytic fungi and its effect on wheat yield and soil health in rice-wheat cropping system. Indian J Agric Sci

Jat ML, Sharma SK, Gupta RK, Sirohi K, Chandana P (2005) Laser land leveling: the precursor technology for resource conservation in irrigated eco-system of India. In: Abrol IP, Gupta RK, Malik RK (eds) Conservation agriculture-status and prospects. CASA, New Delhi, pp 145–154

Kansara NM (1995) Agriculture and animal husbandry in the Vedas, New Delhi: Nag Publishers

Kavikumar KS (2007) Climate change and Indian agriculture: impacts, vulnerability and adaptation assessment, Faculty Seminar at Madras School of Economics, 31 August 2007

Krupa S (2003). Atmosphere and agriculture in the new millennium. Environ Pollut 126, 293–300

Kumar KSK (2007) Climate change studies in Indian agriculture. Economic and political weekly. 42(13):15–18

Kumar KSK, Parikh J (2001). Socio-economic impacts of climate change on Indian agriculture. Int Rev Environ Strateg 2(2), 277–293

Kumar KSK, Parikh J (2001b) Indian agriculture and climate sensitivity. Global Environ Change 11(2):147–154

Ladha JK, Hill J, Gupta RK, Duxbury J, Buresh RJ (eds) (2003) Improving the productivity and sustainability of rice-wheat systems: issues and impact, vol 65, ASA Special Publication. ASA, Madison

Lal OP, Srivastava YN, Sinha SR (2003) Vermicomposting. Indian Farming 52:6–8

Lavalle C, Micale F, Houston TD, Camia A, Hiederer R, Lazar C, Conte C, Amatulli G, Genovese G (2009) Climate change in Europe. 3. Impact on agriculture and forestry. A review. Agron Sustainable Dev 29(3):433–446

Lemon ER (1983) CO2 and plants: the response of plants to rising levels of atmospheric carbon dioxide. West view Press, Boulder

Ligon E, Sadoulet E (2007) Estimating the effects of aggregate agricultural growth in the distribution of expenditures. Background paper for the World Development Report 2007

Lomte MH, Pendke MS (2004) Organic farming inventory. In: Proceedings of the state level seminar on present and future prospects of organic farming, Marathwada Agricultural University, Parbhani, pp 1–133

Mahale P, Sorée H (1999) "Ancient visions and new experiments of rural people". In: Food for thought. India Books for Change, Bangalore

Mall RK, Singh R, Gupta A, Srinivasan G, Rathore LS (2006) Impact of climate change on Indian Agriculture. Clim Change 78:445–478

Malik et al. (2005) Zero Tillage–The voice of farmers. Hisar (Haryana): Directorate of extensive education

McNaughton KG, Jarvis PG (1991) 'Effects of spatial rate on stomatal control of transpiration'. Agric For Meteorol 54:279–302

Ministry of Agriculture (2001b) Statistics at a glance. Government of India

Ministry of Finance, Government of India (2001) Economic Survey 2000–2001. Ministry of Finance, Government of India, New Delhi

Mishra BB, Nayak KC (2004) Organic farming for sustainable agriculture, Orissa Review. Monthly Journal of the Government of Orissa. Published in the month of October 2004. pp 42–46

Misra S, Maikhuri RK, Dhyani D (2008 June) Indigenous soil management to revive below ground biodiversity case of Garhwal. LEISA INDIA, p 13

Morison JIL (1987) Intercellular CO2 concentration and stomatal response to CO2. In: Zeiger E, Cowan IR, Farquhar GD (eds) Stomatal function. Stanford University Press, Stanford, pp 229–251

Padma U, Rao RS, Srinivas N (2002). Eco-friendly disposal of vegetable wastes through vermitechnology. J Ecobiol 14, 155–159

Pereira W (1990) Asking the Earth, Goa: the Other India Press

Pereira W (1993) Tending the Earth. Bombay: Earthcare Books

Planning Commission (2002) Report of the steering group on agriculture and allied sectors for the tenth five year plan 2002–2007, Government of India, New Delhi

Prabha LM, Jayraaj IA, Jeyaraaj R (2005) Macro and Micronutrient changes in vermicomposting of vegetable wastes using Eudrilus eugeniae. South Asian J Socio Political Stud 2:129–130, 156

Randhawa MS (1986) A history of agriculture in India 1980–1986, vol I–IV. Indian Council of Agricultural Research, New Delhi

Rangasamy A, Jayanthi C (2001) Integrated farming systems – a boon to farming community. www.tnau.ac.in/scms/Agronomy/IFS.htm

Rao VB, Ferreira CC, Franchito SH, Ramakrishna SSVS (2008). In a changing climate weakening tropical easterly jet induces more violent tropical storms over the north Indian Ocean. Geophys Res Lett 35, L15710

Ravi A (2008) Climate change risk: an adaptation and mitigation agenda for Indian cities. Environ Urban 20(1):207–229

Reddy PP, Reddy OS (1991) Reproductive performance in population exposed to pesticides in cotton fields in India. Environ Res 55:123

Rotter R, Van de Geijn SC (1999) Climate change effects on plant growth, crop yield and livestock. Climate Change 43(4):651–681

RWC-CIMMYT (2003a) Addressing Resource conservation issues in rice-wheat systems of South Asia: a resource book. Rice-Wheat Consortium for the Indo-Gangetic Plains-International Maize and Wheat Improvement Center, New Delhi, p 305

RWC-CIMMYT (2003b) Agenda notes 11th Regional Technical Coordination Committee

RWC-CIMMYT (2005) Agenda notes 13th Regional Technical Coordination Committee

Sahai VN (1982) Principles and practices of crop production. Inter publication, New Delhi

Sanghi A, Mendelsohn R (2008) The impacts of global warming on farmers in Brazil and India. Global Environ Change 18:655–665

Sarma VV (2008) Multiple cropping to increase agrobiodiversity and sequester carbon dioxide. Curr Sci 95(7), 10 Oct 2008, p 815

Selvaraj S et al. (2002) Use of vulnerability indices for agriculture. Draft Report. www.unep.org/dpdl/indiaworkshop/documents/TS3_2_1.doc

Shukla PR (2003) Climate change and India. Universities Press India, Hyderabad

Sidhu BS, Beri V (1989) Effects of crop residue management on the yields of different crop and on soil properties. Biological wastes 27:15–27

Sinha SK (1997) Global change scenario – current and future with reference to land cover change and sustainable agriculture. Curr Sci 72(11):846–854

Smit B, Smithers J (1994) Sustainable agriculture: interpretations, analyses and prospects. Canadian J Reg Sci 16(3):499–524

Sofia PK, Prasad R, Vijay VK (2006) Organic farming- Tradition reinvented. Ind J Tradit Knowl 5:139–142

Solomon S, Qin D, Manning M, Alley RB, Berntsen T, Bindoff NL, Chen Z, Chidthaisong A, Gregory JM, Hegerl GC, Heimann M, Hewitson B, Hoskins BJ, Joos F, Jouzel J, Kattsov V, Lohmann U, Matsuno T, Molina M, Nicholls N, Overpeck J, Raga G, Ramaswamy V, Ren J, Rusticucci M, Somerville R, Stocker TF, Whetton P, Wood RA, Wratt D (2007) Technical summary. In: Solomon S, Qin D, Manning M, Chen Z, Marquis M, Averyt KB, Tignor M, Miller HL (eds) Climate change 2007: the physical science basis. contribution of working group I to the fourth assessment report of the intergovernmental panel on climate change. Cambridge University Press, Cambridge/New York

Swaminathan MS (2002) Climate change, food security and sustainable agriculture: impacts and adaptation strategies. Shukla PR, Sharma KS, Venkata Ramana P, Eds. In: Climate change and India: issues, concerns and opportunities. Tata McGrawhill, New Delhi, pp 196–216

Thomas R, Rignot E, Casassa G, Kanagaratnam P, Acuna C, Akins T, Brecher H, Frederick E, Gogineni P, Krabill W, Manizade S, Ramamoorthy H, Rivera A, Russell R, Sonntag J, Swift R, Yungel J, Zwally J (2004) Accelerated sea-level rise from West Antarctica. Science 306:255–258

Timsina J, Connor DJ (2001) Productivity and management of rice-wheat cropping systems: issues and challenges. Field Crops Res 69:93–132

Tol RSJ (2002) New estimates of the damage costs of climate change, Part I: Benchmark estimates. Environ Resour Econ 21(1):47–73

Tripathi G, Bhardwaj P (2004) Comparative studies on biomass production, life cycles and composting efficiency of Eisenia fetida (Savigny) and Lampito mauritii (Kinberg). Bioresour Technol 92:275–283

Vandermeer J (1989) The ecology of intercropping. New York: Cambridge University Press

Vani MS (1992) Role of panchayat institutions irrigation management: law and policy: Tamilnadu and Karnataka. Indian Law Institute, New Delhi

Vani MS (2002) "Customary law and modern governance of natural resources in India – conflicts, prospects for accord and strategies". In: Pradhan R (ed) Legal pluralism and unofficial law in social, economic and political development. Papers of the XIIIth international congress, vol I, Chiang Mai, ICNEC, Kathmandu, 7–10 Apr 2002, pp 409–446

Vasudeva G (2002). Environmental security: a South Asian perspective. Tata Energy and Resources Institute, Arlington

Wall E, Smith B (2005) Climate change adaptation in light of sustainable agriculture, Journal of sustainable agriculture 27(1):113–123

Whetton P (1994) Constructing climate scenarios: the practice. In: Jakeman AJ, Pittock AB (eds) Climate impact assessment methods for Asia and Pacific. Australian International Development Assistance Bureau, Canberra, pp 56–92

World Bank (2008) Climate change impacts in drought and flood affected areas: case studies in India. Report No. 43946-IN, India Country Management Unit, New Delhi

World Resources Institute (WRI) in collaboration with United Nations Development Programme, United Nations Environment Programme, and World Bank (2005) World resources 2005: the wealth of the poormanaging ecosystems to fight poverty. WRI, Washington, DC

Zardhari V (2000) "Barah anaaj" – twelve food grains: traditional mixed farming system. LEISA India 2(3):25

Ecological Threats and Agricultural Opportunities of the Aquatic Cane-Like Grass *Phragmites australis* in Wetlands

Andreas P. Mamolos, Anna E. Nikolaidou, Athina K. Pavlatou-Ve, Sofia K. Kostopoulou, and Kiriaki L. Kalburtji

Abstract Wetlands are some of the most biologically productive and dynamic natural ecosystems with multiple value for man and nature. Indeed wetlands provide goods and service such as water storing, floodwater trapping, and trapping of sediment and pollutants. Wetlands also affect climate change by absorbing CO_2, storing and releasing heat, and harnessing sunlight using a rich variety of vegetation that supports animal life. However wetlands can be polluted by industrial and commercial operations, agricultural runoff and storm water. Wetlands are degraded by filling in and drainage for land development. Wetlands are also degraded by dredging for commercial and recreational water traffic. Dam construction and irrigation roads change the hydrological status of wetlands.

Wetlands allow the growth of aquatic macrophytic vegetation such as the emergent *Phragmites australis*. *P. australis* is mainly a clonal plant occurring in natural areas. *P. australis* invasion in wetlands alters the structure and function of the ecosystem by reducing plant and animal biodiversity and changing hydrological regimes and nutrient cycles. This invasion leads to less food or cover for wildlife, decreased use of an area for recreational purposes and decreased availability of drinking and irrigating water. Several methods have been applied in order to control *P. australis* growth. Control methods include non ecological methods such as burning that releases CO_2 in the atmosphere and chemical control with non-specific herbicides. Control methods include also ecological methods such as grazing and removal of above-ground biomass by cutting. Cuttings can in turn be used for on-farm application as green manure or sludge after energy production and compost production.

A.P. Mamolos (✉) · A.E. Nikolaidou · K.L. Kalburtji
Faculty of Agriculture, Laboratory of Ecology and Environmental Protection, Aristotle University of Thessaloniki, 541 24 Thessaloniki, Greece
e-mail: mamolos@agro.auth.gr

A.K. Pavlatou-Ve · S.K. Kostopoulou
Faculty of Agriculture, Laboratory of Soil Science, Aristotle University of Thessaloniki, 541 24 Thessaloniki, Greece

E. Lichtfouse (ed.), *Genetics, Biofuels and Local Farming System*, Sustainable Agriculture Reviews 7, DOI 10.1007/978-94-007-1521-9_9,

An economic and ecological basis for accepting a beneficial role for *P. australis* is lacking. This report focuses on the possibility that people living near wetlands could produce compost using plant material from *P. australis*. Suitable use of compost in agriculture could reduce the fertilizer application and thereby reduce the environmental pollution, improve food security and soil productivity, and increase sustainability in the agroecosystems. It can also play role in the climate change because part of the organic C is released as CO_2 in the atmosphere and the rest is consumed by the decomposers.

Keywords Biodiversity • Common reed • Compost production • Management of reed • Organic farming • Vegetation • Wetlands • Burning • CO_2 • Grazing • Draining • Dredging • Disk harrowing • Compost maturity

1 Introduction

Wetlands allow the growth of natural herbaceous vegetation throughout the growing season or part of it. Studies on wetlands have identified a combination of mechanisms responsible for the commonly observed plant zonation. These mechanisms were reviewed by Keddy (2010) and include ecological succession, physical factors, and biological interactions. Gradients of hydrological conditions are broadly accepted as major factors underlying species distributions (e.g. Grootjans et al. 1996; Wheeler and Proctor 2000; Maltby and Barker 2009). Water regime in wetlands affects the mineralization of organic matter, controlling therefore the nutrient availability and creating a sharp gradient of productivity, which in turn affects the distribution of species by shoot (light) competition (Kotowski et al. 2001; Kotowski and van Diggelen 2004). Changes in species composition along the moisture gradient seem to be determined by the species physiological tolerance limits (Lenssen et al. 1999) as well as by competition (Keddy 1989; Gaudet and Keddy 1995; Lenssen et al. 1999). Disturbances have also a profound effect on species distribution. High water levels in many temperate lakes kill dominant emergent and woody plants and subsequent low water periods allow many plant species and vegetation types to regenerate from buried seeds (Keddy and Reznicek 1986). Thus, fluctuating water levels increase the area of shoreline vegetation and create very diverse conditions, which enhance the diversity of vegetation types and plant species (Keddy and Reznicek 1986; Verhoeven 1992; Hill et al. 1998; Maltby and Barker 2009). On a smaller disturbance scale, grazing is often considered as a major form of disturbance (Grime 2002). Also, mowing affects species richness by changing the competitive relationships between species. Grazing and management alteration, in combination with the intensification of agricultural practices, resulted in considerable changes of the wetland vegetation and, in the long run, a decline in many plant species (Grevilliot et al. 1998; Dupré and Diekmann 2001).

Aquatic macrophytic vegetation consists mainly of the species *Lemna minor*, *Lemna gibba*, *Polygonum amphibium*, and *Potamogeton perfoliatus*, as well as the emergent *Phragmites australis*, *Scirpus lacustris*, *Typha* sp. and *Alisma* sp. (Zalidis and Mantzavelas 1994). *Phragmites australis* [(Cav.) Trin. ex Steudel] (hereafter *P. australis*), is a widely distributed clonal grass species, ranging all over Europe, Asia, Africa, America, and Australia (Holm et al. 1977; Asaeda et al. 2002). It is a tall perennial grass with cane-like stalks that grows in both aquatic and terrestrial environments, and is most commonly found in coastal wetlands. It can reproduce sexually through seeds, but its primary means of reproduction is asexually through rhizomes (Wilcox et al. 2003). Rhizomes are thick stem-like structures that spread underneath the soil. They form a mat of entangled structures that are capable of producing shoots. The thick, tangled mats of rhizomes inhibit other species growing amongst them.

Once established, *P. australis* forms dense monospecific stands, which contribute to clogging and overgrowth of water systems and decrease the use of an area for recreational purposes (Hansson and Fredriksson 2004). This results in economic damage. Chemical control of *P. australis* suggests habitat loss (Tewksbury et al. 2002). Also, *P. australis* invasion alters the structure and function of diverse wetland ecosystems by changing nutrient cycles and hydrological regimes (Benoit and Askins 1999; Meyerson et al. 2000). An economic and ecological basis for accepting a beneficial role for *P. australis* is lacking. Management methods for *P. australis* eradication are common in North America (Marks et al. 1994; Sun et al. 2007), whereas in Europe, scientific efforts are taken, made designed to understand and reverse *P. australis* decline, e.g., van der Putten (1997). In wetlands, *P. australis* is generally a nuisance species. It is an efficient colonizer of areas characterized by soil disturbance (Ailstock et al. 2001), pollution, alteration of the natural hydrologic regime, dredging, increased sedimentation (Marks et al. 1994), and increases in nutrient concentrations due to eutrophication (Hansson and Fredriksson 2004). *P. australis* acts as a climax species thereby forming extensive monocultures and excluding other native plant species (Keller 2000; Silliman and Bertness 2004), which are important as food or cover for wildlife (Ailstock et al. 2001). This means a reduction in plant and animal biodiversity (Russell and Kraaij 2008).

The aim of this review is to explore the possibility of a friendly contribution to environment, by using *P. australis* plant material. People, near wetlands, could produce compost from *P. australis*, and then use it as fertilizer and soil amendment in organic, integrated and conventional farming.

2 Wetlands

2.1 Definitions

One should be careful defining wetlands, since flawed definitions and flawed procedures can lead to unnecessary constrains on land development, or to loss of important wetland habitats. According to Keddy (2010) "*A wetland is an ecosystem that arises when inundation by water produces soils dominated by anaerobic processes and forces the biota, particularly rooted plants, to exhibit adaptations to tolerate flooding*". In an ecological context, wetlands are intermediate between terrestrial and aquatic ecosystems. This definition has a complex structure: there is a cause (inundation by water), a proximate effect (reduction of oxygen levels in the soil) and a secondary effect (the biota must tolerate both the direct effects of flooding and the secondary effects of anaerobic conditions).

Wetlands can also be defined as ecosystems depending on constant or recurrent, shallow inundation or saturation at or near the surface of the substrate (Vymazal et al. 1998). A scientific definition is a tool for the analysis of nature. The definition describes the domain of inquiry of wetland ecology. The prediction of legislation designed to protect wetlands, and regulate their use, therefore requires a different kind of definition. The Ramsar Convention is an international treaty for wetland conservation. According to the 1971 Ramsar Convention, wetlands are defined as areas of marsh, fen, peat land, or water. These wetlands may be natural or artificial, permanent or temporary, with water that is static or flowing, fresh, brackish, or salty including areas of marine water, the depth of which at low tide does not exceed 6 m (Matthews 1993). The Committee on Characterization of Wetlands (1995) developed a definition to stand outside of any single agency, policy or regulation, which is "*A wetland is an ecosystem that depends on constant or recurrent, shallow inundation or saturation at or near the surface of the substrate. The minimum essential characteristics of a wetland are recurrent, sustained inundation or saturation at or near the surface and the presence of physical, chemical, and biological features reflective of recurrent, sustained inundation or saturation. Common diagnostic features of wetlands are hydric soils and hydrophytic vegetation. These features will be present except where specific physiochemical, biotic, or anthropogenic factors have removed them or prevented their development*".

2.2 Functions

Wetlands are dynamic ecosystems of great complexity, which perform a large number of functions, such as water storing, floodwater trapping, trapping of sediment and pollutants, absorbing CO_2, storing and releasing heat, and harnessing sunlight to create a rich variety of vegetation that supports animal life (Greeson et al. 1979; Adamus and Stockwell 1983; Skinner and Zalewski 1995). More analytically, wetlands functions are:

(a) *Water storage*. Wetlands store precipitation and surface water and then slowly release it into associated surface water resources, ground water, and the atmosphere. Wetlands play a critical role in regulating the movement of water within watersheds as well as in the global water cycle (Richardson 1994, 1995; Mitsch and Gosselink 1993).

(b) *Floodwater trapping*. The size, shape, location, and soil type of a wetland determine its capacity to reduce local and downstream flooding. Wetlands can provide flood protection by holding the excess runoff after a storm, and then releasing it slowly (Hollis 1992). While wetlands cannot prevent flooding, they do lower flood peaks by holding water temporarily and slowing its velocity (Hollis 1992).

(c) *Trapping of sediments and pollutants.* Wetlands may act as filters of sediments and organic material. Also, they protect water quality by trapping sediments and retaining excess nutrients and other non point-pollutants such as heavy metals (Leonardsson 1994; Yeh 2008). This function is especially important when a wetland is connected to groundwater or surface water sources (such as rivers and lakes) that are in turn used by humans for drinking, swimming, fishing, or other activities. Sediments, nutrients, and toxic chemicals enter wetlands primarily with "runoff," a term used to describe the rain and storm water that travels over land surfaces on its way to water collectors. Runoff water may carry pesticides and fertilizers. Toxic chemicals reach surface waters in the same way as nutrients, and can have detrimental effects on plants and animals (including humans). Wetlands protect water from the problems of nutrient overload as nutrients are used by wetland plants, and bound or converted to less harmful chemical forms in the soil.

(d) *Carbon dioxide (CO_2) absorption.* Wetlands store carbon in peat and soil organic matter. Storing carbon is an important function within the carbon cycle, particularly given the observations of increasing levels of carbon dioxide in the atmosphere and concerns about global warming. When wetlands are drained, the oxidizing conditions enhance organic matter decomposition and the release of CO_2. When wetlands are preserved or restored, they act as a sink for carbon since organic matter decomposition is stable or slowed down. Climate change threatens the stability of wetlands because temperature rise lowers water tables and consequently increases stored carbon decomposition rates. Increased flux of CO_2 contributes to further amplification of greenhouse gas production (Davidson and Janssens 2006).

(e) *Heat storage and release.* All water reservoirs are storing heat during the warm periods and release it during cool periods. This is related to the high thermal conductivity and specific heat of water. As a consequence the temperature variation between day and night or summer and winter is less near water bodies (Gerakis and Koutrakis 1996).

(f) *Food chains.* Wetlands host photosynthetic organisms and small animals, the latter being essential links at the lowest levels of the food chain, as they support the larger animals that feed on them.

2.3 Values

Wetlands are some of the most biologically productive natural ecosystems in the world comparable to tropical rain forests and coral reefs regarding their productivity and the diversity of species they support (Keddy 2010). The wetlands are known for their multiple values for man and nature (Skinner and Zalewski 1995), meaning the goods and services they supply. The functional analysis of the wetlands is an objective procedure, while the identification of their values can be characterized subjective. Wetland values are not found the same in all wetlands but it is sure that there is no wetland without any value. There is a large number of wetland values, as they are described by Gerakis and Koutrakis (1996) and Schuyt and Brander (2004), such as biological value, drinking-water supply, irrigation, fisheries, grazing, hunting, wood supply, hydroelectricity, salt extraction, sand extraction, scientific value, educational value, cultural value, recreation, flood control, protection against erosion, improvement of water quality and climatic control.

2.4 Threats

Wetlands may be polluted by industrial and commercial operations, agricultural runoff, storm water and other sources (Alho et al. 1988; Harris et al. 2005). Pollutants include sediment, excess nutrients, influx trace metals and organic pollutants.

Wetlands have been destroyed or degraded by filling in or drainage for land development, and by dredging for commercial and recreational water traffic (Best et al. 1993). Mechanical clearance of vegetation in wetland corridors such as ditches can render them impassable to many species and contributes to the isolation of populations, increasing the danger of extirpation. Also, the stabilization of water levels has resulted in decreased productivity and modification of the wetland aquatic plant community structure and diversity.

According to Best et al. (1993) dam construction, irrigation webs etc. change the hydrological status of wetlands. Dam construction to regulate rivers flow provide the appropriate water supply for urban development (for domestic use or irrigation purposes) and prevent water runoff, sediments and nutrients influx (Poff et al. 1997). All the former affect species richness (Ward et al. 1999).

Although *P. australis*, especially in Europe, is beneficial for many human and wildlife needs (Cronk and Fennessy 2001), its emergent vegetation overgrowth that dominates aquatic habitats is often viewed as a potential threat to shallow ecosystems. Therefore, control of dramatic expansion of *P. australis* stands has drawn attention in the recent past (Marks et al. 1994; Asaeda et al. 2002, 2003). The need to control and/or manage the growth of *P. australis* stands arises from several factors. Apart from its robust growth and morphological characteristics, such as longer shoot length and higher density of stems compared to other species, this plant tends to negatively affect the habitat it occupies. The continued spread of *P. australis* reduces local and regional plant diversity, changes the composition and diversity of the fauna associated with these plants, and alters ecosystem function, particularly nitrogen cycling, within coastal marshes (Keller 2000; Meyerson et al. 2000; Warren et al. 2001).

3 *Phragmites australis* [(Cav.) Trin. ex Steudel]

3.1 Description of the Plant

P. australis is a tall (1.5–4 m) perennial rhizomatous emergent macrophyte. It forms two types of roots: (i) thick "soil roots" with few branches and extensive aerenchyma that provide adaptation to anaerobiosis and (ii) "aquatic roots" with numerous finely branched laterals and a less developed airspace system that ensures sufficient nutrient uptake (Koncalová 1990). It spreads by a system of rhizomes and stolons, which expands peripherally and provides a large surface that absorbs plant nutrients from the aquatic environment. The buds of the rhizomes mature and grow about 1 m horizontally and terminate in an upward apex, which grows into a vertical rhizome. The latter produces buds that will form more vertical rhizomes and also horizontal rhizome buds (Marks et al. 1994). The aerial stems arise from rhizomes. *P. australis* may produce great quantities of seeds.

P. australis biomass reaches its maximum in July, August or September. Root growth is more intense in spring (April-May) and autumn (mainly September), while rhizome growth is more intense in summer (June-July) and decreases during spring and winter (Engloner 2009). Individual rhizomes live from 3 to 6 years. Buds develop late in the summer, mature and go dormant until spring. The plants flower and set seed between July and September. After the seed set and physiological maturity, nutrients are translocated down into the rhizomes and the aerial stems die at the end of each growing season (Marks et al. 1994).

P. australis colonizes new areas by seeds or fragments of rhizomes (Marks et al. 1994), while in established stands it is mainly propagated by rhizomes (Engloner 2009). Seed germination increases with increasing temperature from 16°C to 25°C and decreases when salinity is above 20 ppt (2%) and the water depth is more than 5 cm (Marks et al. 1994). Seeds are dispersed by wind during the winter months or carried by birds.

P. australis grows in every continent except Antarctica (Tucker 1990). However, its dynamics differ between wetter and dryer sites (Güsewell and Klötzli 2000). *P. australis* is found both in alkaline and brackish wetlands and in highly acidic wetlands in temperate zones worldwide. It may also grow in tropical wetlands but it is absent from the Amazon Basin and central Africa (Roman et al. 1984). *P. australis* is often found with other wetland species such as *Spartina*, *Carex*, *Nymphaea*, *Typha*, *Glyceria*, *Juncus*, *Myrica*, *Triglochin*, *Calamagrostis*, *Galium* and *Phalaris* (Marks et al. 1994). It forms dense often monospecific stands (Fig. 1).

3.2 Positive Effects

Jiang et al. (2007) reported that *P. australis* has high nutrient uptake ability in aquatic environments. Due to that capacity, it is used as a vegetative filter for wastewater

Fig. 1 *P. australis* in Lake Mikri Prespa (photo by A. Nikolaidou)

treatment (Adler et al. 2008). The dense stands of *P. australis* prevent erosion of the lakeshore (Hawke and José 1996). It also provides habitat for small mammals, insects, reptiles and nesting sites for several species of birds (Marks et al. 1994) contributing in this way to biodiversity.

3.3 Problems

Morphological characteristics favor dominance of *P. australis* in wetlands. It has longer shoot length and higher density of stems compared to other species (Asaeda et al. 2006), as well as a rapidly expanding rhizome system (Haslam 1969). *P. australis* is capable of colonizing rapidly adjacent areas, especially when they are disturbed. Factors that favor invasion are soil disturbance (Ailstock et al. 2001), pollution, alteration of the natural hydrologic regime, dredging, increased sedimentation (Marks et al. 1994), and increases in nutrient concentrations due to eutrophication (Hansson and Fredriksson 2004). Invasion may also be attributed to the introduction of more invasive genotypes of *P. australis* (Sun et al. 2007).

Once established, *P. australis* forms dense monospecific stands. A healthy reedbed in a wetland is defined as "a homogenous, dense or sparse stand with no gaps in its inner parts, with an evenly formed lakeside borderline, without aisles, shaping a uniform fringe or large lobes, stalk length decreasing gradually at the lakeside border, but all stalks of one stand of similar height; at the landside edge the *P. australis* are replaced by sedge or woodland communities or by unfertilized grasslands" (Ostendorp 1989). However, the stands of *P. australis* often spread, excluding other native plant species (Keller 2000; Silliman and Bertness 2004), which are important as food or cover for wildlife (Ailstock et al. 2001). Consequently loss of plant species diversity leads to reduced faunal diversity (Russell and Kraaij 2008).

The dense monospecific stands of *P. australis* contribute to clogging and overgrowth of water systems and decrease the use of an area for recreational purposes (Hansson and Fredriksson 2004). The expansion of *P. australis* has been associated with high sediment accumulation rates (Rooth et al. 2003), decreased dissolved oxygen (Biddlestone et al. 1991), and nitrogen retention (Meyerson et al. 2000).

4 Management of *P. australis*

Several methods have been applied in order to manipulate *P. australis* growth, either separately or in combination, including removal of above-ground biomass by cutting, grazing, mowing or burning (Thompson and Shay 1985, 1989; Lee 1990; van Deursen and Drost 1990; Cowie et al. 1992; Kay 1995), hydrologic control through flooding (Hellings and Gallagher 1992), increasing the salinity, biological control through insect introduction (van der Toorn and Mook 1982; Tscharntke 1989) and chemical control with non-specific herbicides (Riemer 1973, 1976; Kay 1995). Up to date no organisms have been identified that can reduce *P. australis* propagation without damaging other plant species (Marks et al. 1994).

Different responses to the above mentioned management practices have been reported depending on the type, combination, timing, frequency, duration and extent of each method. Moreover, according to Güsewell and Klötzli (2000) *P. australis* response differs in different systems (northern or southern hemisphere, freshwater or saline wetlands etc.).

4.1 *Non-Ecological Management*

4.1.1 Burning

Marks et al. (1994) mentioned that burning does not reduce the growing ability of *P. australis* unless root burning occurs. However, since rhizomes are usually covered by a layer of soil, mud and/or water, root burning is not applicable in most cases. Burning favors the rapid establishment of full vegetative cover by removing leaf litter and therefore is likely to restore wetland functions (Ailstock et al. 2001). On the other hand burning is not always effective in controlling *P. australis* growth. Burning in the spring may stimulate the growth of young shoots if there are active rhizomes, due to the increased exposure to light. Thompson and

Shay (1985) found that shoot biomass was greater in plots burned in spring and in fall than in control plots and less in plots burned in summer. Burning resulted in higher total shoot densities and lower mean shoot weights than in control plots because of greater densities of shorter and thinner shoots. Belowground production also increased following spring and fall burns but not following summer burning. In winter, removal of litter had no effect or doubled above ground biomass (van der Toorn and Mook 1982), probably due to increased light availability and better aeration of the soil around the rhizomes (Granéli 1989).

Burning is not the appropriate management method in wetlands with a high conservation status or in wetlands close to urban areas. It temporarily leaves soil surface without vegetation cover, with a risk of nutrient release and loss from ash and surface erosion, depending on a site's topography and hydrology. Burning also reduces vegetation cover for faunal habitat during winter and may result in lower long-term plant biodiversity. It is also dangerous; *P. australis* can cause spot fires over 328 m away (Marks et al. 1994). Burning releases CO_2 in the atmosphere (Janhäll et al. 2010), so it contributes to global worming.

4.1.2 Herbicide Application

The most commonly used herbicide for *P. australis* control in aquatic ecosystems is Rodeo [Monsanto Co., St. Louis, MO, active ingredient: isopropylamine salt of glyphosate, N-(phosphonomethyl) glycine] (Ailstock et al. 2001). Glyphosate has a half life of 60 days. If applied properly, it has a minimum impact on other plant species and wildlife. However, according to Marrs et al. (1993) in areas where glyphosate is sprayed, no-spray buffer zones of 6–10 m and 20 m wide must be established for the protection of semi-natural vegetation and of seedlings, respectively. Jiraungkoorskul et al. (2003) found that long-term exposure to glyphosate at sublethal concentrations had deleterious effects on the fish *Oreochromis niloticus*. Often glyphosate does not kill below ground reserves, therefore *P. australis* may return after glyphosate application.

After the herbicide application the dead aboveground biomass of *P. australis* may increase invertebrate populations that feed on the detritus (Weis and Weis 2003). Also O_2 may reach the soil through stems of dead biomass leading to increased decomposition rates and providing suitable conditions for the growth of other species (Gries et al. 1990). One-time application of glyphosate or glyphosate application followed by burning reduced the abundance of *P. australis* and increased plant species abundance and diversity in a non-tidal wetland for 2–3 years (Ailstock et al. 2001). One-time application of herbicide led to reduced re-growth of other plant species and wetland functions in the first year due to the shading effect of *P. australis* dead shoots and to the biomass that accumulated on the soil surface. Plant diversity increased linearly during the next 2 years due to the diverse microhabitats that were created and to differences in light and soil moisture (Ailstock et al. 2001). One-time herbicide application followed by burning increased plant diversity and wetland functions in the first year. Later on plant diversity was reduced due to the expansion of a new plant species at the area. There were no changes in the invertebrate groups that inhabited the area due to the different management practices that were used (Ailstock et al. 2001).

4.1.3 Management of Geochemical Conditions

In coastal wetlands, periodic tidal enhancement has been effective in reducing *P. australis* within one growing season in degraded salt marshes (Roman et al. 1984). Sun et al. (2007) reported a minimum soil water salinity level of 280 cmol/l for a permanent reduction of *P. australis* and a significant negative relationship between Mg^{2+} or K^{+} and *P. australis* height, indicating that NaCl may not be the only salt in tidal salt water responsible for growth suppression.

Cutting followed by inundation with moderately saline water (5.0–7.5 g kg^{-1}) has also eliminated *P. australis* (Russell and Kraaij 2008). The effectiveness of cutting and inundation depends mainly on flooding depth and duration. Weisner and Ekstam (1993) concluded that after *P. australis* seedling establishment has occurred, the water level needed to eliminate the plants in the following year may be estimated from the size of the juvenile plants. According to Russell and Kraaij (2008) continuous inundation of emerging shoots for at least 1 month is required, in order to suppress *P. australis* growth.

4.2 Ecological Management

4.2.1 Grazing, Dredging and Draining

Grazing, dredging and draining have been used in order to reduce reed beds. Grazing may trample the rhizomes, but in the long-term *P. australis* is not significantly affected. van Deursen and Drost (1990) found that cattle consumed 67–98% of aboveground biomass; however, in the next 4 years, *P. australis* reached a new equilibrium under grazing regimes. Dredging and draining are not appropriate methods for most wetlands.

4.2.2 Cutting and Disk-Harrowing

By controlling the cutting time or cutting height (Fig. 2), it is possible to manipulate subsequent *P. australis* growth (Rolletschek et al. 2000; Mauchamp et al. 2001; Russell and Kraaij 2008). In general, cutting increases *P. australis* density (Granéli 1989; Buttler 1992; Ostendorp 1995, 1999; Warren et al. 2001), while it decreases shoot length (Ostendorp 1999; Warren et al. 2001) and diameter (Buttler 1992; Ostendorp 1995). It also accelerates organic matter decomposition and total N release due to improved transparency and aeration thus leading to a decrease in

Fig. 2 Cutting *P. australis* in Lake Mikri Prespa (photo by Y. Kazoglou)

sediments (Jiang et al. 2007). There is conflicting information about the best season to cut (Gryseels 1989a,b; Asaeda and Karunaratne 2000; Asaeda et al. 2002, 2003). Granéli (1990) concluded that winter cutting has no effect or may increase biomass due to better spring light conditions. Asaeda and Karunaratne (2000) showed that cutting the aboveground biomass in late June led to decreased re-growth of *P. australis* plants, because at that period shoot growth is rapid and carbohydrate supplies in the rhizomes are at a minimum. Weisner and Granéli (1989) and Asaeda et al. (2006) attained to similar results, while no significant difference was observed after reed cutting in July or August. Cut shoots must be removed to prevent sprouting and forming stolons. Disk-harrowing removes rhizomes; however, it could potentially result in an increase of *P. australis* since pieces of the rhizome can produce new plants (Marks et al. 1994).

On-Farm Application of Cuttings

P. australis has a high N and P uptake ability (Jiang et al. 2007). The above ground biomass of *P. australis* has the highest nutrient content towards the end of the vegetative period, i.e. in August in the northern hemisphere and in February in the southern hemisphere (Haslam 1969; van der Toorn and Mook 1982). During the next period the major proportion of nutrients is transported down to rhizomes (Granéli et al. 1992; Gessner 2001). By cutting the biomass in summer the material can be applied in agriculture to improve soil structure and to increase soil organic matter and soil fertility.

P. australis cuttings are usually considered a waste product and they are left at the site. The easiest and most cost-effective way of disposal of the cuttings is to chop the material and spread it on farmland as green manure. The application can improve soil structure and increase soil organic matter but nutrient availability to the crop may be considerably low, mainly due to the high C/N ratio of the cuttings (Hansson and Fredriksson 2004). Cheshire et al. (1999) reported reduced maize growth due to N immobilization as a result of a high C/N ratio of straw decomposing in soil; however, Huijser et al. (2004) reported that maize growth and yield were not affected by N immobilization. Additionally the seeds, stolons and other diasporas that may be present in the cuttings may lead to greater weed species richness, which could reduce crop yield (Prew et al. 1995).

Energy Production

Winter harvested *P. australis* biomass has been long used as raw material in energy production (Allirand and Gosse 1995). This material can also be anaerobically digested for the production of biogas, which consists mainly of methane (CH_4) and CO_2. In this case *P. australis* biomass should not be harvested too late in the growing season, since an increased lignin content would decrease gas yield (Hansson and Fredriksson 2004). Other organic materials (i.e. manure) can be used in combination with *P. australis* to stabilize the biogas production process and to improve biogas production (Hansson and Fredriksson 2004).

Sludge Application

The harvested *P. australis* biomass can be used for biogas production and the sludge that is a by-product of the process can be used as an organic fertilizer. In the biogas production process nutrients are not lost; moreover they are more easily available to plants since most of the mineralization has already taken place. According to Hansson and Fredriksson (2004) approximately 62% of the N in the biogas sludge is in the form of ammonium while the rest of N is organically bound. Unless sealed storage containers are used, considerable amounts of N may be lost during storage of the sludge. Also when spreading the sludge on agricultural soil, approximately up to 15% of the N may be lost through ammonia volatilization (Hansson and Fredriksson 2004).

5 Compost Production

Cuttings from *P. australis* can be used to produce compost as fertilizer or soil amendment. Compost production can take place near the wetland in order to minimize energy consumption and also provide an extra income to nearby communities.

5.1 Process

Composting is a process that aims to control the conditions under which decomposition of organic materials takes place. The final product can be applied as soil conditioner and amendment. Passive composting refers to the construction of a pile with different organic materials. Passive composting usually takes place under anaerobic conditions and that leads to loss of N as ammonia and emissions of methane and hydrogen sulfide. Most large on-farm and commercial composting facilities manufacture compost using turned windrow systems, passively aerated windrows, or forced-aeration static piles. Windrows (long and narrow piles; Fig. 3) are generally 3–6 m wide and 3–4 m tall. Initial mixing and frequent turning (a) allow the uniform distribution of particles of different sizes, (b) improve aeration, (c) prevent decomposition under anaerobic conditions in areas of the pile that are too moist or rich in nitrogenous substrates, (d) allow the release of hot gases and water vapor that are produced in the middle of the windrow and (e) material of the exterior is moved to the middle. In passively aerated windrows O_2 enters the pile (with height up to 1 m) through perforated pipes placed at the base of the pile. For adequate aeration the windrow should not be higher than 1 m. Organic materials must be well-mixed prior to placement, while they should be placed on an absorbent layer of mature straw or peat to absorb moisture. In forced aeration static piles the producer can control temperature and aeration. Air is pumped through ventilation pipes placed underneath the pile of organic materials and controlled either by timers or by thermostats. Pile height should not exceed 3 m (de Bertoldi et al. 1984).

Initially mesophilic microflora consumes the readily degradable C fractions of the composting substrate. At that phase most N of the substrate is in the organic

Fig. 3 Compost windrow (photo by K.L. Kalburtji)

form as proteins and simple peptides. At temperatures higher than 40°C or if pH rises above 7.5, N generally volatilizes as NH_3^+. During the thermophilic and bio-oxidative second phase (4–6 weeks), O_2 is consumed, CO_2 is produced, and temperature increases as high as 70°C, while thermophilic bacteria consume the rest of the easily degradable C. During the maturation phase microbial activity slows down, temperature decreases and fungi populate the compost substrate and slowly consume the remaining recalcitrant forms of C, which are lignins and cellulose. When temperature is lower than 40°C, chemoautotrophic bacteria (*Nitrosomonas* spp.) transform NH_4^+ to NO_2^-. Due to lack of easily degradable C, O_2 concentrations in the substrate increase leading to increases in *Nitrobacter* spp. activity that transforms NO_2^- to NO_3^- (Bernal et al. 1998a; Eiland et al. 2001). Factors that contribute to the decomposition of organic matter during the composting process are moisture, microbial populations, O_2 and a balance of C and N. Average O_2 concentration inside the pile is 15–20%. When O_2 content falls below these levels, populations of anaerobic microorganisms are increased and methane and malodorous fatty acids are produced.

The authors carried out an experiment in 2008, which aimed to produce compost from: (a) 70% cuttings of young *P. australis* amended with 30% cow manure or (b) cuttings of young *P. australis* alone based on the turned windrow system. Plant material from *P. australis* was collected in the wetland of Lake Mikri Prespa, Northern Greece. Composting lasted for 7 months. Parameters that were evaluated were C/N ratio and C, N, P and K contents. By the end of the composting process N, P and K concentrations were higher in compost derived from *P. australis* cuttings amended with cow manure than from *P. australis* cuttings alone (Fig. 4). The C/N ratios were initially 32 and 65 for the compost manufactured from *P. australis* amended with cow manure and for the compost manufacture from *P. australis* alone, respectively, while at the maturity phase C/N ratios were 17 and 23, respectively (Fig. 4). Similar results have been reported by Toumpeli (2009) for compost derived from *P. australis* plant material collected in the wetland of Lake Koronia, Northern Greece. She found that in the beginning of the composting period, the C/N ratio was 24.7 and 43.6 for the compost manufactured from *P. australis* amended with sheep and goat manure and for the compost manufactured from *P. australis* alone, respectively, while by the end, C/N ratios were 15.6 and 19.5, respectively.

5.2 *Compost Maturity and Stability Indices*

The successful use of compost depends on the degree of maturity and stability (Fig. 5). Compost maturity refers to the degree of decomposition of phytotoxic organic substances produced during the active composting stage, while compost stability refers to the level of activity of the microbial biomass (Benito et al. 2005). If an unstable or immature compost is applied to the soil, microorganisms can utilize soil O_2 to decompose the organic material, thus leading to anaerobic soil conditions. Also organic acids that are present in immature composts can cause phytotoxicity.

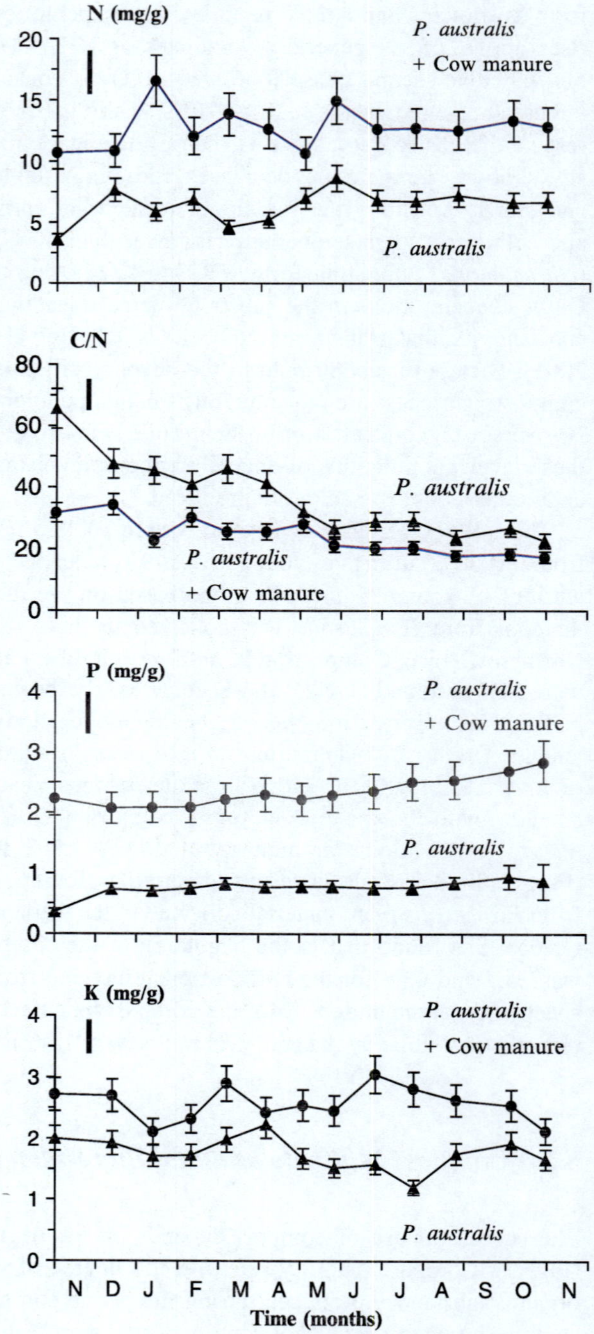

Fig. 4 Changes over time in the Nitrogen (N) concentration, C/N, Phosphorus (P) and Potassium (K) concentrations during the composting. Two types of compost were studied: 70% cuttings of young *P. australis* and 30% cow manure (circles) and cuttings of young *P. australis* (triangles). Bars on the top left of each subfigure represent the standard error of the difference of the means. Bars of each mean represent the standard deviation

Fig. 5 *P. australis* compost (photo by A. Nikolaidou)

Indices that are used to evaluate the degree of maturity of composts are:

(a) Physical parameters, e.g. temperature, odour and colour,
(b) Chemical parameters, e.g. C/N ratio, CEC (Cation Exchange Capacity), humification index, water soluble C and NH_4^+/NO_3^- ratio. According to Sullivan and Miller (2001), a C/N ratio for composts ranging from 20/1 to 15/1 is a good indicator for their maturity, but this is the case when the initial C/N ratio of compost is not higher than 25/1. For materials with higher initial C/N ratio, such as pruning waste, Benito et al. (2005) considered that the value of final C/N around 30/1 should be adequate. Regarding *P. australis* cuttings, despite their high (about 40/1) initial C/N ratio (Hansson and Fredriksson 2004; Toumpeli 2009) after composting alone or with other organic material rich in N, the C/N ratio was lower than 20/1 (Toumpeli 2009). Similar results were reached in the experiment mentioned in the previous subchapter, as C/N ratios ranged from 17 to 23 for the *P. australis* composts at the maturity phase. The CEC increases during humification due to the formation of carboxyl and phenolic functional groups (Vincelas-Akpa and Loquet 1997; Tomati et al. 2000). Bernal et al. (1998a) reported rapid increases in CEC from 36.4–119.9 to 95.1–236.3 cmol kg^{-1}. Humification index, which is the ratio of humic/fulvic acids, increases as compost matures due to loss of readily degraded fulvic acids (Bernal et al. 1998a). The CEC and humification often do not correlate well with microbial respiration (Eggen and Vethe 2001). Water soluble C declines with composting time mainly during the bio-oxidative phase. Researchers have established the values of 0.5%, 1% and 1.7% of water soluble C as a maximum content below

which compost could be considered mature (Bernal et al. 1998a; Belete et al. 2001; Eggen and Vethe 2001). Temperatures greater than 40°C inhibit nitrifiers, therefore nitrification generally begins at the end of the thermophilic phase and leads to the production of increasing NO_3^--N quantities. Moreover during the final phases of composting, mineralization of organic N is limited (Sánchez-Monedero et al. 2001) leading to decreased phytotoxic NH_4^+ quantities. Values smaller than 0.16 of NH_4^+/NO_3^- ratio indicate that compost is mature enough (Bernal et al. 1998a; Eiland et al. 2001).

(c) Biological parameters, e.g. seed germination indexes in compost or compost extracts, microbial counts, ATP (adenosine triphosphate) content, microbial biomass or enzyme activities. A germination index greater than 50% indicates phytotoxin-free composts (Benito et al. 2005). Belete et al. (2001) estimated microbial populations in a compost manufactured by household and yard waste and wood chips in a forced-aeration windrow, by measuring colony forming units (CFUs) and cell counts using epifluorescence microscopy. They found a significant negative correlation between compost age and microbial CFUs, respiration and bacterial cell counts. Benito et al. (2005) suggested that organic matter decomposition during the composting of pruning waste was adequately described by dehydrogenase activity (refers to a group of mostly endocellular enzymes that catalyze the oxidation of organic matter) and the potential metabolic index (dehydrogenase/water soluble C).

Compost stability can be determined by O_2 uptake rate, CO_2 production rate or by the heat released as a result of microbial activity (Benito et al. 2005). Microbial respiration decreases with the loss of the readily biodegradable C fractions. Brewer and Sullivan (2003) reported high respiration rates during the first 27 days of aerated yard waste compost and stable respiration during the period of 70–133 days. Several researchers have attempted to correlate various physical and chemical parameters to microbial respiration (Bernal et al. 1998a; Belete et al. 2001; Eggen and Vethe 2001). C/N ratio has been well correlated with stability indices (Bernal et al. 1998a, b; Eiland et al. 2001).

5.3 *Agricultural Applications of Compost*

Composts may contain more than 50% organic matter. Sufficient applications of composts ($\geq$34 Mg ha^{-1}) can increase soil organic matter and improve soil structure and soil physical properties, since polysaccharides and other polymeric substances of organic matter (Bronick and Lal 2005) act as aggregating compounds (Hansen et al. 2001). Household waste compost application at a rate of 17 Mg ha^{-1}increased aggregate stability of a sandy loam soil up to 88% (Debosz et al. 2002). Increased rates of compost application resulted in increased hydraulic conductivity (33–95% in a loamy soil, 55–168% in a clay soil), total porosity (up to 33%) and water holding capacity compared to no compost application (Aggelides and Londra 2000).

Bulluck et al. (2002) reported that bulk density in soils amended with composts was 1.01 versus 1.17 g cm^{-3} in soils that received synthetic fertilizers. *P. australis* compost addition decreased clay dispersion and increased wet aggregate stability of a sandy soil (Toumpeli 2009).

Composts may improve soil chemical properties too. Generally soil organic matter is increased during the second year or after successive applications of composts. Compost addition may initially enhance soil organic matter mineralization rates due to increased microbial activity (Min et al. 2003). Tester (1990) observed that the application of 134 Mg sludge compost ha^{-1}resulted in a 24% loss of C at a depth of 13 cm in the first year. Grandy et al. (2002) reported a 28% increase in soil total C in the second year of compost application at a rate of 22 Mg ha^{-1}. Differences in the organic compounds present in composts probably affect soil organic matter change after compost application. In the experiment carried out by the authors, the soil application of compost manufactured from 70% cuttings of young *P. australis* amended with 30% cow manure resulted in a 9% increase in soil total C, while that of compost manufactured from cuttings of young *P. australis* alone resulted in a 16% loss of C (unpublished data). Soil CEC may also be increased due to increased soil organic matter (Stockdale et al. 2002), thereby increasing nutrient availability for plant uptake (Schlegel 1992; Altieri and Nicholls 2003) and reducing the leaching of fertilizer nutrients.

The authors found out during their experimentation that the application of compost manufactured from 70% cuttings of young *P. australis* and 30% cow manure icreased soil N and P compared to the application of compost manufactured from cuttings of young *P. australis* and to the application of synthetic fertilizers (unpublished data). Toumpeli (2009) reported higher N, P, K, Ca and Mg concentrations in tomato plants that received compost derived from *P. australis* plant material compared to those that received synthetic fertilizers. The use of compost can reduce Cu, Cd, Pb and B plant uptake in soils with high concentrations of these elements, since they are immobilized on the increased compost adsorption sites (Yermiyahu et al. 2001; Zhou and Wong 2001; Shuman et al. 2002). Neutral to slightly alkaline composts can increase pH of acid soils, thus reducing the potential for Al and Mn toxicity (McConnell et al. 1994).

The application of mature compost can increase biotic diversity in the soil due to the presence of mesophilic bacteria and fungi. Microbial biodiversity can mitigate colonization by pathogenic populations in soil (Altieri 1995). In compost amended soils, beneficial fungal and bacterial propagules cover densely pathogen spores, parasitize on pathogenic fungi and consume resources needed by the pathogenic populations.

Compost is applied on farms as a soil conditioner and organic fertilizer, replacing synthetic fertilizer, sequestering carbon, improving soil fertility and increasing the water holding capacity of soils (IPCC 2007). Long-term compost application can play a positive role in climate change mitigation by soil carbon sequestration, which in turn can reverse the process of soil degradation (Diacono and Montemurro 2010). Compost C is released partly as CO_2 in the atmosphere and the rest C is consumed by the decomposers (IPCC 2007).

6 Conclusion

Phragmites australis is a clonal plant of wetlands. Its invasion creates problems since it alters the structure and function of wetland ecosystems by changing nutrient cycles and hydrological regimes. Ecological management such as grazing, dredging and cutting is a friendly contribution to environment. Cuttings of the aboveground biomass can be used for on-farm application as green manure or as sludge after energy production. Also, people near wetlands can produce compost by using plant material from *P. australis*, and use it as fertilizer and soil amendment in farming systems. Suitable use of compost in agriculture could reduce the fertilizer application and thereby reduce the environmental and food pollution, improve food security and soil productivity, and increase sustainability in the agroecosystems. It can also play role in the climate change because it releases partly CO_2 in the atmosphere and the rest of C is consumed by the decomposers.

Acknowledgements We would like to extend our gratitude to the Greek Biotope, Wetland Centre for the useful information and to the emeritus professor A. Gatzianas for his constructive criticism and helpful suggestions, allowing us to improve on the paper's earlier version.

References

Adamus PR, Stockwell LT (1983) A method for wetland functional assessment. US Transp. Fed. Highway Admin., Washington DC

Adler A, Karacic A, Weih M (2008) Biomass allocation and nutrient use in fast-growing woody and herbaceous perennials used for phytoremediation. Plant Soil 305:189–206

Aggelides SM, Londra PA (2000) Effects of compost produced from town wastes and sewage sludge on the physical properties of a loamy and a clay soil. Bioresource Technol 71:253–259

Ailstock MS, Norman CM, Bushmann PJ (2001) Common reed *Phragmites australis*: Control and effects upon biodiversity in freshwater non-tidal wetlands. Restor Ecol 9: 49–59

Alho CJR, Lacher TE, Goncalves HC (1988) Environmental degradation in the Pantanal ecosystem of Brazil. BioScience 38:164–171

Allirand JM, Gosse G (1995) An above-ground biomass production model for common reed (*Phragmites communis Trin.*) stand. Biomass Bioenerg 9:441–448

Altieri M (1995) Agroecology: The Science of Sustainable Agriculture. In: Altieri M (ed) Plant Disease Ecology and Management. Westview Press, Boulder, CO, pp 307–320

Altieri MA, Nicholls CI (2003) Soil fertility management and insect pests: harmonizing soil and plant health in agroecosystems. Soil Till Res 72(2):203–211

Asaeda T, Karunaratne S (2000) Dynamic modeling of the growth of *Phragmites australis*: model description. Aquat Bot 67:301–318

Asaeda T, Nam LH, Hietz P, Tanaka N, Karunaratne S (2002) Seasonal fluctuations in live and dead biomass of *Phragmites australis* as described by a growth and decomposition model: implications of duration of aerobic conditions for litter mineralization and sedimentation. Aquat Bot 73:223–239

Asaeda T, Manatunge J, Fujino T, Sovira D (2003) Effects of salinity and cutting on the development of *Phragmites australis*. Wetl Ecol Manag 11:127–140

Asaeda T, Rajapakse L, Manatunge J, Sahara N (2006) The effect of summer harvesting of *Phragmites australis* on growth characteristics and rhizome resource storage. Hydrobiologia 553:327–335

Belete L, Egger W, Neunhauserer C, Insam H, Caballero B (2001) Can community level physiological profiles be used for compost maturity testing? Compost Sci Util 9:13–18

Benito M, Masaguer A, Moliner A, Arrigo N, Palma RM, Effron D (2005) Evaluation of maturity and stability of pruning waste compost and their effect on carbon and nitrogen mineralization in soil. Soil Sci 170:360–370

Benoit LK, Askins RA (1999) Impact of the spread of *Phragmites* on the distribution of birds in Connecticut tidal marshes. Wetlands 19:194–208

Bernal MP, Paredes C, Sánchez-Monedero MA, Cegarra J (1998a) Maturity and stability parameters of composts with a wide range of organic wastes. Bioresource Technol 63: 91–99

Bernal MP, Sánchez-Monedero MA, Paredes C, Roig A (1998b) Carbon mineralization from organic wastes at different composting stages during their incubation with soil. Agr Ecosyst Environ 69:175–189

Best EPH, Verhoeven JTA, Wolff WJ (1993) The ecology of the Netherland's wetlands: characteristics, threats, prospects and perspectives for ecological research. Hydrobiologia 265:305–320

Biddlestone AJ, Gray KR, Thorairagan K (1991) A botanical approach to the treatment of wastewaters. J Biotechnol 17(3):209–220

Brewer LJ, Sullivan DM (2003) Maturity and stability evaluation of composted yard trimmings. Compost Sci Util 11(2):96–112

Bronick CJ, Lal R (2005) Soil structure and management: a review. Geoderma 124:3–22

Bulluck III LR, Brosius M, Evanylo GK, Ristaino JB (2002) Organic and synthetic fertility amendments influence soil microbial, physical and chemical properties on organic and conventional farms. Appl Soil Ecol 19:147–160

Buttler A (1992) Permanent plot research in wet meadows and cutting experiment. Vegetatio 103:113–124

Cheshire MV, Bedrock CN, Williams BL, Chapman SJ, Solntseva I, Thomsen I (1999) The immobilization of nitrogen by straw decomposing in soil. Eur J Soil Sci 50:329–341

Committee on Characterization of Wetlands (1995) Wetlands. Characteristics and Boundaries, National Academy Press, Washington, DC

Cowie NR, Sutherland WJ, Ditlhogo MKM, James R (1992) The effects of conservation management of reed beds. II. The flora and litter disappearance. J Appl Ecol 29:277–284

Cronk J, Fennessy MS (2001) *Phragmites australis* as a declining species in Europe. In: Cronk J, Fennessy MS (eds) Wetland Plants. CRC Press, Boca Roton, pp 317–320

Davidson EA, Janssens IA (2006) Temperature sensitivity of soil carbon decomposition and feedbacks to climate change. Nature 440:165–173

de Bertoldi M, Vallini G, Pera A (1984) Technological aspects of composting including modeling and microbiology In: Gasser JKR (ed) Composting of Agricultural and Other Wastes. Elsevier Applied Science Publishers, London, pp 27–41

Debosz K, Petersen SO, Kure LK, Ambus P (2002) Evaluating effects of sewage sludge and household compost on soil physical, chemical and microbiological properties. Appl Soil Ecol 19:237–248

Diacono M, Montemurro F (2010) Long-term effects of organic amendments on soil fertility. A review. Agron Sustain Dev 30:401–422

Dupré C, Diekmann M (2001) Differences in species richness and life-history traits between grazed and abandoned grasslands in southern Sweden. Ecography 24:275–286

Eggen T, Vethe O (2001) Stability indices for different composts. Compost Sci Util 9:19–26

Eiland F, Lind AM, Leth M, Iversen JJL, Klamer M, Jensen HEK (2001) C and N turnover and lignocellulose degradation during composting of *Miscanthus* straw and liquid pig manure. Compost Sci Util 9:186–197

Engloner AI (2009) Structure, growth dynamics and biomass of reed (*Phragmites australis*) – A review. Flora 204:331–346

Gaudet CL, Keddy PA (1995) Competitive performance and species distribution in shoreline plant communities: a comparative approach. Ecology 76:280–291

Gerakis PA, Koutrakis ET (1996) Greek Wetlands. Commercial Bank of Greece, Goulandris Natural History Museum, Greek Biotope/Wetland Centre, English edn

Gessner MO (2001) Mass loss, fungal colonisation and nutrient dynamics of *Phragmites australis* leaves during senescence and early aerial decay. Aquat Bot 69:325–339

Grandy AS, Porter GA, Erich MS (2002) Organic amendment and rotation crop effects on the recovery of soil organic matter and aggregation in potato cropping systems. Soil Sci Soc Am J 66:1311–1319

Granéli W (1989) Influence of standing litter on shoot production in reed, *Phragmites australis* (Cav.) Trin.ex Steudel. Aquat Bot 35:99–109

Granéli W (1990) Standing crop and mineral content of reed, *Phragmites australis* (Cav.) Trin. Ex Studel, in Sweden–Management of reed stands to maximize harvestable biomass. Folia Geobot Phytotaxon 25:291–302

Granéli W, Weisner SEB, Sytsma MD (1992) Rhizome dynamics and resource storage in *Phragmites australis*. Wetl Ecol Manag 1:239–247

Greeson PE, Clarke JR, Clark JE (1979) (eds) Wetland functions and values: the state of our understanding. American Water Resources Association, Minneapolis

Grevilliot F, Broyer J, Muller S (1998) Phytogeographical and phenological comparison of the Meuse and the Saone valley meadows (France). J Biogeogr 25:339–360

Gries C, Kappen L, Losch R (1990) Mechanism of flood tolerance in reed, *Phragmites australis* (Cav.) Trin. Ex Steudel. New Phytol 114:589–593

Grime JP (2002) Plant Strategies, Vegetation Processes, and Ecosystem Properties, 2nd edn, Wiley, Chichester, UK

Grootjans AP, van Wirdum G, Kemmers R, van Diggelen R (1996) Ecohydrology in The Netherlands: principles of an application-driven interdiscipline. Acta Bot Neerl 45:491–516

Gryseels M (1989a) Nature management experiments in a derelict reed marsh. I. Effects of winter cutting. Biol Conserv 47:171–193

Gryseels M (1989b) Nature management experiments in a derelict reed marsh. II. Effects of summer mowing. Biol Conserv 48:85–99

Güsewell S, Klötzli F (2000) Assessment of aquatic and terrestrial reed (*Phragmites australis*) stands. Wetl Ecol Manag 8:367–373

Hansen B, Alrøe HF, Kristensen ES (2001) Approaches to assess the environmental impact of organic farming with particular regard to Denmark. Agr Ecosyst Environ 83:11–26

Hansson PA, Fredriksson H (2004) Use of summer harvested common reed (*Phragmites australis*) as nutrient source for organic crop production in Sweden. Agr Ecosyst Environ 102:365–375

Harris MB, Thomas W, Mourão G, Da Silva CJ, Guimarãews E, Sonoda F, Fachium E (2005) Safe quarding the Pantanal Wetland: threats and conservation initiatives. Conserv Biol 19:714–720

Haslam S (1969) The development of shoots in *Phragmites communis* Trin. Ann Bot 33:695–709

Hawke CJ, José PV (1996) Reedbed management for commercial and wildlife interests. The Royal Society for the Protection of Birds, Sandy, UK

Hellings SE, Gallagher JL (1992) The effects of salinity and flooding on *Phragmites australis*. J Appl Ecol 29:41–49

Hill NM, Keddy PA, Wisheu IC (1998) A hydrological model for predicting the effects of dams on the shoreline vegetation of lakes and reservoirs. Environ Manage 22:723–736

Hollis GE (1992) The hydrological functions of wetlands and their management. In: Gerakis PA (ed) Conservations and Management of Greek Wetlands. Proceedings of a Greek Wetlands workshop held in Thessaloniki, Greece, IUCN, Gland Switzerland, pp 9–60

Holm LG, Plucknett DL, Pancho JV, Herberger JP (1977) The World's Worst Weeds: Distribution and Biology. The University Press of Hawaii, Honolulu, Hawaii, USA

Huijser MP, Meerburg BG, Holshof G (2004) The impacts of ditch cuttings on weed pressure and crop yield in maize. Agr Ecosyst Environ 102:197–203

IPCC (2007) Agriculture. In: Climate change 2007: Mitigation. Contribution of Working Group III to the Fourth Assessment Report of the Intergovernmental Panel on Climate Change. Cambridge University Press, Cambridge, UK

Janhäll S, Andreae MO, Pöschl U (2010) Biomass burning aerosol emissions from vegetation fires: particle number and mass emission factors and size distributions. Atmos Chem Phys 10:1427–1439

Jiang C, Fan X, Cui G, Zhang Y (2007) Removal of agricultural non-point source pollutants by ditch wetlands: implications for lake eutrophication control. Hydrobiologia 581:319–327

Jiraungkoorskul W, Upatham ES, Kruatrachue M, Sahaphong S, Vichasri-Grams S, Pokethitiyook P (2003) Biochemical and histopathological effects of glyphosate herbicide on Nile tilapia (*Oreochromis niloticus*). Environ Toxicol 18:260–267

Kay S (1995) Efficacy of wipe-on applications of glyphosate and imazapyr on common reed in aquatic sites. J Aqua Manage 33:25–26

Keddy PA (1989) Effects of competition from shrubs on herbaceous wetland plants: a 4-year field experiment. Can J Bot 67:708–716

Keddy PA (2010) Wetland Ecology: Principles and Conservation, 2nd edn. Cambridge Univ. Press, Cambridge UK

Keddy PA, Reznicek AA (1986) Great lakes vegetation dynamics: the role of fluctuating water levels and buried seeds. J Great Lakes Res 12:25–36

Keller BEM (2000) Plant diversity in *Lythrum*, *Phragmites* and *Typha* marshes, Massachusetts, USA. Wetl Ecol Manag 8:391–401

Koncalová H (1990) Anatomical adaptations to water logging in roots of wetland graminoids: limitations and drawbacks. Aquat Bot 38:127–134

Kotowski W, van Diggelen R (2004) Light as an environmental filter in fen vegetation. J Veg Sci 15:583–594

Kotowski W, van Andel J, van Diggelen R, Hogendorf J (2001) Responses of fen plant species to groundwater level and light intensity. Plant Ecol 155:147–156

Lee SY (1990) Net aerial primary productivity, litter production and decomposition of the reed *Phragmites communis* in a nature preserve in Hong Kong: management implications. Mar Ecol Prog Ser 66:161–173

Lenssen JPM, Menting FBJ, Van der Putten WH, Blom CWPM (1999) Control of plant species richness and zonation of functional groups along a freshwater flooding gradient. Oikos 86:523–534

Leonardsson L (1994) Våtmarker som kvävefällor, Svenska och internationella erfarenheter. Swedish Environmental Protection Agency, Solna, Sweden (in Swedish)

Maltby E, Barker T (2009) The Wetlands Handbook. Wiley-blackwell, UK

Marks M, Lapin B, Randall J (1994) *Phragmites australis* (*P. communis*): threats, management and monitoring. Nat Area J 14:285–294

Marrs RH, Frost AJ, Plant RA, Lunnis P (1993) Determination of buffer zones to protect seedlings of non target plants from the of glyphosate spray drift. Agr Ecosyst Environ 45:283–293

Matthews GVT (1993) The Ramsar Convention on wetlands: its history and development, Ramsar Convention Bureau; Gland

Mauchamp A, Blanch S, Grillas P (2001) Effects of submergence on the growth of *Phragmites australis* seedlings. Aquat Bot 69:147–164

McConnell D, Shiralipour A, Smith W (1994) Compost impact on soil/plant properties, in Composting source separated organics. JG Press, Inc., Emmaus, PA

Meyerson LA, Saltonstall K, Windham L et al. (2000) A comparison of *Phragmites australis* in freshwater and brackish marsh environments in North America. Wetl Ecol Manag 8:89–103

Min DH, Islam KR, Vough LR, Weil RR (2003) Dairy manure effects on soil quality properties and carbon sequestration in alfalfa- orchardgrass systems. Commun Soil Sci Plan 34:781–799

Mitsch WJ, Gosselink JG (1993) Wetlands, 2nd edn. Van Nostrand Reinhold NY

Ostendorp W (1989) "Die-back" of reeds in Europe- a critical review of literature. Aquat Bot 35:5–26

Ostendorp W (1995) Effect of management on the mechanical stability of lakeside reeds in Lake Constance-Untersee. Acta Oecol 16:77–294

Ostendorp W (1999) Management impacts on stand structure of lakeshore *Phragmites* reeds. Int Rev Hydrobiol 84:33–47

Poff NL, Allen JD, Bain MB, Karr JR, Prestegaard KL, Richter BD, Sparks RE, Stromberg JC (1997) The natural flow regime: a paradigm for conservation and restoration of riverine ecosystems. BioScience 47:769–784

Prew RD, Ashby JE, Bacon ETG, Christian DG, Gutteridge RJ, Jenkyn JF, Powell W, Todd AD (1995) Effects of incorporating or burning straw, and of different cultivation systems, and of different cultivation systems, on winter wheat grown on two soil types 1985–1991. J Agric Sci 124:173–194

Richardson CJ (1994) Ecological functions and human values in wetlands: A Framework for assessing forestry impacts. Wetlands 14(1):1–9

Richardson CJ (1995) Wetlands ecology, in Encyclopedia of Environmental Biology, Vol. 3, Academic Press, Inc., New York

Riemer DN (1973) Effects of rate, spray volume, and surfactant on the control of *Phragmites* with glyphosate. Proc Annu Meet Northeastern Weed Sci Soc 27:101–104

Riemer DN (1976) Long-Term Effects of Glyphosate Applications to *Phragmites*. J Aquat Plant Manag 14:39–43

Rolletschek H, Rolletschek A, Hartzendorf T, Kohl J (2000) Physiological consequences of mowing and burning of *Phragmites australis* stands for rhizome ventilation and amino acid metabolism. Wetl Ecol Manag 8(6):425–433

Roman CT, Niering WA, Warren RS (1984) Salt marsh vegetation change in response to tidal restriction. Environ Manag 8:141–150

Rooth JE, Stevenson JC, Cornwell JC (2003) The influence of 5 and 20 yr-old *Phragmites* populations on rates of accretion in an oligohaline tidal marsh of Chesapeake Bay. Estuaries 26(2B):475–483

Russell IA, Kraaij T (2008) Effects of cutting *Phragmites australis* along an inundation gradient, with implications for managing reed encroachment in a South African estuarine lake system. Wetl Ecol Manag 16:383–393

Sánchez-Monedero MA, Roig A, Paredes C, Bernal MP (2001) Nitrogen transformation during organic waste composting by the Rutgers system and its effects on pH, EC and maturity of the composting mixtures. Bioresource Technol 78:301–308

Schlegel AJ (1992) Effect of composted manure on soil chemical properties and nitrogen use by grain sorghum. J Prod Agric 5(1):153–157

Schuyt K, Brander L (2004) The economic values of the world's wetlands. Swiss Agency for the Environment, Forest and Landscape (SAEFL), Amsterdam

Shuman LM, Dudka S, Das K (2002) Cadmium forms and plant availability in compost-amended soil. Commun Soil Sci Plant Anal 33:737–748

Silliman BR, Bertness MD (2004) Shoreline development drives invasion of *Phragmites australis* and the loss of plant diversity on New England salt marshes. Conserv Biol 18(5):1424–1434

Skinner J, Zalewski S (1995) Functions and values of Mediterranean wetlands, Station Biologique de la Tour du Valat, Arles. A Med-Wet Publication, France

Stockdale EA, Shepherd MA, Fortune S, Cuttle SP (2002) Soil fertility in organic farming systems-fundamentally different? Soil Use Manage 18:301–308

Sullivan DM, Miller RO (2001) Compost quality attributes, measurements, and variability. In: Stoffella PJ, Kahn BA (eds) Compost utilization in horticultural cropping systems. CRC Press LLC, Boca Raton Florida, pp 95–120

Sun H, Brown A, Coppen J, Steblein P (2007) Response of *Phragmites* to environmental parameters associated with treatments. Wetl Ecol Manag 15:63–79

Tester CF (1990) Organic amendment effects on physical and chemical properties of a sandy soil. Soil Sci Soc Am J 54:827–831

Tewksbury L, Casagrande R, Blossey B, Hafliger P, Schwarzlander M (2002) Potential for biological control of *Phragmites australis* in North America. Biol Control 23(2):191–212

Thompson DJ, Shay JM (1985) The effects of fire on *Phragmites australis* in the Delta Marsh, Manitoba. Can J Bot 63:1864–1869

Thompson DJ, Shay JM (1989) First-year response of a *Phragmites* marsh community to seasonal burning. Can J Bot 67:1448–1455

Tomati U, Madejon E, Galli E (2000) Evolution of humic acid molecular weight as an index of compost stability. Compost Sci Util 8:108–114

Toumpeli A (2009) Compost production from *Phramites australis* and its use in tomato cultivation. MS thesis, School of Agriculture, Aristotle University of Thessaloniki (In Greek with an English summary)

Tscharntke T (1989) Attach by stem-boring moth increases susceptivility of *Phragmites australis* to gall-making by a midge: mechanisms and effects of midge population dynamics. Oikos 55:92–100

Tucker GC (1990) The genera of *Arundinoideae* (*Gramineae*) in the southeastern United States. J Arnold Arboretum 71:145–177

van der Putten WH (1997) Die-back of *Phragmites australis* in European wetlands: an overview of the European research programme on reed die-back and progression (1993–1994). Aquat Bot 59:263–275

van der Toorn J, Mook JH (1982) The influence of environmental factors and management on stands of *Phragmites australis*. I. Effect of burning, frost and insect damage on shoot density and shoot size. J Appl Ecol 19:477–500

van Deursen EJM, Drost HJ (1990) Defoliation and treading by cattle of reed *Phragmites australis*. J Appl Ecol 27:284–297

Verhoeven JTA (1992) Vegetation as a resource in wetlands with special reference to the wetlands in Greece. In: Gerakis PA (ed) Conservations and Management of Greek Wetlands, Proceedings of a Greek Wetlands Workshop held in Thessaloniki, Greece, IUCN, Gland, Switzerland, pp 9–60

Vincelas-Akpa M, Loquet M (1997) Organic matter transformations in lignocellulosic waste products composted or vermicomposted (*Eisenia fetida andrei*): chemical analysis and ^{13}C CPMAS NMR spectroscopy. Soil Biol Biochem 29:751–758

Vymazal J, Brix H, Cooper PF, Green MB, Haberl R (1998) (eds) Constructed Wetlands for Wastewater Treatment in Europe. Backhuys Publishers, Leiden, The Netherlands

Ward JV, Tockner K, Schiemer F (1999) Biodiversity of floodplain river ecosystems: ecotones and connectivity. Regul River 15:125–139

Warren RS, Fell PE, Grimsby JL, Buck EL, Rilling GC, Fertik RA (2001) Rates, patterns, and impacts of *Phragmites australis* expansion and effects of experimental *Phragmites* control on vegetation, macroinvertebrates, and fish within Tide lands of the Lower Connecticut River. Estuaries 24:90–107

Weis JS, Weis P (2003) Is the invasion of the common reed, *Phragmites australis*, into tidal marshes of the eastern US an ecological disaster? Mar Pollut Bull 46:816–820

Weisner SEB, Ekstam B (1993) Influence of germination time on juvenile performance of *Phragmites australis* on temporarily exposed bottoms – implications for the colonization of lake beds. Aquat Bot 45(2–3):107–118

Weisner SEB, Granéli W (1989) Influence of substrate conditions on the growth of *Phragmites australis* after a reduction in oxygen transport to below ground parts. Aquat Bot 35:71–80

Wheeler BD, Proctor MCF (2000) Ecological gradients, subdivisions and terminology of north-west European mires. J Ecol 88:187–203

Wilcox KL, Petrie SA, Maynard LA, Meyer SW (2003) Historical distribution and abundance of *Phragmites australis* at Long Point, Lake Erie, Ontario. J Great Lakes Res 29:664–680

Yeh TY (2008) Removal of metals in constructed wetlands: Review. Pract Periodical of Haz Toxic and Radioactive Waste Mgmt 12(12):96–101

Yermiyahu U, Keren R, Chen Y (2001) Effect of composted organic matter on boron uptake by plants. Soil Sci Soc Am J 65:1436–1441

Zalidis GC, Mantzavelas AL (eds) (1994) Inventory of Greek Wetlands as Natural Resources (first approximation), Greek Biotope/Wetland Centre (EKBY), Thessaloniki, Greece

Zhou LX, Wong JWC (2001) Effect of dissolved organic matter from sludge and sludge compost on soil copper sorption. J Environ Qual 30:878–883

Effects of Genotype, Environment and Management on Yields and Quality of Black Tea

P. Okinda Owuor, David M. Kamau, Samson M. Kamunya, Solomon W. Msomba, Marie A. Uwimana, Amos W. Okal, and Bowa O. Kwach

Abstract The tea beverages processed from the young tender shoots of *Camellia sinensis* (L.) O. Kuntze, are claimed to be the most widely consumed fluids after water. The tea plant originates from the point of confluence of Northeast India, North Burma, Southwest China and Tibet. Its production has spread and economic production has been reported in between 49°N in Outer Carpathians to 33°S in Natal, South Africa, at altitudes ranging from sea level to 2,700 m above mean sea level. The adaptability of the plant to areas with large variations in geographical, climatic and environmental factors can cause changes in growth patterns in different genotypes leading to variations in yields and black tea quality. Tea producers usually import genotypes, management and production techniques suitable for optimal production in one region in the hope that beneficial attributes observed at source shall be maintained in the new areas. But the tea plant responses in new environments have not always yielded the desired results. Here we review the effects of genotypes, environment and management on the yields and quality of black tea.

Previous investigations demonstrated that black tea yields and quality changes are due to environmental factors like soil type, altitude, seasons, weather factors,

P.O. Owuor (✉) · A.W. Okal · B.O. Kwach
Department of Chemistry, Maseno University, P.O. Box 333, Maseno 40105, Kenya
e-mail: owuoropo@africaonline.co.ke

D.M. Kamau
Department of Chemistry, Tea Research Foundation of Kenya, P.O. Box 820, Kericho 20200, Kenya

S.M. Kamunya
Department of Botany, Tea Research Foundation of Kenya, P.O. Box 820, Kericho 20200, Kenya

S.W. Msomba
Crop Improvement Programme, Tea Research Institute of Tanzania, P.O. Box 2177, Dar Es Salaam, Tanzania

M.A. Uwimana
Office Des Cultures Industrielles du Rwanda The' (OCIR The'), P.O. Box 1344, Kigali, Rwanda

E. Lichtfouse (ed.), *Genetics, Biofuels and Local Farming System*, Sustainable Agriculture Reviews 7, DOI 10.1007/978-94-007-1521-9_10,

geographical areas of production, agronomic inputs, processing technologies and management. Black tea quality and yields of similar genotypes grown on different soils vary. The extent and patterns of the variations change with varieties. High altitude grown teas are more aromatic than low altitude grown black teas, implying that the low grown teas are plain in character. Thus producers at high altitudes should aim at producing aromatic black teas, although yields will be lower than same genotypes at lower altitudes. Producers growing teas at low altitudes should focus on high output and ensure optimal conditions for production of plain black teas. There are seasonal black tea quality and yield variations. Cold seasons lead to slow growth resulting in low yields, but high black tea quality. Provided soil moisture and temperatures are adequate, warm temperatures lead to fast growth, leading in turn to high yields, but low black tea quality. It is therefore not possible to have uniform production or to produce the same black tea quality throughout the year. The situation is adverse further away from the equator with no production in winter as the labour management can be critical during the long cold seasons, necessitating long labour layoffs. Many genotypes have been developed, some with very high yields and quality. As a result, producers continuously try to access the good varieties into new geographical areas in the hope the genotypes would retain their economic advantages. While some genotypes are stable to locational changes, most show wide variations due to planting in the new areas.

Management policies induce yield and black tea quality differences. Imported management policies should be domesticated and modified to suit the new environments. Harvesting, by hand plucking, the young tender shoots is done when they are of the right size. Delayed harvesting leads to shoot overgrowth and crop loss. Whereas plucking two leaves and a bud is a compromise between yields and black tea quality, some growers practice coarser plucking standards. Black tea quality declines with coarse plucking standards. Short plucking rounds lead to high production and high quality black teas. When a plucking standard is preset, growers in a location need to establish the shortest harvesting interval for realization of good yields and quality. Fertilizers are essential for establishment and growth. Varying results have been recorded on yield and quality responses to NPK application. For potassium and phosphorus, evaluations are necessary in different regions because where there is no beneficial effects their application can be reduced to decrease costs. High rates of nitrogen reduce black tea quality and do not increase yields. Nitrogen fertilizers need to be applied at rates that are a compromise between yields and black tea quality. Such rates vary with regions and genotypes. To reach high production and quality, region- and genotype-specific fertilizer rates are needed.

In regions producing relatively inferior black teas producers try to import processing technologies from other areas. These efforts may not improve quality due to variations of environmental conditions. Indeed, for the same genotype grown in different regions and processed under identical conditions, differences in quality and chemical composition have been reported. This was due to variations in the leaf biochemical constituents composition caused by the environment in which the plant was grown. Different regions must therefore optimize their processing conditions to realize high quality.

Keywords Black tea • *Camellia sinensis* • Environment • Genotypes • Cultural and agronomic practices • Yields • Quality

1 Introduction

Tea (*Camellia sinensis* (L.) O. Kuntze) is believed to have originated in Southeast Asia around the intersection of altitude 29°N and longitude 98°E, the point of confluence of the lands of Northeast India, North Burma, Southwest China and Tibet (Mondal 2007). The perennial tree crop is now grown in many parts of the world for production of various tea beverages which are the most widely consumed fluids after water. Commercially, three varieties of tea: – the China type (*Camellia sinensis* var. *sinensis*), the Assam type (*Camellia sinensis* var. *assamica*) and the hybrid (*Camellia sinensis* var. *assamica* ssp *lasiocalyx*) are extensively exploited (Banerjee 1992). However recently, two new varieties, *Camellia sinensis* var. *pubilimba*, and *Camellia sinensis* var. *Kucha* have been recognized to have commercial potential in China (Yao et al. 2008). The China type consists of small semi-erect leaves, while, the Assam types has relatively larger horizontally held leaves and the hybrid type has characteristics in between the China and Assam types (Banerjee 1992). If left to grow freely, the plant can grow to over 30 m (Fig. 1). The classification of tea beverages is based on the processing methods. Black tea is prepared by biochemical oxidation (or fermentation) of crushed fresh tender leaves, involving multi-step enzyme-mediated oxidation of polyphenols that is terminated by firing (drying); while green tea is prepared by steaming withered or unwithered leaves to inactivate oxidative enzymes, prior to crushing before drying. Processing conditions for oolong are intermediate between black and green teas, whereby, in oolong tea macerated leaves undergo short fermentation durations.

Tea is an important economic crop grown in many countries for production of various tea beverages. As at the year 2008, the ten leading world tea producers were India, China, Kenya, Sri Lanka, Turkey, Indonesia, Japan, Iran and Argentina (Anon 2008). In Africa, tea is mainly grown in Kenya, Malawi, Uganda, Tanzania, Zimbabwe, Rwanda, South Africa, Burundi and Mauritius, although some tea is also grown in Ethiopia, Nigeria and Cameroon. Tea is a major economic crop for some developing countries. In Kenya, for example, tea is a key player within the agro-industrial crops and is the single commodity leading foreign exchange earner accounting for about 26% of the total export earnings and 4% of the gross domestic product (GDP), and is a source of livelihood to over 3 million people (Mbadi and Owuor 2008). Over 62% of the Kenya tea is produced by the smallholder growers, living in the rural set ups where industrialisation is low and economic activities are rare. The crop is viewed as a source of rural development in many developing countries.

Being an antioxidant, tea has been reported to have ability to manage several diseases including colon, oesophageal, and lung cancer, urinary stone, dental caries, etc (Sharma et al. 2007). For black tea, there is clear evidence that intake of $\geq$3 cups per day reduces the risk of coronary heart disease (Gardner et al. 2007). It has been

Fig. 1 Tea plants left to grow in a seed barie

claimed that tea beverages are the most widely consumed fluids after water (Agarwal 1989; Sharma et al. 2007). World tea production has grown rapidly, causing over supply of the commodity, resulting in stagnation or reduction in prices (Anon 2008). As a result tea growers strive to improve productivity and profits through optimising agronomic in-puts and cultural practices to realise highest production per unit area and best quality at the lowest cost of production. Tea yields are controlled by yield components which include; harvestable shoot size, number of shoots per unit area and rate of shoot growth (Odhiambo 1989; Mathews and Stephens 1998a). The expressions of these components are controlled by environments, management practices and the genotype. Similarly, black tea quality is influenced by agronomic in puts (Ravichandran 2004) and cultural practices (Cloughley 1983; Owuor and Othieno 1991).

As a results of high demand, commercial production has been reported under diverse environments from as 49°N, Outer Carpathians, to 33°S, Natal, South Africa (Shoubo 1989), and from altitudes ranging from sea level in Japan and Sri Lanka (Anandacoomaraswamy et al. 2000) to 2,700 m above mean sea level (amsl) in Kenya and Rwanda (Owuor et al. 2008b) and under varying topographies (See Fig. 2). The plant is adaptable to environments with large climatic variations. The variations in the environment and growing conditions can cause large differences in yields and black tea quality. But despite the variations, tea farmers usually import genetic materials and production technologies across the borders. In countries where tea plant growth parameters are different, usually experimental testing for genetic materials or production technologies are centralised. It is usually assumed the genotype with good yield and/or quality attributes or a successful pro-

Fig. 2 Some geographical features of areas where tea grows (**a**) tea on hill topography, (**b**) tea in a valley, (**c**) tea in plain field, (**d**) tea on water logged flat land

duction/manufacturing technology developed in one location maintains the status irrespective of where the plant is grown or technology used. Similarly, tea planters tend to import production technologies and management systems, sometimes even without re-evaluating the technologies in the new environments. The assumption is that proven technology in one region works in all regions. However, tea growers have not managed to replicate yields or black tea quality using production technologies imported from different regions. The lack of replication of the results can be due to many factors including, but not limited to low genetic stability of the genotype over wide localities, climatic changes, agronomic/cultural practices and processing techniques which can cause differences in growth and chemical composition and hence black tea quality (Owuor et al. 2010a,b, 2009, 2008a).

Use of suitable genotypes under optimal management in different environments can lead to most economic yields and realisation of best black tea quality. Despite the large variations in the environmental conditions under which tea is grown, leading to differences in yields and quality, uniform/standard agronomic recommendations/practices are used (Anon 2002; Ranganathan and Natesan 1987) over wide regions. In this review, the variations in black tea yields and/or quality due to growing the plant in different locations, changing management system and use of same production technologies in several locations are reported.

2 Effect of Soil Types on Yields and Black Tea Quality

Environment is a major factor that affects tea yields (Carr and Stephens 1992) and black tea quality (Babu et al. 2007; Owuor et al. 2008a). In particular, soil is an essential environmental factor affecting growth of plants. Soil characteristics in different parts of the world vary (Wallis 1997). However, tea plant prefers deep acidic soils with pH below 5.6 (Othieno 1992). The variation in other soil characteristics in different parts of the world where tea is grown is large. As a result, tea productivity and quality of similar genotypes grown in different soil types vary (Fung et al. 2003; Jin et al. 2008; Li et al. 2007). Caffeine and polyphenol contents of tea vary with geographical origin of the leaf and type of soil (Baptista et al. 1998, 1999). The variations in tea responses, particularly yields due to general agro ecological conditions, and farm-specific soil characteristics have been reported in China (Han et al. 2007), India (Venkatesan et al. 2004) and Uganda (Carter 2007), although within Kericho in Kenya, soil characteristics did not affect productivity (Kamau 2008). This lack of response in Kericho, Kenya could in part be as result of uniformity of soils characteristics within a small region. However, even within some small regions, where tea has been grown for a long time, soil quality deteriorates, especially when the soil nutrients content become depleted (Dang 2002; Kamau et al. 2008a). This stagnates and/or reduces yields (Kamau 2008; Kamau et al. 2008a,b; Zeiss and denBraber 2001). The variations in black tea quality due to soil deterioration after long term tea plantation are not documented. However, the reported effects of soil type on tea quality and yields are normally confounded by non uniformity of tea genotypes and management in the studies.

These results suggest that studies should be conducted to evaluate the effects of soil deterioration due to long term tea plantation on black tea quality. The noted variations in tea yields and black tea quality due to soil types indicate that to maximise yields and quality, it is necessary to evaluate new genotypes on the soils of intended growth, so that only cultivars that produce potential optimal yields and black tea quality in a particular soil type are cultivated. Again cultivars should be tested in different soil types to establish stable varieties that are widely adaptable and those with limited stability that perform well only on specific soil types.

3 Altitudes

Successful economic production of tea has been reported from sea level to over 2,700 m amsl (Anandacoomaraswamy et al. 2000; Owuor et al. 2008b). The variations in the altitude cause large differences in temperatures resulting in changes in growth rates and patterns (Burgess and Carr 1997; Obaga et al. 1989; Squire et al. 1993). An increase in mean air temperature with decrease in altitude lowers the shoot population density (SPD) but increases the mean dry weight of the harvested shoot in same clones (Balasuriya 1999; Squire et al. 1993) (Table 1). Consequently, tea yields decrease with rise in altitude (Anandacoomaraswamy et al. 2000; Balasuriya 1999; Obaga and Ng'etich 1989; Obaga et al. 1989; Squire et al. 1993). Decreased tea yields at rate of 1 kg per 100m rise in altitude have been reported in Kericho, Kenya (Othieno et al. 1992). Tea suffers less from water deficit at higher altitudes since 65% of available soil water has to be depleted before a significant decrease of transpiration occurs (Anandacoomaraswamy et al. 2000). However,

Table 1 Effect of altitude on shoot population density measured in the centre of the bush (SPD), density of shoots harvested from the centre of the bush [HSD(centre)], and density of shoots harvested from the entire plot [HSD(plot)] of clones TRI 2023 and TRI 2025

		Number of shoots per m^2								
		SPD			HSD (centre)			HSD (plot)		
Site	Altitude (m)	TRI 2023	TRI 2025	Mean	TRI 2023	TRI 2025	Mean	TRI 2023	TRI 2025	Mean
Glassaugh	1,859	5,115	4,603	4,859	1,063	2,613	1,838	783	1,078	931
Talawakele	1,382	4,867	3,974	4,421	1,424	2,258	1,741	814	965	890
Vellai-oya	1,300	4,711	3,747	4,229	1,859	1,917	1,888	824	656	740
Strathdon	914	3,854	3,601	3,728	2,086	2,173	2,130	784	708	746
Kottawa	30	3,877	2,794	3,336	2,428	1,821	2,125	788	704	746
Mean		4,485	3,744		1,772	2,116		799	822	
CV		4.4%			1.3%			3.1%		
Clone		***			***			*		
Altitude		***			***			***		
Clone × altitude		**			***			***		

Source: Balasuriya (1999)

*, **, *** LSD values significantly different at P = 0.05, 0.01 and 0.001, respectively

the response to environment can be variable in different genotypes (Ng'etich and Stephens 2001a). In young clonal teas, whereas clone, EPK TN14-3, produced the highest matter at the low altitude, clone AHP S15/10 produced the least (Ng'etich and Stephens 2001b). Analysis of genotype and environment interactions showed that clone EPK TN14-3 has above average stability in dry matter production at various altitudes, but below average stability in yield (Ng'etich and Stephens 2001a). Clone AHP S15/10 showed above average stability for tea yield, but was below average stability in dry matter production (Ng'etich and Stephens 2001a). The harvest index of the clones also varied at the different altitudes (Ng'etich and Stephens 2001b), under uniform management. Overall, yields and dry matter production of tea genotypes differed at different sites situated at different altitudes (Ng'etich et al. 2001) despite uniform management. This suggests that different tea genotypes require different management strategies to realise optimal yields and black tea quality.

The composition of volatile compounds of black tea varied with altitude (Mick and Schreier 1984). Black tea quality increased as the altitude increased (Mahanta et al. 1988) more so in the same genotypes under uniform management (Owuor et al. 1990a). Caffeine (Owuor et al. 1990a) and flavour index increase with altitude (Mahanta et al. 1988; Owuor et al. 1990a); verifying the superiority of quality of higher grown over those teas grown in lower altitudes (Table 2). The slow rate of shoot growth due to cool temperatures at high altitudes (Obaga and Ng'etich 1989; Obaga et al. 1989; Squire et al. 1993) is the source of frequently recorded best quality for high grown black teas. These results demonstrate that black tea quality will improve with rise in altitude particularly in the same genotypes when under uniform management. The extent of changes however, vary with genotypes. Thus there are genotypes that are less susceptible to quality changes as altitude rises and *vice versa.* The quality (Moreda-Pineiro et al. 2003) and mineral composition (Street et al. 2006) of tea from diverse sources also vary with altitude. In Turkey, the iron and manganese levels in tea shoot change with altitude (Sahin et al. 1991). However, it was not documented whether the studies used same tea genotypes or received uniform management.

In conclusion, for production of high quality black teas, especially aromatic black teas, growth at high altitude is essential. However production of tea at low attitudes leads to high yields, provided temperatures and soil moisture are adequate and well distributed. Some cultivars show reasonable stabilities across different altitudes. New genotypes should be evaluated for suitability/stability for production of high yields and black tea quality at different altitudes.

4 Geographical Locations and Seasonal Variability

The weather and seasonal fluctuations in variables such as rainfall, temperature and humidity, and soil water deficits influence annual yield distribution, and hence annual yield (Mathews and Stephens 1998a; Uddin et al. 2005) and black tea quality (Owuor 1992, 1994; Owuor et al. 1991a). Performances of most crops,

Table 2 Effects of altitude on chemical quality parameters of CTC black tea

		Clone S15/10				Clone TN 14-3			
Site	Altitude	Theaflavins (μ mol g^{-1})	Caffeine (%)	Flavour index	Sensory evaluation	Theaflavins (μ mol g^{-1})	Caffeine (%)	Flavour index	Sensory evaluation
Timbilil	2,180 m	25.14	3.57	3.56	37.30	28.97	3.93	1.79	
Chepgoiben	2,120 m	21.93	3.41	3.32	34.33	29.84	3.53	1.51	40.67
Cheptabes	1,940 m	21.19	3.27	2.46	32.97	31.14	3.57	1.41	39.33
Kaproret	1,860 m	20.62	3.10	2.13	30.37	31.54	3.63	1.16	38.00
CV (%)		5.96	4.53	6.62	3.85	2.42	7.24	16.17	37.33
LSD, P = 0.05		2.64	0.30	0.38	1.91	1.47	NS	NS	Ns
r^2		0.85	0.97*	0.997*	0.95*	−0.98*	0.53	0.94*	0.98*

Source: Owuor et al. (1990a)

* Significant, P = 0.05, NS = Not significant

Table 3 Mean yields in tons made tea per hectare (x) and stability parameter estimates (bi, S^2d, σ^2i and SE^2i) for tea yields of 20 genotypes tested in 12 (i.e. 6 years in two sites) environments

Genotype	x	bi	Cluster group	SE^2i	S^2d	σ^2i
1. STC 5/3	1.200	0.80	1	1.052	0.414**	0.258
2. 6/8	1.567	0.84**	2	0.830	0.114***	0.036
3. 7/3	1.786	0.99	2	1.179	0.174**	0.045
4. 7/9	1.763	0.88	2	1.032	0.284***	0.112
5. 2X1/4	1.644	1.16	2	2.032	0.426*	0.364
6. 11/26	1.290	0.78***	1	0.723	0.130**	0.064
7. 12/12	1.616	0.91	2	1.014	0.178**	0.058
8. 12/19	1.650	0.96	2	1.061	0.082***	0.009
9. TN 14-3	1.788	1.00	2	1.233	0.219***	0.061
10. S15/10	2.051	1.20**	3	1.665	0.136**	0.056
11. 31/8	2.277	1.25	4	1.972	0.326***	0.186
12. 31/27	1.632	0.93**	2	0.984	0.048**	0.008
13. BB35	1.994	1.11	3	1.448	0.141**	0.039
14. 54/40	1.933	1.09	3	1.412	0.185***	0.049
15. 56/89	1.542	0.94	2	1.140	0.252**	0.089
16. 57/15	1.738	1.06	2	1.312	0.126***	0.023
17. 303/259	1.659	0.96	2	1.069	0.095***	0.015
18. 303/577	1.960	1.06	3	1.389	0.187***	0.048
19. 303/999	1.925	1.02	3	1.255	0.178**	0.049
20. 303/1199	1.882	1.06	3	1.335	0.153**	0.037

Source: Wachira et al. (2002)
*, **, *** indicate significantly different from 1.0 for the regression coefficients (bi) and from 0.0 for the deviation mean squares (S^2d) at 0.05; 0.01 and 0.001 levels of probability.
Cluster based on non-weighted values of mean yield (x) and regression coefficient (bi).

tea inclusive, vary from locality to locality and season to season (Wachira et al. 2002). These variations arise from differences in growth parameters (Ng'etich and Stephens 2001a,b) leading to changes in economic yields (Wachira et al. 1990, 2002; Wickremaratne 1981) (Table 3), chemical composition and overall black tea quality (Owuor et al. 2008a) (Table 4). Yield is primarily determined by shoot numbers, shoot weight and the rate of shoot growth (Carr and Stephens 1992; Odhiambo 1989). These components vary with environment, management practices and the genotype (Burgess 1992). The variations in tea productivity and quality due to weather factors have been reported in several studies (Carr and Stephens 1992; Ng'etich 1995; Odhiambo 1989; Wachira et al. 2002). In particular, tea yields vary with temperature, the saturation of water vapour pressure of air, rainfall and evapotranspiration (Stephens and Carr 1990) which vary with locations and seasons. The components of tea yield (i.e. number of harvested shoots per unit area, their rate of growth and the average weight of shoots at harvest) are largely influenced by weather factors (Burgess 1992). Yields are usually low during the dry or very cold seasons (Nixon et al. 2001).

Table 4 Impact of geographical area of production on fermentation and quality of cultivar SFS 150

Parameter	Source	Fermentation time (min)					Mean source
		30	50	70	90	110	
Theaflavins	Kenya	12.86	18.86	21.55	23.05	22.20	19.90
(μ mol/g)	Malawi	15.01	22.11	21.78	21.29	21.20	20.28
	Mean time	13.94	20.49	21.67	22.17	21.70	
	C.V. (%)			8.69			
	LSD, ($P < 0.05$)			1.78			NS
	Interactions			2.52			
Thearubigins	Kenya	9.61	11.47	11.73	13.02	12.98	11.76
(%)	Malawi	6.97	7.83	7.89	8.56	8.87	8.02
	Mean time	8.29	9.65	9.81	10.79	10.92	
	C.V. (%)			5.08			
	LSD, ($P < 0.05$)			0.80			0.51
Total colour	Kenya	2.51	3.48	3.89	4.47	4.53	3.77
(%)	Malawi	3.14	4.21	4.58	5.22	5.34	4.49
	Mean time	2.83	3.84	4.23	4.84	4.93	
	C.V. (%)			13.35			
	LSD, ($P < 0.05$)			0.88			0.56
Brightness	Kenya	34.32	32.91	32.70	31.01	29.64	32.11
(%)	Malawi	30.11	29.74	27.16	22.28	21.67	26.19
	Mean time	32.21	31.32	29.93	26.64	25.66	
	C.V. (%)			21.80			
	LSD, ($P < 0.05$)			NS			NS

Source: Owuor et al. (2008a)

Tea is grown from the equator to sub tropical environments (Owuor et al. 2008b; Shoubo 1989) where seasonal variations can be very large. For example, at the Tea Research Foundation of Kenya (TRFK) (latitude 0° 22′S, longitude 35° 21′E, altitude 2,200 m amsl) in the Kericho District of Kenya mean seasonal temperatures vary from 15–17°C, substantially less than 18–24°C at the Tea Research Foundation of Central Africa (TRFCA), (latitude 16° 05′S, longitude 35° 37′E, altitude 650 m amsl) Mulanje, Southern Malawi (Carr and Stephens 1992). Tea yields also vary with seasons where significant reductions are observed during the cool season (Tanton 1979). For example, in Malawi, more than 70% of the annual crop is harvested during a 5 month wet warm season (Cloughley 1983; Fordham and Palmer-Jones 1977). Within-season yield variability is directly related to base temperature for shoot extension and shoots population density, which affect management in terms of labour planning, transport requirements and factory capacity (Carr and Stephens 1992; Gulati and Ravindranath 1996; Nixon et al. 2001). Near the Equator, the seasonal changes in temperature are minimal, and the monthly crop distribution can be relatively smooth (Carr and Stephens 1992), but at high latitudes, it can be very uneven as low temperatures and/or droughts restrict shoot extension for part of the year (Burgess and Carr 1997; Stephens and Carr 1990). Away from the equator, large yield peaks often occur following a cool or dry season, with subsequent oscillations which may continue throughout the remainder of the season (Matthews and Stephens 1998b). The yield variations in a single clone can range from about 18% during the rainy season to 30% during the cool, dry weather (Burgess et al. 2006). However, there are genotypic differences in the shoot base temperatures and extension rates resulting in differences in the seasonal distributions of yield in different cultivars with some tea clones undergoing less seasonal variations than the others (Burgess and Carr 1997).

The adverse effects of drought can be partially mitigated by use of shade trees (McCulloch et al. 1965), or irrigation (Kigalu et al. 2008). However, shade trees reduce tea yields (Obaga and Othieno 1987; Othieno and Ng'etich 1992) and nutrient availability (Othieno 1983) but improve black tea quality (Owuor et al. 1989a). The tea yield response to irrigation can be large during dry weather (De Costa et al. 2007). Different tea genotypes differ in responses to irrigation even when they are grown at a single site under uniform management (Kigalu 2007; Kigalu et al. 2008).

The quality of tea is a polygenically controlled trait and is directly or indirectly influenced by various traits (Kamunya et al. 2010) and environments (Babu et al. 2007; Owuor et al. 2008a). Commercial tea of variable genotypes, including seedling and clonal teas show seasonal quality variations even close to the equator (Owuor 1994). Volatile aroma compounds levels and compositions also vary with location of production (Owuor et al. 2008a) and seasons (Cloughley et al. 1982). The variation is more noticeable under temperate or sub-tropical condition (Cloughley et al. 1982; Robinson and Owuor 1992). Dry and cool seasons cause slower shoot growth rate leading to high black tea quality; while, wet season especially rainy period leads to fast tea flush (growth) which reduces black tea quality (Odhiambo et al. 1988). Provided there is adequate moisture in the soil, high temperatures favour fast shoot growth thus increasing yield but lowering the black tea quality. On the contrary, low temperatures in cold seasons cause slow shoot growth rates leading to reduced yields with improved black tea quality. Growth under dry weather with cooler nights and desiccating winds favour the biogenesis of flavour compounds leading to production of flavoury black teas (Rawat and Gulati 2008). Thus, better quality and higher valuation are obtained from teas manufactured from shoots plucked during slow growth conditions (Hilton et al. 1973; Rawat and Gulati 2008).

Seasonal responses to some tea quality parameters with agro-inputs such as fertilizers have been reported in several studies (Mahanta 1988; Mahanta et al. 1995; Ozdemir et al. 1993; Ravichandran and Parthiban 2000; Turkmen and Velioglu 2007; Yao et al. 2005). These occur in different genotypes like in seedling (Sud and Baru 2000) and clonal (Owuor 1994, 2001; Rawat and Gulati 2008) teas due to nitrogenous fertilizer rates; indicating that regardless of season, the use of high nitrogen fertiliser rates reduces black tea quality. The extent of the variations may vary with tea genotypes. Other nutrients of tea also influence the seasonal quality (Ruan et al. 1998, 1999) and extent of seasonal variations in black tea quality vary with location of production.

Locality and seasons affect crop performance. In Malawi, the size of harvested shoots and number of shoots harvested, accounts for 11% and 89%, respectively of the total seasonal yield variations (Tanton 1981). Reduced rate of shoot extension is evident during period of water stress (Ng'etich 1995) and low temperatures (Burgess and Carr 1997). High rate of shoot growth is recorded in warmer conditions at lower altitude (Anandacoomaraswamy et al. 2000; Balasuriya 1999; Obaga and Ng'etich 1989; Obaga et al. 1989; Squire et al. 1993) provided water and temperatures are not limiting. Such fluctuations in harvestable yields due to seasonal variations affect the optimal efficiency of both tea growers and the processors.

Seasonal variations in length of shoot replacement cycle cause variability in yield distribution during the year (Tanton 1982). The shoot replacement cycles vary with geographical region. In the Kenya highlands, along the equator shoot regeneration takes from 80 to 120 days (Odhiambo et al. 1993) unlike further away from the equator in Malawi where shoot regeneration can be as short as 42 days during the favourable growing seasons (Smith et al. 1993; Tanton 1982). Such large differences in growth patterns cause differences in crop distribution, total yields and black tea quality even in the same genotype leading to the need for different management strategies. For example, harvesting intervals vary from 6 to 8 days and 10 to 19 days under warm wet and dry cool seasons in southern Tanzania (Burgess and Carr 1998), respectively.

In conclusion, the seasonal variations in yields and black tea quality dictate that management must make seasonal adjustments. The cool seasons characterised by slow growth usually translate into low production but high quality black teas. There should therefore be efforts to emphasise management to produce high quality black teas. When crop is low due to dry and cool seasons, farm in puts like plucking labour deployment, should change to reflect the low production. Warm wet seasons are characterised by fast growth, high production and low quality black teas. Management should ensure availability of labour or ability to remove all the crop and adequate factory processing capacity to handle all the leaf. These seasonal changes will affect growers further away from the equator much more than those close to the equator where seasonal variations are minimal.

5 Genotypes

Many tea genotypes have been developed in the various tea producing countries. Remarkable achievements have been demonstrated and yields up to 10,995 kg made tea ha^{-1} $year^{-1}$ have been realised under commercial production from selected genotype (Oyamo 1992). This arguably is the highest recorded tea yield per unit area under commercial production. To realise maximum benefits from tea growing, farmers are continuously looking for such high yielding and good quality

Table 5 Comparison of the variations in the theaflavins levels of Cultivar 6/8 grown in Kenya and Malawi due to fermentation duration

Location	Fermentation time (min)	Total TF (Flavognost)	TF	TF-3-G	TF-3'-G	TFDG	TFDG equiv.
Malawi	30	18.18	8.45	4.29	3.11	2.33	6.22
	50	17.30	8.09	4.48	2.75	1.98	5.74
	70	17.30	7.79	4.51	2.79	2.21	5.97
	90	15.10	5.96	4.10	2.56	2.49	5.73
	110	15.28	5.87	4.19	2.64	2.59	5.88
Kenya	30	11.38	7.51	1.79	1.65	0.42	2.21
	50	13.78	8.41	2.51	2.03	0.83	3.72
	70	18.91	9.90	4.56	2.80	1.65	5.75
	90	17.75	9.16	4.53	2.52	1.56	5.44
	110	17.82	8.70	5.11	2.41	1.60	5.57

Source: Owuor et al. (2008b), Total TF = Total theaflavins, TF = Theaflavin, TF-3-G = Theaflavin-3-monogallate, TF-3'-G = Theaflavin-3'-monogallate, TFDG = Theaflavin-3,3'-digallate, TFDG equiv. = Theaflavin-3,3'-digallate equivalent

genotypes. Although tea growers believe that such high yielding cultivars retain their yield potentials wherever they are planted, several studies have demonstrated wide response ranges in yield (Ng'etich et al. 2001; Wachira et al. 1990; Wickremaratne 1981), yield partitioning (Ng'etich et al. 2001), growth (Ng'etich and Stephens 2001a,b), shoot population density (Balasuriya 1999) and dry matter partitioning (Ng'etich and Stephens 2001b) of tea genotypes to different environments (Carr and Stephens 1992; Wachira et al. 1990, 2002; Wickremaratne 1981). The noted variations are due to several factors including water stress (Carr and Stephens 1992), temperatures (Tanton 1982) and altitudes (Obaga et al. 1989; Squire et al. 1993). Such variations are observed even in same genotype planted in various regions (Wachira et al. 1990, 2002; Wickremaratne 1981), although there are also genotypes which are more adaptable in several regions (Wachira et al. 2002) (Table 3).

In terms of black tea quality, variations in chemical composition and quality of same tea genotypes due to geographical influence have been documented (Owuor et al. 2008a) (Table 5). But the management of the cultivars might have been different making comparison difficult. Limited data exist on use of same genotype in different geographical areas due to restricted exchange of genotypes. Where such data exist, management practices are not uniform making comparison difficult. When same genotypes were subjected to same management in different locations, there were variations in yields (Ng'etich and Stephens 2001a,b; Ng'etich et al. 2001; Obaga et al. 1989; Squire et al. 1993; Wachira et al. 2002) and black tea quality (Tables 6 and 7) (Owuor et al. 2010a) even within a radius of only 10 km (Owuor et al. 1990a; Ng'etich and Stephens 2001a,b; Ng'etich et al. 2001; Obaga et al. 1989; Squire et al. 1993). These results demonstrate that the variations in the yields and quality of genotypes can be large when environmental conditions vary widely. However, there are genotypes that are very adaptable over wide regions

Table 6 Response of clonal black tea total theaflavins (μ mol/g) and relative ranking based on theaflavins levels to growing environment

Site	Timbilil		Kipkebe		Kangaita		Mean clones	
Clone	Theaflavins (μ mol/g)	Rank	Theaflavins (μ mol/g)	Rank	Theaflavins (μ mol/g)	Rank	Theaflavins (μ mol/g)	Rank
TRFK 6/8	24.03	7	26.33	4	30.14	1	26.83	4
TRFK 31/8	21.75	14	25.71	7	23.44	16	23.64	12
AHP S15/10	18.06	19	22.48	16	21.19	18	20.58	18
EPK TN 14-3	28.02	1	25.46	9	29.32	4	27.60	1
BBK 35	24.49	5	25.14	10	27.34	8	25.66	9
TRFK 54/40	24.15	6	27.15	2	29.37	3	26.85	3
TRFK 12/12	23.77	8	24.97	11	29.89	2	26.21	5
TRFK 12/19	19.58	16	20.89	18	23.50	15	21.32	17
TRFK 31/27	18.50	17	19.01	20	21.17	19	19.56	19
TRFK 11/26	23.05	10	25.48	8	28.97	6	25.84	8
TRFK 57/15	25.62	2	26.17	5	25.92	10	25.90	7
TRFK 7/3	22.97	11	22.78	15	23.06	17	22.94	14
TRFK 7/9	21.96	13	24.64	13	24.69	11	23.76	11
TRFK 56/89	17.58	20	20.52	19	18.63	20	18.91	20
STCK 5/3	19.76	15	22.98	14	23.58	13	22.10	16
TRFK 303/259	18.48	18	27.30	1	24.15	12	23.31	13
TRFK 303/577	25.29	3	24.97	11	28.23	7	26.16	6
TRFK 303/999	24.86	4	26.84	3	29.05	5	26.92	2
TRFK 303/1199	23.49	8	25.80	6	26.74	9	25.34	10
TRFK 2XI/4	22.07	12	22.20	17	23.54	14	22.60	15
Mean site	22.38		24.34		25.59			
CV (%)			13.27					
LSD, ($P \leq 0.05$)			1.16				2.99	

Source: Owuor et al. (2010a)

(Wachira et al. 2002; Owuor et al. 2010a).Thus new genotype imported into an area may need to be re-evaluated for the yields and quality potentials in the new regions.

The conclusions derived from the observations are that there are cultivars which produce superior yields and black tea. To maximise profits from tea cultivation, growers need to invest in the search for high yielding and black tea quality genotypes. The black tea yields and quality of different genotypes in one location vary suggesting a need for proper cultivar selections and evaluation before wide spread cultivation. Responses of selected cultivars in an area vary when grown in new environments/locations. Thus superior cultivar in one location may not maintain the superior attributes in new locations. But there are genotypes which are suitable for particular geographical areas while some genotypes perform poorly when cultivated away from the locations they were developed. Genotypes should be re-evaluated in new areas before they are availed for wide spread production in areas where they were not developed.

Table 7 Regression coefficients (R^2) of linear regression analyses between same parameters in different regions

Item		Timbilil	Kipkebe	Kangaita
	Timbilil	–		
Theaflavins	Kipkebe	0.3149	–	
	Kangaita	0.5250	0.5665	–
	Mean	0.6864	0.7261	0.9180
Thearubigins	Timbilil	–		
	Kipkebe	0.4340	–	
	Kangaita	0.2329	0.1303	–
	Mean	0.7537	0.7074	0.5383
Total colour	Timbilil	–		
	Kipkebe	0.3768	–	
	Kangaita	0.3996	0.5863	–
	Mean	0.6928	0.8292	0.8203
Chemical Brightness	Timbilil	–		
	Kipkebe	0.2126	–	
	Kangaita	0.4668	0.0562	–
	Mean	0.7760	0.5281	0.6247
Tasters B evaluations	Timbilil	–		
	Kipkebe	0.3425	–	
	Kangaita	0.2563	0.4886	–
	Mean	0.6272	0.7341	0.5905
Tasters A evaluations	Timbilil	–		
	Kipkebe	0.0120	–	
	Kangaita	0.3002	0.0009	–
	Mean	0.6964	0.1234	0.7001

Source: Owuor et al. (2010a)

6 Management

Tea genotypes vary widely in their growth and nutrients uptake responses even in the same environment. This implies that the management and agronomic inputs need to be adjusted with the changes in genotypes and environments to ensure production of optimised tea yields and quality. However, in many regions tea agronomic practices such as fertilizer rates and plucking intervals are similar. In Eastern Africa, for, example, most of the agronomic recommendations originated from the Tea Research Foundation of Kenya, in Kericho (Anon 2002) that was originally part of the Tea Research Institute of East Africa (1951–1980) prior to its dissolution after the break-up of the East Africa Community in 1977. Up to this period, most tea plantations in East Africa were of seedling plants. Such plants have variable yields and quality as individual plants are unique. Usually the yields and black tea quality from seedling plantations are low. Most tea plantations are now planted with clonal materials which have been selected for high yields and/or black tea quality. These plants are sensitive to changes in environmental and management factors. In eastern Africa, there has been lack of research body in Uganda, Rwanda, and Burundi and only recently did Tanzania initiate the Tea Research Institute of Tanzania. It is doubtful whether the recommendations then or those developed in Kenya (Anon 2002) suit all environments in the region for the production of high yields and black tea quality from the clonal teas. Indeed, even, within Kenya yields (Wachira et al. 1990, 2002) and quality (Owuor et al. 2010a) of tea genotypes change with region of production suggesting it is necessary to apply different management practices that would promote realisation of high yields and quality in different regions. The optimal plucking interval (Owuor et al. 2009) and nitrogenous fertiliser rates (Owuor et al. 2010b) for realisation of high quality in same genotype vary with geographical area of production further emphasising the need to fine tune agronomic recommendations for different regions. The optimal fermentation durations to realise high quality black tea also change in the same genotype with geographical area of production. The optimal fermentation duration is attained much faster at TRFCA (altitude 650 m amsl, latitude 16° 05′S, longitude 35° 37′E) than at TRFK (altitude 2,200 m amsl, latitude 0° 22′S, longitude 35° 21′E) (Owuor et al. 2008a). These results suggest that it is necessary to adjust cultural practices and agronomic inputs to suit different geographical areas of tea production for realisation of high yields and quality of black tea.

In conclusion, different regions need to develop region specific production technologies suitable for the particular region. In single locations different genotypes grow at different rates leading to variations in yields and black tea quality. Each genotype should have unique management technologies to enhance their values. The different growth patterns in different cultivars suggest that planting mixed cultivars should be discouraged to ease management. Genotypes should be planted in pure stands or blocks to ease management.

7 Plucking

Tea yields and black tea quality are influenced by agronomic practices. One such agronomic practice is harvesting or plucking (Owuor 1996). Tea plucking costs amount to up to 80% of the total field production costs in manual harvesting (hand plucking) based-estates (Sharma 1987; Willson 1992). To reduce costs of harvesting, several methods are used to pluck tea (Fig. 3). In different genotypes black tea quality and yields vary with plucking intervals (Baruah et al. 1986; Owuor et al. 1997, 2000) (Table 8), standards (Table 9) (Asil 2008; Mahanta et al. 1988; Obanda et al. 2002; Owuor et al. 2000) and methods (Burgess et al. 2006; Owuor et al. 1991b; Ravichandran and Parthiban 1998); while tea yields are affected by plucking intensity (Chandra-Mouli et al. 2007). These results demonstrate that uniform plucking policies may not be appropriate for different tea genotypes even when they are grown in same environment. There are also tea genotypes that are more resistant to quality changes due to plucking standard (Obanda et al. 2002).

Fig. 3 (**a**) A method of mechanical tea harvesting. (**b**) Hand plucking of tea

Table 8 Effects of plucking intervals and nitrogen fertiliser on yields and black tea quality

Item	Plucking intervals (days)	Rates of nitrogen (kg N ha^{-1} year^{-1}) 100	200	300	400	500	600	Mean plucking intervals
Yield (kg black tea/ha)	7	5,076	7,117	6,787	7,432	7,407	7,388	6,868
	14	5,118	6,491	6,766	6,812	7,449	7,206	6,715
	21	4,607	6,383	5,226	6,254	6,464	6,882	6,186
	Mean rates	4,934	6,813	6,693	6,893	7,107	7,158	
	CV (%)			4.15				7.72
	LSD,$_{0.05}$			713				297
Theaflavins (mol/g)	7	12.80	13.32	11.75	11.21	10.84	12.32	12.04
	14	11.67	10.79	10.53	9.74	9.51	9.87	10.35
	21	10.39	10.03	9.26	9.52	8.59	8.69	9.41
	Mean rates	11.62	11.38	10.51	10.16	9.64	10.29	
	CV (%)			5.84				7.95
	LSD,$_{0.05}$			1.62				0.49
Flavour Index	7	1.13	0.99	0.89	0.80	0.77	0.73	0.89
	14	0.91	0.75	0.72	0.67	0.64	0.61	0.72
	21	0.77	0.66	0.66	0.59	0.57	0.53	0.63
	Mean rates	0.94	0.76	0.80	0.69	0.66	0.62	
	CV (%)			4.67				8.42
	LSD,$_{0.05}$			0.11				0.04
Sensory evaluation	7	44	43	42	40	39	38	41
	14	40	38	37	36	35	36	36
	21	37	32	33	31	33	33	33
	Mean rates	40	38	37	36	36	34	
	CV (%)			1.77				8.80
	LSD,$_{0.05}$			2				2

Source: Owuor et al. (1997)

The plucking rounds vary among tea growing countries in the eastern Africa region. In Kenya, plucking rounds range from 7 to 10 days (Anon 2002), while, in Rwanda it ranges from 9 to 14 days (Uwimana, personal communication) and in Southern Tanzania, the practice is to pluck after 13–14 and 27–30 days under normal long rain season and adverse conditions, respectively (Burgess 1992) in clone TRFK 6/8. These plucking policies demonstrate that there may be a strong plucking interval and environment interactions which could affect profits in tea production in the different regions. Where it is warmer, shoot growth is much faster (Anandacoomaraswamy et al. 2000; Balasuriya 1999; Squire et al. 1993), thus requiring shorter plucking intervals to realise optimal yields and quality (Owuor et al. 2009). As in black tea quality (Obanda et al. 2002), there are differences in the extent of yield variations from region to region in different genotypes. It is necessary to use different plucking management practices to optimise both yield and black tea quality in different cultivars and environments.

Table 9 The influence of plucking standard on quality

Plucking standard	Clone 6/8 Theaflavins (μ mol/g)	Clone 6/8 Thearubigins (%)	Clone 6/8 Theaflavin digallate equivalent	Clone 6/8 Sensory evaluation	S15/10 Theaflavins (μ mol/g)	S15/10 Thearubigins (%)	S15/10 Theaflavin digallate equivalent	S15/10 Sensory evaluation
1 + bud	25.01	14.47	4.28	11.0	17.70	13.09	4.16	11.0
2 + bud	29.28	18.04	3.10	9.7	19.16	14.37	3.21	9.7
3 + bud	23.52	18.25	1.44	7.7	16.58	13.67	1.84	8.3
4 + bud	21.86	17.39	0.93	5.7	14.51	13.66	1.33	7.0
5 + bud	19.10	16.29	0.70	5.7	13.40	13.28	0.95	7.0

Source: Obanda et al. (2002)

In conclusion, plucking younger shoots (fine plucking) ensures production of high yields and black tea quality, and this can be realised through short plucking intervals. However, yield response to plucking intervals varied with locations and possibly duration of the trials or genotypes being evaluated. There is need to develop region specific plucking management technologies to enhance yields and black tea quality.

8 Fertilisers and Nutrients Management

Although fertilizer is one of most expensive agronomic inputs in tea plantations (Bonheure and Willson 1992), it is essential for economic tea production. Fertilizers increase yields (Bonheure and Willson 1992; Kamau et al. 2008a; Owuor et al. 2008b; Venkatesan et al. 2003, 2004) through increased growth rate and density of harvested shoots (Odhiambo 1989) (Fig. 4). The recommended agro-inputs and management on tea such as rates of nitrogen fertilizer in the eastern Africa region are widely adopted from Kenya (Anon 2002). But in practice, the rates, and sources vary from country to country (Bonheure and Willson 1992) and the popular formulation is NPK 25:5:5 or NPK 20:10:10 (Bonheure and Willson 1992). In recent years, there

Fig. 4 Typical fertiliser field trials

have been efforts to use other NPK formulations as tea growers have realised that yield responses vary widely. For example, in Kenya, nitrogen rate between 100 and 250 kg N ha^{-1} $year^{-1}$ is used under rain-fed conditions (Anon 2002); and rates up to 300 kg N ha^{-1} $year^{-1}$ is considered a normal practice for high yielding clones (Owuor et al. 2008b). In Tanzania, 150–250 kg N ha^{-1} $year^{-1}$ rates are applied under rain-fed regime (Anonymous 2004). Similarly, in Rwanda between 100 and 150 kg N ha^{-1} $year^{-1}$ is used (Uwimana, pers. Com). Despite the almost similar nitrogenous fertiliser rates used, the yields in the three east African countries per unit area are very different, even when the genotype is uniform especially for clone TRFK 6/8 that is widely grown in the region. In Iran, Sulphate of Ammonia produced higher yields than Urea in the same genotypes at two locations (Salardini 1978).

There are variations in tea yields (Han et al. 2008) and black tea quality (Venkatesan and Ganapathy 2004; Venkatesan et al. 2003) among tea cultivars with increase in nitrogenous fertilizer. Increasing nitrogen fertilizer rate in combination with rise in tea base temperature increases yield by increasing shoot population (Obaga and Ng'etich 1989). Excessive use of nitrogenous fertilizer reduces the quality of made tea (Cloughley 1983; Owuor and Othieno 1996; Owuor and Wanyoko 1996). There is a decline in black tea total colour with increase in nitrogen rate (Owuor et al. 1997, 2000). Also there is increase of caffeine and decrease of flavour index with higher rates of nitrogenous fertilizers (Owuor et al. 1997, 2000). The extent of these variations were however dependent on the genotypes (Owuor and Othieno 1996; Owuor and Wanyoko 1996) and the environment (Owuor et al. 2010b) (Table 10). But, high rates of nitrogen do not increase tea yields economically when applied beyond 300 kg N ha^{-1} $year^{-1}$and reduce black tea quality (Table 8) (Owuor et al. 1997, 2000; Venkatesan and Ganapathy 2004; Venkatesan et al. 2004). The yields and quality of black tea show dissimilar response patterns to increased nitrogen fertilizer rate, suggesting the importance to optimize the two parameters (Ravichandran 2002) for economic production.

While tea yield responds to potassium in India (Ranganathan and Natesan 1985; Ranganathan et al. 1988; Verma 1997) and Iran (Salardini 1978), tea does not respond to potash application in Kenya (Wanyoko and Othieno 1987), possibly due to the high levels of potassium in east African soils (Bonheure and Willson 1992). Tea quality responses to potassium fertiliser change with the environment. In Southern India applying potassium at high levels lowers the overall tea quality (Venkatesan and Ganapathy 2004), but in Kenya potassium did not affect tea quality (Owuor et al. 1998). In yet other studies in southern India, potassium improved black tea quality (Venkatesan et al. 2004, 2005, 2006). The conflicting responses recorded in Indian teas could be due to differences in genotypes and/or environments of the trials. While further studies may be necessary to explain the basis of these conflicting results, it is likely the differences arise from the native potassium levels in the soil. In young tea, yields have been reported to increase with phophatic fertiliser application (Othieno 1980), while there was no yields response in replanted tea (Othieno et al. 1997). However, in black tea quality there was no response to phophatic fertilisers (Owuor et al. 1998).

Table 10 Effects of rates on nitrogenous fertilisers on mean (1998–2007) tea yields in different regions

Parameter	Location	Rate of nitrogen (kg N ha^{-1} $year^{-1}$)					Mean
		0	75	150	225	300	location
Mean yield (kg black tea/ha/year) (1998–2007)	Kipkebe	1,947	2,977	3,534	3,657	3,859	3,194
	Sotik Highlands	3,151	4,034	5,112	5,720	5,942	4,792
	Karirana	3,379	3,393	3,693	3,647	3,635	3,549
	Changoi	3,952	4,595	4,757	4,925	4,928	4,632
	Timbilil	2,359	2,982	3,532	3,309	3,451	3,126
	Mean rate	2,957	3,595	4,126	4,252	4,363	
	CV (%)			5.55			
	LSD, ($P \leq 0.05$)			217			217
	Interactions			371			
Theaflavins (μ mol/g)	Kipkebe	26.21	24.09	23.31	22.69	21.65	23.59
	Sotik Highlands	22.47	19.85	21.11	19.97	17.05	20.09
	Karirana	26.41	25.21	24.83	23.63	25.35	25.09
	Changoi	26.71	25.85	24.48	25.56	24.58	25.44
	Timbilil	25.04	24.04	23.68	23.49	22.02	23.66
	Mean rate	25.37	23.81	23.48	23.07	22.13	
	CV (%)			13.50			
	LSD, ($P \leq 0.05$)			3.23			3.23
Thearubigins (%)	Kipkebe	17.86	17.94	16.75	15.00	14.67	16.44
	Sotik Highlands	16.82	17.91	16.67	16.87	15.05	16.67
	Karirana	16.53	16.88	15.82	15.64	15.09	15.99
	Changoi	17.68	17.87	16.41	16.67	15.99	16.93
	Timbilil	18.38	16.23	15.82	15.44	14.46	16.07
	Mean rate	17.46	17.37	16.29	15.93	15.05	
	CV (%)			9.51			
	LSD, ($P \leq 0.05$)			1.58			NS
Sensory evaluations	Kipkebe	78	67	73	73	62	71
	Sotik Highlands	107	110	104	59	55	87
	Karirana	104	90	84	76	78	86
	Changoi	85	74	72	62	55	70
	Timbilil	105	98	94	96	79	94
	Mean rate	96	88	86	73	66	
	CV (%)			24.38			
	LSD, ($P \leq 0.05$)			20			20

Source: Owuor et al. (2010b)

Leaf and chemical soil analysis are two methods routinely used in the determination of nutrients availability to plants (Kamau et al. 2005; Venkatesan et al. 2004). Plant analysis elucidates how much of the available nutrients are taken up by the plant (Nathan and Warmund 2008). Thus, chemical analysis details the potential of plant nutrients in the soil and the ability of plant to extract those nutrients (Anon 2002; Kamau et al. 2008a). However, even when planted in one field, different tea genotypes have varying ability to extract nutrients from the soil resulting in variations in leaf nutrients content of different clones grown at the same site (Nyirenda 1991; Wanyoko and Njuguna 1983). The levels of nutrients in different parts of tea bushes also vary (Dang 2005; Kamau 2008; Ruan et al. 2003). This suggests that nutrients removal from the soil by tea plant is dependent on both genotypes and harvesting management which could lead to variations in amounts of fertiliser needed to optimise yields and quality.

In conclusion, different regions need to adopt use of nitrogenous fertiliser rates that are a compromise between yields and quality for the particular region. Use of high rates of nitrogenous fertilisers reduce black tea quality and do not enhance yields, while low rates of nitrogen give high quality black tea but low production. In eastern Africa, for example, it may not be justifiable to use more than 200 kg N ha^{-1} $year^{-1}$, even for high yielding tea cultivars.

9 Processing

The biochemical processes that occur during black tea processing have impact on the resultant quality (Baptista et al. 1998, 1999). The processes are largely influenced by both processing conditions (management) (Obanda et al. 2004) and growing environment (Owuor et al. 2008a, 2009, 2010b). For same genotypes, grown in different environments, optimal fermentation duration, hence quality changes with geographical area of production (Tables 4 and 6) (Owuor et al. 2008a). Thus to produce similar quality tea at different environments, it is necessary to adjust processing conditions. Similarly, maceration methods (Mahanta 1988; Owuor and Obanda 1994a; Owuor et al. 1989b) or sequence (Owuor and Obanda 1994b) cause conditions which alter the fermentation processes resulting in difference in black tea quality. Indeed, the extent of moisture removal during the withering phase (physical wither) (Obanda et al. 1997; Tomlins and Mashingaidze 1997; Ullah et al. 1986), wither duration (chemical wither) (Owuor et al. 1995b; Tomlins and Mashingaidze 1997; Ullah 1984) and withering temperature (Owuor and Obanda 1996) cause differences in black tea quality. The extent of such variations depend on the genotype (Obanda and Owuor 1994) but generally hard physical wither produce more aromatic black teas with low plain black tea quality parameters, while softer physical wither produce black teas with higher plain tea quality parameters but less aromatic black teas. Chemical withers beyond 20 h (Owuor et al. 1990b) and below 8 h (Owuor and Orchard 1992) impair quality of plain teas. Indeed, different genotypes under the same wither conditions have different optimal fermentation durations (Wright et al. 2002). The extent of these variations depend on clones/genotypes, demonstrating that there is strong genotype x environment x management interactions effects in black tea processing. For different genotypes, it is necessary to develop site specific processing conditions (Wright et al. 2002) to realise high black tea quality.

In general, environmental conditions of growth influence black tea processing such that suitable conditions for producing high quality tea in one environment may not be ideal in different regions. Processing technologies should be developed that are region/location specific.

10 Conclusion

Yields and black tea quality vary largely due to genotypes, environment and management. For realisation of high yields and black tea quality, it is necessary that optimal production technologies are developed that are suitable for different genotypes in various environments. Development of genotype and region specific technologies may be a viable way for tea farmers realising better profits from tea growing. There should be specific attention to the effects of soil type, altitude, weather factors, seasons and cultivars on management of tea enterprises, especially on agronomic practices like plucking and fertiliser use and processing technologies.

Acknowledgement Financial support for this review from the Inter-University Council of East Africa as VicRes Research Grant is gratefully acknowledged.

References

Agarwal B (1989) Factors affecting quality of tea during processing. Sri Lanka J Tea Sci 58:64–72

Anandacoomaraswamy A, De Costa WAJM, Shyamalie HW, Campbell GS (2000) Factors controlling transpiration of mature field-grown tea and its relationship with yield. Agri For Meteor 103:375–386

Anonymous (2002) Tea growers handbook, 5th edn. Tea Research Foundation of Kenya, Kericho

Anonymous (2004) Annual report. Tea Research Institute of Tanzania (TRIT), Tanzania

Anonymous (2008) Annual bulletin of statistics. International Tea Committee, London

Asil MH (2008) Effects of plucking methods on yield and quality of black of tea (*Camellia sinensis L.*). J Food Agric Environ 6(2):337–341

Babu SM, Saravanan S, Murugesan S (2007) Association analysis of greenleaf quality parameters in tea: UPASI Tea Research Foundation Newsletter 17:1, ISSN 0972–3145

Balasuriya J (1999) Shoot population density and shoot weight of clonal tea (*Camellia sinensis*) at different altitudes in Sri Lanka. Eur J Agron 11:123–130

Banerjee B (1992) Botanical classification of tea. In: Willson KC, Clifford MN (eds) Tea: cultivation to consumption, 1st edn. Chapman & Hall, London, pp 227–263, Chapter 8

Baptista JAB, Tavares JFP, Carvalho RCB (1998) Comparison of catechins and aromas among different green teas using HPLC/SPME-GC. Food Res Intern 31:729–736

Baptista JAB, Tavares JFP, Carvalho RCB (1999) Comparative study and partial characterization of Azorean green tea polyphenols. J Food Comp Anal 12:273–287

Baruah S, Hazarika M, Mahanta PK, Horita H, Murai T (1986) Effect of plucking intervals on the chemical constituents of CTC black teas. Agric Biol Chem 50:1039–1041

Bonheure D, Willson KC (1992) Mineral nutrition and fertilisers. In: Willson KC, Clifford MN (eds) Tea: cultivation to consumption. Chapman & Hall, London, pp 269–329, Chapter 9

Burgess PJ (1992) Responses of tea clones to drought on Southern Tanzania. Ph.D. thesis, Silsoe College, Cranfield University, Bedford

Burgess PJ, Carr MKV (1997) Responses of young tea (*Camellia sinensis*) clones to drought and temperature. 3. Shoot extension and development. Expl Agri 33:367–383

Burgess PJ, Carr MKV (1998) The use of leaf appearance rates estimated from measurements of air temperature to determine harvest intervals for tea. Expl Agric 34:207–218

Burgess PJ, Carr MKV, Mizambwa FCS, Nixon DJ, Lugusi J, Kimambo EI (2006) Evaluation of simple hand-held mechanical systems for harvesting tea (*Camellia sinensis*). Expl Agric 42:165–187

Carr MKV, Stephens W (1992) Climate, weather and the yield of tea. In: Willson KC, Clifford MN (eds) Tea: cultivation to consumption. Chapman & Hall, London, pp 87–135

Carter R (2007) Water management and smallholder fairtrade tea producers in South Western Uganda. MSc thesis, School of Applied Sciences, Cranfield University, Bedford

Chandra-Mouli MR, Onsando JM, Corley RHV (2007) Intensity of harvesting in tea. Expl Agric 43:41–50

Cloughley JB (1983) The effect of harvesting policy and nitrogen fertilizer application rates on the production of tea in Central Africa. II, Quality and total value of crop. Expl Agric 19:47–54

Cloughley JB, Ellis RT, Pendlington S, Humphrey P (1982) Volatile flavour constituents of some Central African black tea clones. J Agric Food Chem 30:342–345

Dang MV (2002) Effects of tea cultivation on soil quality in the northern mountainous zone, Vietnam. Ph.D. thesis, University of Saskatchewan, Canada

Dang MV (2005) Soil–plant nutrient balance of tea crops in the northern mountainous region, Vietnam. Agric Ecosys Environ 105:413–418

De Costa WAJM, Mohotti AJ, Wijeratne AJ (2007) Ecophysiology of tea. Braz J Plant Phys 19(4):299–332

Fordham R, Palmer-Jones RW (1977) Simulation of intra-seasonal yield fluctuations of tea in Malawi. Expl Agric 13:33–42

Fung KF, Zhang ZQ, Wong JWC, Wong MH (2003) Aluminium and fluoride concentrations of three tea varieties growing at Lantau Island, Hong Kong. Environ Geochem Health 25:219–232

Gardner EJ, Ruxton CHS, Leeds AR (2007) Black tea, helpful or harmful? A review of the evidence. Eur J Clin Nutr 61:3–18

Gulati A, Ravindranath SD (1996) Seasonal variations in quality of Kangra Tea (*Camellia sinensis* (L.) *0* Kuntze) in Himachal Pradesh. J Sci Food Agric 71:231–236

Han W, Kemmitt SJ, Brookes PC (2007) Soil microbial biomass and activity in Chinese tea gardens of varying stand age and productivity. Soil Biol Biochem 39:1468–1478

Han W-Y, Ma L-F, Shi Y-Z, Ruan J-Y, Kemmitt SJ (2008) Nitrogen release dynamics and transformation of slow release fertiliser products and their effects on tea yield and quality. J Sci Food Agric 88:839–846

Hilton PJ, Palmer-Jones R, Ellis RT (1973) Effects of season and nitrogen fertilizer upon the flavanol composition in tea making quality of fresh shoots of tea (*Camellia sinensis*) in Central Africa. J Sci Food Agric 24:819–826

Jin CW, Du ST, Zhang K, Lin XY (2008) Factors determining copper concentration in tea leaves produced at Yuyao County, China. Food Chem Toxic 46:2054–2061

Kamau DM (2008) Productivity and resource use in ageing tea plantations. Ph.D. thesis Wageningen University, Wageningen. ISBN: 978-90-8504-808-4

Kamau DM, Wanyoko JK, Owuor PO (2005) Mature leaf nutrient levels as a diagnostic tool for making fertilizer recommendations in mature tea: the case of clone BBK 35. Tea 26:57–69

Kamau DM, Spiertz JHJ, Oenema O (2008a) Carbon and nutrient stocks of tea plantations differing in age, genotype and plant population density. Plant Soil 307:29–39

Kamau DM, Spiertz JH, Oenema O, Owuor PO (2008b) Productivity and nitrogen use of tea in relation to plant age and genotype. Field Crops Res 108:60–70

Kamunya SM, Wachira FN, Owuor PO, Pathak RS, Wanyoko JK, Sharma RK, Muoki RC (2010) Quantitative genetic parameters for yield, drought tolerance and some quality traits in tea (*Camellia sinensis* (L.) O. Kuntze). Res J Agric Sci 1(2):53–65

Kigalu JM (2007) Effects of planting density and drought on the productivity of tea clones (*Camellia sinensis* L.): yield responses. Phys Chem Earth 32:1098–1106

Kigalu JM, Kimambo EI, Msite I, Gembe M (2008) Drip irrigation of tea (*Camellia sinensis* L.). 1. Yield and crop water productivity responses to irrigation. Agric Water Manag 95:1253–1260

Li T, Yu L-J, Li M-T, Li W (2007) Comparative studies on the qualities of green teas in Karst and non-Karst areas of Yichang, Hubei Province, PR China. Food Chem 103:71–74

Mahanta PK (1988) Biochemical analysis as a measure of dynamic equilibrium in genomic setup during processing of tea. J Biosci 13:343–350

Mahanta PK, Baruah S, Owuor PO, Murai T (1988) Flavour volatile of Assam black teas manufactured from different plucking standards and orthodox teas manufactured from different altitudes of Darjeeling. J Sci Food Agric 45:317–324

Mahanta PK, Tamuly P, Bhuyan LP (1995) Comparison of lipoxygenase activity and lipid composition in various harvests of north-eastern Indian tea. J Agric Food Chem 43:208–214

Matthews RB, Stephens W (1998a) Cuppa-Tea: a simulation model describing seasonal yield variation and potential production of tea. 1. Shoot development and extension. Expl Agric 34:345–367

Matthews RB, Stephens W (1998b) The role of photoperiod in regulating seasonal yield variation in tea (*Camellia sinensis* L.). Expl Agric 34:323–340

Mbadi PK, Owuor PO (2008) The role of Kenya Tea Development Agency Limited in the small-scale tea holder development in Kenya. Intern J Tea Sci Spec Issues 7(3&4):7–15

McCulloch JSG, Pereira HC, Kerfoot O, Goodchild NA (1965) Effect of shade trees on tea yields. Agric Meteor 2:385–399

Mick W, Schreier P (1984) Additional volatiles of black tea aroma. J Agric Food Chem 32:924–929

Mondal TK (2007) Tea. In: Pua EC, Davey MR (eds) Biotechnology in agriculture and forestry, 60th edn, Transgenic Crops V. Springer, Berlin, pp 519–535. ISBN ISBN 354049160

Moreda-Pineiro A, Fisher A, Hill SJ (2003) The classification of tea according to region of origin using pattern recognition techniques and trace metal data. J Food Comp Anal 16:195–211

Nathan MV, Warmund M (2008) Use of plant analysis for evaluating the nutrients status of perennial fruit. Mo Environ Gard Newsl 14:64–64

Ng'etich WK (1995) An evaluation of the responses of some tea clones to environment in Kenya. Ph.D. thesis, Silsoe College, Cranfield University, Bedford

Ng'etich WK, Stephens W (2001a) Responses of tea to environment in Kenya. 1. Genotype×Environment interactions for total dry matter production and yield. Expl Agric 37:333–342

Ng'etich WK, Stephens W (2001b) Responses of tea to environment in Kenya. 2. Dry matter production and partitioning. Expl Agric 37:343–360

Ng'etich WK, Stephens W, Othieno CO (2001) Responses of tea to environment in Kenya. 3. Yield and yield distribution. Expl Agric 37:361–372

Nixon DJ, Burgess PJ, Sanga BNK, Carr MKV (2001) A comparison of the responses of mature and young clonal tea to drought. Expl Agric 37:391–402

Nyirenda HE (1991) Use of growth measurements and foliar nutrient content as criteria for clonal selection in tea (*Camellia sinensis*). Expl Agric 27:47–52

Obaga SO, Ng'etich WK (1989) Thermal-time and growth of shoots of tea clones. Tea 10:80–89

Obaga SO, Othieno CO (1987) Effect of different types of shade and nitrogen rates on tea yields: preliminary results. Tea 8:57–62

Obaga SO, Othieno CO, Lang'at JK (1989) Observations on the effect of altitude on yield attributes of some tea clones – growth and density of shoots. Tea 10:73–79

Obanda M, Owuor PO (1994) The effect of wither and plucking standard on the biochemical and chemical parameters of selected Kenya tea. Discov Innov 6:190–197

Obanda M, Owuor PO, Bore JK (1997) Effects of moisture loss and temperature of leaf during withering on black tea quality parameters. Tea 18:45–50

Obanda M, Owuor PO, Wekesa H (2002) The influence of clone and plucking standards on the formation of black tea chemical and sensory parameters. Tea 23:81–90

Obanda M, Owuor PO, Mang'oka R, Kavoi MM (2004) Changes in thearubigin fractions and theaflavin levels due to variations in processing conditions and their influence on black tea liquor brightness and total colour. Food Chem 85:163–173

Odhiambo HO (1989) Nitrogen rates and plucking frequency on tea. The effect of plucking frequency and nitrogenous fertilizer rates on yield components of tea (*Camellia sinensis*) in Kenya. Tea 10:90–96

Odhiambo HO, Owuor PO, Othieno CO (1988) Factors affecting tea quality: I. Agronomic practices: a review. Tea 9:36–40

Odhiambo HO, Nyabundi JO, Chweya J (1993) Effects of soil moisture and vapour pressure deficits on shoot growth and the yield of tea in the Kenya highlands. Expl Agric 29:341–350

Othieno CO (1980) Effects of mulches on phosphorus uptake by tea. Tea 1(2):9–11

Othieno CO (1983) Studies on the use of shade in tea plantations in Kenya. I. Effects on nutrient uptake and yield of tea: preliminary results. Tea 4(2):13–20

Othieno CO (1992) Soils. In: Willson KC, Clifford MN (eds) Tea: cultivation to consumption. Chapman & Hall, London, pp 241–258

Othieno CO, Ng'etich WK (1992) Studies on the use of shade in tea plantations in Kenya. II. Effect on yield and their components. Tea 13(2):100–110

Othieno CO, Stephens W, Carr MKV (1992) Yield variability at the Tea Research Foundation of Kenya. Agric For Meteor 61:237–252

Othieno CO, Wanyoko JK, Owuor PO, Lang'at JK (1997) Response of replanted young clonal tea to different rates and sources of phosphatic fertilizers. Tea 18:32–41

Owuor PO (1992) Changes in quality parameters of commercial black seedling tea due to time of the year in the eastern highlands of Kenya. Food Chem 45:119–124

Owuor PO (1994) Clonal variations in the response of black tea (*Camellia sinensis* (L.)) quality parameters to time of the year in the western Kenya highlands. Trop Sci 34:225–230

Owuor PO (1996) Improving black tea quality through optimum plucking: a review. J Plant Crops Suppl 24:697–702

Owuor PO (2001) Effects of fertilisers on tea yields and quality: a review with special reference to Africa and Sri Lanka. Intern J Tea Sci 1:1–11

Owuor PO, Obanda M (1994a) Impact of maceration method on black tea quality: CTC vs. LTP. Tea 15:50–55

Owuor PO, Obanda M (1994b) Changes in CTC black tea quality due to variations of maceration-fermentation sequence. Tea 15:113–118

Owuor PO, Obanda MA (1996) The impact of withering temperature on black tea quality. J Sci Food Agric 70:288–292

Owuor PO, Orchard JE (1992) Effects of storage time in a two-stage withering process on the quality of made seedling tea. Food Chem 45:45–49

Owuor PO, Othieno CO (1991) Response of black tea quality to locality in smallholder tea farms. Tea 12:41–45

Owuor PO, Othieno CO (1996) Optimizing nitrogen fertilizer application rates to different tea cultivars. Trop Sci 36:211–223

Owuor PO, Wanyoko JK (1996) Rationalisation of nitrogen fertiliser use in tea production. Tea 17:53–59

Owuor PO, Othieno CO, Howard GE, Robinson JM, Cooke RD (1989a) Studies on the use of shade in tea plantations in Kenya. Effect on chemical composition and quality of CTC black tea. J Sci Food Agric 46:63–70

Owuor PO, Othieno CO, Takeo T (1989b) Effects of maceration method on the chemical composition and quality of black tea. J Sci Food Agric 49:87–94

Owuor PO, Obaga SO, Othieno CO (1990a) The effects of altitude on the chemical composition of black tea. J Sci Food Agric 50(1):9–17

Owuor PO, Orchard JE, Robinson JM, Taylor SJ (1990b) Variations in the chemical composition of clonal black tea due to delayed withering. J Sci Food Agric 52:55–61

Owuor PO, Othieno CO, Robinson JM, Baker DM (1991a) Response of tea quality parameters to time of the year and nitrogen fertilisers. J Sci Food Agric 55:1–11

Owuor PO, Othieno CO, Robinson JM, Baker DM (1991b) Changes in the quality parameters of seedling tea due to height and frequency of mechanical plucking. J Sci Food Agric 55:241–249
Owuor PO, Obanda M, Edmonds CJ, Rono MKA (1995) Evaluation of tank withering method in processing plain black tea. Trop Sci 35:167–170
Owuor PO, Othieno CO, Odhiambo HO, Ng'etich WK (1997) Effect of fertilizer levels and plucking intervals of clonal tea *Camellia sinensis* L. O. Kuntze. Trop Agric 74:184–191
Owuor PO, Wanyoko JK, Obanda M, Othieno CO (1998) Potash and phosphorus fertilisers on black tea quality in the western Kenya highlands. Tea 19:43–48
Owuor PO, Ng'etich WK, Obanda M (2000) Quality response of clonal black tea to nitrogen fertilizer, plucking interval and plucking standard. J Sci Food Agric 80:439–446
Owuor PO, Othieno CO, Kamau DM, Wanyoko JK, Ng'etich WK (2008a) Effects of long term fertilizer use on a high yielding tea clone AHP S15/10: yields. Intern J Tea Sci 7:19–31
Owuor PO, Obanda M, Nyirenda HE, Mandala WL (2008b) Influence of region of production on clonal black tea chemical characteristics. Food Chem 108:263–271
Owuor PO, Kamau DM, Jondiko EO (2009) Response of clonal plain black tea quality parameters and yields to geographical region of production and plucking frequencies. Food Chem 115:290–296
Owuor PO, Kamau DM, Jondiko EO (2010a) The influence of geographical area of production and nitrogenous fertiliser on yields and quality parameters of clonal tea. J Food Agric Environ 8:682–690
Owuor PO, Wachira FN, Ng'etich WK (2010b) Influence of region of production on relative clonal plain tea quality parameters in Kenya. Food Chem 119:1168–1174
Oyamo JR (1992) The golden clone in a golden field. Tea 13:1
Ozdemir F, Gokalp HY, Nas S (1993) Effects of shooting period, times within shooting periods and processing systems on the extract, caffeine and crude fiber contents of black tea. Z Lebensm Unters Forsch 197:358–362
Ranganathan V, Natesan S (1985) Potassium nutrition of tea. In: Potassium in agriculture. ASA-CSSA-SSSA, Wisconsin, pp 981–1022
Ranganathan V, Natesan S (1987) Manuring of tea – recommendations for high yielding fields. In: Handbook of tea culture 1/1983. UPASI Tea Science Department, Tea Research Institute, Coimbatore, 12 pp
Ranganathan V, Natesan S, Palani N (1988) Response of tea to potassium application in the pruned year and effect of manuring on yield trend. Planters' Chron 83:373–378
Ravichandran R (2002) Carotenoid composition, distribution and degradation to flavour volatiles during black tea manufacture and the effect of carotenoid supplementation on tea quality and aroma. Food Chem 78:23–28
Ravichandran R (2004) The impact of pruning time from pruning on quality and aroma constituents of black tea. Food Chem 84:7–11
Ravichandran R, Parthiban R (1998) The impact of mechanization of tea harvesting on the quality of south Indian CTC teas. Food Chem 63:61–64
Ravichandran R, Parthiban R (2000) Lipid occurrence, distribution and degradation to flavour volatiles during tea processing. Food Chem 68:7–13
Rawat R, Gulati A (2008) Seasonal and clonal variations in some major glycosidic bound volatiles in Kangra tea (*Camellia sinensis* (L.) O. Kuntze). Eur Food Res Technol 226:1241–1249
Robinson JM, Owuor PO (1992) Tea aroma. In: Willson KC, Clifford MN (eds) Tea: cultivation to consumption, 1st edn. Chapman & Hall, London, pp 603–647, Chapter 18
Ruan J, Wu X, Ye Y, Hardter R (1998) Effect of potassium, magnesium and sulphur applied in different forms of fertilisers on free amino acid content in leaves of tea (*Camellia sinensis* L). J Sci Food Agric 76:389–396
Ruan J, Wu X, Hardter R (1999) Effects of potassium and magnesium nutrition on the quality components of different types of tea. J Sci Food Agric 79:47–52
Ruan J, Ma L, Shi Y, Han W (2003) Uptake of fluoride by tea plant (*Camellia sinensis* L) and the impact of aluminium. J Sci Food Agric 83:1342–1348

Sahin Y, Nas S, Gokalp HY (1991) Effect of shooting period, region of growth and processing method on the Fe and Mn content of tea determined by X-ray fluorescence. Intern J Food Sci Technol 26:485–492

Salardini AA (1978) Yield response of tea (*Camellia sinensis*) to fertilizer and irrigation in two locations in Iran. Plant Soil 50:113–123

Sharma VS (1987) Harvest in tea. Planters Chron 82(8):261–266

Sharma VK, Bhattacharya A, Kumar A, Sharma HK (2007) Health benefits of tea consumption. Trop Pharmaceutical Res 6(3):785–792

Shoubo H (1989) Meteorology of the tea plant in China. a review. Agric For Meteor 47:19–30

Smith RI, Harvey FJ, Cannell MGR (1993) Clonal responses of tea shoot extension to temperature in Malawi. Expl Agric 29:47–60

Squire GR, Obaga SM, Othieno CO (1993) Altitude, temperature and shoot production of tea in Kenyan highlands. Expl Agri 29:107–120

Stephens W, Carr MKV (1990) Seasonal and clonal differences in shoot extension rates and numbers in tea (*Camellia sinensis*). Expl Agric 26:83–98

Street R, Száková J, Drábek O, Mládková L (2006) The status of micronutrients (Cu, Fe, Mn, Zn) in tea and tea infusions in selected samples imported to the Czech Republic. Czech J Food Sci 24:62–71

Sud RG, Baru A (2000) Seasonal variation on theaflavins, total color and brightness of Kangra orthodox tea (*Camellia sinensis* (L.) O. Kuntze) in Himachal Prandesh. J Sci Food Agric 80:1291–1299

Tanton TW (1979) Some factors limiting yields of tea (*Camellia sinensis*). Expl Agric 15:187–191

Tanton TW (1981) Crop growth and yields of the tea bush. Expl Agric 17:323–331

Tanton TW (1982) Environmental factors affecting the yield of tea (*Camellia sinensis*). I Effect of air temperature. Expl Agric 18:53–63

Tomlins KI, Mashingaidze A (1997) Influence of withering, including leaf handling, on the manufacturing and quality of black teas—a review. Food Chem 60:573–580

Turkmen N, Velioglu YS (2007) Determination of alkaloids and phenolic compounds in black tea processed by two different methods in different plucking seasons. J Sci Food Agric 87:1408–1416

Uddin J, Hoque R, Ahmed M, Saha JK (2005) Studies on the effect of saturation deficit on the yield of tea. Pak J Meteor 2:5–10

Ullah MR (1984) A reappraisal of withering process in black tea manufacture. 11. Chemical wither in current practices of artificial withering and its effects in tea liquors. Two Bud 31:24–27

Ullah MR, Bordoloi AK, Gogoi N (1986) Hastening of leaf senescence in the processing of green tea leaf to black tea. J Sci Food Agric 37:1042–1044

Venkatesan S, Ganapathy MNK (2004) Impact of nitrogen and potassium fertilizer application on quality of CTC teas. Food Chem 84:325–328

Venkatesan S, Verma DP, Ganapathy MNK (2003) Targeted yield equations of nitrogen for clonal teas under southern India conditions. J Indian Soc Soil Sci 51:178–183

Venkatesan S, Murugesan S, Ganapathy MNK, Verma DP (2004) Long-term impact of nitrogen and potassium fertilisers on yield, soil nutrients and biochemical parameters of tea. J Sci Food Agric 84:1939–1944

Venkatesan S, Murugesan S, Senthur Pandian VK, Ganapathy MNK (2005) Impact of sources and doses of potassium on biochemical and green leaf parameters of tea. Food Chem 90:535–539

Venkatesan S, Senthurpandian VK, Murugesan S, Maibuam W, Ganapathy MNK (2006) Quality standards of CTC black teas as influenced by sources of potassium fertiliser. J Sci Food Agric 86:799–803

Verma DP (1997) Potassium nutrition of tea. J Potassium Res 13:93–100

Wachira FN, Ng'etich WK, Obaga SMO, Othieno CO (1990) Genotype x environment interactions and genotypic stability in tea. A preliminary indication. Tea 11:51–57

Wachira FN, Ng'etich WK, Omolo J, Mamati G (2002) Genotype × Environment interactions for tea yields. Euphytica 127:289–296

Wallis JAN (1997) Intensified systems of farming in the tropics and subtropics. World Bank discussion paper no. 364. The International Bank for Reconstruction and Development/The World Bank, Washington, DC

Wanyoko JK, Njuguna CK (1983) Variation of foliar nutrients of some Kenyan tea and their relationship with yield. Tea 4(1):37–43

Wanyoko JK, Othieno CO (1987) Rates of potassium fertilizer on clone 6/8 tea: effects on soil extractable potassium and leaf nutrient contents, yield and plant water status. Tea 8(1):14–20

Wickremaratne MT (1981) Genotype-environment interactions in tea (*Camellia sinensis* L.) and their implications in tea breeding and selection. J Agric Sci 28:556–580

Willson KC (1992) Field operations. In: Willson KC, Clifford MN (eds) Tea: cultivation to consumption, vol 2, 1st edn. Chapman & Hall, London, pp 227–267, Chapter XXX

Wright LP, Mphangwe NI, Nyirenda HE, Apostolides Z (2002) Analysis of the theaflavin composition in black tea (*Camellia sinensis*) for predicting the quality of tea produced in Central and Southern Africa. J Sci Food Agric 82:517–525

Yao L, Caffin N, D'Arcy B, Jiang Y, Shi J, Singanusong R, Liu X, Datta N, Kakuda Y, Xu Y (2005) Seasonal variations of phenolic compounds in Australia-grown tea (*Camellia sinensis*). J Agric Food Chem 53:6477–6483

Yao MZ, Chen L, Liang YR (2008) Genetic diversity among tea cultivars from China, Japan and Kenya as revealed by ISSR markers and its implication for parental selection in tea breeding programmes. Plant Breed 127:166–172

Zeiss M, den Braber K (2001) Growth and physiology of the tea plant. In: Tea, IPM ecological guide. CIDSE, Ha Noi, pp 30–37

Biotechnology and Agroforestry in Indian Arid Regions

Varsha Sharma, Shaily Goyal, and K.G. Ramawat

Abstract Agroforestry evolved with agriculture to make the best use of the land and to maintain equilibrium between man, land, livestock and plants. In recent years, agroforestry has been developed as an autonomous science that can help farmers to increase profitability and land sustainability. Tree growing in combination with agriculture, including individual farms, watersheds and regional landscape can be integrated to take advantage of the benefits provided by adjacent natural, semi-natural or restored ecosystems. Trees in a variety of agroforestry systems enhance nutrient cycling, improve soil conservation and soil faunal activities and contribute to food security. Trees increase rural income through a diversity of products and services. Tree planting reduces salinity and prevents soil erosion, increases local biodiversity and reduces the greenhouse effect.

Multiplication and establishment by conventional propagation of some of the arid and semi-arid zone trees are difficult. The plants showing high endemism need to be conserved and improved as they provide useful products to support the life system of arid and semi-arid zones of India. In arid and semi-arid environments, agroforestry systems support livelihood improvement through simultaneous production of food, fodder and firewood without much affecting climate change. The important trees of Indian arid region agroforestry are *Acacia nilotica*, *Acacia senegal*, *Capparis deciduas*, *Phoenix dactylifera*, *Prosopis cineraria*, *Salvadora oleoides*, *Salvadora persica*, *Tecomella undulata*, and *Zizyphus mauritiana*. There is also a potential to utilize agroforestry to solve global problems like desertification and to conserve biodiversity.

Biotechnology has the potential to ameliorate agroforestry plants. Recently developed technology in molecular biology and tissue culture play an increasing role in the choice of genotype for successful establishment and improvement of

V. Sharma · S. Goyal · K.G. Ramawat (✉)
Laboratory of Bio-Molecular Technology, Department of Botany,
M.L. Sukhadia University, Udaipur 313001, India
e-mail: kg_ramawat@yahoo.com

E. Lichtfouse (ed.), *Genetics, Biofuels and Local Farming System*, Sustainable Agriculture Reviews 7, DOI 10.1007/978-94-007-1521-9_11,

agroforestry plants. Agroforestry tree improvement intends to identify and improve several important trees attributes like growth rates, disease and pest resistance and adaptability, tree form, e.g. straightness, branch number and size, and wood fiber quality. Biotechnological approaches to agroforestry plants of arid region have produced excellent results in micropropagation, e.g. for *Prosopis* species and *Tecomella undulata*, production of useful metabolites, e.g., by *Commiphora wightii*, genomics and genetic markers, e.g. for *P. cineraria, T. undulata, C. wightii*, gene mapping, e.g. in *P. dactylifera*, and transformed plants, e.g. *Tecomella undulata*. Some of the future directions for arid and semi-arid zone agroforestry need to focus on the evaluation and improvement of plants which will make a profound impact on the agroforestry.

Keywords Agroforestry • Arid zone • Germplasm • Micropropagation • Plant biotechnology • Secondary metabolites

1 Introduction

Agroforestry is as old as the origin of agriculture. It makes maximum use of the land and helps in maintaining equilibrium between man, land, livestock and plants. In recent years, agroforestry has been developed as an autonomous science that aims at helping farmers to increase profitability and sustainability of their land. Tree growing in combination with agriculture as well as numerous vegetation management regimes in cultural landscape, including individual farms, watersheds and regional landscape can be integrated to take advantage of the benefits provided by adjacent natural, semi-natural or restored ecosystems (Harsh and Tewari 2007). Trees in a variety of agroforestry systems enhance nutrient cycling, improve soil conservation and soil faunal activities and contribute to food security, rural income generation through diversity of products and services. Therefore, the agroforestry systems have potential option for livelihood improvement and biodiversity conservation in agroecosystems as well as yield of goods and services to the society (Pandey 2007). In an agroforestry systems trees are grown for the essential purposes as they improve crop productivity by reducing the wind flow and thereby reducing water loss through evapotranspiration (Zinkhan and Mercer 1997). The main component of the production technology are proper tree density, grazing, lopping and application of biofertilizers, use of biotechnology for genetic improvements of existing tree species for better yields and the selection of the most suitable species for the region (Tewari and Singh 2006). Agroforestry systems can considerably increase the rainfall utilization as compared to annual cropping systems, largely due to the subsequent utilization of system components (Ong et al. 2002).

(a) Role of trees in agroforestry: Various trees species of arid and semi-arid regions have the dominant role to play in all agroforestry systems for sustainable agriculture and environmental protection. Tree planting has beneficial effects like-reduces salinity and prevents soil erosion, increases local bio-diversity and reduces the green house effect. This is well illustrated by *Prosopis cineraria* and *Zizyphys mauritiana* plantations in fields of farmers of the western Rajasthan. When the nutrient supply in the soil is limited, agroforestry and other tree-based systems are more efficient than herbaceous and monocropping systems in the utilization of nutrients to sustain the required levels of agricultural production (Pandey 1998). Agroforestry systems in India include trees in farms, community forestry, and a variety of local forest management and ethnoforestry practices. Overall India is estimated to have between 14,224 million and 24,602 million trees outside forests, spread over an equivalent area of 17 million hectares, supplying 49% of the 201 million tones of fuel wood and 48% of the 64 million m^3 of timber consumed annually by the country (Pandey 2007; Rai and Chakrabarti 2001).

(b) Role of Biotechnology: Biotechnology is a promising, fast emerging, powerful technology, which can create a balance between agriculture and environment. The production of improved crop varieties and agroforestry plant species of arid and semi-arid regions would require application of new innovative biotechnological approaches, where conventional propagation is poor to get over from the field and other intrinsic barriers (Rathore et al. 2007; Nandwani et al. 2005). Recent developments in plant science, allow scientists to modify DNA, to enhance plant's tolerance against pest and diseases, and to increase yield and improve quality and nutritional value of plant products. It has a wide range of possible applications in agriculture, food biotechnology, medicine, environment and industry (Arora et al. 2010; Ramawat and Goyal 2008). The economically important trees of arid and semi-arid regions, which are over-exploited, threatened and endangered, need to be conserved and improved as they provide many useful products to support the life system. The modern plant biotechnology in agroforestry system provides classical approaches to achieve desirable characteristics and better productivity of food, fodder, fuel, timber and other useful non-edible materials including resins, tannins, insecticides and medicinal compounds. Thus, the scope of research and development in agroforestry are much higher, e.g., forest biotechnology to identify desirable characteristics and then use vegetative propagation and somatic embryogenesis to rapidly develop stocks for planting of superior agroforestry trees (Rathore et al. 2007).

Use of various biotechnological approaches such as micropropagation, somatic embryogenesis, genetic engineering, marker-aided selection and molecular diagnostics can be merged with traditional forest biological studies to identify and produce better suited tree for agroforestry (Rathore et al. 2007; Nandwani et al. 2005). Therefore, serious efforts are needed to develop technology for agroforestry plants, like tissue culture technique, wonderful for multiplication of millions of plants from a single mother plant (Raj Bhansali 2010). Agroforestry requires multi-disciplinary approach of biotechnology for amelioration of plants and offers the opportunities to enhance research in agriculture and forestry. The recent advances in plant biotechnology have tremendous scope in agroforestry for future research.

The distant promise of agroforestry biotechnology is that designer trees that are perfect for specific wood, paper or environmental remediation applications can be a possibility. Consequently, understanding of the ecophysiological aspects of forests and agroforestry systems is vital for setting better management guidelines, because it provides opportunities for intercropping components with management adaptation to the environment. In the recent times, large tree plantations of multipurpose agroforestry tree species have been made to improve the shortage of fruit, shade, timber and fauna protection (Raj Bhansali 2010; Pasiecznik et al. 2001; Singh 1994).

The objective of this review is to present at one place scattered information about biotechnological efforts made on arid region agroforestry plants. This information will be helpful in deciding further course of action towards conservation and micropropagation of useful plants, produce bioactive molecules and ameliorate the plants utilizing biotechnological methods. Biotechnological developments, viz., regeneration, production of useful metabolites, transgenic development, disease resistance, genomic analysis and nitrogen fixation have been presented about the plants of agroforestry of arid regions.

2 Suitable Tree Species for Agroforestry

The selection of suitable tree species is made on the basis of their ability to generate higher cash income and multiple uses. In arid and semi-arid environments, good management and crop associations are required in agroforestry systems to provide greater insurance against weather abnormalities. Many multipurpose and fruit trees such as *Acacia nilotica*, *Acacia senegal*, *Capparis decidua*, *Phoenix dactylifera*, *Prosopis cineraria, Tecomella undulata*, *Salvadora oleoides*, *Salvadora persica* and *Zizyphus mauritiana* (Figs. 1–3) thrive in arid and semi-arid areas (Kumar and Singh 2009; Jain et al. 2005). Crops accompanying these trees may not show any significant reduction in grain yield. Though the worldwide list of such trees is long, some of them have received more attention with respect to arid and semi-arid regions. *Prosopis cineraria* is the most accepted tree of arid region and provides food, fodder, wood, shelter and increases soil fertility (Nandwani et al. 2005). *Salvadora* based agroforestry system is very common where soil is clay loam or sandy loam. These trees have multipurpose uses such as its seeds contain 50% non-edible oil and native people eat fruits and it does not have any adverse effect on the companion crops (Harsh and Tewari 2007). Similarly, *Azadirachta indica* and *Albizia lebbek* have been traditionally planted on the field borders in the semi-arid areas. It is vital to select an appropriate tree density, so as to minimize the effect of root competition and to reduce the trade-offs between crop and tree productivity. Trees that shed its leaves during the dry season and are adapted to harsh environmental conditions, the accumulation of organic carbon and nitrogen in the topsoil may be a result of leaf litter decomposition. Soil organic carbon storage and distribution is controlled by the balance of carbon inputs from plant production and outputs through decomposition (Burke et al. 1998). In an

Fig. 1 Comparative morphological characters *of Prosopis cineraria* (*Pc*), *P. juliflora* (*Pj*) and *P. tamarugo* (*Pt*)

Fig. 2 A branch showing fruits and flowers of *Salvadora oleoides*

agroforestry system, the yield and physiological behavior of agricultural crops are strongly affected by different intercropping designs. Besides fuel fodder and shade, some of agroforestry trees have potential biological activities (Table 1). These trees can provide biomass for extraction of bioactive molecules or value added products. The biological activities range from antimicrobial to hepatoprotective, and cancer prevention, antiprotozoan and hypolipidemic. Some of the promising herbs can

Fig. 3 *Capparis decidua* plant with flowers

also be cultivated for better revenues in mix cropping (Arora et al. 2010). In arid regions, agroforestry plants have been selected on the basis of long traditional uses of these plant. This paper gives an overview of the area of plant biotechnology, which has now become an important integral constituent of agroforestry research (Rhouma et al. 2010; Rathore et al. 2007).

3 Biotechnological Applications

Biotechnology in agroforestry offers related scientific means to increase forest productivity and conservation of multipurpose trees. Its application will be another 'hot' area for research in agroforestry; ranging from vegetative propagation and tissue culture to genome analysis and gene cloning, DNA recombination gene transfer to DNA-based selection, biological nitrogen fixation to low cost plant production measures, from propagation of rare germplasm to breeding of desirable plant ideotypes and from use of plant hormones for a variety of purposes to processing of agroforestry products, the potential application of biotechnology in agroforestry research are unlimited (Raj Bhansali 2005, 2010).

Use of tissue culture and genetic engineering for tree improvement is approaching reality. Modern molecular techniques in the characterization of agroforestry trees of arid and semi-arid area can be used to study gene diversity, phylogeny and fingerprinting as well as to construct physical genetic maps in agroforestry plants. AFLP markers are now being routinely employed for assessment of genetic makeup in various crop plants (Narula et al. 2004). The plants like *Dalbergia, Prosopis, Eucalyptus, Azadirachta* and *Tectona* can now be successfully micropropagated

Table 1 Isolated bioactive compounds of important agroforestry trees and their biological activity

Plant species	Bioactive compounds	Plant part used	Biological activities	References
Acacia nilotica	Kaempferol (AN-5), D-pinitol, a sex hormone viz 3 β-acetoxy-17-hydroxy-androst-5-ene	Stem bark, gum, flower, leaves	Antioxidant	Singh et al. (2008)
			Immunosuppressive	Aderbauer et al. (2008)
			Anticancer and antimutagenic	Meena et al. (2006)
			Antiplasmodial	Kirira et al. (2006)
			Antileshmanial	Fatima et al. (2005)
			Antidiarrheal	Agunu et al. (2005)
			Antimicrobial activity against multi-drug resistant *Salmonella typhi*	Rani and Khullar (2004)
			Inhibitory effect on hepatitis C virus (HCV) protease	Hussein et al. (2000)
Capparis decidua	Triacontanol (C1), 2-carboxy-1, 1dimethylpyrrolidine (C2)	Stem, flower, fruit	Insecticidal and oviposition inhibitory activity	Upadhyay et al. (2006)
			Hypolipidemic	Purohit and Vyas (2005)
			Antidiabetic, antioxidative	Yadav et al. (1997)
Phyllanthus emblica syn Emblica officinalis	Emblicanin-A, B, gallic acid, ellagic acid, pyrogallol, apigenin 7-0-(6″ butyryl-β-glucopyranoside), quercetin, putranjivain A	Fruit	Antimicrobial, virucidal action against HIV-1NL4.3 and HPV infections	Talwar et al. (2008)
			Hepatoprotective and anti- tuberculor	Panchabhai et al. (2008)
			Cancer chemopreventive	Pinmai et al. (2008), Arulkumaran et al. (2007), and Sandhya and Mishra (2006)

(continued)

Table 1 (continued)

Plant species	Bioactive compounds	Plant part used	Biological activities	References
			Potent antioxidant	Deep et al. (2005)
			Memory improvement and reversal of memory deficits	Saito et al. (2008)
			Radioprotective effect	Vasudevan and Parle (2007a, b)
			Antihypercholesterolemic, antiatherosclerotic	Singh et al. (2006), Kim et al. (2005)
Salvadora persica	Four benzylamides of which N-benzyl-2-phenylacetamide is pharmacologically important	Stem	Carries prevention	Sofrata et al. (2007), Khalessi et al. (2004), and Darmani et al. (2006)
			Antiplasmodial	Ali et al. (2002)
			Anticonvulsant and sedative effects	Monforte et al. (2002)
			Antiulcer	Sanogo et al. (1999)
			Hypolipidemic	Galati et al. (1999)
Tecomella undulata	Bsitosterol, triacontanol, cirsimaritin, cirilineol, pentatriacontanol and 4,5-dihydroxy-3,6,8-trimethoxy flavone oleanolic acid, ursolic acid and betulinic acid	Leaves	HIV prohibition	Azam (1999)
			Iinhibition of typhoid fever	Dushyent and Bohra (2000)
			Hepatoprotective	Rana et al. (2008)
Zizyphus mauritiana	Betulinic acid	Fruits	Hepatoprotective and immunomodulatory	Adhvaryu et al. (2007)
			Anticacerous activity	Mahajan and Chopda 2009, Mukherjee et al. (2006)

either through multiple shoot induction or somatic embryogenesis (Nandwani et al. 2005). Among the agroforestry trees of the arid and semi-arid regions, Department of Biotechnology, New Delhi has supported scaling up of protocols for forestry species such as *Eucalyptus* and teak used in energy plantation and reforestation (Muralidharan and Kallarackal 2004). There is need for taking up projects on *Capparis* species, *P. cineraria*, *S. oleoides*, *T. undulata* and *Zizyphus* species. These species bear important traits and products, that can be characterize and identified for future prospecting in plant biotechnology (Looy et al. 2008). Various applications of biotechnology have been presented in the following text.

3.1 Micropropagation

Micropropagation is defined as an artificial asexual propagation technique. It is a rapid method for producing plants of particular germplasm within short span of time. Micropropagation also provides an alternative approach of cloning which has vast potential in forestry science and is playing a key role in commercial application for quick multiplication of plantation and fruit crops (Raj Bhansali 2005, 2010). It not only provides economy of time and space but also gives greater output and allows further augmentation of elite disease free propagule. There are two main objectives for developing clonal techniques for forest trees. Firstly, viable large-scale clonal systems can be accepted for selected trees and, secondly, reliable protocols are necessary for genetic manipulation. Progress of micropropagation involves several stages, each of which may be critical to decide its success for particular plant species.

The conventional method of propagation of most of tree species is through seeds. This has disadvantage as progeny produced from sexually produced seeds has large number of variables, as in *Acacia* species, *P. cineraria*, *T. undulata*, while some species seeds have poor viability, e.g., *Commiphora wightii*. The forest vegetation, especially agroforestry species such as *Prosopis cineraria*, *Prosopis tamarugo* (Fig. 4), *Tecomella undulata*, *Salvadora oleoides*, *Salvadora persica* and *Acacia nilotica*, *Acacia senegal*, all are well adapted, shows high endemism and need to be conserved and improved (Raj Bhansali 2010; Kumar and Singh 2009; Khalafalla and Daffalla 2008; Beck and Dunlop 2001; Nandwani et al. 2005). The *in vitro* plant regeneration may be direct from the explants such as nodal segments, buds and meristems, and indirect via initiation and differentiation of callus. Success in tree regeneration greatly depends on the choice of tissue source, age of the tree and species or genotype chosen (Nandwani et al. 2005). Whatever system is chosen for tissue culture of a given species, the ability to regenerate intact plants is essential; finally trees regenerated from a propagation system must perform satisfactory field tests to prove their practical value in tree improvement programs.

Superior germplasm of Date palm was introduced in Indian desert region and micropropagation and agrotechnology were developed (Raj Bhansali 2010). *P. cineraria* has been tried to propagate through tissue culture by using selected

Fig. 4 Highly regenerative cultures of *Prosopis tamarugo* showing large number of shoots

seeds, buds and root by various workers (Kumar and Singh 2009; Nandwani et al. 2005). The number of plants produced from each explants from mature selected tree was quite low and inconsistent for the purpose of micropropagation. Serious efforts are needed to develop technology for these plants to multiply on large scale and for field transplantation. *T. undulata*, known as teak of Marwar, is quickly depleted due to extensive cutting of trees. The quality of wood is good and fairly resistant to termite and borer attack (Singh et al. 2004). The tissue culture technology has been developed to multiply the plant of *T. undulata*. Similarly, large numbers of shoots were regenerated from nodal buds of *S. oleoides* in tissue culture within 2 months (Raj Bhansali 2005, 2010; Nandwani et al. 2005). Micropropagation technology has been used for medicinal plants grown under agroforestry trees, e.g., *Curculigo orchioides*. In this tuberous root crop, which is used medicinally (Jain et al. 2010), high frequency regeneration of plantlets (Figs. 5–9) through bulbils formation from leaf explants have been reported in this plant (Nema et al. 2008; Suri et al. 2000).Thus well established micropropagation methods are available in several plants like *Prosopis* species, *T. undulata*, *P. dactylifera* and several others.

3.2 Somatic Embryogenesis

Somatic or asexual embryogenesis is the process by which somatic cells develop into plants. Somatic embryogenesis is the production of embryo-like structures from somatic cells without gametes fusion. During their development, somatic embryos pass through stages similar to those observed in zygotic embryogenesis (Dodeman et al. 1997). They arise naturally in some species in a process known as direct

Fig. 5 A plant of *Curculigo orchioides*

Fig. 6 Leaf explants of *Curculigo orchioides* in liquid medium for *in vitro* culture

somatic embryogenesis; directly from the cultured explants. In contrast, somatic embryos arise from *in vitro* cultured cells in the process called indirect somatic embryogenesis. The rapid improvement in somatic embryogenesis methods allows the use of somatic embryos in plant micropropagation as synthetic seeds. Somatic embryogenesis allows regeneration of whole plants from single cells (Toonen and de Vries 1996), making possible its use in cell selection programs. The special interests for agroforestry are the use of somatic embryos as synthetic seeds via matrix coating or encapsulation of somatic embryos and their scale-up production in bioreactors in which extensive multiplication can be achieved for commercialization. However, somatic embryogenesis has other practical applications in agroforestry, including crop improvement (Vicient and Martínez 1998). Somatic embryogenesis has now become one of the most potential *in vitro* methods for

Fig. 7 Bulbils regeneration from leaf explants of *Curculigo orchioides*

Fig. 8 Large number of plantlets produced by germinating bulbils of *Curculigo orchioides*

Fig. 9 Plantlets of *Curculigo orchioides* produced from *in vitro* germination of bulbils

efficient plant regeneration in many of arid zone taxa, since literally million of bipolar embryos and complete plants could be produced from a relatively small amount of cultured cells or tissues (Raj Bhansali and Singh 2000).

Somatic embryogenesis have been reported in many arid and semi-arid zone agroforestry tree species, such species are *Commiphora wightii, Azadirachta indica, Acacia nilotica, Zizyphus mauritiana* and *Eucalyptus camadulensis* (Nandwani et al. 2005; Muralidharan and Kallarackal 2004; Kumar et al. 2003). Plant regeneration via somatic embryogenesis was achieved in embryogenic callus cultures derived from immature zygotic embryos 40 days after anthesis in *A. indica*. Somatic embryo-derived plantlets were acclimatized in a greenhouse and subsequently showed normal growth. This efficient protocol will be helpful for propagating elite clones on a mass scale and will also be useful for genetic transformation study (Rout 2005). Somatic embryo proliferation in *C. wightii* and evidence for guggulsterone production in the cultures has been reported (Kumar et al. 2003, 2006). Somatic embryos of *C. wightii* were obtained from immature zygotic embryos (Fig. 10) and synthetic seeds (Fig. 11) were developed from such embryos and used for direct germination in soil. In *Acacia nilotica* somatic embryogenesis has been reported using immature endosperm (Garg et al. 1996).

Although the list of arid and semi-arid zone plant species that can induce somatic embryo is short, however, few species can regenerate complete plants and is not restricted to few taxa. Somatic embryogenesis has been proved as potentially stable, true-to-type source of plant multiplication system for many plants including arid medicinal zone plant species. Plant regeneration via somatic embryogenesis (Arora et al. 2006, 1999) has been reported in *Chlorophytum borivilianum* (Figs. 12 and 13), a medicinal herb grown under trees for its tuberous roots (Jain et al. 2010). Some benefits are that somatic embryos are bipolar structure having both root and shoot

Fig. 10 Somatic embryos of *Commiphora wightii* on hormone free medium

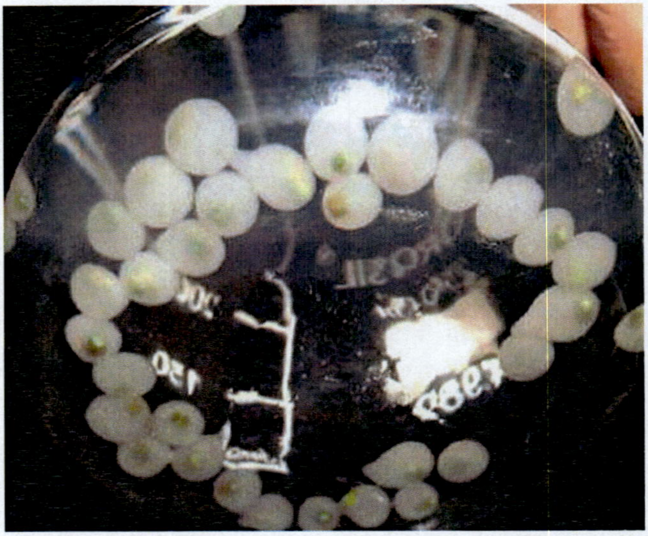

Fig. 11 Synthetic seed of *Commiphora wightii* obtained by encapsulation of somatic embryos in calcium alginate beads

meristems, and are capable of regenerating in to complete plant that too in large number, e.g., in *P. dactylifera* (Raj Bhansali and Singh 2000). Embryonic callus of *P. dactylifera* produce enormous numbers of somatic embryos in which a few embryos germinate on the same medium (Fig. 14). So this technique offers an inexpensive, large-scale propagation system for superior genotypes (Vicient and

Fig. 12 A flowering plant of *Chlorophytum borivilianum*

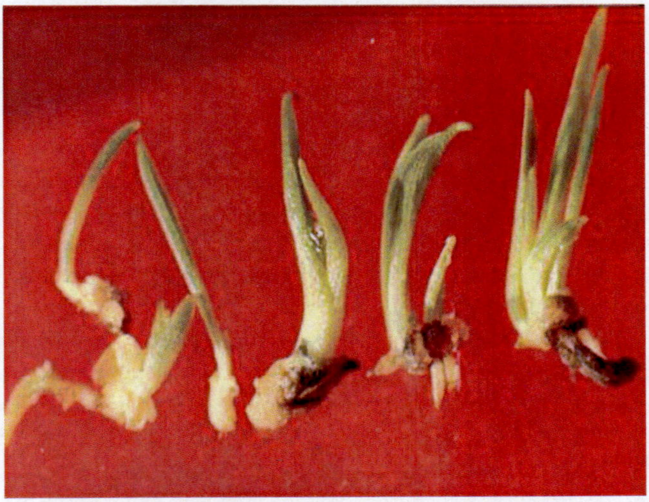

Fig. 13 Germinated somatic embryos of *Chlorophytum borivilianum*

Martínez 1998). Though somatic embryogenesis in trees is a difficult task, somatic embryogenesis has been achieved in some of the important plants of arid region agroforestry.

Fig. 14 Embryonic callus and a few germinating somatic embryos of *Phoenix dactylifera*

3.3 Germplasm Preservation

Genetic diversity in primitive cultivars and wild relatives of crop plants constitute an invaluable gene pool for future breeding programs. There are many species of agroforestry, which are rare or threatened with environmental changes. Agroforestry trees are also threatened in a way that is not problematic for most other wild species. Because trees are managed in or near their native forests, the potential for gene-pool disruption due to production forestry and horticulture is high. An important challenge to conservation is to conserve genetic diversity in ways that protect the germplasm in its desired form. Conservation strategies range from preservation of germplasm in cloned DNA sequences and genomic libraries, through storing genetic samples as tissue cultures, seeds, seedling and clonal plantations. Germplasm for tree improvement is most often conserved in plantations and other kinds of gene banks. An important phase of the conservation process is to evaluate the germplasm being protected in natural forest, plantations, tissue culture or as DNA. Evaluating the performance of accessions makes the germplasm collections valuable and useful (Singh 2004).

Tissue culture methods for storage of germplasm are thus of great practical significance and could be a boon for long-term storage of germplasm. Storage of genetic materials has become increasingly important for developing new cultivars as well as preserving old varieties, rare or endangered species. Two tissue culture approaches have been developed for germplasm preservation, *in vitro* conservation and cryopreservation. These approaches contribute to the establishment of an *in vitro* gene bank of plant germplasms. *In vitro* conservation maintains plant cultures at low temperature and low light intensity. This *in vitro* long-term storage of plant cells, tissues and organs is often favored because it can greatly reduce the labor and space requirements of traditional storage methods. Cryopreservation, freezing *in vitro* cultures with liquid nitrogen and storing at −30°C to −196°C for years, would be an appropriate method to preserve these valuable germplasms for the relatively long-term. It is a secure and reliable conservation method with a high level of genetic stability (Caswell and Kartha 2009; Huang et al. 1993).

Cryopreservation of embryogenic cell cultures, explants or somatic embryos is a good alternative to cell culture maintenance (Ahmed et al. 2001). Cryopreservation minimizes the necessity of the establishment and maintenance of the embryogenic cell cultures, reducing manipulations of the cultures, genetic variation and risks of loss. Hence, cryopreservation should be regarded as a strategically an important support to traditional in-field banks. Indeed, the combination of traditional conservation approaches with the potential of cryogenic technology represents an important step forward in minimizing the risks of accidental loss of woody plant genetic resources. The optimal composition of the cryoprotectant solution depends on the plant material and species (Vicient and Martínez 1998). Use of this technology requires well established protocols for regeneration and need to preserve germplasm. Hence this technology has not been explored for arid region plants.

3.4 Regeneration of Transformed Plants

Gene transfer methods are an essential part of the new technologies that are altering conventional plant breeding and have become indispensable tools. Most of the reports on regeneration of genetically transformed plants correspond to tests or optimizations of the methods using either genes conferring resistance to antibiotics or reporter genes (Vicient and Martínez 1998). The modern techniques developed for genetic engineering include vector-mediated systems based on *Agrobacterium* plasmids or viral vectors, and non-vector systems (direct gene transfer), such as biolistics (particle bombardment), microinjection, electroporation (electrical pulse treatment), chemical poration (polyethyleneglycol-mediated osmotic DNA insertion), liposome fusion, and protoplast fusion. In order to develop the *Agrobacterium* plasmid as a gene vector for genetic engineering of plants, researchers have genetically engineered the T-DNA by replacing the oncogenic genes (tumor causing) with useful genes. DNA transfer using the *Agrobacterium* vector has been demonstrated in many woody plants but without subsequent plant regeneration. *A. tumefaciens* and *A. rhizogenes* have proven to be excellent vector systems for the production of transgenic plants (Ahmed et al. 2001). In comparison to normal roots (Fig. 15) grown in the medium, transformed roots (Figs. 16 and 17) grow fast and do not require auxin for their grown are negatively geotropic. Such roots form excellent system to study various biological problems like growth, regeneration and production of useful compounds. Successful systems have been developed for a wide range of plant species, including some agroforestry tree species. However, the ease with which production of genetically engineered plants can be accomplished varies between species (Brink et al. 1998).

An efficient system for *Agrobacterium*-mediated transformation in *Eucalyptus camaldulensis* and production of transgenic plants was developed and five

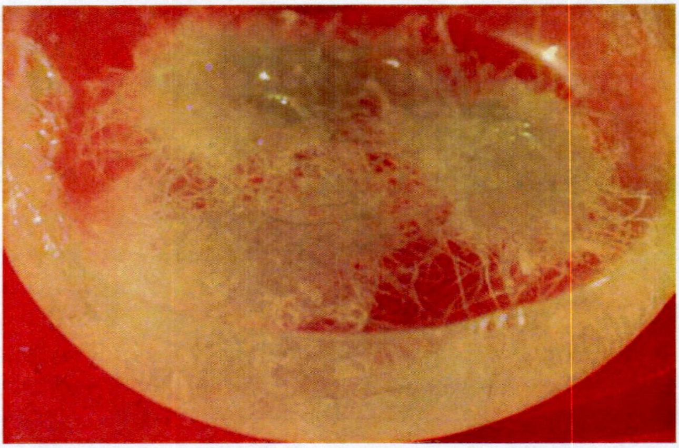

Fig. 15 Normal root culture in liquid medium

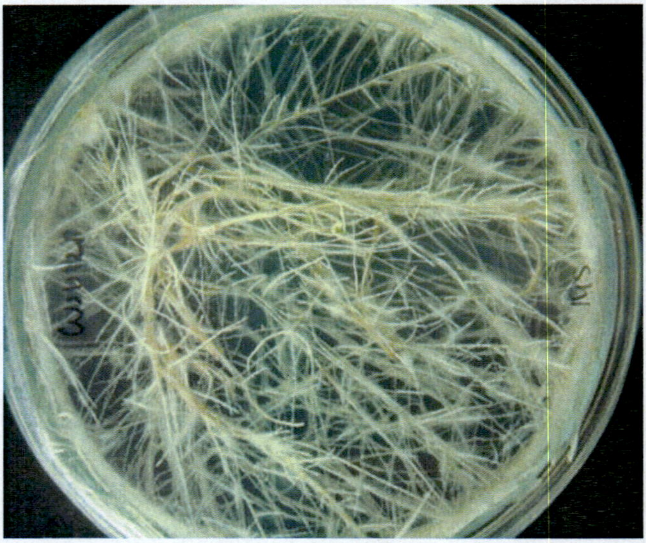

Fig. 16 Initial growth of transformed root

distinct genomic DNA fragments encoding phosphate transpoter were isolated from *E. camadulensis* (Koyama et al. 2006). Transformation was accomplished by cocultivation of hypocotyl segments with *A. tumefaciens* containing a binary Ti-plasmid vector harboring chimeric neomycin phosphotransferase and β-glucuronidase (GUS) genes (Ho et al. 1998). The tree-legume *Leucaena leucocephala* is used as a perennial fodder because of its fast-growing foliage, which is high in protein content. Improvements in the nutritional contents as

Fig. 17 Transformed roots grown in liquid medium showing negative geotropism

well as other agronomic traits of *Leucaena* can be accomplished through genetic transformation. The transformation protocol for *Leucaena* using phosphinothricin resistance as the plant selectable marker was developed. Explants obtained from immature zygotic embryos infected with the *A. tumefaciens* strain C58C1 containing the binary plasmid pCAMBIA3201 produced four putative transformed *Leucaena* plants. Transformation was confirmed by PCR, RT-PCR, Southern blot, Western analyses, GUS-specific enzyme activity and herbicide leaf spraying assay (Jube and Borthakur 2009). Similarly, recent protocol was optimized for genetic transformation of *T. undulata* from cotyledonary node tissues using *A. tumefaciens* strain GV2260 harboring binary vector pBinAR containing osmotin and *nptII* gene under control of CaMV35S promoter. This was the first report on the transformation of *T. undulata* in which PCR and Southern blot confirmed integration of osmotin gene into genome of the plant (Aslam et al. 2009). Azadirachtin is one of the most potent biopesticides so far developed from a plant sources. Influence of different culture media and elicitation on growth and production of azadirachtin by hairy root cultures of *A. indica* was reported (Satdive et al. 2006). Efforts have been started in this direction for arid region plants with improvement in facilities for transgenics.

3.5 Development of Technology for Disease Resistance

The importance of a more systematic collection of pest and disease related information for agroforestry in a central database, and of the development of strategies for reducing pest and disease risks in agroforestry in cooperation with farmers will be useful (Staver et al. 2001; Rao et al. 2000). Many biotechnological methods are now available and useful in the production of disease resistant plants in forestry system. Many arid and semi-arid zone plants, especially *Z. mauritiana, S. oleoides, C. wightii* traditionally being vegetatively propagated are systemically infected with viruses, mycoplasma and other unknown pathogens carried over from infected mother plants which greatly reduce yield and quality of agriculture products (Schroth et al. 2000). When *S. persica* occurs on river terraces, it is a preferred host of *Cistanche tubulosa,* an obligate phanerogamic root parasite. The mite *Eriophyes* causes leaf gall on it while a number of fungi such as *Cercospora udaipurensis, Placosoma salvadorae* and *Sephogloeum salvadorae* damage the leaves. Eradication of viruses and other pathogens is extremely important in order to optimize the yield and also to facilitate the researchers to use pure and disease-free plants for developing new varieties.

The various approaches to *in vitro* selection for disease resistance are known. Cultures can be exposed to a toxin, toxin analogues, filtrate or the pathogen itself. The use of purified toxins as selection agents in culture is potentially effective when symptoms of the disease are caused by a toxin produced by the pathogen, and where the toxin operates at the level of the explant cultured (e.g. at the cell level). Viruses have been eliminated from a number of economically important plant species, which has resulted in a significant increase in the yield and production. The technology has wide applications in producing disease-free plants, as there is no chemical therapy available to control such pathogens (Ahmed et al. 2001). Tissue culture method has been used to understand seed born nature of viruses, as these pathogens are non-cultivable on artificial media. These techniques permit the rapid screening of large amounts of germplasm in limited space and facilitate quick release of disease-resistant cultivars. Disease-resistant crop plants have been produced through *in vitro* selection procedures in *Zizyphus* against *Alternaria alternate*. The technique offers considerable opportunities of selecting disease-resistant genotypes in most of crops including fruit and forest trees (Raj Bhansali 2005, 2010).

3.6 Protoplast Culture and Fusion

Somatic cell hybridization via protoplast fusion is another alternative for agroforestry tree improvement. In this process of parasexual hybridization, two protoplasts (with their cell walls removed by enzymes) are fused in culture medium, and then the resultant hybrid cells are regenerated into whole plants *in vitro*. In order to facilitate the use of protoplasts in tree breeding, efficient and reliable methods must be developed for regeneration of trees from protoplasts or cell cultures (Huang 1998). Moreover, the regeneration of somatic hybrids could be difficult in some species, even using embryo rescue techniques. The protoplast fusion technique overcomes part of these problems and also allows new combinations of nuclear and organelle DNA, but a plant regeneration system is first required. The regeneration of fused protoplasts can be done via organogenesis or via somatic embryogenesis. As has been mentioned previously, somatic embryogenesis has the advantage of regenerating plants from single cells, and for these reason it is the best regeneration method in many somatic hybridization projects (Vicient and Martínez 1998). Protoplast fusion can be used to produce unique nuclear-cytoplasmic combination and to transfer cytoplasmically controlled male sterility between breeding lines. Population of regenerated plants arising from protoplast fusion contains more variability than comparable populations of plants produced by sexual hybridization (Ahmed et al. 2001).

Using somatic hybridization the genes conferring tolerance for cold, frost and salt were introduced (Muralidharan and Kallarackal 2004). Using cybridization method, it was possible to transfer cytoplasmic male sterility. During the process of cybridization and heterokaryon formation, the nuclei are stimulated to segregate so that one protoplast contributes to the cytoplasm while the other contributes nucleus alone. The irradiation with gamma rays and X-rays and use of metabolic inhibitors makes the protoplasts inactive and non-dividing. Some of the genetic traits in certain plants are cytoplasmically controlled. This includes certain types of male sterility, resistance to certain antibiotics and herbicides. Therefore cybrids are important for the transfer of cytoplasmic male sterility (CMS), antibiotic and herbicide resistance in agriculturally useful plants. Many hybrids have been tested for establishment of industrial plantation taxa such as *Eucalyptus* (Mullins et al. 1997). A number of these hybrids are in commercial use, and it seems likely that hybrids will become of wider importance in industrial plantation forestry. Freshly isolated protoplasts are shown in the Fig. 18. This technology will be effective only after developing regenerative protocols for arid region plants.

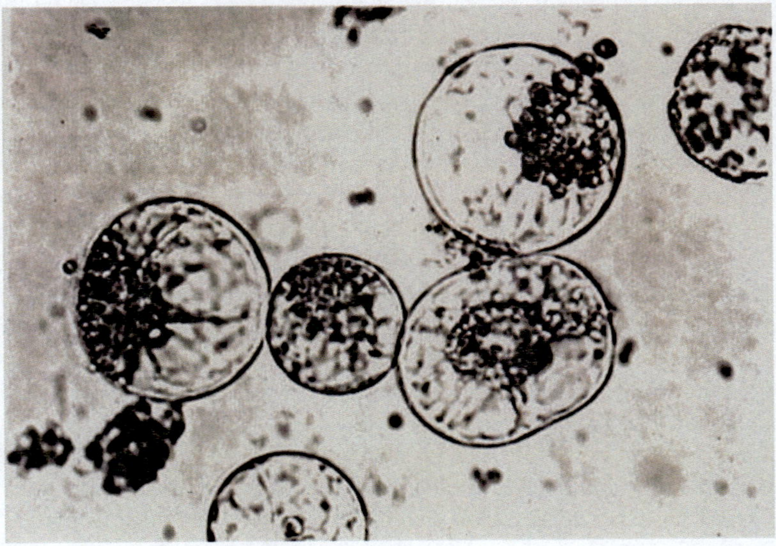

Fig. 18 Freshly isolated protoplasts

3.7 Genome Analysis and Gene Mapping

Trees and shrubs constitute an important part of the hot and arid ecosystem. Recent advances in molecular genetics and recombinant DNA technology offer novel tools to explore genetic organization and gene expression in tree species. Biotechnology has also introduced new ways to study genes and their functions through the use of genetic markers. A genetic marker can be a whole gene, part of a gene, and a sequence in non-coding DNA between genes, or an enzyme produced by a gene. Allozymes have been used for several decades to illustrate differences in genetic structure among individuals and within and among populations. Molecular markers such as mini- and microsatellites, Random Amplified Polymorphic DNA (RAPD), Restricted Fragment Length Polymorphism (RFLP), Sequence Tagged Microsatellite Sites (STMS), Amplified Fragment Length Polymorphism (AFLP), and Sequence Amplified Microsatellite Polymorphic Loci (SAMPL) provide more comprehensive genetic information due to either higher polymorphism information content or multiplex ratio. Among these, AFLP seems to be the most promising, especially for genetic diversity analysis. AFLP is an extremely reproducible and robust technique that may be applied to DNA of any origin or complexity. AFLP is an excellent tool for genetic diversity analysis as it generates a large number of bands per assay. However, the foremost disadvantage of this technique is that it is a dominant marker and cannot detect heterozygous alleles and simply measures DNA marker diversity, which assesses DNA sequence variation, and does not account for allelic interactions within or among different loci (Bhau et al. 2007). Currently, these are powerful tools for assaying genetic variation and developing genetic linkage maps. Genetic analyses with RFLP and RAPD markers have quickly been extended to tree species to develop genetic linkage maps. RAPD profile of nine different accessions of *Commiphora wightii* is shown in the Fig. 19. This marker showed high variability in these accessions (Suthar et al. 2008). RFLP markers have numerous additional advantages over most other types of markers (such as morphological and isozyme). They are developmentally stable, display normal Mendelian inheritance and generally exhibit multiple, co-dominant alleles. Molecular level markers also do not display environmental effects. RFLP markers are easily detected and relatively easy to map. The major disadvantage of RFLP markers is that they are expensive and time-consuming to generate. RAPD markers, on the other hand, are fairly simply, quickly and inexpensively detected. RAPD markers can be produced with relatively simple protocols and equipment and without the use of ^{32}P which gives them an advantage over RFLPs as far as application in field labs with limited facilities. They are, however, dominant markers, thus they provide much less information for mapping than do RFLPs (Huang et al. 1993).

Date palm, one of the oldest domesticated fruit crop, is the most adapted tree growing in desert areas. The Date Palm is extending over an area from the Indus Valley (now Pakistan) to Mesopotamia (now Iraq), the Nile Valley, Southern Persia, Eastern Mediterranean and the Horn of Africa and was under cultivation for the last 5,000 years. It has always been looked upon as a key source of stability,

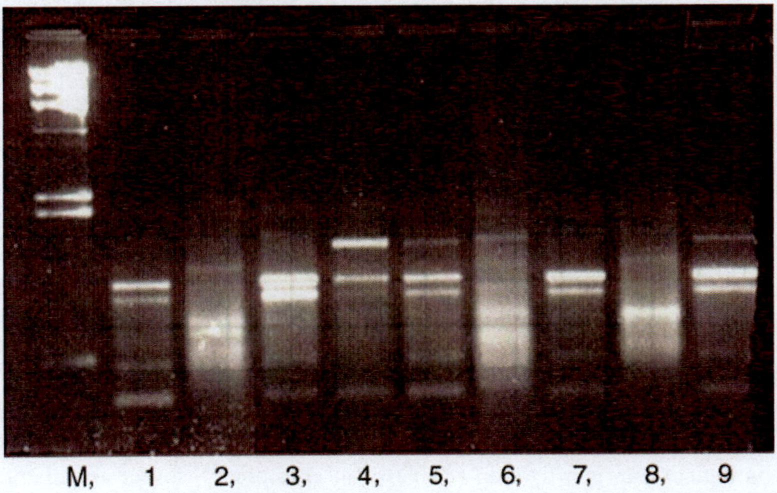

Fig. 19 RAPD (Random amplified polymorphic DNA) profile of *Commiphora* accessions, where M = λ DNA marker digested with Hind III, 1–9 accessions collected from different localities

survival and evolution of the oasis agro system since it constitutes the basic features for ecological pyramid in desert regions. Tunisian date palm germplasm is characterised by a high genetic diversity because more than 250 varieties have been identified. Efforts were focussed on development of phenotypic, biochemical and DNA-based markers useful in characterizing the genetic diversity of date palm populations and establishment of relationships between the different cultivars (Rhouma et al. 2010). Such as molecular phylogenetics of *Acacia* based on the choloroplast *matsK* coding sequence and flanking *trnK* introne spacer regions have been reported (Miller and Bayer 2001). Quantification of genetic diversity in the existing populations of *Tecomella undulata* prevalent in India was undertaken to investigate the utility of AFLP markers in the assessment of genetic diversity and phenetic relationship among 42 accessions collected from different regions of India. Based on molecular study, it was concluded that a high level of genetic diversity exists in these accessions, which may be attributed to its out-crossing nature. Eight AFLP amplification primer combinations were tested across 42 accessions of *T. undulata* leading to the amplification of a total of 338 AFLP bands. On an average, the AFLP technique yielded on an average 42 fragments per assay (Bhau et al. 2007).

Analysis of variations by RAPD profiles among 32 accessions of *Prosopis* has been reported. When 101 primers were used, 47 produced some RAPD patterns while the rest of the primers resulted in either no amplification or smeared profiles in as many as two out of the three DNAs tested. A total of 357 bands were generated by these 14 primers (Goswami and Ranade 1999). In a study, a RAPD-based approach was used to develop sequence-characterized regions, from the nuclear genome, with levels of DNA sequence variation appropriate for resolving relationships of *Leucaena*. RAPDs were used to amplify a set of seven *Leucaena* accessions. Equal length fragments amplified in two or more species were sequenced and compared (Baily et al. 2004). There is a need for taking up projects on Expressed Sequence Tags (ESTs) and microsatellite as molecular markers for *Capparis* spp., *P. cineraria*, and *T. undulata*. These species bear important traits and genes that can be characterize and identified for future prospecting (Rathore et al. 2007). Molecular markers have been used in a few cases to assess variability in arid region plants.

3.8 Improvement for Nitrogen Fixation

The biological nitrogen fixation (BNF) is an enormous resource for maintaining and improving soil fertility. There is a wide range of nitrogen fixing bacteria consisting of both free-living and symbiotic types, which fix and provide nitrogen to crops and also greater extent agroforestry system. The wild legumes in arid zones harbor diverse and promiscuous rhizobia in their root-nodules. Based on phenotypic characteristics and molecular techniques (protein profiles, polysaccharides, plasmids, DNA–DNA hybridization, 16srRNA, etc.), the root-nodule bacteria that were isolated from wild legumes had been classified into four genera (*Rhizobium, Bradyrhizobium, Mesorhizobium* and *Sinorhizobium*). The rhizobia of wild legumes in arid zones, exhibit higher tolerance to the prevailing adverse conditions, e.g. salt stress, elevated temperatures and desiccation. These rhizobia may be used to inoculate wild, as well as, crop legumes, cultivated in reclaimed desert lands. Recent reports indicated that the wild-legume rhizobia formed successful symbioses with some grain legumes. Moreover, intercropping of some N_2-fixing tree legumes (e.g. *Lablab, Leucaena, Sesbania,* etc.) to pasture grasses improved biomass yield and herb quality (Toky et al. 1995).

Nitrogen fixing trees and shrubs offer a clear opportunity to increase yield of subsequent crops in agroforestry system. Nitrogen-fixing tree species, which used in agroforestry are *A. senegal, A. nilotica,* and *L. leucocephala*. *A. senegal* nodulates only with fast-growing strains of *Rhizobium*. *L. leucocephala* generally nodulates with *Rhizobium* and occasionally nodulates with *Bradyrhizobium.*

Several genomic approaches have been used to define and understand the involvement of whole bacterial genomes in the symbiotic process. Genomic analyses of the model symbiotic bacterial species *Sinorhizobium meliloti*, *Rhizobium leguminosarum*, and *Bradyrhizobium japonicum* have revealed a few surprises concerning genome evolution and structure, how plant and microbes communicate, and physiological diversity among the microsymbionts of legumes (MacLean et al. 2007). These bacteria may have specific traits that can be transferred to other rhizobia through genetic engineering tools or used to produce industrially important compounds. Therefore, these bacteria are very important from both economic and environmental points of view (Zahran 2001). There is still much work ahead to collect *Rhizobium* strains for leguminous nitrogen-fixing trees and then screen them for genetic compatibility, nitrogen-fixation effectiveness, and tolerance to environmental stresses, especially soil acidity.

3.9 Production of Secondary Metabolites

Secondary metabolites are compounds biosynthetically derived from primary metabolites but more limited in occurrence in the plant kingdom and may be restricted to a particular taxonomic group. Secondary metabolites are considered as end product of primary metabolites and in general are not involved in metabolic activity. Plant cell and tissue are highly sophisticated chemical factories in which a large variety of chemical compounds are manufactured with great precision from simple raw materials (Ramawat et al. 2009; Merillon and Ramawat 2007). The production of metabolites in traditional agriculture is based on the growing of large quantities of plants, collection of certain organs or tissues that accumulate the product, and a chemical extraction. Sometimes the product is only accumulated in very specific parts of the plants such as fruits, seeds, flowers or embryos and sometimes very large quantities of plants are necessary for the production of small quantities of metabolite (Ramawat et al. 2004). Agroforestry plants require more biotechnological inputs to maintain their quality. In arid and semi-arid region number of such agroforestry trees products have high commercial value and demand. These trees species are *C. wightii* (Fig. 20), *A. indica, S. oleoides* and *Phyllanthus emblica* (Fig. 21). These plants contain number of compounds of economic value such as non-edible oils, pesticides and insecticidal compounds (Azadirachtin), alkaloids, steroids and many anticancerous compounds (Table 1, Ramawat et al. 2009; Ramawat and Goyal 2008). Various secondary products of these plants are extracted through complicated processes of extraction from different plant parts.

Some agroforestry plants identified as they have high medicinal value, crude extract of *Salvadora persica* twigs and roots have demonstrated *in vitro* antimicrobial effects. The compounds identified are esters of fatty acids and aromatic acids and some terpenoids (Abdel Rahman et al. 2003). Azadirachtin from *Azadirachta indica,* is a triterpenoid isolated from fruits in addition to several other valuable constituents of high commercial value and two novel azadirachtin derivatives were isolated. The first 29-oxymethylene azadirachtin analogue, 29-oxymethylene-11-demethoxycarbonyl-11α-hydroxyazadirachtin (azadirachtin M) and 22,23-dihydro-23α-hydroxy-3-tigloyl-11-deoxyazadirachtinin (azadirachtin N), together with known compound 11-*epi*-azadirachtin H were isolated from a methanolic extract of the seed kernels of *A. indica* (Luo et al. 1999). Successful scale-up of *A. indica* suspension culture for azadirachtin production was done in stirred tank bioreactor with two different impellers (Prakash and Srivastava 2007;

Fig. 20 A plant of *Commiphora wightii*

Fig. 21 A fruiting branch of *Phyllanthus emblica*

Srivastava and Srivastva 2008). Flavonoids isolated from *P. emblica* showed hypolipidemic and hypoglycemic activities (Anila and Vijaylaxmi 2000). Cinnamic acid and 5-hydroxymethylfurfural were identified as components of *Emblica* fruit for the first time (Luo et al. 2009).

The gum resin of *C. wightii* has been used as an Ayurvedic medicine for over 2,000 years to treat obesity, liver disorders, internal tumors, malignant sores and ulcers, urinary complaints, intestinal worms, leucoderma, sinus, edema and sudden paralytic seizures. Research over several decades has shown that gum guggul has antioxidant, hepatoprotective, cardioprotective, neuroprotective and thyroid stimulatory effects. It has been effective in the treatment of inflammation, obesity, arthritis, and lipid disorders. Although the exact mechanism of lipid lowering is far from clear, guggulsterone has been shown to modulate the nuclear receptors, farnesoid X receptor (FXR), pregnane X receptor (PXR), CYP 2b10 gene expression, and the bile salt export pump for cholesterol elimination (Ramawat et al. 2008). The active component of gum guggul, E- and Z-guggulsterone has been shown to suppress the pro-inflammatory transcription factor, NF-κB and NF-κB-regulated gene products involved in anti-apoptosis (IAP1, xIAP, Bfl-1/A1, Bcl-2, cFLIP, and survivin), proliferation (cyclin D1 and c-Myc), and metastasis (MMP-9, COX-2, and VEGF) of tumor cells. Guggulsterone inhibits the growth and induces apoptosis in a variety of tumor cells through the suppression of Akt and activation of JNK and the caspase pathway (Shishodia et al. 2008). Modern science has revealed that this ancient medicine mediates its effects by modulation of several important molecular targets, including transcription factors, enzymes, mitogen activated protein kinases, cell cycle proteins, cytokines, and receptors (Shishodia et al. 2008). Several attempts have been made to ameliorate guggulsterone production by alternative technological methods such as optimizing the production of guggulsterone in callus culture (Mathur et al. 2007) and in cell suspension culture, grown in shake flasks and bioreactor (Figs. 22–26, Mathur and Ramawat 2007). Improved guggulsterone production by sugars and morphactin in the cell cultures (Mathur and Ramawat 2008) and by plant gum elicitation (Dass and Ramawat 2009).

Two spermidine alkaloids (14-N-acetylisocodonocarpine and 15-N-acetylcapparisine) were isolated from the root bark of *Capparis decidua* (Ahmad et al. 1992). Several groups of piperidine alkaloids (juliflorine, juliprosine, juliprosopine, julifloricine and julifloridine) were isolated from *Prosopis juliflora*

Fig. 22 Cell cultures grown in liquid medium on a rotary shaker

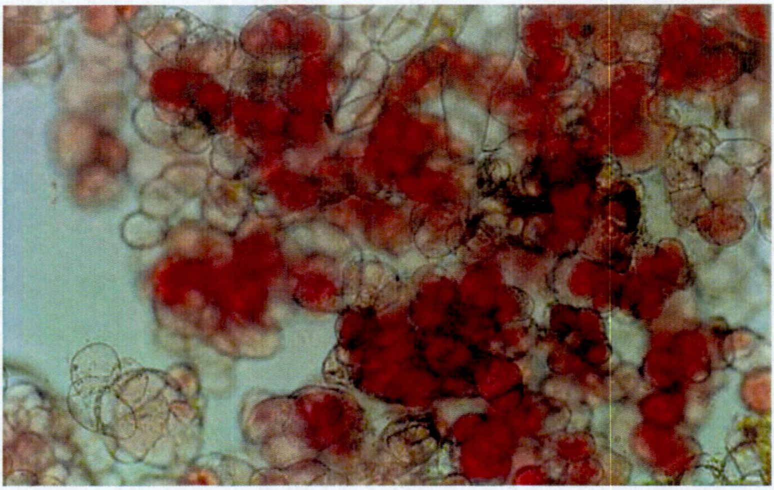

Fig. 23 Cells of *Commiphora wightii* callus cultures under microscope showing anthocyanin

Fig. 24 A 3 L aerated bioreactor containing cell suspension culture

Fig. 25 A 2 L stirred tank bioreactor with cells

(Nakano 2010; Vimal and Tyagi 1986). A mixture of alkaloids from *P. juliflora* has significant inhibitory effects on gram-positive bacteria. Constraints to the use of *P. juliflora* extracts as a source of medicinal compounds are the presence of such irritant and potentially toxic chemicals, and further work would be required before making any recommendations for the use of *Prosopis* plant extracts for medicinal use (Pasiecznik et al. 2001). Through it has been possible to recover some drugs and chemicals, there is need to further perfect the procedures to make technique applicable to all types of agroforestry plants products and to use at commercial scale. Another alternative is to grow organized cultures, such as normal roots or hairy roots in shake flasks (Figs. 15–17) and bioreactors. Production of useful secondary metabolites from arid region plants has been explored and technology has been developed to grow cells in bioreactor. However, there are many more plants yet to be explored growing in this harsh environment for novel bioactive molecules.

Fig. 26 Stirred tank bioreactor showing growth of cells for harvest

4 Conclusion

The preceding discussion highlighted some of the promising aspects of biotechnology in agroforestry systems. In agroforestry systems, numerous issues are involved with livelihood improvement because agroforestry practices are one option with multifunctional value. Biotechnology provided the basic tools to develop entirely new processes and products of agroforestry trees for the benefit of society. Biotechnology is helpful in traditional tree improvement activities by providing valuable information to tree breeders and supplementing the production of high quality seeds. An interrelationship of biotechnological inputs and plant improvement is presented in Fig. 27.

The biotechnology provides an opportunity to convert bioresources into economic wealth. A number of biotechnology and molecular tools are now being introduced into agroforestry research and improvement programs which have a direct bearing on lively hood improvement. Remarkable technological developments have been achieved inmicropropagation, e.g. in *Prosopis tamarugo*, *Tecomella undulata*, *Zizyphus mauritiana* and *Phoenix dactylifera*. Adequate research facilities, technical information system and funding are key elements need to be added for effective and realistic productive output of this technology in agroforestry in developing countries.

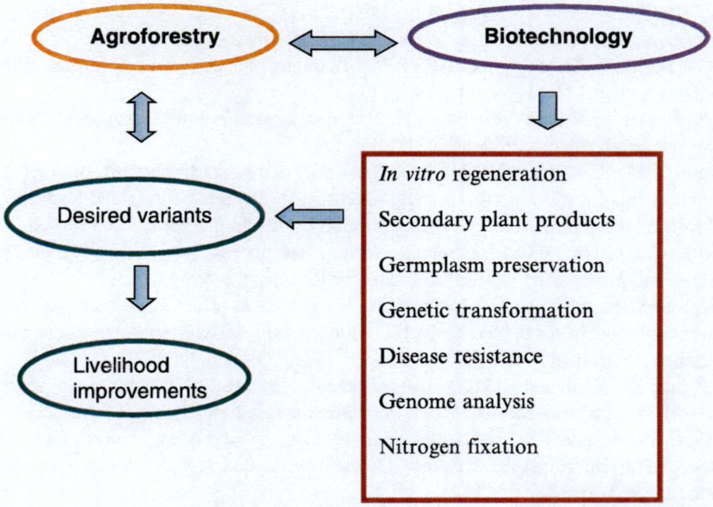

Fig. 27 Schematic representation of role of biotechnology in amelioration of tree species for agroforestry. Biotechnological inputs help in developing desirable varieties and consequently improvement in quality of life by providing better finance

Molecular markers are indispensable tools to understand the genetic structures of populations. Genetic engineering may enable forest researchers to grow agroforestry trees with new and desirable traits as well as selection of multi-layered cropping system to grow trees-bushes-herbs-tubers. Significant technical challenges must be overcome before genetically engineered trees could be used for operational reforestation. Biotechnological inputs are helpful in rapid reforestation program. Plant biotechnology can therefore, be an effective catalyst in commercializing crops, which can create jobs, earn foreign exchange and ensure a better quality of life for all kinds.

Acknowledgements This work was supported by financial assistance from UGC-DRS under special assistance program for medicinal plant research to KGR. VS and SG thanks UGC and CSIR, New Delhi for financial assistance in the form of JRF and RA, respectively.

References

Abdel Rahman HF, Skaug N, Whyatt AM, Francis GW (2003) Volatile compounds in crude *Salvadora persica* extracts. Pharmaceut Biol 41:399–404

Aderbauer B, Clausen PH, Kershaw O, Melzig MF (2008) In vitro and in vivo trypanocidal effect of lipophilic extracts of medicinal plants from Mali and Burkina Faso. J Ethnopharmacol 119:225–231

Adhvaryu MR, Reddy N, Parabia MH (2007) Effects of four Indian medicinal herbs on Isoniazid-, Rifampicin- and Pyrazinamide-induced hepatic injury and immunosuppression in guinea pigs. World J Gastroenterol 13:3199–3205

Agunu A, Yusuf S, Andrew GO, Zezi AU, Abdulrahman EM (2005) Evaluation of five medicinal plants use in diarrhea treatment in Nigeria. J Ethnopharmacol 101:27–30

Ahmad VU, Ismail N, Arif S, Amber AR (1992) Two new N-acetylated spermidine alkaloids from *Capparis decidua*. J Nat Prod 55:1509–1512

Ahmed Z, Akhter F, Haque MS, Banu H, Rahman MM, Faruquzzaman AKM (2001) Novel micropropagation system. J Biol Sci 11:1106–1111

Ali H, Konig GM, Khalid SA, Wright AD, Kaminsky R (2002) Evaluation of selected Sudanes medicinal plants for their in vitro activity against hemoflagellates, selected bacteria, HIV-1-RT and tyrosine kinase inhibitory and for cytotoxicity. J Ethnopharmacol 83:219–228

Anila L, Vijaylaxmi NR (2000) Beneficial effects of flavonoids from *Sesasum indicum, Emblica officinalis* and *Momordica charantia*. Phytother Res 14:592–595

Arora DK, Suri SS, Ramawat KG, Merillon JM (1999) Factors affecting somatic embryogenesis in long term callus cultures of Safed Musli (*Chlorophytum borivilianum*), an endangered wonder herb. Indian J Exp Biol 37:75–82

Arora DK, Suri SS, Ramawat KG (2006) Assessment of variability in regenerants from long term cultures of Safed Musli (*Chlorophytum borivilianum*). Indian J Biotech 5:527–534

Arora J, Goyal S, Ramawat KG (2010) Biodiversity, Biology and conservation of medicinal plants of Thar desert. In: Ramawat KG (ed) Desert plants: biology and biotechnology. Springer, Heidelberg, pp 3–36

Arulkumaran S, Ramprasath VR, Shanthi P, Sachdanandam P (2007) Alteration of DMBA-induced oxidative stress by additive action of a modified indigenous preparation-Kalpaamruthaa. Chem Biol Interact 167:99–106

Aslam M, Singh R, Anandhan S, Pande V, Ahmed Z (2009) Development of transformation protocol for *Tecomella undulata* (Smith) Seem from cotyledonary node explants. Sci Hort 121:119–121

Azam MM (1999) Anti-HIV agents and other compounds from *Tecomella undulata*. Orient J Chem 15:375–377

Baily CD, Hughes CE, Harris SA (2004) Using RAPDs to identify DNA sequence loci for species level phylogeny reconstruction: an example from *Leucaena* (Fabaceae). Syst Bot 29:4–14

Beck SL, Dunlop RW (2001) Micropropagation of *Acacia* species – a review. In Vitro Cell Dev Biol Plant 37:531–538

Bhau BS, Negi MS, Jindal SK, Singh M, Lakshmikumaran M (2007) Assessing genetic diversity of *Tecomella undulata* (Sm.) – an endangered tree species using amplified fragment length polymorphisms-based molecular markers. Curr Sci 93:67–72

Brink JA, Woodward BR, Dasilva EJ (1998) Plant biotechnology: a tool for development in Africa. Electron J Biotechnol 1:1–10

Burke IC, Lauenroth WK, Vinto MA, Hook PB, Kelly RH, Epstein HE, Aguiar MR, Robles MD, Aguilera MO, Gill RA (1998) Plant-soil interactions in temperate grasslands. Biogeochem 42:121–143

Caswell KL, Kartha KK (2009) Recovery of plants from pea and strawberry meristems cryopreserved for 28 years. Cryo Lett 30:41–46

Darmani H, Nasayr T, AL-Hiyasat AS (2006) Effects of extracts of miswak and derum on proliferation of Balb/C 3 T3 fibroblasts and viability of cariogenic bacteria. Int J Dent Hyg 4:62–66

Dass S, Ramawat KG (2009) Elicitation of guggulsterone production in cell cultures of *Commiphora wightii* by plant gums. Plant Cell Tiss Org Cult 96:349–353

Deep G, Dhiman M, Rao AR, Kale RK (2005) Chemopreventive potential of Triphala (a composite Indian drug) on benzo(a)pyrene induced forestomach tumorigenesis in murine tumor model system. J Exp Clin Cancer Res 24:555–563

Dodeman VL, Ducreux G, Kreis M (1997) Zygotic embryogenesis versus somatic embryogenesis. J Exp Bot 48:1493–1509

Dushyent G, Bohra A (2000) Toxic effects of various plant part extracts on the causal organism of typhoid fever. Curr Sci 78:780–781

Fatima F, Khalid A, Nazar N, Abdalla M, Mohomed H, Toum AM, Magzoub M, Ali MS (2005) In vitro assessment of anti-cutaneous leishmaniasis activity of some Sudanese plants. Turk Parazitol Derg 29:3–6

Galati EM, Monforte MT, Forestieri AM, Miceli N, Bader A, Trovato A (1999) *Salvadora persica* L. hypolipidemic activity on experimental hypercholesterolemia in rat. Phytomed 6:181–185

Garg L, Bhandari NN, Rani V, Bhojwani SS (1996) Somatic embryogenesis and regeneration of triploid plants in endosperm cultures of *Acacia nilotica*. Plant Cell Rep 15:855–858

Goswami M, Ranade SA (1999) Analysis of variations in RAPD profiles among accessions of *Prosopis*. J Genetics 78:141–147

Harsh LN, Tewari JC (2007) Agroforestry systems in arid regions of India. In: Puri S, Panwar P (eds) Agroforestry: systems and practice. New India Publisher, New Delhi, pp 175–189

Ho CK, Chang SH, Tsay JY, Tasai CJ, Chiang VL, Chen ZZ (1998) *Agrobacterium tumefaciens*-mediated transformation of *Eucalyptus camaldulensis* and production of transgenic plants. Plant Cell Rep 17:675–680

Huang W (1998) Modelling the coexistence gain and interactions of populations in *Taxodium ascendens*-intercrop systems. Ecol Model 107:189–212

Huang Y, Karnosky DF, Tauer CG (1993) Applications of biotechnology and molecular genetics to tree improvement. J Arboricult 19:84–98

Hussein G, Miyashiro H, Nakamura N, Hattori M, Kakiuchi N, Shimotohno K (2000) Inhibitory effects of Sudanese medicinal plant extracts on hepatitis C virus (HCV) protease. Phytother Res 14:510–516

Jain SH, Kumar H, Arya R, Srivastava RL (2005) Protection and efficient utilization of plantation grown lesser-known timbers of arid region in India-*Acacia tortilis, Prosopis juliflora* and *Prosopis cineraria*. IRG Secretariat, Stockholm, pp 1–9

Jain N, Goyal S, Ramawat KG (2010) Biotechnological approaches to aphrodisiac plants of Rajasthan, India. In: Ramawat KG (ed) Desert plants-biology and biotechnology. Springer, Berlin/Heidelberg, pp 479–495

Jube S, Borthakur D (2009) Development of an *Agrobacterium* –mediated transformation protocol for the tree-legume *Leucaena leucocephala* using immature zygotic embryos. Plant Cell Tiss Org Cult 96:325–335

Khalafalla MM, Daffalla HM (2008) In vitro micropropagation and micrografting of gum Arabic tree [*Acacia senegal* (L.) Wild]. Int J Sustain Crop Prod 3:19–27

Khalessi AM, Pack AR, Thomson WM, Tompkins GR (2004) An in vivo study of the plaque control efficacy of Persica: a commercially available herbal mouthwash containing extracts of *Salvadora persica*. Int Dent J 5:279–283

Kim HJ, Yokozawa T, Kim HY, Tohda C, Rao TP, Juneja LR (2005) Influence of amla (*Emblica officinalis* Gaertn.) on hypercholesterolemia and lipid peroxidation in cholesterol-fed rats. J Nutr Sci Vitaminol Tokyo 51:413–418

Kirira PG, Rukunga GM, Wanyonyi AW, Muregi FM, Gathirwa JW, Muthaura CN, Omar SA, Tolo F, Mungai GM, Ndiege IO (2006) Anti-plasmodial activity and toxicity of extracts of plants used in traditional malaria therapy in Meru and Kilifi Districts of Kenya. J Ethnopharmacol 106:403–407

Koyama T, Kato N, Hibino T, Kawazu T, Kimura T, Sakka K (2006) Isolation and expression analysis of phosphate transporter genes from *Eucalyptus camadulensis*. Plant Biotechnol 23:215–218

Kumar S, Singh N (2009) Micropropagation of *Prosopis cineraria* (L.) Druce-A Multipurpose Desert Tree. Researcher 1:28–32

Kumar S, Suri SS, Sonie KC, Ramawat KG (2003) Establishment of embryonic cultures and somatic embryogenesis in callus cultures of guggul- *Commiphora wightii* (Arnott.) Bhandari. Indian J Exp Biol 41:69–77

Kumar S, Mathur M, Jain AK, Ramawat KG (2006) Somatic embryo proliferation in *Commiphora wightii* and evidence for guggulsterone production in culture. Indian J Biotechnol 5:217–222

Looy TV, Carrero GO, Mathijs E, Tollens E (2008) Underutilized agroforestry food products in Amezonas (Venezuala): a market chain analysis. Agrofores Syst 74:127–141
Luo X, Ma Y, Wu S, Wu D (1999) Two novel azadirachtin derivatives from *Azadirachta indica*. J Nat Prod 62:1022–1024
Luo W, Zhao M, Yang B, Shen G, Rao G (2009) Identification of bioactive compounds in *Phyllanthus emblica* L. fruit and their free radical scavenging activities. Food Chem 14:499–504
MacLean AM, Finan TM, Sadowsky MJ (2007) Genomes of the symbiotic nitrogen-fixing bacteria of legumes. Plant Physiol 144:615–622
Mahajan RT, Chopda MZ (2009) Phyto-pharmacology of *Zizyphus jujube* Mill – a plant review. Pharmacog Rev 3:320–329
Mathur M, Ramawat KG (2007) Guggulsterone production in cell suspension cultures of the guggul tree, *Commiphora wightii*, grown in shake-flasks and bioreactor. Biotech Lett 29:979–982
Mathur M, Ramawat KG (2008) Improved guggulsterone production by sugars, precursors and morphactin in cell cultures of *Commiphora wightii* grown in shake flasks and bioreactor. Plant Biotech Rep 2:133–136
Mathur M, Jain AK, Dass S, Ramawat KG (2007) Optimization of guggulsterone production in callus culture of *Commiphora wightii* (Arnott.) Bhandari. Indian J Biotechnol 6:525–531
Meena PD, Kaushik P, Shukla S, Soni AK, Kumar M, Kumar A (2006) Anticancer and antimutagenic properties of *Acacia nilotica* (Linn.) on 7,12-dimethylbenz(a)anthracene-induced skin papillomagenesis in Swiss albino mice. Asian Pac J Cancer Prev 7:227–232
Merillon JM, Ramawat KG (2007) Biotechnology for medicinal plants: research need. In: Ramawat KG, Merillon JM (eds) Biotechnology Secondary metabolites. Science Publication, Enfield, pp 1–20
Miller JT, Bayer RJ (2001) Molecular phylogenetics of *Acacia* (Fabaceae: Mimosoideae) based on the chloroplast *matsK* coding sequence and flanking *trnK* introne spacer regions. Am J Bot 88:697–705
Monforte MT, Trovato A, Rossitto A, Forestieri AM, DAquino A, Miceili N, Galati EM (2002) Anticonvulsant and sedative effects of *Salvadora persica* L. stem extracts. Phytoether Res 16:395–397
Mukherjee R, Khattar D, Jaggi M, Singh AT, Kumar M, Bala H (2006) Method for treating cancer using betulinic acid rich herbal extract. US Patent No. 20,060,159,783, http://www.freshpatents.com
Mullins KV, Llewellyn DJ, Hartney VJ, Strauss S, Dennis ES (1997) Regeneration and transformation of *Eucalyptus camaldulensis*. Plant Cell Rep 16:787–791
Muralidharan EM, Kallarackal J (2004) Current trends in forest tree biotechnology. In: Srivastava PS, Narula A, Srivastava S (eds) Plant biotechnology and molecular markers. Anamaya Publisher, New Delhi, pp 169–182
Nakano H (2010) Plant growth inhibitors from mesquite (*Prosopis juliflora*). In: Ramawat KG (ed) Desert plants: biology and biotechnology. Springer, Heidelberg, pp 342–354
Nandwani D, Purohit SD, Jain SM, Ramawat KG (2005) Propagation technique in woody plants with special reference to arid and semiarid zone trees. In: Shanmughavel P, Ignacimuthu S (eds) Tree improvement and biotechnology. Pointer Publisher, Jaipur, pp 16–52
Narula A, Kumar S, Bansal KC, Srivastava PS (2004) Biotechnology approaches towards improvement of medicinal plants. In: Srivastava PS, Narula A, Srivastava S (eds) Plant biotechnology and molecular markers. Anamaya Publisher, New Delhi, pp 78–116
Nema RK, Dass S, Mathur M, Ramawat KG (2008) Morphactin and cytokinin promote high frequency bulbil formation from leaf explants of *Curculigo orchioides* grown in shake flasks cultures. Indian J Biotech 7:520–525
Ong CK, Wilson J, Deans JD, Mulayta J, Raussen T, Wajja-Musukuee N (2002) Tree-crop interactions: manipulation of water use and root function. Agric Water Manag 53:171–186

Panchabhai TS, Ambarkhane SV, Joshi AS, Samant BD, Rege NN (2008) Protective effect of *Tinospora cordifolia, Phyllanthus emblica* and their combination against antitubercular drugs induced hepatic damage: an experimental study. Phytother Res 22:646–650

Pandey DN (1998) Ethnoforestry: local knowledge for sustainable forestry and livelihood security. Himanshu/AFN, New Delhi

Pandey DN (2007) Multifunctional agroforestry system in India. Curr Sci 92:455–463

Pasiecznik NM, Felker P, Harris PJC, Harsh LN, Cruz G, Tewari JC, Cadoret K, Maldonado LJ (2001) The *Prosopis juliflora – Prosopis pallida* Complex: a monograph. HDRA, Coventry, p 172

Pinmai K, Chunlaratthanabhorn S, Ngamkitidechakul C, Soonthornchareon N, Hahnvajanawong C (2008) Synergistic growth inhibitory effects of *Phyllanthus emblica* and *Terminalia bellerica* extracts with conventional cytotoxic agents: doxorubicin and cisplatin against human hepatocellular carcinoma and lung cancer cells. World J Gastroenterol 14:1491–1497

Prakash G, Srivastava AK (2007) Azadirachtin production in stirred tank reactors by *Azadirachta indica* suspension culture. Process Biochem 42:93–97

Purohit A, Vyas KB (2005) Hypolipidaemic efficacy of *Capparis decidua* fruit and shoot extracts in cholesterol fed rabbits. Indian J Exp Biol 43:863–866

Rai SN, Chakrabarti SK (2001) Demand and supply of fuel wood and timber in India. Indian Forester 127:263–279

Raj BR (2005) Micropropagation of woody plants of arid zone. In: Kukreja AK, Mathur AK, Banerjee S, Mathur A, Sharma A, Khanuja SPS (eds) Proceedings of National Symposium on Plant Biotechnology. CIMAP, Lucknow, pp 339–348

Raj BR (2010) Biology and multiplication of *Prosopis* species grown in Thar Desert. In: Ramawat KG (ed) Desert plants: biology and biotechnology. Springer, Heidelberg, pp 371–406

Raj BR, Singh M (2000) Somatic embryogenesis in fruit and forest trees of arid zone. In: Mohan S, Gupta PK, Newton RJ (eds) Somatic embryogenesis in woody plants. Kluwer, Dordrecht/Boston/London, pp 141–168

Ramawat KG, Goyal S (2008) Indian herbal drugs scenario in global perspectives. In: Ramawat KG, Merillon JM (eds) Bioactive molecules and medicinal plants. Springer, Heidelberg, pp 323–345

Ramawat KG, Sonie KC, Sharma MC (2004) Therapeutic potential of medicinal plants. In: Ramawat KG (ed) Biotechnology: medicinal plants. Science Publishers Inc., Enfield, pp 1–18

Ramawat KG, Mathur M, Dass S, Suthar S (2008) Guggulsterone: a potent natural hypolipidemic agent from *Commiphora wightii* – problems, perseverance and prospects. In: Ramawat KG, Merillon JM (eds) Bioactive molecules and medicinal plants. Springer, Heidelberg, pp 101–121

Ramawat KG, Dass S, Mathur M (2009) The chemical diversity of bioactive molecules and therapeutic potential of medicinal plants. In: Ramawat KG (ed) Herbal drugs: ethnomedicine to modern medicine. Springer, Heidelberg, pp 7–321

Rana MG, Katbamna RV, Dudhrejiya AV, Sheth NR (2008) Hepatoprotection of *Tecomella undulata* against experimentally induced liver injury in rats. Pharmacol 3:674–682

Rani P, Khullar N (2004) Antimicrobial evaluation of some medicinal plants for their anti-enteric potential against multi-drug resistant *Salmonella typhi*. Phytother Res 18:670–673

Rao MR, Singh MP, Day R (2000) Insect pest problems in tropical agroforestry systems: contributory factors and strategies for management. Agroforest Syst 50:243–277

Rathore MS, Singh M, Rathore JS, Panwar D, Shekhawat NS (2007) Molecular tools for improvement of forest trees. Am Eurasian J Agric Environ Sci 2:545–551

Rhouma S, Zehdi-Azouzi S, Dakhlaoui-Dkhil S, Ould Mohamed Salem AOB, Othmani A, Cherif E, Marakachi M, Trifi M (2010) Genetic variation in Tunisian date-palm (*Phoenix dactylifera* L.). In: Ramawat KG (ed) Desert plants: biology and biotechnology. Springer, Heidelberg, pp 355–370

Rout GR (2005) *In vitro* somatic embryogenesis in callus cultures of *Azadirachta indica* A. Juss.–a multipurpose tree. J For Res 10:263–267

Saito K, Kohno M, Yoshizaki F, Niwano Y (2008) Extensive screening for edible herbal extracts with potent scavenging activity against superoxide anions. Plant Foods Hum Nutr 63:65–70

Sandhya T, Mishra KP (2006) Cytotoxic response of breast cancer cell lines, MCF 7 and T 47 D to triphala and its modification by antioxidants. Cancer Lett 238:304–313

Sanogo R, Monforte MT, Daquino A, Rossitto A, Maur DD, Galati EM (1999) Antiulcer activity of *Salvadora persica* L. structural modifications. Phytomed 6:363–366

Satdive RK, Fulzele DP, Eapen S (2006) Enhanced production of azadirachtin by hairy root cultures of *Azadirachta indica* A. Juss by elicitation and media optimization. J Biotechnol 128:281–289

Schroth G, Krauss U, Gasparotto L, Aguilar JAD, Vohland K (2000) Pests and diseases in agroforestry system of the humid tropics. Agroforest Syst 50:199–241

Shishodia S, Harikumar HB, Dass S, Ramawat KG, Aggarwal BB (2008) The guggul for chronic diseases: ancient medicine. Mod targets Anticancer Res 28:3647–3664

Singh SP (1994) Handbook of agroforestry. Agrotech Publishing Academy, Udaipur

Singh AK (2004) Endangered economic species of Indian desert. Genet Resou Crop Evol 51: 371–380

Singh G, Bala N, Rathore TR, Limbza NK (2004) Biomass production of *Tecomella undulata* agroforestry systems in India. Biol Agric Hort 22:205–216

Singh I, Soyal D, Goyal PK (2006) *Emblica officinalis* (Linn.) fruit extract provides protection against radiation-induced hematological and biochemical alterations in mice. J Environ Pathol Toxicol Oncol 25:643–654

Singh R, Singh B, Singh S, Kumar N, Kumar S, Arora S (2008) Anti-free radical activities of kaempferol isolated from *Acacia nilotica* (L.) Willd. Ex Del Toxicol In Vitro 22:1965–1970

Sofrata A, Lingstrm P, Baljoon M, Gustafsson A (2007) The effect of miswak extract on plaque pH, an in vivo study. Caries Res 41:451–454

Srivastava S, Srivastva AK (2008) *In vitro* azadiractin production. In: Ramawat KG, Merillon JM (eds) Bioactive molecules and medicinal plants. Springer, Heidelberg, pp 101–121

Staver C, Guharay F, Monterroso D, Muschler RG (2001) Designing pest- supressive multistata pernnial crop systems: shade-grown coffee in Central America. Agroforest Syst 53:151–170

Suri SS, Arora DK, Ramawat KG (2000) A method for large-scale multiplication of *Curculigo orchioides* through bulbil formation from leaf explants in shake flask cultures. Indian J Exp Biol 38:145–148

Suthar S, Thul S, Kukreja AK, Ramawat KG (2008) Assessment of genetic diversity by RAPD marker and guggulsterone content in *Commiphora wightii*, an endangered medicinal tree. J Cell Tiss Res 8:1477–1480

Talwar GP, Dar SA, Rai MK, Reddy KV, Mitra D, Kulkarni SV, Doncel GF, Buck CB, Schiller JT, Muralidhar S, Bala M, Agrawal SS, Bansal K, Verma JK (2008) A novel polyherbal microbicide with inhibitory effect on bacterial, fungal and viral genital pathogens. Int J Antimicrob Agents 32:180–185

Tewari VP, Singh M (2006) Tree-crop interaction in the That Desert of Rajasthan (India). Scheresse 17:326–332

Toky OP, Kaushik N, Sharma PK (1995) Genetic variability in progenies of *Acacia nilotica* (L.) ex Del. ssp. indica (Benth.) Brenan for nitrogen fixing ability. Silvae Genet 44:161–165

Toonen MAJ, De Vries SC (1996) Initiation of somatic embryos from single cells. In: Wang TL, Cuming A (eds) Embryogenesis: the generation of a plant. Bios Scientific, Oxford, pp 173–189

Upadhyay RK, Rohatgi L, Chaubey MK, Jain SC (2006) Ovipositional responses of the pulse beetle, *Bruchus chinensis* (Coleoptera: Bruchidae) to extracts and compounds of *Capparis decidua*. J Agric Food Chem 54:9747–9751

Vasudevan M, Parle M (2007a) Effect of Anwala churna (*Emblica officinalis* Gaertn.): an ayurvedic preparation on memory deficit rats. Yakugaku Zasshi 127:1701–1707

Vasudevan M, Parle M (2007b) Memory enhancing activity of Anwala churna (*Emblica officinalis* Gaertn.): an Ayurvedic preparation. Physiol Behav 91:46–54

Vicient CM, Martnez FX (1998) The potential uses of somatic embryogenesis in agroforestry are not limited to synthetic seed technology. Rev Bras Fisiol Veg 10:1–12

Vimal OP, Tyagi PD (1986) *Prosopis juliflora*: chemistry and utilization. In: Patel VJ (ed) The role of *prosopis* in wasteland development. Javrajbhai Patel Agroforestry Center, Surendrabag, pp OVP1–OVP8

Yadav P, Sarkar S, Bhatnagar D (1997) Action of *Capparis decidua* against alloxan-induced oxidative stress and diabetes in rat tissues. Pharmacol Res 36:221–228

Zahran HH (2001) Rhizobia from wild legumes: diversity, taxonomy, ecology, nitrogen fixation and biotechnology. J Biotechnol 91:143–153

Zinkhan FC, Mercer DE (1997) An assessment of agroforestry systems in the southern USA. Agroforest Syst 35:303–321

Mutual Legume Intercropping for Forage Production in Temperate Regions

Branko Ćupina, Aleksandar Mikić, Frederick L. Stoddard, Đorđe Krstić, Eric Justes, Laurent Bedoussac, Joële Fustec, and Borivoj Pejić

Abstract Carefully designed intercropping systems can have many advantages in comparison to monocropping such as increased forage yield, enhanced weed control, reduced soil erosion and, in the case of legumes, improved soil fertility due to their symbiosis with nitrogen-fixating bacteria. In addition the use of forage legumes is increasing for the rations of ruminants because legumes supply animal husbandry with protein-rich diets. Due to lower forage yield from perennial legumes in the first planting year and a critical standing ability of annual forage legumes, farmers tend to establish these crops with a companion crop.

The first trials in Serbia studied the role an annual legume in the establishment of a perennial legume. Field pea cultivars with reduced plant height, semi-leafless leaf types and improved lodging tolerance were included together with a pure red clover stand and its mixture with oats as controls. When sown as the companion crop an annual forage legume can provide an economic yield during the perennial forage

B. Ćupina (✉) · Đ. Krstić · B. Pejić
Faculty of Agriculture, University of Novi Sad, Trg Dositeja Obradovića 8, 21000 Novi Sad, Serbia
e-mail: cupinab@polj.uns.ac.rs; djordjek@polj.uns.ac.rs; pejic@polj.uns.ac.rs

A. Mikić
Institute of Field and Vegetable Crops, M. Gorkog 30, 21000 Novi Sad, Serbia
e-mail: aleksandar.mikich@gmail.com

F.L. Stoddard
Department of Agricultural Sciences, University of Helsinki, Helsinki, Finland
e-mail: frederick.stoddard@helsinki.fi

E. Justes · L. Bedoussac
INRA Toulouse, UMR 1248 AGIR Equipe VASCO, Castanet-Tolosan, France
e-mail: Eric.Justes@toulouse.inra.fr; laurent.bedoussac@toulouse.inra.fr

J. Fustec
UP-SP Laboratoire d'Écophysiologie Végétale et Agroécologie, Ecole Supérieure d'Agriculture – PRES Université Nantes Angers Le Mans, Angers, France
e-mail: j.fustec@groupe-esa.com

E. Lichtfouse (ed.), *Genetics, Biofuels and Local Farming System*, Sustainable Agriculture Reviews 7, DOI 10.1007/978-94-007-1521-9_12,

crop establishment. In average, field pea as a companion crop increases forage annual dry matter yield by 2.56 t ha^{-1} and reduces weeds in red clover stand by 29%.

Another group of trials involved mixtures of autumn-sown cool season, spring-sown cool season and warm season annual legumes for forage production. Here one plant had good and another poor standing ability and with concurring development stages and similar growth habit. There were economically justified intercrops with Land Equivalent Ratio (LER) values of forage dry matter yield higher than 1, such as winter faba bean with winter common vetch (1.42), spring faba bean with spring grass pea (1.44) and pigeon pea with lablab bean (1.12). The achieved results in the mutual legume intercropping research in Serbia encourage the similar research in the neighbouring West Balkan Countries and other European temperate regions.

Keywords Agronomic performance • Annual legumes • Forage yield • Intercropping • Land equivalent ratio • Perennial legumes establishment

1 Introduction

Cropping systems based on carefully designed species mixtures reveal many potential advantages such as forage yield increase, weed control (Avola et al. 2008), reduce soil erosion (Wiersma et al. 1999), lower pests and diseases (Trenbath 1993; Altieri 1999) under various conditions (Malézieux et al. 2009). According to Willey (1979) intercropping is the simultaneous growing of two or more crop species in the same field. Intercropping may often be considered a practical application of ecological principles based on biodiversity, plant interactions and other natural regulation mechanisms (Vandermeer et al. 1998). This is a result of differences in competitive ability for growth factors between intercrop components (Anil et al. 1998) which thereby tend to use resources in a complementary way and then more efficiently than sole crops. In particular, intercropping non fixing species with legumes are often mentioned as resulting in higher yields than monocultures (Corre-Hellou et al. 2006). Because of their ability to biological nitrogen fixation, legumes are largely involved in N facilitation and N dynamic in the plant community and in agrosystems (Hauggaard-Nielsen and Jensen 2005; Temperton et al. 2007; Zarea et al. 2008; Fustec et al. 2010).

Intercropping of two or more crop species for food production tends to be limited to the tropics as a result of economic and mechanical reasons as opposed to physiological or agronomic ones, while in temperate regions it is receiving increased attention as a means of efficient forage production (Anil et al. 1998). These tend to be restricted to four forms (Koivisto 2002): (1) two or more annual forages sown together, (2) companion cropping in establishing perennial forage crops, (3) annual forages drilled into an existing stand to boost short-term yields (Thorsted et al. 2002), (4) perennial legumes sown in between the rows of an arable crop (Zemenchik et al. 2000).

Although efficient utilisation of available growth resources is fundamental in achieving sustainable systems of agricultural production, arable intercropping for feed and human consumption is presently not so common in Europe (Hauggaard-Nielsen et al. 2001). In these systems such mixture are important for increasing forage yield and enhanced weed control (Avola et al. 2008). In addition, companion cropping could reduce soil erosion on sites prone to either wind or water erosion (Wiersma et al. 1999) with is of particular interest for both conventional and low inputs systems.

This review comprises two parts focused successively on the following issues with benefit for forage production in temperate regions such as West Balkan Countries: (a) use of annual legumes as cover crops for establishing perennial forage legumes and (b) mutual intercropping of annual legumes.

2 Annual Legumes: Cover Crops for Establishing Perennial Legumes

Perennial forage legumes can be established either in summer/autumn or in spring. Since a small seed size, perennial legumes are sown shallowly and are especially vulnerable to drought. A spring-sown crop has often a significantly lower yield in the year of establishment than a crop sown in the previous autumn. This could be partially explained by weeds which are a much greater problem in a spring-sown crop (Ćupina et al. 2000, 2004). The difficulty in obtaining an economically significant yield from perennial forage legumes in the year of establishment has forced the farmers to establish these crops with a companion crop (Tesar and Marble 1988). Zollinger and Meyer (1996) reported that the weed control during perennial forage legume establishment without a companion crop is critical in obtaining a productive stand. Also, Klesnil (1980) and Matejkova (1982) claimed that pure crops of perennial legumes produced significantly lower forage yields than intercrops, as well as that their slow growth during establishment increased their vulnerability to weed invasion. However, intercropping with short-season crops, such as field pea (*Pisum sativum* L.) and vetches (*Vicia* spp.), significantly increased forage yield and quality and reduced incursion of weeds. Vandermeer et al. (1998) presumed the mechanism of weed reduction, where one of the crops in the mixture provided an environment that reduced weed biomass. This would thereby allow the other component species to out-compete the weeds. In those situations, where companion cropping is used to establish perennial forages, this is a frequent case. However, it must be clear that a companion crop does not have to compete too long with the undersown crop in order to avoid damage to the latter (Tesar and Marble 1988).

In general, intercropping with perennial legumes is deemed to be an effective method of agricultural and, more specifically, forage production because it offers greater yield stability, higher yields, reduced weed competition, an increase in protein content within a mixed diet and higher land-use efficiency (Anil et al. 1998). An annual forage crop, when sown as the companion, can provide an economic yield during the establishment of the perennial forage crop (Lanini et al. 1991;

Joanne et al. 2001; Tan et al. 2004) and so producing a return in the seeding year (Sturgul et al. 1990; Chapko et al. 1991). The undersown crop refers to the perennial forage crop established with the annual companion crop (Koivisto 2002). In the West Balkan Countries and beyond, it is small grains, primarily oats (*Avena sativa* L.) and barley (*Hordeum vulgare* L.), that are traditionally intercropped with perennial legumes, although these tend to be too fast-growing and therefore too competitive for the legume component (Ćupina et al. 2010b). Such companion crops compete with the undersown species, reducing yield and in some cases the persistence of the stand (Tesar and Marble 1988; Nickel et al. 1990). However, information on alternative companion crops is limited (Sule 1993).

Annual legumes such as field pea and vetches are suitable for intercropping with perennial forage legumes because the crop can be harvested quickly and the canopy structure is not dense enough to cause suppressive shading. Because of its short growing season, field pea and vetches are suited to be harvested when the first (establishment) cut of perennial legumes is due, increasing protein yield significantly in the harvested crop (Koivisto 2002). Cultivars of field pea currently in use differ in morphology, primarily in leaf type and plant height. The semi-leafless field pea cultivars with short stems could be more adapted for intercropping since the light penetration is much stronger, providing better conditions for the initial growth of the undersown crop (Ćupina et al. 2006). These results are coherent with those of Heath and Hebblethwaite (1985) who found that the semi-leafless, as well as the field pea with normal leaf type but with reduced leaflet area, allowed a greater amount of light through their canopy as a companion crop than the cultivars with normal leaf type and large leaflet area. They also found that semi-leafless peas were more efficient in using available light than the normal-leafed cultivars. This ability to allow more light through the canopy may have benefits for an undersown crop. These two points suggest that they would be better companion crops in establishing perennial forages. Cultivating semi-leafless field pea cultivars may increase the weeds, but an appropriate choice of pea cultivars could decrease weeds amount and enable the undersown crop to reach light (Fig. 1). In addition, including peas in this arable mixture makes possible to improve forage quality and digestibility, that is, to increase the forage crude protein content and to decrease both neutral and acid detergent fibre (Chapko et al. 1991).

2.1 Agronomic Performance

In addition to the selection of a suitable field pea cultivar as a companion crop, an appropriate production technology also needs to be developed. To mitigate the effect of competition among the intercropped plants, it is recommended that the normal seeding rate of the companion crop be reduced (Smith et al. 1972; Lanini et al. 1991; Vough et al. 1995), what requests that the optimum stand density, i.e., the number of plants of the companion crop per unit area, needs to be determined (Tan et al. 2004). Once these criteria are met, it should be possible to produce a crop that will be high yielding, nutritious and palatable to most livestock (Ćupina et al. 2009a).

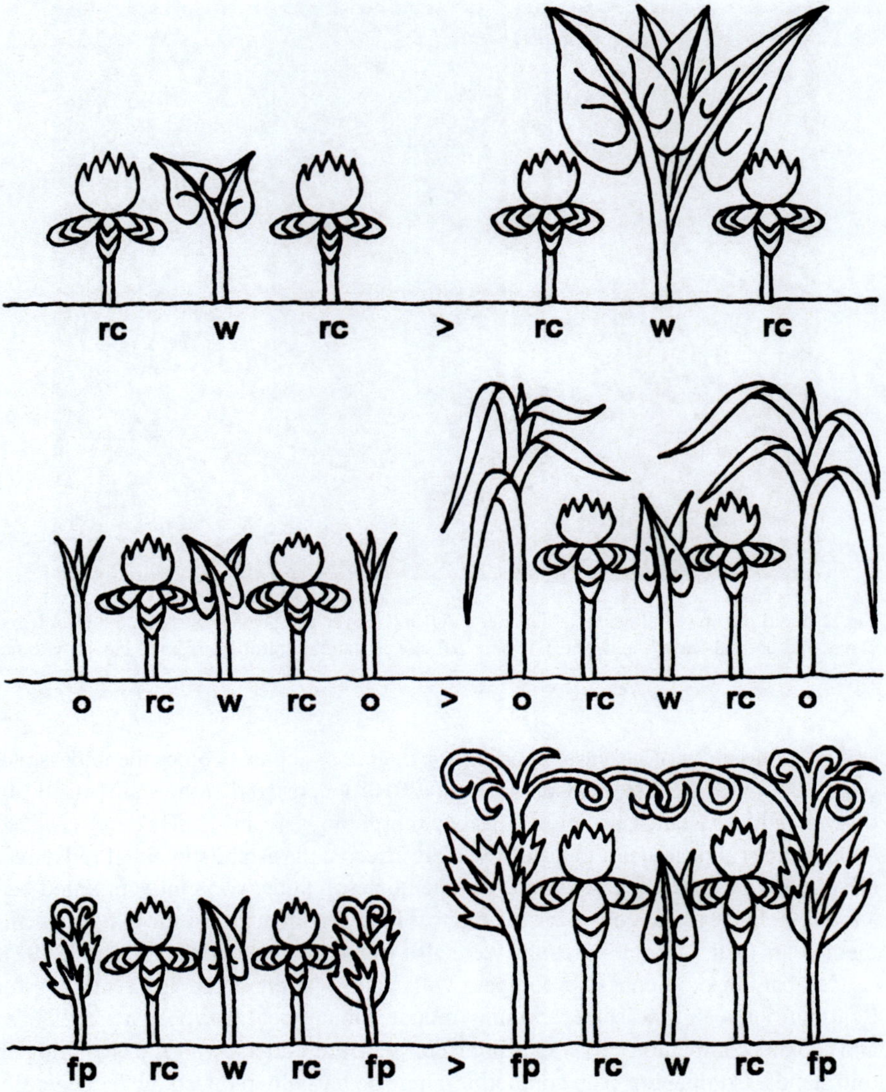

Fig. 1 Different ways of the perennial legumes establishment. Red clover (rc) is easily matched by weeds (w) in its pure stand (*upper row*). Oats (o) decreases weeds (w) but also red clover (rc) (*middle row*). Field pea (fp) decreases weeds (w) but enhances the growth of red clover (rc) (*lower row*).

The choice of pea cultivar as a companion with perennial legumes is of particular importance. Faulkner (1985) and Gilliland and Johnston (1992) found fully leafed pea cultivars prone to lodging, suppressing growth of the undersown species. Therefore, appropriate field pea cultivars should be chosen with care for intercropping since this will affect the risk of lodging, which in turn will affect the likelihood of

Fig. 2 Field pea in red clover establishment. A1: red clover and semi-leafless field pea; A2: red clover and normal-leafed field pea; B1: pure red clover stand (control 1); B2: red clover with oat (control 2)

soil contamination of harvested herbage or degree of suppression of the undersown crop. In a 3 year field study, Ćupina et al. (2010b) reported that the standing ability of two field pea cultivars with different morphology (semi-leafless and reduced leaflets type) appeared not to have adversely affected the establishment of red clover (*Trifolium pratense* L.) and that the companion crop under two cultivars tested had similar yields, with no consistent differences between them across the two establishment years (Fig. 2). Similar results were also found in the work of Koivisto (2002).

Companion crop can also compete with young perennial forage seedlings for light, nutrients and moisture, and may reduce annual yield and persistence. Dense and lodged companion crops can interfere with the undersown crop, resulting in thin stands. Companion crop competition may be partially reduced by the generally recommended cultural practices, such as reducing the companion crop seeding rate and cutting as early as possible (Tan et al. 2004). Tesar and Marble (1988) and Horrocks and Vallentine (1999) recommend a compromise between high sowing rates, which maximise the total yield but are too aggressive for the undersown crop, and low rates which decreases yield and are insufficient to suppress weeds. To avoid too severe competition with the undersown legume, it has been recommended that the sowing rate of the companion crop should be reduced by up to half (Vough et al. 1995). Also, if the cost of pea seed is taken into account, it is reasonable to question the use of the highest seeding rate (Faulkner 1985; Ćupina et al. 2000). Vough et al. (1995) also recommended cutting the companion crop early for forage,

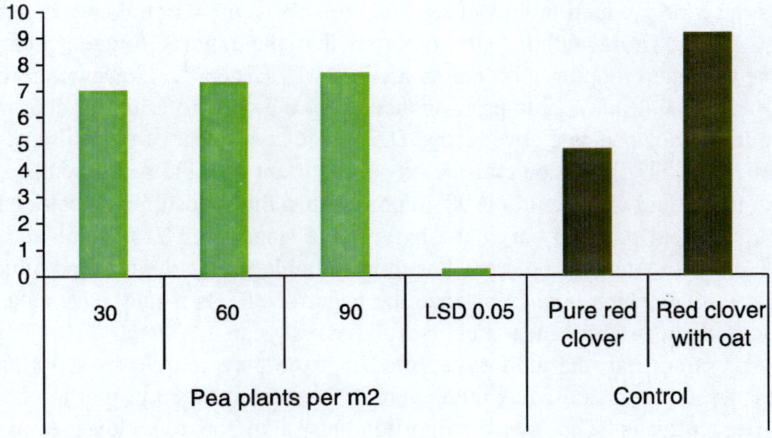

Fig. 3 Effect of field pea plant population density on annual dry matter yield (t ha^{-1}) when intercropped with red clover in comparison to pure stand of red clover and red clover with oat (Ćupina et al. 2010b)

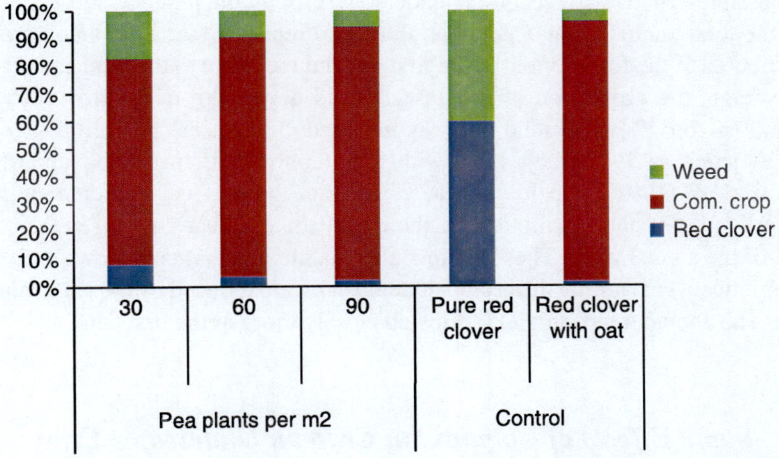

Fig. 4 Weed proportion (%) in the first cutting of red clover intercropped with field pea in comparison to pure stand of red clover and red clover with oat (Ćupina et al. 2010b)

as opposed to its use for grain, to reduce its competition with the undersown species. The second reason for using the companion crop for forage is to reduce the risk of lodging prior to a grain harvest (Tesar and Marble 1988). Ćupina et al. (2010b) reported that the highest annual forage dry matter yield (7.66 t ha^{-1}) and the lowest weed proportions in the first cut (5.9%) were obtained by the highest plant number (90 plants per m^{-2}) of field pea as a cover crop (Figs. 3 and 4). In the second and third red clover cuttings, the difference decreases, while the yield was equalized in the first year of utilization. Regarding the economic side, it could be recommended that in the wide production a lower seed rate (60 plants m^{-2}) of field pea companion crop is used. Those authors also reported that the highest forage yields were obtained in intercropping with oat as a control (9.15 t ha^{-1}). However, according to Ćupina et al. (2006), yield performance is not the only criterion for determining the suitability of pea as a cover crop. The forage coefficient of digestibility in the diet of ruminants should be considered as important as yield performance. Smith et al. (1972) and Obračević (1990) reported that the digestibility coefficients of red clover, field pea and oat at the harvestable stage were 71%, 79% and 54%, respectively. In addition, field pea has more suitable morphological and biological characteristics, which tend to balance the negative effects it may have as a cover crop in both the establishment and first full harvest year.

In the same trial, the authors reported that the pure red clover stand had the highest weed proportion while there were differences in weed proportion between field pea cultivars. The weed proportion, just like the red clover proportion, decreases with increasing the plant number of the companion crop (Fig. 4). Lanini et al. (1991) also found that using an oat companion crop also reduced the number of weeds in the alfalfa (*Medicago sativa* L.) by up to 50%, over a solo seeded control.

For intercropping, particularly with perennial legumes, it is important to check for possible negative effects on the undersown crop in the first cut, subsequent cuts and the total annual yield. Ćupina et al. (2006) reported that companion cropping contributed to the forage yield in the first cut and therefore to the total annual yield. In average, the proportion of the annual yield derived from the first cut ranged from 50% to 69% when field pea was included, 58% in the pure crop (two cuts) and as much as 79% when sown with oats (control variant). It is supported by Lanini et al. (1991) who found that yield from the first cut of perennial forage legume (alfalfa) and oat mixture in the establishment year comprised from 71 to 98% of the annual yield. These authors also maintained that undersown crop yields at subsequent cuts during the establishment year were reduced by the oat companion crop. The annual trends in yield followed closely those at the first cut.

2.2 *Some Effects of Companion Crop on Undersown Crop*

The success of companion cropping depends on the capacity of the undersown legume to develop in the shade of the cover crop, because the competition for light, nutrients and water may reduce the yield and resistance of the undersown crop (Tan et al. 2004). Of all the major environmental factors that contribute to reported multispecies system merits, the capture and use of solar radiation is the one that has received the most attention (Keating and Carberry 1993). The competition of cover and undersown crops for light directly affects the morphological and physiological parameters and their dynamics in the latter. Such results have also been demonstrated recently on durum wheat-winter pea annual intercrops (Bedoussac and Justes 2010b). According to Simmons et al. (1995), the light intensity at the level of the perennial legume within the semi-dwarf pea companion crop canopy was consistently higher than with the conventional-stature companion crops. The improved photosynthetically active radiation (PAR) allows the semi-leafless pea cultivars to convert solar radiation into dry matter more effectively than normal-leafed cultivars. Thus such types are more appropriate (Heath and Hebblethwaite 1985).

The light intensity and quality are crucial for the formation of optimum leaf area index (LAI). Reduced light intensity retards the processes of growth, bioproduction and leaf area development (Heichel et al. 1988). Krstić et al. (2005a,b) reported a perceptible reduction in the leaf area index of the alfalfa and red clover as undersown crop as the number of plants of the field pea companion crop increased, regardless of its cultivar.

According to Tsialtas and Maslaris (2008), the shaded leaves and plants, receiving only diffuse light, typically have higher chlorophyll content than those exposed to direct light. This is supported by Krstić et al. (2005a,b) who found that the lowest average contents of both chlorophyll types (a and b) were obtained in pure alfalfa ($9.65\,\text{mg g}^{-1}$) and red clover ($12.2\,\text{mg g}^{-1}$) stands comparing to intercropping ($18.10\,\text{mg g}^{-1}$), with both increasing with stand density. In the treatments with higher yields, that is, with higher number of the companion cover plants, the shading was increased. The shaded plants produced more chlorophyll and used the nutrients in a more economic way in comparison to the plants in pure stand. The plants provided with better conditions had higher LAI and due to an appropriate amount of light do not have to increase their pigment contents.

In the alfalfa and red clover establishment in the case of field pea as the cover crop, which has one reduced and one metamorphosed leaf, light penetration is probably sufficient to meet photosynthetic maximum. According to Krstić et al. (2005b) in companion cropping the red clover plants remaining in the field are capable of unimpeded photosynthesis and that the average potential intensity of net photosynthesis of intercropped red clover and sole crop are $341.80\,\mu\text{mol}\,0_2\,\text{g}^{-1}\,\text{h}^{-1}$ and $454.67\,\mu\text{mol}\,0_2\,\text{g}^{-1}\,\text{h}^{-1}$, respectively. Dennis and Woledge (1982) reported that in a mixture of white clover and perennial ryegrass the shaded leaves of white clover did not differ significantly regarding the intensity of photosynthesis from the fully exposed leaves. In addition, it should be taken into account that the first cut of such intercrop leaves a large part of the red clover rosette not harvested. This un-harvested part could represent about over two thirds of the aboveground plant part, and the plant takes a short time to recover and continue to do photosynthesis, grow and develop, preparing itself for the second cut. Thus, the photosynthetic apparatus of intercropped red clover is capable of unimpeded photosynthesis after first cutting, which primarily removes the taller crop (Krstić et al. 2005b).

The shade-grown plants in intercropping usually show some anatomical modifications of vegetative organs, especially leaves (Arora and Gupta 1996; Dickison 2000). In the preliminary research on alfalfa histology as affected by intercropping with field pea, carried out as a part of the trials in Novi Sad, Serbia, it was shown that the leaf structural adjustments to differences in light availability affected mostly photosynthetic apparatus. Although alfalfa is regarded as a species with high light demands, the same analysis showed that companion cropping did not significantly affect the leaf structure, proportion of leaf tissues or the size of the cells, as well as that the plants of all treatments showed low variability of measured parameters and high level of homogeneity, which was proved by principal component analysis.

3 Mutual Intercropping of Annual Legumes

Growing annual legume crops in mixed systems can be beneficial as compared to sole crops. The superiority of intercropped plants over pure stands has been attributed generally to the significance variations of morphological characteristics including plant height and leaf architecture which results in efficient exploitation of environmental resources (Biabani et al. 2008; Ćupina et al. 2010a).

The cultivation of annual legumes in mixtures may also have many advantages in comparison to pure stands, especially in a more efficient exploitation of environmental resources. In a similar way, the efficiency of the intercrops of legumes with cereals (Mariotti et al. 2009; Bedoussac and Justes 2010a) and forage brassicas depend mostly on the use of complementary N resources. Mutual intercrops of annual legumes improve soil characteristics (Ghosh et al. 2006), increase availability of nitrogen (Abu-Gyamfi et al. 2007), phosphorus (Cu et al. 2005) and other nutrients, facilitate nodulation (Li et al. 2009) and increase yield (Roy et al. 2003).

A series of trials with mutual intercrops of several annual legume species has been carried out in the conditions of Serbia during past few years. It consisted of three groups of crops, namely winter cool season, spring cool season and warm season annual legumes (Table 1).

The basic principles used in making a mixture of two annual legumes were:

- The same time of sowing;
- The similar growing habit (Fig. 5);
- The similar time of maturing for cutting;
- That one had good standing ability (supporting crop) and that another had poor standing ability (supported crop).

The primary goal of this complex study was to find out if any of these combinations may mean an increase in forage yield, with subsequent research aimed at forage quality, tolerance to abiotic and biotic stress and other agronomic and physiological aspects. As a main indicator of the reliability and profitability, the land equivalent ratio (LER), defined as the relative land area required when growing sole crop to produce the aerial biomass dry weight, yield or N accumulated achieved in intercrop (Willey 1979), was determined for each combination.

The field experiments were conducted mostly at the Experimental Field of the Institute of Field and Vegetable Crops at Rimski Šančevi (45°20′N, 19°51′E and 84 m asl) and prevailing soil and weather conditions given in Tables 2 and 3.

Table 1 Three main groups of the mutual annual legume intercropping in the trials carried out in Serbia in 2008 and 2009

Group	Supporting crop	Supported crop
Winter cool season	Faba bean	Forage pea
		Dual-purpose pea
		Common vetch
		Hungarian vetch
		Hairy vetch
	Semi-leafless field pea	Normal-leafed field pea
		Bitter vetch
Spring cool season	Faba bean	Forage pea
		Common vetch
		Grass pea
	White lupin	Forage pea
		Common vetch
		Grass pea
	Semi-leafless field pea	Normal-leafed pea
		Lentil
Warm season	Soybean 00	Mung bean
	White lupin	
	Soybean I	Lablab bean
	Pigeon pea	

3.1 Mutual Intercropping of Winter Cool Season Annual Legumes

The research on the forage yields in winter cool season annual legumes has begun in autumn 2008 (Mikić et al. 2010a).

The intercropping of winter cool season annual legumes included diverse mixtures of faba bean and semi-leafless field pea, playing the role of a supporting crop, and forage and dual-purpose peas and common, Hungarian, hairy and bitter vetches that acted as supported crops (Table 1). In all intercrops, both supporting and supported crops were sown with 50% of their pure stand sowing rate in early October.

The preliminary results of the trial reveal that majority of the intercrops were more reliable than pure stands, especially of faba bean with dual-purpose pea and common vetch (Table 4).

3.2 Mutual Intercropping of Spring Cool Season Annual Legumes

The most advanced research on the mutual intercrops of annual legumes was done with spring cool season ones. A field trial was carried out at Rimski Šančevi, Novi

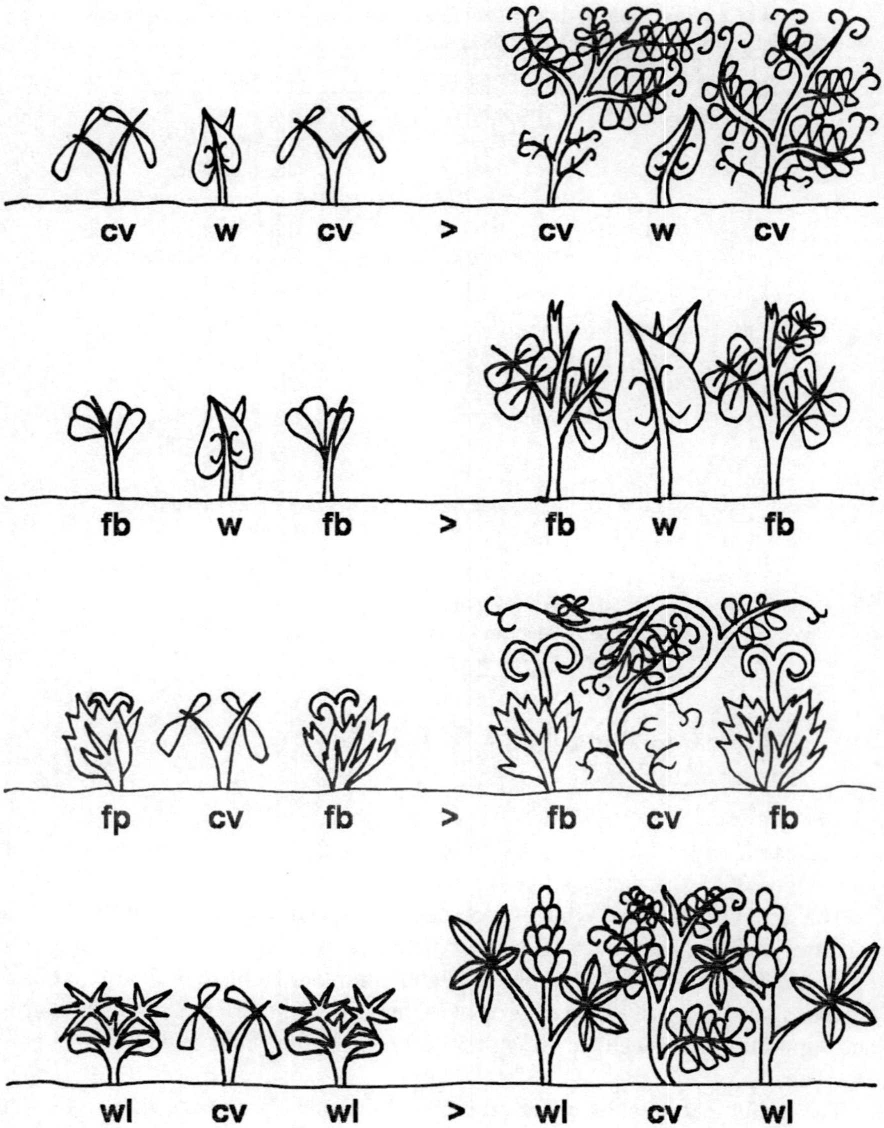

Fig. 5 Different aspects of the mutual intercropping of annual legumes. In the pure stand of a lodging-susceptible crop (common vetch, cv), weeds (w) are suppressed but lower leaves are easily lost (*first row*); in the pure stand of a lodging-resistant crop (faba bean, fb), weeds (w) have favourable conditions (*second row*); semi-leafless field pea (fp) and common vetch (cv) are less-beneficial choice since different growing habit (*third row*); white lupin (wp) and common vetch (cv) are appropriate choice respecting all basic principles (*fourth row*)

Table 2 Basic chemical characteristics of the chernozem soil at Rimski Šančevi

Depth (cm)	pH KCl	pH H_2O	$CaCO_3$ (%)	Humus (%)	N total (%)	Al-P_2O_5 mg 100 g^{-1}	Al-K_2O mg 100 g^{-1}
0–30	7.41	7.90	5.61	2.97	0.196	17.99	21.00

Table 3 A long-term average of the monthly precipitation sums and monthly temperatures at Rimski Šančevi

Year/month	I	II	III	IV	V	VI	VII	VIII	IX	X	XI	XII
Precipitation (mm)												
Long term average	38.4	32.2	36.7	48.3	58.3	85.4	69.4	56.5	46.8	47.3	50.9	47.7
Temperature (°C)												
Long term average	−0.6	1.6	6.2	11.4	16.8	19.9	21.2	21.0	16.9	10.4	8.4	2.5

Table 4 Average LER values of forage dry matter yield in the intercrops of winter cool season annual legumes in the trial at Rimski Šančevi in 2008/2009

Mixture	Forage dry matter yield (supporting crop)	Forage dry matter yield (supported crop)	LER
Faba bean + forage pea	1.95	8.55	1.05
Faba bean + dual-purpose pea	3.9	3.6	1.23
Faba bean + common vetch	3.0	3.3	1.42
Faba bean + Hungarian vetch	2.7	3.3	1.06
Faba bean + hairy vetch	2.55	2.25	0.95
Protein afila pea + bitter vetch	1.65	4.2	0.91
$LSD_{0.05}$	1.2		0.334

Sad in 2008 and 2009 to study various combinations of faba bean and white lupin, as supporting crops, and forage pea, common vetch and grass pea, as supported crops (Table 1). In both combinations, there were pure stands of all three crops and the three mixing ratios, namely 75% : 25%, 50% : 50% and 25% : 75% (Table 5).

The green forage yields were at the same level with a long-term average for majority of the intercropped components, such as common vetch grass pea, in the same conditions. With the LER values higher than 1 in most of the mixtures (Table 5), the intercropping of spring cool season annual legumes, especially grass pea, proved economically reliable and deserving attention (Ćupina et al. 2009b).

3.3 *Mutual Intercropping of Warm Season Annual Legumes*

The research of the intercropping warm season annual legumes has been started in 2009 with a trial carried out in two locations in the vicinity of Novi Sad and

Table 5 Forage dry matter yields (t ha^{-1}) in the pure stands and the intercrops of spring cool season annual legumes in the trial at Rimski Šančevi in 2008 (Ćupina et al. 2010a)

Supporting crop	Supported crop	Ratio	Total forage yield	LER
Faba bean	Common vetch	25:75	4.4	0.96
		50:50	4.5	0.91
		75:25	5.1	1.04
	Forage pea	25:75	4.4	1.02
		50:50	4.4	0.93
		75:25	4.6	0.98
	Grass pea	25:75	6.7	1.27
		50:50	8.5	1.44
		75:25	6.8	1.02
White lupin	Common vetch	25:75	7.1	1.29
		50:50	6.7	1.28
		75:25	6.7	1.27
	Forage pea	25:75	6.9	1.28
		50:50	6.9	1.24
		75:25	4.9	0.95
	Grass pea	25:75	6.1	1.14
		50:50	6.0	1.04
		75:25	6.8	1.21
$LSD_{0.05}$			1.0	0.14

Belgrade. It comprised the cultivars of soybean belonging to 00 and I maturity group, white lupin and pigeon pea, as supporting crops, and mung bean and lablab bean, as supported crops (Table 1).

As may be seen in Table 6, the average LER values for forage dry matter yield varied between 1.01 in the mung bean – soybean 00 intercrop and 1.13 in the lablab bean – pigeon pea intercrop. Although preliminary, these results provide solid grounds for the possibility of intercropping warm season annual legumes for forage production in the agro-ecological conditions of Serbia (Mikić et al. 2010b).

4 Conclusion

So far, few studies have dealt with the mutual legume intercropping systems in temperate agro-ecosystems. This could be explained by a limited interest of both researchers and farmers, considering them significant to low-input farming and due to the supposed complexity of intercropping systems.

An annual legume as the companion crop can provide an economic yield during the establishment of the perennial forage crop. The superiority of intercropped

Table 6 Forage dry matter yields (t ha^{-1}) in the pure stands and the intercrops of mid- to late spring-sown annual legumes in the trial at Rimski Šančevi and Zemun Polje in 2009 (Mikić et al. 2010b)

			Pure stand		Intercropping			
Location	Supported crop	Supporting crop	Supported	Supporting	Supported	Supporting	Total	LER
Rimski Šančevi	Mung bean	Soybean 00	6.3	13.8	3.1	7.0	10.1	1.01
		White lupin		7.6	3.6	4.2	7.8	1.12
	Lablab bean	Soybean I	6.7	13.9	3.6	7.3	10.9	1.06
		Pigeon pea		3.9	4.1	1.8	5.9	1.07
Zemun Polje	Mung bean	Soybean 00	9.4	7.7	4.2	4.3	8.5	1.01
		White lupin		3.7	6.1	1.5	7.6	1.06
	Lablab bean	Soybean I	10.5	10.4	6.3	4.6	10.9	1.05
		Pigeon pea		4.8	7.3	2.1	9.3	1.13
Average	Mung bean	Soybean 00	7.9	10.7	3.6	5.7	9.3	1.01
		White lupin		5.7	4.8	2.9	7.7	1.09
	Lablab bean	Soybean I	8.6	12.2	5.0	5.9	10.9	1.05
		Pigeon pea		4.4	5.7	2.0	7.7	1.10
$LSD_{0.05}$			1.4					0.05

plants over pure stands has been attributed generally to the significance variations of morphological characteristics resulting in efficient exploitation of environmental resources. In addition, field pea companion crop contributes to improving forage quality and digestibility. The plants provided with better conditions had higher LAI and due to an appropriate amount of light do not have to increase their pigment contents, while the shaded plants of the undersown crop produce more chlorophyll, have more intensive photosynthesis and use the nutrients in a more economic way.

The first results of the mutual intercropping of annual legumes prove its reliability for forage production in the temperate regions such as West Balkan Countries and Serbia as one of their representatives. The future research needs to answer many details of the other aspects, such as:

- Should there be a limit to two components only?
- What is the impact of such intercrops on soil properties? Could positive changes in nitrogen content be expected? Could they be risky for nitrate leaching, especially in nitrate-sensitive areas?
- Will there be a difference in forage yield components, forage quality and stress resistance between the legume crops in pure stands and in their intercrops?

Re-introducing intercropping in European agriculture to a greater extent should not be reversion to an abandoned methodology, but rather a consideration of the usefulness of this old and sustainable cropping practice in a modern, innovative and technology-oriented low input and organic agriculture.

References

Abu-Gyamfi JJ, Myaka FA, Sakala WD, Odgaard R, Vesterager JM, Høgh-Jensen H (2007) Biological nitrogen fixation and nitrogen and phosphorus budgets in farmer-managed intercrops of maize–pigeonpea in semi-arid southern and eastern Africa. Plant Soil 295:127–136

Altieri M (1999) The ecological role of biodiversity in agroecosystems. Agric Ecosyst Environ 74:19–31

Anil L, Park J, Phipps RH, Miller FA (1998) Temperate intercropping of cereals for forage: review of the potential for growth and utilisation with particular reference to the UK. Grass Forage Sci 53:301–317

Arora DK, Gupta S (1996) Advances in plant morphology and anatomy, vol 5, Morphology and anatomy of leaf. Anmol Publications Ptv Ltd, New Delhi

Avola G, Tuttbene R, Gresta F, Abbate V (2008) Weed control strategies for grain legumes. Agron Sustain Dev 28:389–395

Bedoussac L, Justes E (2010a) The efficiency of a durum wheat-winter pea intercrop to improve yield and wheat grain protein concentration depends on N availability during early growth. Plant Soil 330:19–35

Bedoussac L, Justes E (2010b) Dynamic analysis of competition and complementarity for light and N use to understand the yield and the protein content of a durum wheat–winter pea intercrop. Plant Soil 330:37–54

Biabani A, Hashemi M, Herbert SJ (2008) Agronomic performance of two intercropped soybean cultivars. Int J Plant Prod 2(3):215–221

Chapko LB, Brinkman MA, Albrecht KA (1991) Oat, oat-pea, barley, and barley-pea for forage yield forage quality and alfalfa establishment. J Prod Agric 4:360–365

Corre-Hellou G, Fustec J, Crozat Y (2006) Interspecific competition for soil N and its interaction with N-2 fixation, leaf expansion and crop growth in pea-barley intercrops. Plant Soil 282:195–208

Cu STT, Hutson J, Schuller KA (2005) Mixed culture of wheat (*Triticum aestivum* L.) with white lupin (*Lupinus albus* L.) improves the growth and phosphorus nutrition of the wheat. Plant Soil 272:143–151

Ćupina B, Mihailović V, Erić P (2000) Management practices for yield and quality of field pea. Period Sci Res Field Veg Crops 33:91–102

Ćupina B, Erić P, Mihailović V, Mikić V (2004) The role and importance of cover crops in sustainable agriculture. Period Sci Res Field Vëg Crops 40:419–430

Ćupina B, Erić P, Krstić Đ, Mihailoić V, Mikić A, Vučković S (2006) Feed peas: a cover crop in establishing red clover. Bibl Fragm Agron 11:521–522

Ćupina B, Mikić A, Krstić Đ (2009a) Field pea: a cover crop in establishing perennial legumes. Grain Legum 52:9

Ćupina B, Mikić A, Mihailović V, Krstić Đ, Đuric B (2009b) Intercropping of grass pea (*Lathyrus sativus*) with other grain legumes for forage production. Grain Legum 54:32

Ćupina B, Mikić A, Krstić Đ, Antanasović S, Erić P (2010a) Agronomic performance of intercropped annual legumes. Book of abstracts, 5th international food legumes research conference (OFLRC V) and 7th European conference on grain legumes (AEP VII), Antalya, 26–30 Apr 2010, p 295

Ćupina B, Krstić Đ, Mikić A, Erić P, Vučković S, Pejić B (2010b) Field pea companion crop management on red clover establishment and productivity. Turk J Agric For 34:275–283

Dennis WD, Woledge J (1982) Photosynthesis by white clover leaves in mixed clover/ryegrass swards. Ann Bot 49:627–635

Dickison WC (2000) Integrative plant anatomy. Academic, San Diego

Faulkner JS (1985) A comparison of faba beans and peas as whole-crop forages. Grass Forage Sci 40:161–169

Fustec J, Lesuffleur F, Mahieu S, Cliquet JB (2010) N rhizodeposition of legumes. A review. Agron Sustain Dev 30:57–66

Ghosh PK, Mohanty M, Bandyopadhyay KK, Painuli DK, Misra AK (2006) Growth, competition, yield advantage and economics in soybean/pigeonpea intercropping system in semi-arid tropics of India: I. Effect of subsoiling. Field Crops Res 96:80–89

Gilliland TJ, Johnston J (1992) Barley/pea mixtures as cover crops for grass reseed. Grass Forage Sci 47:1–7

Hauggaard-Nielsen H, Jensen ES (2005) Facilitative root interactions in intercrops. Plant Soil 274:237–250

Hauggaard-Nielsen H, Ambusa P, Jensen ES (2001) Interspecific competition, N use and interference with weeds in pea-barley intercropping. Field Crops Res 70:101–109

Heath MC, Hebblethwaite PD (1985) Solar radiation interception by leafless, semi-leafless and leafed peas (*Pisum sativum* L.) under contrasting field conditions. Ann Appl Biol 7:309–318

Heichel GH, Delaney RH, Cralle HT (1988) Carbon asimilation, partioning and utilization. In: Hanson AA, Barnes DK, Hill RR (eds) Alfalfa and alfalfa improvement. Medison, Visconsin, pp 195–222

Horrocks RD, Vallentine JF (1999) Establishment of forage species. In: Horrocks RD, Vallentine JF (eds) Harvested forages. Academic, London, pp 135–154

Joanne R, Thiessen M, Hoeppner JW, Entz MH (2001) Legume cover crops with winter cereals in southern Manitoba: establishment, productivity, and microclimate effects. Agron J 93:1086–1096

Keating BA, Carberry PS (1993) Resource capture and use in intercropping – Solar radiation. Field Crops Res 34:273–301

Klesnil J (1980) Contribution to the problem of establishing of clover crop stands. Rostl Vyr 26:6–10

Koivisto JM (2002) Semi leafless pea: a cover crop for establishing lucerne or red clover. PhD thesis, Royal Agricultural College & Coventry University, Cirencester

Krstić, Ćupina B, Erić P, Mihailović V (2005a) Fodder pea as companion crop in alfalfa establishment-Effect on alfalfa morphological and physiological parameters. XXXV ESNA annual meeting, Amiens, pp 40–41

Krstić Đ, Ćupina B, Erić P, Mihailović V (2005b) Fodder pea as companion crop in red clover establishment – effects on red clover morphological and physiological parameters. Book of abstracts, XVI symposium of Society of Plant Physiology of SCG, Bajina Bašta, 13–16 Jun 2005, p 11

Lanini WT, Orlof SB, Vargas RN, Orr JP, Marable VL, Grattan SR (1991) Oat companion crop seeding rate effect on alfalfa establishment, yield and weed control. Agron J 83:330–333

Li Y, Ran W, Zhang R, Sun S, Xu G (2009) Facilitated legume nodulation, phosphate uptake and nitrogen transfer by arbuscular inoculation in an upland rice and mung bean intercropping system. Plant Soil 315:285–296

Malézieux E, Crozat Y, Dupraz C, Laurans M, Makowski D, Ozier-Lafontaine H (2009) Mixing plant species in cropping systems: concepts, tools and models. A review. Agron Sustain Dev 29:43–62. doi:10.1051/agro:2007057

Mariotti M, Masoni A, Ercoli L, Arduini I (2009) Above- and below-ground competition between barley, wheat, lupin and vetch in a cereal and legume intercropping system. Grass Forage Sci 64:401–412

Matejkova O (1982) Limited ways of the soil cultivation to summer sowing of alfalfa. Rostl Vyr 28:233–248

Mikić A, Angelova S, Vishnyakova M, Đurić B, Burstin J, Marget P, Delaitre C, Lejeune I, Sabeva M, Bulyntsev S, Ćupina B, Vasić M, Mihailović V, Krstić Đ, Duc G (2010a) Project ECO-NET 18817: In search for new genetic variability in Pea and Faba Bean to support the development of the sustainable agriculture. Book of abstracts, 5th International Food Legumes research Conference (IFLRC V) and 7th European conference on grain legumes (AEP VII), Antalya, 26–30 Apr 2010, p 295

Mikić A, Đorđević V, Peric V, Ćupina B, Mihailović V, Srebrić M, Krstić Đ (2010b) Preliminary report on forage yields in mid- to late spring-sown annual legume intercrops. Biotechnol Anim Husb 26(special issue), 2:269–275

Nickel SE, Simmons SR, Sheaffer CC, Radosevich SR (1990) Addition series approach to assessing competition in a small grain-alfalfa companion crop community. Crop Sci 30:1139–1141

Obračević Č (1990) Tablice hranjivih vrednosti stočnih hraniva i normativa u ishrani preživara [Tables of feed nutritional value and normatives in ruminants nutrition]. Naučna knjiga, Belgrade

Roy AK, Sikdar B, Rahman M, Rahman MA, Haque ME (2003) Lablab bean based intercropping system in northwest region of Bangladesh. Pak J Biol Sci 6:948–951

Simmons SR, Craig S, Sheffer DC, Rasmusson DD, Stuthman S, Nickel E (1995) Alfalfa establishment with barley and oat companion crops differing in stature. Agron J 87:268–272

Smith LW, Goering HK, Gordon CH (1972) Relationships of forage compositions with rates of cell wall digestion and indigestibility of cell walls. J Dairy Sci 55:1140–1147

Sturgul SJ, Daniel TC, Mueller DH (1990) Tillage and canopy cover effects on iterrill erosion from first-year alfalfa. Soil Sci Am J 54:1733–1739

Sule RM (1993) Ryegrass companion crops for alfalfa establishment: I Forage yield and alfalfa suppression. Agron J 85:67–74

Tan M, Serin Y, Erkovan HI (2004) Effect of barley as companian crop on the hay yield and plant density of red clover and botanical composition of hay. Turk J Agric For 28:35–41

Temperton VM, Mwangi PN, Scherer-Lorenzen M, Schmid B, Buchmann N (2007) Positive interactions between nitrogen-fixing legumes and four different neighbouring species in a biodiversity experiment. Oecologia 151:190–205

Tesar MB, Marble VL (1988) Alfalfa establishment. In: Hanson AA, Barnes DK, Hill RR (eds) Alfalfa and alfalfa improvement. ASA, CSSA, SSSA, Madison, pp 303–332

Thorsted MD, Søegaard K, Koefoed N (2002) Yield and quality of oat/white clover intercrops. Grassl Sci Eur 7:94–95

Trenbath BR (1993) Intercropping for the management of pests and diseases. Field Crops Res 34:381–405

Tsialtas JT, Maslaris N (2008) Seasonal trends and relationships of light, temperature and leaf physiological traits of sugar beets (*Beta vulgaris* L.) grown under semi-arid, Mediterranean conditions. Int J Plant Prod 2(3):223–242

Vandermeer J, Van Noordwijk M, Anderson J, Ong C, Perfecto I (1998) Global change and multi-species ecosystems: concepts and issues. Agric Ecosyst Environ 67:1–22

Vough LR, Decker AM, Taylor TH (1995) Forage establishment and renovation. In: Barnes RF, Miller DA, Nelson J (eds) The science of grassland agriculture. Iowa State University Press, Ames, pp 29–43

Wiersma DW, Hoffman PC, Mlynarek MJ (1999) Companion crops for legume establishment: forage yield, quality, and establishment success. J Prod Agric 12:116–122

Willey R (1979) Intercropping – its importance and research needs. 1. Competition and yield advantages. Field Crop Abstr 32:1–10

Zarea MJ, Ghalavand A, Jamshidi E (2008) Role of forage legumes mixed cropping on biomass yield and bacterial community composition. Proceedings of the second scientific conference of the International Society of Organic Agriculture Research, Modena, 18–20 Jun 2008, pp 180–183

Zemenchik RA, Albrecht KA, Boerboom CM, Lauer J (2000) Corn production with kura clover as a living mulch. Agron J 92:698–705

Zollinger RK, Meyer DV (1996) Imazethapyr for weed control in alfalfa establishment. Weed Sci 49:19–21

Alternative Farming Techniques for Sustainable Food Production

K. Padmavathy and G. Poyyamoli

Abstract Sustainability and food security are the major challenges faced by third world countries for the past several decades. Most of the third world countries are also facing problems of climate change, increasing population, overexploitation of natural resources and resource degradation associated with rapid economic growth. Among the scientific and policy circles there are controversies in using inorganic chemicals and biotechnology for sustaining the agricultural production. There is no critical comprehensive review on sustainability of alternative farming systems and their relative advantages over conventional, chemicalized and hi-tech agriculture for decision making at various levels. This review tries to fulfill the knowledge gap in this vital sector.

The first part of the review discuss the current status of agroecosystems, with emphasis on their threats in terms of food security, long term sustainability, impacts on ecosystem services and climate change. We also evaluate the ecological, economic, social and cultural sustainability of inorganic agriculture. This analysis points emerging issues such as environmental degradation, loss of ecosystem services, non-sustainability and threats to food security in the context of global population growth and climate change. Hence there is an urgent need for identifying potential alternative farming strategies to achieve long term sustainability and food security as indicated by several leading workers in the field. The next section traces the background and evolution of alternative farming systems with their scope and importance. Then we classified potential sustainable farming techniques practiced in various parts of the world. For that we review potentials, constraints, strategies and case studies for ten alternatives farming techniques and four innovative endogenous farming techniques from India.

K. Padmavathy · G. Poyyamoli (✉)
Department of Ecology and Environmental Sciences, Pondicherry University, Puducherry 605014, India
e-mail: ecopadma@gmail.com; gpoyya9@gmail.com

E. Lichtfouse (ed.), *Genetics, Biofuels and Local Farming System*, Sustainable Agriculture Reviews 7, DOI 10.1007/978-94-007-1521-9_13,

The alternative farming techniques that were field tested and perfected over several generations in the past portrayed the following advantages over chemical farming: (1) eco-friendly by protecting and revving life support systems and ecosystem services, (2) higher cost benefit ratio, benefiting the farmers as well as the consumers, (3) control and reduction of bioaccumulation and biomagnification, (4) reduction in air, water and soil pollution caused by various pesticides and other chemicals, (5) control of health hazards in humans and livestock, and (6) conservation and sustainable use of on-farm biodiversity, including traditional cultivated germplasm and natural resources in agrosystems.

Keywords Alternative farming techniques • Bio dynamic farming • Ecofarming • HOMA • India • LEISA • Integrated farming systems • No tillage farming • Natural farming • Organic farming • Permaculture • Polyculture • Sustainability • Urban and Peri-Urban farming • Zero budget natural farming

Abbreviations

BD	Bio dynamic farming
CBO	Community based organizations
CSD	Convention on biological diversity
CT	Conventional tillage
EM	Effective Microorganisms
FAO	Food and Agriculture organization
FAOSTAT	Food and Agriculture Organization statistics
GIAHS	Globally Important Agriculture Heritage Systems
GM	Genetically modified
GRA	Green Revolution Agriculture
IAASTD	International Assessment of Agricultural Knowledge Science and Technology for Development
ICCA	Indigenous Community Conserved Areas
IEA	International Energy Agency
IFOAM	International Federation of Organic Agriculture Movements
IFS	Integrated Farming Systems
IPCC	Intergovernmental Panel on Climate Change
IUCN	International Union for Conservation of Nature
LEISA	Low External Input Sustainable Agriculture
NF	Natural Farming
NGO	Non Governmental organizations
NT	No tillage
USAID	United States Agency for International Development
USDA	United States Department of Agriculture
WBCSD	World Business Council for Sustainable Development

1 Introduction

Agriculture has been the basic source of subsistence for human societies, over thousands of years and it provides a livelihood to half of the world's population, even today. Food and Agricultural Organization, 2007 reports that the quantity of food produced per capita has been declining since 1984 as the population is increasing in alarming rate. In 1960 when the world population numbered only 3 billion, approximately 0.5 ha of cropland per capita was available and considered as the minimum area essential for the production of a diverse, healthy, nutritious diet of plant and animal products (Keyzer et al. 2005; Southgate 2009). Globally available per-capita cropland is now about 0.23 ha (FAO 2007; LaSalle et al. 2008). For the world as a whole, per-hectare output of cereals, which account for more than half the food people eat if the grain fed to livestock is factored in, had risen by the late 1990s to 3.0 metric tons, which was double the average yield in the early 1960s (Southgate et al. 2007; Southgate 2009).

The global population is predicted to increase to 9.5 billion people in the year 2050 (U.S. Census Bureau 2008). Total food requirements will increase by 100% (Tilman et al. 2002) as a function of both the 50% increase in population and the additional global demand for animal protein as people in developing countries become more affluent (Keyzer et al. 2005). The ever-growing population rates coupled with the decreasing per capita availability of natural resources, the associated negative impacts on the environment and the consequent unplanned developmental activities have stretched the resilience of the natural resources and socioeconomic systems to a level of catastrophe. Thus compulsion to produce more has further compounded the problems leading to non sustainability of the agricultural production system all over the world in general, and the developing countries in particular, necessitating a paradigm shift towards a holistic agroecosystem management in an integrated manner for development and renewal of eco-friendly sustainable agricultural technologies (Kiers et al. 2008).

1.1 Issues and Challenges in Agriculture

Currently global agriculture faces several critical issues in terms of food security, climate change, ecosystem degradation, biodiversity loss and various effects on ecosystem services. Hence, an in depth analysis of each of these issues is warranted.

1.1.1 Food Security

Household and national food security are complex and complicated goals influenced by many factors such as technologies, human capacities, policies, prices, trade and infrastructural context. Demand for food is certain to increase with increasing population pressure and income, even though this demand and ability to supply the demand are not equal in all communities. Indeed, today's total global agricultural production is sufficient to feed the current world population and both necessary technologies and multilateral environmental agreements are available to help meet development and conservation needs (FAO 2007).

However, hunger, poverty and environmental degradation persist even as concerns about global human food security issues continue to increase. Moreover, the last decades provide uncompromising evidence of diminishing returns on grains despite the rapid increases of chemical pesticide and fertilizer applications (Sanders 2006) questioning the sustainability of these high input technologies for equitable household and national food security in the next decades. Overall global cereal output is declining, mainly among the major producing and exporting countries, despite the recent advances in high–tech agriculture (FAO 2007; Pretty 2008; Robinson 2008).

The projections by the Food and Agriculture Organization for the period of 1999 to 2030 estimate an increase of global agricultural production by 56% with arable land expansion accounting for 21% of production growth in developing countries. For this same period, the share of irrigated production in developing countries is projected to increase from 40 to 47% (FAO 2006). Arable land expansion and large-scale irrigation may be a cause of concern for the loss of ecosystem services. Although the number of undernourished people will decline i.e. from more than 850 million at present to about 300 million by 2050, high rates of poverty and food insecurity are expected to continue with the present models of food production and consumption, along with further natural resource degradation (Lee 2005; Lal 2008; FAO 2008).

Worldwide undernourishment is not explained only by a lack of food availability as several causes of hardship lie outside the agricultural sector. However, there is need to seek new solutions to address the problems posed by growing populations and environmental degradation due to inorganic agriculture through new paradigms for sustainable agriculture and food supply chains. The multidimensional nature of food security includes food availability, food access, stability and utilization (Bradford and Wichner 2009; Dreyfus and McIntyre 2009). **Food availability** refers to having sufficient quantities of food of appropriate quality, supplied through domestic production or inputs, food aid and net imports. It comprises of global supply, yields, energy use, nutrient inputs, urban food supply and food import. The **food access** dimension of food security refers to the access, by individuals, to adequate resources and entitlements for acquiring appropriate foods for a nutritious diet. It comprises of agricultural inputs, farming viability, knowledge, quality of the market and multifunctional farms. **Food stability** is to food secure a population, household or individual must have access to adequate food at all times. They should not risk losing access to food as a consequence of sudden shocks e.g. an economic or climatic crisis or cyclical events e.g. seasonal food insecurity. It refers to resilience, soil stability, water use efficiency, agrobiodiversity, risk mitigation and climate change. **Food Utilization** aspect of food security refers to ways in which food contributes to an adequate diet, clean water, sanitation and health care, and in turn, to a state of nutritional well-being where all physiological needs are met. It includes diversifying diets, quality, nutrition, safety, health, water quality, handling and biosecurity (FAO 2007; Dreyfus and McIntyre 2009)

1.1.2 Climate Change and Agriculture

Models predict that average global temperatures will increase between 1.8° and 4.0°C at the end of the twenty-first century (IPCC 2007) and will be accompanied by random alterations i.e. both temporal and geographic, in precipitation patterns (Mendelsohn and Williams 2006). Agriculture is a key sector of global greenhouse gas emissions (14% or 6.8 Gt of CO_2eq) but with a high technical mitigation potential (5.5–6 Gt of CO_2eq. per year by 2030). Seventy-four percent of emissions from agriculture are in developing countries (Von Koerber and Kretschmer 2006). Agriculture is a sector where mitigation action has strong potential co-benefits for sustainable development i.e. food security, poverty reduction among the 70% of the poor living in rural areas, environmental services and climate change adaptation i.e. improving agro-ecosystem resilience. Most of the mitigation potential from agriculture could be achieved through soil carbon sequestration (89%) and roughly 70% could be realized in developing countries (Fischer et al. 2006a and b; FAO 2009).

Agriculture is extremely vulnerable to climate change. Higher temperatures eventually reduce yields of desirable crops while encouraging weed and pest proliferation. Changes in precipitation patterns increase the likelihood of short-run crop failures and long-run production declines. Although there will be gains in some crops in some regions of the world, the overall impacts of climate change on agriculture are expected to be negative, threatening global food security (Mendelsohn and Williams 2006). Populations in the developing world, which are already vulnerable and food insecure, are likely to be the most seriously affected. In 2005, nearly half of the economically active population in developing countries –2.5 billion people relied on agriculture for its livelihood and 75% of the world's poor live in rural areas. The following are the predicted impacts of climate change in agriculture and human well-being (Haas and Köpke 1994; FAO 2007; Hassan 2010).

- In developing countries, climate change will cause yield declines for the most important crops; particularly South Asia will be affected hardly
- Climate change will have varying effects on irrigated yields across regions, but irrigated yields for all crops in South Asia will experience large declines
- Climate change will result in additional price increases for the most important agricultural crops–rice, wheat, maize, and soybeans. Higher feed prices will result in higher meat prices. As a result, climate change will reduce the growth in meat consumption slightly and cause a more substantial fall in cereals consumption.

- Calorie availability in 2050 will not only be lower than in the no–climate-change scenario – it will actually decline relative to 2000 levels throughout the developing world
- By 2050, the decline in calorie availability will increase child malnutrition by 20% relative to a world with no climate change. Climate change will eliminate much of the improvement in child malnourishment levels that would occur with no climate change

Thus, aggressive agricultural productivity investments of US$7.1–7.3 billion are needed to raise calorie consumption enough to offset the negative impacts of climate change on the health and well-being of children.

1.1.3 Ecosystem Degradation and Biodiversity Loss in Agroecosystems

Humans have converted an estimated 38.2% or 4973 million ha of the Earth's land surface area to agriculture i.e. temporary or permanent crops and pastures, at the expense of natural habitat and this figure is forecast to reach 60% in the next 100 years (FAOSTAT 2008). The ecosystem services and bio diversity are severely affected by chemical intensive agricultural and high tech practices. There are several studies to prove how the farming practices influence the species richness and abundance of taxa in agro ecosystems (Fuller et al. 2005; Firbank et al. 2007; Araujo et al. 2009) about the threats posed by agricultural change to biodiversity (Tilman et al. 2001; Firbank et al. 2007) how farming practices can be modified to mitigate these threats and generate benefits (Godwin et al. 2003; Araujo et al. 2009). There are several reports indicating that biodiversity changes are due to agricultural changes and its intensification (Chamberlain et al. 2000; Cooke et al. 2009). The human appropriation of terrestrial net primary production is the ideal indicator of the agricultural intensification on biodiversity. Currently, it is 30% and can reach to 72%, as a result of population and economic growth (Imhoff et al. 2004; Firbank et al. 2007; Hobbs et al. 2008).

Large-scale agriculture leads to ecosystem simplification and loss of biodiversity, thus reducing the potential to provide ecosystem services other than food production. Of some 270,000 known species of higher plants, about 10,000–15,000 are edible and only about 7,000 are used in agriculture (UNEP/GRID-Arendal 2009) (http://www.grida.no/publications/rr/food-crisis/page/3569.aspx).

However globalization, and agricultural intensification have diminished the varieties traditionally used, with only 30% of the available crop varieties dominating global agriculture. These, together with only 14 animal species, provide an estimated 90% of the worlds consumed calories (FAO 2008). Globally, over 1,000 (87%) of a total of 1,226 threatened bird species are impacted by agriculture. More than 70 species are affected by agricultural pollution, 27 of them seriously. Pesticides and herbicides pose a threat to 37 threatened bird species globally, in addition to deleterious effects of agricultural chemicals on ground water (BirdLife 2008; Hazell and Wood 2008). Such evidences indicate that small scale, diversified farming systems are preferable in terms of food security and sustainability. Even for farming systems in the developed countries where farms are larger in size, it would be worthwhile, if some comparative studies are made on the pros and cons of agro ecology. Considering the higher cattle population and the consequent problem of excess dung and urine production, they can explore the possibilities for recycling these "wastes" more usefully at the farm and the collective levels.

1.1.4 Impacts on Ecosystem Services

Agriculture produces more than just crops. Agricultural practices have environmental impacts that affect a wide range of ecosystem services, including the maintenance of water cycle and water quality, pollination, nutrient cycling, soil retention, carbon sequestration, and biodiversity conservation. In turn, the impacted ecosystem services affect agricultural productivity. Understanding the contribution of various agricultural practices to the range of ecosystem services would help inform choices about the most beneficial agricultural practices. Measuring the impact of alternative agricultural practices on ecosystem services and of ecosystem services on agricultural production is a big challenge (Byerlee and Alex 2005; Badgley et al. 2007).

The following are reported as the major impacts of agricultural practices on Ecosystem services (Tilman et al. 2002; Dalea and Polasky 2007; Sandhua et al. 2008):

- The use of resources such as food, water, and timber has increased rapidly continues to grow, sometimes unsustainably
- Fossil fuel use has intensified, thus leading to greater green house gas emission
- Human interventions have led to changes in the regulation of climate, disease, and other ecosystem processes
- Loss of agrobiodiversity especially the pollinators and decomposers is accelerated, due to the indiscriminate use of agro chemicals
- The use of ecosystems for recreation, spiritual enrichment, and other cultural purposes is growing. However the capacity of ecosystems to provide these services has declined significantly.

A doubling of global food production would have major impacts on the ability of nonagricultural ecosystems to provide services (Daily 1997) vital to humanity. Existing nonagricultural ecosystems provide, at no cost, pure, drinkable water. In contrast, the groundwater associated with intensive agricultural ecosystems often contains sufficiently high concentrations of nitrite and nitrates or of pesticides and their residues as to be unfit for human consumption. Expensive treatment is required to make it potable. The biodiversity of non agroecosystems provides many services to agriculture. For instance, the genetic diversity of both wild relatives of crop plants and unrelated organisms is used to increase yields and to reduce impacts of agricultural pests and pathogens. However the maintenance of the wild biodiversity needed for future development of crops and medicines occurs mainly in nonagricultural ecosystems and these ecosystems are threatened by agricultural expansion and nutrient release as well as pesticide accumulation (Kremen and Ricketts 2000; Scialabba and Williamson 2004).

Agriculture depends on soil fertility and the natural fertility created by the ecosystem services are destroyed when lands are converted to intensive chemical agriculture. Especially on sandy soils, the best way to regain soil fertility lost because of tilling is to allow re-establishment of the native ecosystems. Many agricultural crops depend on the pollination services provided by insects, birds, or mammals that live in nearby nonagricultural ecosystems (Dalea and Polasky 2007; Carpenter et al. 2009). Similarly, agricultural crops benefit from biocontrol agents, such as parasitic and predatory insects, birds, bats that live in neighboring nonagricultural ecosystems and that decrease outbreaks of agricultural pests (Mertz et al. 2007; Badji et al. 2007).

A brief overview on ecosystem services by Daily (1997) demonstrates that society and agriculture depend on many services provided by nonagricultural ecosystems. Although it is difficult to establish economic values for such services (Costanza et al. 1997), it is clear that, when possible, technological substitutes for lost ecosystem services can be extremely expensive. This highlights the need for public policy to consider the short-term and long-term costs of actions that decrease the ability of nonagricultural ecosystems to provide vital ecosystem services to society.

Current challenges in agriculture sector in developing countries include (FAO 2006, 2007):

- Increasing population and economic growth demanding more food for consumption and decreasing availability of cultivable land rural
- Depopulation (world's urban population exceeded rural population in 2006), posing questions on availability of food
- Degradation of soil quality and water quality, loss of bio diversity and disruption of ecosystem services as a direct consequence of chemical intensive agriculture
- Introduction of hybrid mono cultures replacing the traditional land races and crop genetic diversity
- Decline in pollinators and biological control agents as a result of extensive use of pesticides
- Current social and economic inequities and political uncertainties about war and conflicts
- Uncertainties about the ability to sustainably produce and access sufficient food
- Uncertainties about the future of world food prices
- Changes in the economics of fossil based energy use
- The emergence of new competitors for natural resources
- Increasing chronic diseases that are partially a consequence of poor nutrition and poor food quality as well as food safety
- Changing environmental conditions and the growing awareness of human responsibility for the maintenance of global ecosystem services (provisioning, regulating, cultural and supporting).

1.1.5 Energy Use in Agriculture

Agriculture is a modest user of energy relative to other economic sectors accounting for an estimated 3.5% of commercial energy use in developed countries and 4.5% in developing countries. Energy demand is projected to increase by 45% between 2006 and 2030, based on the IEA's (2008) reference scenario (Fig. 1). Biofuels for transport and biomass for heat and electricity will be used to meet some of this demand, leading to greater competition for land and crops between energy and food markets (Mitchell 2008). In addition, the production of the mineral fertilizers on which modern, intensive agriculture has come to rely to replenish nutrient-depleted soils requires significant energy. Up to fivefold increase in fertilizer prices between 2005 and 2008 was strongly influenced by the soaring oil price during this period (Piesse and Thirtle 2009), alongside production capacity constraints linked to the availability of phosphates, potash and other mineral ores.

In developing countries, fertilizers require the highest commercial energy inputs, followed by machinery and irrigation. The Agriculture: Toward 2000 study (FAO 1981) assumed that significant increases in commercial energy application will be needed to boost agricultural yields and farm earnings. Its two scenarios for the year 2000 project an average increase of 7.5% per year in commercial energy use in agriculture in 90 developing countries, resulting in more than a quadrupling of energy use between 1980 and 2000. Growth rates in fertilizer consumption in the early 1980s were somewhat below this rate of increase, averaging 6.2% per year between 1981–1982 and 1985–1986. Given the high costs of importing fertilizers and fossil fuels, it seems likely that developing countries may increasingly turn to alternatives such as biogas and animal power. Biogas consists mostly of methane and hydrogen gases produced by anaerobic fermentation of crop and animal wastes. Usable nitrogen, phosphorus, and potassium are the byproducts (Mitchell 2008; Piesse and Thirtle 2009; Beddington 2010).

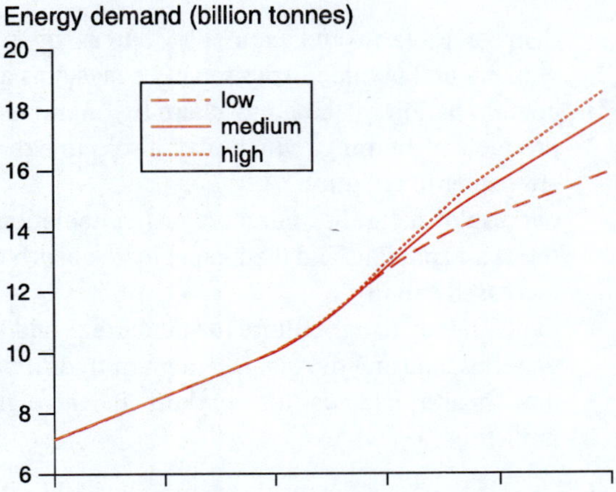

Fig. 1 Global energy demand in billion tones oil equivalent (1980–2030, from IEA 2008) (Beddington 2010)

1.2 Green Revolution Agriculture (GRA)

Green Revolution is the transformation of agriculture that began in 1945. Mexican government established an agricultural research station to develop more varieties of wheat that could be used to feed the rapidly growing population of the country. In 1943, Mexico imported half its wheat but by 1956 the Green Revolution had made Mexico self-sufficient and by 1964 Mexico exported half a million tons of wheat. The associated transformation has continued as the result of programs of agricultural research, extension, and infrastructural development. These programs were instigated and largely funded by the Rockefeller Foundation, along with the Ford Foundation and among other major agencies. The term "Green Revolution" was first used in 1968 by former USAID director William Gaud. The main objective of GRA is to increase production to feed the growing population by new hybrid varieties for after varieties, more pesticides, fertilizers and mechanization (Ruttan 1977; Chapman 2002; Ponting 2007). GRA was introduced and transferred to all the countries with a wrong conviction that organic farming alone will not be able to meet the growing global population and their needs (LaSalle et al. 2008; Gianessi 2009). This has unwittingly culminated in a myriad of negative impacts to the environment and to the society.

1.2.1 Negative Impacts of Chemical Farming

The adverse environmental and social impacts of modern agriculture are summarized by several workers (for instance - Lichtenberg 1992; Pretty 1995; Svotwa et al. 2008; LaSalle et al. 2008; Ericksen et al. 2009) as:

- contamination of water by pesticides, nitrates, soil and livestock wastes, causing harm to wildlife, disruption of ecosystems and possible health problems in drinking water
- contamination of food and fodder by residues of pesticides, nitrates and antibiotics
- damage to farm and natural resources by pesticides, causing harm to farm workers and public, disruption of ecosystems and harm to wildlife
- contamination of the atmosphere by ammonia, nitrous oxide, methane and the products of burning, which play a role in ozone depletion, global warming and atmospheric pollution
- overuse of natural resources, causing depletion of groundwater and loss of wild foods and habitats and their capacity to absorb wastes, causing water-logging and increased salinity
- the tendency in agriculture to standardize and specialize by focusing on modern varieties, causing the displacement of traditional varieties and breeds
- new health hazards for workers in the agrochemical and food-processing industries

GRA is not cost-efficient. Industrial agricultural methods can reduce labor costs by substituting herbicides, insecticides, synthetically-produced fertilizers as well as farm machinery for application and crop maintenance. However, the energy costs, health costs of the environment and people are much lower in alternative farming systems than the intensive chemical farming (Liebhardt 2001; Baker et al. 2002; Delate et al. 2003; Macgregao and Hill 2006). A study of Rodale Institute's Farming Systems Trial from 1981 to 2002 shows that fossil energy inputs for organic corn production were about 30% lower than for conventionally produced corn (Pimentel et al. 2005; Pimentel 2006). GRA practices have been shown to reduce returns by 25–30% in few 5 years by increasing the amounts of costly and toxic artificial inputs. Unfortunately, once started, the intensive chemical system of farming is very difficult to escape the vicious cycle of non sustainability (USDA 2003). The negative and serious consequences of the Green Revolution led researchers to question the industrial farming as a viable approach to address problems of soaring food prices, hunger, social injustice and environmental degradation in the developing world (IAASTD 2008).

Subsequently, the negative impacts of chemical farming and its un-sustainability motivated the innovative farmers all over the world to rediscover the traditional self reliant and ecofriendly alternative sustainable farming techniques (Bockstaller et al. 1997; Dore et al. 2007). However, they are mostly regionalized and localized and poorly documented in scientific journals (Scialabba 2007; Svotwa et al. 2008; Babou et al. 2009; Bhaskaran et al. 2009).

1.3 Evolution of Alternative Farming Strategies

The interest in the sustainability of agricultural and food systems can be traced to environmental concerns that began to appear in the 1950s–1960s. However, ideas about sustainability date back at least to the oldest surviving writings from India, China, Greece and Rome (Pretty 2006; Dhama et al. 2005; Sofia et al. 2006; Balasubramanian et al. 2009).

Today there is an urgent need of sustainable agricultural technologies and practices that (1) do not have adverse effects on the environment i.e. partly because the environment is an important asset for farming (2) are accessible to and effective for farmers (3) lead to both improvements in food productivity and have positive side effects on environmental goods and services. Sustainability in agricultural systems incorporates concepts of both resilience i.e. the capacity of systems to buffer shocks and stresses, and persistence i.e. the capacity of systems to continue over long periods. This culminates in many wider economic, social and environmental outcomes (Pretty 2008; Bradford and Wichner 2009).

1.4 Scope of Alternative Farming Practices

The majority of sustainable alternatives farming techniques indicate a return of time tested traditional and eco-friendly practices. The ultimate goals of sustainable agriculture stated by several researches like Vandermeer (1998), Xu et al. (2000), Greene and Kremen (2003) are: to

- maintain or improve the natural resource base
- protect the environment
- ensure profitability to the farmers
- conserve energy
- increase the farm productivity
- improve food quality, security and safety and
- create more viable and vibrant socio-economic infrastructure for farms and rural communities

As Green revolution agriculture is detrimental to the environment as well as the local communities, there is an urgent need to find more sustainable farming methods (Mader et al. 2002; Thamaga-Chitja and Sheryl 2008).

1.5 Urgent Need for Adopting Sustainable Alternative Farming Techniques

The Green Revolution agriculture aimed at stimulating agriculture production primarily by replacing traditional varieties of crops by high response varieties-the hybrids. Green revolution agriculture has vast and adverse impacts on earth. During the late 1970s efforts were initiated to bring the traditional practices into the realm of modern agricultural science (Bene et al. 1977; LaSalle et al. 2008). These initiatives arose from the frustrations arising from failure of the green revolution (Evenson and Gollin 2003; FAO 2007) to benefit poor farmers and those in less-productive agroecological environments. More than two-thirds of human water use is for agriculture and crop and livestock production. They are the main source of water pollution by nitrates, phosphates and pesticides thus impacting agriculture, forestry and fisheries leading to loss of the world's biodiversity (Robertson and Swinton 2005; Nair 2008). The sustainable alternative organic farming practices play a vital role in maintaining biodiversity, regulating climate change and energy use, conserving ecosystem services, optimizing productivity and ensuring food security.

1.5.1 Biodiversity and Agriculture

Focus on a few selected species is the key feature of modern agriculture; these breeds give maximum quantities of preferred commodities only when they are in hybrid monocultures. In this single-commodity paradigm, the traditional practices of growing multi crops and trees together are ignored leading to serious biodiversity decline. The Convention on Biological Diversity (CBD) has estimated that human activity is causing species extinction at a rate of 100–1,000 times the natural rate of extinction (CBD 2006). Accelerating acculturation in several traditional communities throughout the world is a matter of serious concern for the loss of agricultural biodiversity – the diversity of traditional cultivated germplasm (Eyzaguirrea et al. 2007; Nautiyal et al. 2008). The Bio-cultural diversity, Bio-cultural heritage areas, Globally Important Agricultural Heritage Systems (GIAHS) and Indigenous and Community Conserved Areas (ICCA) are emerging priorities for conservation in traditional agricultural landscapes that are still intact (Chigora et al. 2007; http://www.fao.org/nr/giahs/en/; Bisong and Andrew-Essien 2010).

The indiscriminate application artificial chemical fertilizers, pesticides and other agrochemicals degrade the natural fertility of the soil through changes in soil structure, aeration, water holding capacity (Wen and Liang 2004), poisoning and killing of beneficial non target flora and fauna such as earth worms, symbiotic bacteria, pollinators and seed dispersal agents (Pimentel 1995; Wilson 2000; Horrigan et al. 2002). The simplification of agro-ecosystems to monoculture production and the removal of non-crop vegetation from the farm unit has contributed to the homogeneity of agricultural landscapes by reducing botanical and structural variation, resulting in both a reduced capacity of agricultural areas to serve as habitat for wild species as well as to effectively internally regulate populations of pests and disease causing organisms which affect crop productivity (Defra 2003). This has resulted in a widespread decline in farm species abundance and diversity across many taxonomic groups which are essential for various agro-ecosystem functions such as purification of water, internal regulation of pests and diseases, carbon sequestration and degradation of toxic compounds (Altieri 1999; Bugg and Trenham 2003; Benton et al. 2003).

Over several hundred million birds are exposed to pesticides each year in the world with an estimated 10% of these animals dying with chronic symptoms that affecting their behavior and reproductive success (Kegley 1999; Robinson et al. 2001). Pesticides are also known to negatively affect insect pest-predator population dynamics in agroecosystems (Landis et al. 2000; Brickle and Harper 2002) resulting in pest population resurgences and the development of genetic resistance of pests to pesticides (Flint 1998; Macgregao and Hill 2006).

1.5.2 Pollution in Agroecosystems

Synthetically compounded fertilizer poses multiple risks to extensive eutrophication in freshwater and marine ecosystems in many areas, where agriculture is concentrated (Drinkwater et al. 1998; Bugg and Trenham 2003; Pimentel et al. 2005). It is estimated out of 70–90% of ground and aerially applied pesticides only 25–50% reaches their target and the remaining amount is released into surrounding ecosystems and enters the food chain (FAO 2007; LaSalle et al. 2008). This has the potential to interfere with the reproductive success of wildlife, reduce the habitat quality, biodiversity of agricultural and surrounding ecosystems (Edge 2000; Liebman and Davis 2000; Mader et al. 2002).

1.5.3 Ecosystem Services

Agroecosystem services (Carpenter et al. 2006, 2009) such as clean water, carbon regulation and nutrient cycling or soil maintenance are equally important in sustaining agricultural ecosystems (WBCSD 2008) and they are seriously impacted by intensive chemical farming (Matson et al. 1997; Sandhua et al. 2008). There are several recent papers that prove the positive and reinforcing linkages between ecosystem services and various sustainable farming techniques (Milestad and Hadatsch 2003; Letourneau and Bothwell 2008; Sandhua et al. 2008, 2010).

Both organic and low-input systems resulted in increase of soil organic carbon content and larger pools of stored nutrients which are needed for long-term fertility maintenance of the land (Clark 1998). The role of organic agriculture and renewable resource management in reducing greenhouse gas emissions and increasing sinks is an important subject of research in the recent years (Bradford and Wichner 2009; Khanal 2009; Muller 2009; Soil Association 2009; Sandhua et al. 2010).

1.5.4 Climate Change

Sustainable Organic agriculture systems contribute to reduced consumption of fossil fuel energy (especially nitrogen fertilizers), reduced carbon dioxide emissions (48–60% less, except for very intensive crops), reduced nitrous dioxide (due to less mobile nitrogen concentrations and good soil structure), reduced soil erosion and increased carbon stocks, especially in already degraded soils. Energy consumption in organic systems is reduced by 10–70% in European countries and 28–32% in the USA as compared to high-input systems, except for difficult crops such as potatoes or apples where energy use is equal or even higher. Greenhouse warming potential in organic systems was 29–37% lower on per ha basis, because of omission of synthetic fertilizers and pesticides as well as less use of high energy feed (Fischer et al. 2005; Steinfeld et al. 2006). Methane emissions of organic rice and ruminants are equal to conventional systems but the increased longevity of organic cattle is favorable on methane emissions. Carbon sequestration efficiency of organic systems in temperate climates is almost double (575–700 kg carbon per ha per year) as compared to conventional soils, mainly due to use of grass clovers for feed and of cover crops in organic rotations (Fischer et al. 2005; Steinfeld et al. 2006; FAO 2007).

1.5.5 Energy Use

Inputs in organic management replace fossil fuel elements (e.g. highly soluble fertilizers, pesticides, machines) with lower impact, often locally accessed inputs and management skills. Higher labour input decreases expenses on purchased inputs by some 40% but labour costs increase by 10–15%. The main benefit of organic systems is energy efficiency in natural resource use, 33% less energy per ha in organic maize and 56% in biodynamic systems in temperate areas, as well as reduced irrigation requirements (Wetterich and Haas 1999; Kus and Stalenga 2000; Geier et al. 2001).

1.5.6 Productivity

Productivity in organic production systems is management specific. Studies suggest that switching to organic management commonly results in yield reduction in perennial crops (up to 50%) and during the conversion period for high external input systems in areas with favorable crop growth conditions (up to 40%). But in regions with medium growth conditions and moderate use of synthetic inputs, organic productivity is comparable to conventional systems (92%) and in subsistence agricultural systems, it results in increased yields up to 180% (Hepperly et al. 2006; Badgley et al. 2007). Overall, the world average organic yields are calculated to be 132% more than current food production levels (FAO 2007).

Nevertheless, demand for food will increase in the future so there are reasons why production issues and the relevance of sustainable agriculture are needed to be addressed (World Food Summit 1996; LaSalle et al. 2008):

- The main strategy for increasing both food production as well as ensuring food security and access to food is through sustainable production by farmers in developing countries
- Conventional agriculture give short-term gains in production, but in most cases it is not sustainable in the long term nor does it guarantee safe food and it also result in inadequate disadvantaged farming communities
- Organic production has the potential to produce sufficient food of a high quality. In addition organic agriculture is particularly well suited for those rural communities that are currently most exposed to food shortages and food insecurity.

1.5.7 Sustainable Agriculture and Food Security

Sustainable agriculture contributes to food security by a combination of many features, most notably by (World Food Summit 1996; FAO 2007; Beddington 2010):

- Increases yields in low-input areas
- Conserving bio-diversity and natural resources on the farm and in the surrounding area
- Increasing income and reducing costs
- Producing safe and varied food
- Long term sustainability

Understandably all these benefits cannot be expected from high tech chemicalised agriculture.

1.6 Types of Alternative Farming Techniques

Environmental sustainability is a major driving force for the development and adoption of sustainable farming practices where monoculture production of agriculture and forestry commodities has led to reduced biodiversity and loss of wildlife habitat, increased non-point source pollution of ground and surface water, and deterioration of family farms (Jose and Gordon 2008; Rigueiro et al. 2008).

The different forms of integrated land-use systems that embrace the concepts of sustainable agriculture include (Fukuoka 1985; Sachchidananda and Rajiv 1999):

- Organic farming
- Bio-dynamic farming
- No tillage farming
- Urban and Peri-urban Farming
- Natural farming
- Eco-farming
- Permaculture
- Polyculture
- Integrated farming system
- Floating Farming

These are the predominant potential sustainable farming techniques practiced in various parts of the world. This article reviews the potentials, constraints, strategies and case studies for these ten alternative farming techniques. Based on an extensive critical review of literature, it may be concluded that these farming techniques have demonstrated their ecological, economic, social and cultural sustainability.

The last section for the review comprises of some innovative endogenous farming techniques practices in India, as there is a rapid revival of alternative farming in recent decades. This part include

- Nammazhvar's Organic farming techniques from Tamilnadu
- Subhash Palekar's Zero Budget Natural farming from Maharashtra
- HOMA – Agnihotra – Vedic science of bio-energy denoting the process healing and purifying the atmosphere by removing the toxins through the agency of fire
- LEISA- Low Energy Input Sustainable Agriculture

2 Alternative Farming Techniques

2.1 Organic Farming

The principles of organic farming is the maintenance of soil fertility by bio-intensive nutrient management, recycling of agricultural wastes, vermicomposting, avoidance or reduction of external inputs, use of natural forms of pest management and weed control (Goldsmith and Hildeyard 1996; Hansen et al. 2006). The organic movement began in the 1930s and 1940s as a reaction to the growing reliance of agriculture on synthetic fertilizers. Organic farming is a form of agriculture which excludes the use of synthetic fertilizers and pesticides; plant growth regulators, livestock feed additives, and genetically modified organisms. Organic agriculture can be considered a subset of sustainable agriculture, the difference being that organic implies certification in accordance with legal standards.

Sir Albert Howard was widely considered to be the father of modern organic farming worked as an agricultural adviser in Pusa, Bengal from 1905 to 1924. He documented traditional Indian farming practices and came to regard them as superior to conventional agriculture, Rudolf Steiner a German philosopher made influential strides in the earliest organic theory with his biodynamic agriculture. More work was done by Rodale in the United States, Lady Eve Balfour in the United Kingdom and many others across the world-such as Masanobu Fukuoka, Aldo Leopold, William Albrecht, Louis Bromfield, Edward Faulkner, Ehrenfried Pfeiffer, Alan Chadwick, Wes Jackson, Garth Youngberg (http://www.ifoam.org/growing_organic/definitions/pioneers/index.php).

One of the earliest recorded examples of research on organic farming is the comparison of organic and conventional farming systems at Haughley, Suffolk, England initiated by Eve Balfour (Balfour 1943). But the practice of organic farming is as old as the early history of agriculture-if we track back some 12,000 years and beyond into prehistory. Organic agriculture is distinct from conventional agriculture through alternative agricultural practices, in their view and values (Lotter 2003; Watson et al. 2006). Organic agricultural methods are standard, internationally regulated and legally enforced International Federation of Organic Agriculture Movements (IFOAM) an international umbrella organization for organic organizations established in 1972 (Lockeretz 2007). This is known as certification. Certification of organic food products is advantageous for both producers as well as consumers (IFOAM 2006). Farmers fallowing certification are rewarded with eliminating the risk of exposure to toxic agrochemicals, premium prices and better market access. Several countries have already adopted community certification of organic food (Setboonsarng 2008).

Organic agriculture world over involves certain basic steps as like:

- Green manuring
- Bio fertilizers
- Crop rotation
- Cover cropping
- Soil Health Management

2.1.1 Green Manuring

A green manure is a type of cover crop grown primarily to add nutrients and organic matter to the soil for soil improvement and soil protection. Typically a green manure crop is grown for a specific period, plowed and incorporated into the soil.

- Leguminous green manures contain nitrogen-fixing symbiotic bacteria in root nodules that fix atmospheric nitrogen in a form that plants can use.
- Green manures increase the percentage of organic matter (biomass) in the soil, thereby improving water retention, aeration, and other soil characteristics.
- The root systems of some varieties of green manure grow deep in the soil and bring up nutrient resources unavailable to shallower-rooted crops.
- Common cover crop functions of weed suppression and prevention of soil erosion and compaction are often also taken into account when selecting and using green manures.
- Some green manure crops, when allowed to flower, provide forage for pollinating insects.

The green manure crops could contribute 30–60 kg nitrogen per hectare annually to the subsequent crop and is an inexpensive source of organic fertilizer to build up or maintain soil fertility (Amanullah 2008). For instance, the rice yield could be significantly improved by incorporating green manure and stem nodulating green manure has the capacity to fix approximately 150–220 kg N ha^{-1}in 50–60 days (Shivay and Rahal 2008). Green manuring alone (without fertilizer nitrogen) manifested an yield increase of toria by 122% equivalent to solitary application of 60 kg N ha^{-1} and the residual effect of green manuring on the following sunflower crop resulted in an additional yield of 317 kg ha^{-1} (Bahi and Pasricha 2001).

Organic agriculture is no longer a phenomenon of developed countries. It is now commercially practiced in 120 countries, representing 31 million ha of certified croplands and pastures (~0.7% of global agricultural lands and an average of 4% in the European Union), 62 million ha of certified wild lands (for organic collection of bamboo shoots, wild berries, mushrooms and nuts) and a market of US$40 billion in 2006 (~2% of food retail in developed countries) (Willer and Youssefi 2007). Although difficult to quantify, non-certified organic systems e.g. indigenous models that follow organic principles by intent or by default) of several million small farmers may represent at least an equivalent share in subsistence agriculture of developing countries (FAO 2007).

2.1.2 Biofertilizers

Biofertilizers are the substance which contains symbiotic nutrients fixating living microbes which are capable of colonizing in rhizosphere and enhances plant growth by increasing the availability of primary nutrients or by synthesizing growth promoting. The plant inoculation with Azospirillum promoted the uptake of K^+, NO_3^- and $H_2PO_4^-$, releases various metabolites such as auxines, cytokines, riboflavin and vitamins leading to higher growth in various legume and non-leguminous plant (Saubidet et al. 2000; Matriu and Dakora 2004).

Azospirillum and Pseudomonas fluorescens colonize plant roots and exert beneficial effects on plant growth and development (Bashan et al. 2004; Choong et al. 2005). Rhizobium, Azospirillum and phosphobacteria encourage plant growth by producing growth regulators, facilitating nutrient uptake, accelerating mineralization, reducing plant stress, stimulating nodulation and promoting nitrogen fixation (Somers et al. 2005; Remans et al. 2006; Mallik and Williams 2008).

2.1.3 Crop Rotation

Crop rotations serve to provide new above-and below-ground habitats as each new crop has a distinct chemical and biological make-up, introducing new vegetation types to the landscape eventually increases crop residues to the soil ecosystem (Magdoff and Harold 2000). Different crop residues promote or inhibit different soil organisms which may have inhibitory or growth promoting effects to subsequent crops. By interrupting the continuous presence of a crop host, crop rotation serves to break the build-up in the cycles of weeds and insects and diseases, thus eliminating the need for pesticide application. Fallow periods i.e. ground left uncultivated for an extended period of time, allow a limited amount of secondary succession to advance and hence the recovery of the diversity of both terrestrial and below-ground species are possible (Ditsch et al. 1993; Svotwa et al. 2008; LaSalle et al. 2008).

2.1.4 Cover Cropping

Cover cropping is an ideal cropping pattern adopted specifically for soil improvement purposes. Both annual and perennial cover crops used to harnessing natural resources effectively in above-and below-ground biodiversity (Burgos and Talbert 1996; Anonymous 2001). Cover crops may provide a physical temporary habitat for many different species of ground-nesting birds, small mammals as well as nectar and pollen sources for many species of insects. The habitat value of cover crops varies by species and variety therefore cover crops must be carefully selected to meet specific management objectives. Cover crops root system improves water penetration and prevents soil erosion (Roberson et al. 1991; Tyler et al. 1994; Sainju and Singh 1997).

Cereal cover prevents, excessive water consumption, nutrient leaching into sensitive water ways and can be an important source of organic matter when incorporated into the soil (Holderbaum et al. 1990; Sullivan et al. 1991). The use of perennial cover crops in farms is an effective means of enhancing the biodiversity and productive capacity of cropping systems by minimizing the environmental risks associated with chemical use (Creamer and Bennett 1997; Costello 1999).

2.1.5 Soil Health Management

Enhancing soil quality is essential for maintaining agricultural productivity and minimizing environmental degradation. Organic farming plays a key role in maintaining soil quality (Lal 2008). Intensive chemical agricultural practices either depletes soil nutrients or resulting in over-reliance of inorganic fertilizers leading to nutrient build up can be harmful to yields and the environment (Craswell 1998; Limpinuntana et al. 2001; Noble and Ruaysoongnern 2002). The use of naturally occurring soil mineral amendments e.g. rock phosphate, sulphate of potash, serves to supply essential plant nutrients and reduces nutrient leaching and runoff. Compost is used to improve and maintain soil organic matter levels. The Higher soil biodiversity in organic farms has shown to increase the rate of nutrient cycling, improve soil aggregation and aggregate stability and improve the disease suppression of agricultural soils (Tugel et al. 2000; Mader 2002; Fuller et al. 2005; Bhaskaran et al. 2009). Additionally in order to work effectively, manures, vermicompost and enriched phospho-composts must be incorporated into the soil. This provides the double benefit of increasing availability for crop use and decreasing potential for runoff (Mader 2002; Bhaskaran et al. 2009).

2.1.6 Integrated Pest Management

Since 1960s, chemical pesticides have been the dominant approach in controlling and eliminating pests resulting in more consistent crop yields as well as a reduction in labor needed to manage the crops. Crops grown in healthy soils tend to be more resistant and resilient to pest and pathogens which require little to no applications of pest control materials (Flint 1998). Integrated Pest Management addresses both the concern of the farmer by regarding the increasing amounts of pesticide to maintain the same effectiveness on insects and human as well as ecosystem health (Anonymous 2001). However, there is an imminent danger that it can unwittingly promote the corporate interest by unnecessarily emphasizing the use of avoidable pesticides and fungicides against the locally available more economically, ecologically and culturally appropriate pest control solutions. For instance the effectiveness and relevance of panchagavya, biofertilisers and bio-pesticides have been rediscovered recently by the scientific community (Dhama et al. 2005; Sofia et al. 2006; Babou et al. 2009; Balasubramanian et al. 2009).

2.1.7 Potentials

Production of safe and healthy products (Makatouni 2001; Rigby et al. 2001), increasing preference for environmentally friendly processes and products (Hellin and Higman 2002; Scialabba 2007), growing demand for organic products in the markets and reduced input cost with return of diverse income (Blowfield 2001; Lotter 2003) are the significant benefits.

2.1.8 Constrains

The following are some of the notable constraints-poor access to productive land, inputs and credit; poor access to water and resources especially for small-scale farmers; lack of awareness of niche markets for organic produce; problems with accessing local, national and international markets; dependence on standards set by northern hemisphere countries, which limits the development of local standards; lack of technical skills by farmers in organic production; and lack of efficient extension service in organic production systems (Niemeyer and Lombard 2003; Quansah 2003; Poulton 2004; Walaga 2006).

2.1.9 Case Studies

There are several case studies on Organic Farming throughout the world. Notable among them are the following: studies on Mid-Atlantic region by Cavigelli et al. (2009) that describes how the organic farming is environmentally and economically stable for years together. Svotwa et al. (2008) carried out a 5 years field comparison of socio economic benefits for smallholders between the conventional and organic farmers in Zimbabwe. Joyce Thamaga-Chitja and Hendriks (2008) gave a synopsis of emerging issues in smallholder organic production and marketing in South Africa.

2.2 *Bio-Dynamic Farming*

Biodynamic agriculture was the first ecological farming systems arise in response to commercial fertilizers and specialized agriculture after the turn of the century yet it remains largely unknown to the modern farmer. Biodynamic farming places great importance on the rhythmic positions of moon, sun and planets when sowing seeds, transplanting, applying liquid manures or spraying fruit trees and crops.

Biodynamic (BD) agriculture is an advanced organic farming system which gains increased attention of farmers and consumers for its emphasis on food quality and soil health. There are about 4,200 Biodynamic certified farms in 43 countries over 128,000 ha, according to Demeter standards (Demeter 2008; Turinek et al. 2009). Biodynamic agriculture developed out of eight lectures on agriculture given in 1924 by Rudolf Steiner (1861–1925) an Austrian scientist and philosopher to a group of farmers near Breslau (which was then in the eastern part of Germany and is now Wroclaw in Poland). Biodynamic farming is a combination of biological and dynamic practices; it also involves animal manures, crop rotations, care for animal welfare, looking at the farm entity and local distribution systems. BD farming practices are also gaining importance in the face of increasing climate change, energy scarcity and population growth, where they indicate a more resilient, diverse and efficient system (Turinek et al. 2009).

2.2.1 Biodynamic Preparations

The original biodynamic preparations are numbered 500–508. The BD 500 horn-manure is made by fermenting cow manure in a cow horn, that is buried in the soil for 6 months through autumn and winter) and is used as a soil spray to stimulate root growth and humus formation. The BD 501 preparation (horn-silica) is made from powdered quartz i.e. packed inside a cow horn and buried in the soil for 6 months through spring and summer and applied as a foliar spray to stimulate and regulate growth. The next six preparations of BD 502–507 are used in making compost.

Some important BD compost preparations are

- No. 502 Yarrow blossoms (Achillea millefolium)
- No. 503 Chamomile blossoms (Chamomilla officinalis)
- No. 504 Stinging nettle (whole plant in full bloom) (Urtica dioca)
- No. 505 Oak bark (Quercus robur)
- No. 506 Dandelion flowers (Taraxacum officinale)
- No. 507 Valerian flowers (Valeriana officinalis)
- No. 508 silica-rich horsetail plant (Equisetum arvense) and used as a foliar spray to suppress fungal diseases in plants.

Biodynamic preparations are intended to help moderate and regulate biological processes as well as enhance and strengthen the life forces on the farm. The preparations are used in homeopathic quantities, meaning they produce an effect in extremely diluted amounts. As an example, just 1/16th ounce-10 ml- of each compost preparation is added to 7–10-ton piles of compost.

2.2.2 Biodynamic Farming Practice

It involves the fallowing Biological and Dynamic practices (Reganold 1995; Turinek et al. 2009)

Biological Practices

Green manures, Cover cropping, Composting, Companion planting, Integration of crops and livestocks, Tillage and Cultivation.

Dynamic Practices

Special compost preparation, Special foliar spray, Planting by calendar, peppering for pest control, Homeopathy and Radionics.

2.2.3 Potentials

Biodynamic (BD) farming is a diversified method which could provide ecological, economical and physical long-term sustainability for humankind

(Turinek et al. 2009; Reeve et al. 2005; Carpenter-Boggs et al. 2000). Energy efficiency (Pimentel et al. 2005) use of animal manures, crop rotations, care for animal welfare, looking at the farm as an organism/entity and local distribution system which contribute toward the protection of the environment, safeguard biodiversity and improve livelihoods of farmers (Reganold 1995; Carpenter-Boggs et al. 2000).

2.2.4 Constrains

Need for optimal temperature management, that could promotes quick compost preparation (Carpenter-Boggs et al. 2000), need of certain specific flora species, lack of public awareness and training capacity, undiscovered perspectives (Reganold 1995), lack of comparable subsidies for organic inputs (in comparison to artificial agrochemicals), lack of support price to produce, lack of wide spread on-farm research, trails and awareness on high-value crops (Ryan and Ash 1999; Turinek et al. 2009) are some of the important constraints.

2.2.5 Case Studies

Among the several available case studies on Bio-dynamic farming the following are notable. Reganold et al. (1993) study compared soil quality and financial performance of Biodynamic and conventional farms in New Zealand. Their study reported that the Biodynamic farms proved in most enterprises to have soils of higher biological and physical quality: significantly greater in organic matter, content and microbial activity, more earthworms, better soil structure, lower bulk density, easier penetrability, and thicker topsoil. The biodynamic farms were just as financially viable on a per hectare basis.

Mäder et al., from FiBL Institute in Switzerland (2002) in their 21-year study compared the agronomic and ecological performance of biodynamic, organic and two conventional systems. The study found that nutrient input in the biodynamic and organic systems was 34–51% lower than in the conventional systems but crop yield was only 20% lower on average, indicating more efficient production. The total energy (for fuel, production of mineral fertilizer and pesticides, etc.) to produce a dry-matter unit of crop was 20–56% lower for the biodynamic and organic systems, and pesticide input was reduced by 97% (by 100% for the biodynamic system). With regard to soil aggregate stability, soil pH, humus formation, soil calcium, microbial biomass, and faunal biomass (earthworms and arthropods), the biodynamic system was superior even to the organic system, which in turn had superior results over the conventional systems. With the significant increase in microbial diversity in the biodynamic and organic systems, there was a significant associated decrease in metabolic quotient, indicating a greater ability to use organic material for plant growth.

2.3 *No Tillage Farming*

Tillage is a critical soil management practice used for weed control, seed bed preparation, root growth stimulation, soil moisture control, soil temperature control, soil compaction alleviation and incorporation of crop residues and manure (Lal 1997; Basamba et al. 2006). Masanobu Fukuoka wrote a book "The One-Straw Revolution" was one of the pioneers work about No tillage farming or Fukuoka Farming. Producing crops usually undergo regular tilling that agitates the soil in various ways and it's usually done by tractor-drawn implements. Usually this tilling practice affects the soil compaction, loss of organic matter, degradation of soil aggregates, death or disruption of soil microbes including mycorrhiza, arthropods, and earthworms (Chan 2001; Hussain et al. 1999) and soil erosion.

No-till farming avoids these effects by excluding the use of tillage practice. By this way of farming, crop residues or other organic amenities are retained on the soil surface, sowing and fertilizing is done with minimal soil disturbance (Chan 2001). It sometimes involves in problems like residue management, increased weed and disease infestations and this can be avoided by crop rotations and cover crops (Fukuoka 1978). No-tillage farming is effective in terms of economic and soil erosion control, than any other cost effective practices which are commonly used (King 1983; Sorrenson and Montoya 1984). Tebrügge and Böhrnsen (1997) reported that No-tillage is a very profitable cultivation system compared to conventional tillage because of the cost savings from lower machinery costs and lower operation costs.

2.3.1 Farming Practice

No-till farming i.e. zero tillage is a way of growing crops from year to year without disturbing the soil through tillage practices. No-till is an emergent agricultural technique which increases the amount of water in the soil, soil biodiversity and decrease erosion. It includes certain management strategies like crop rotation, cover crops, disease and pathogens control though herbicides and removal of residues.

2.3.2 Potentials

Environmentally friendly no-tillage farming is being recognized as a best management practices all over the world. This farming practice improves soil health by increasing earthworms, diversity of decomposer organisms, organic matter content, minimizes soil erosion (Halvorson 2006), low production costs due to less fuel consumption and labor input (Uri 1999; Elmi et al. 2009). They also produce products that are safe and healthy for consumers (Basamba et al. 2006). Several other workers have reported the multiple benefits of no tillage farming (Magdoff and Harold 2000; LaSalle et al. 2008).

2.3.3 Constrains

Impacts on non targeted biota by the type and quantity of herbicide used, lack of wide spread on-farm research and trails, awareness on the choice, sequencing of cash, green manure crops in rotations, managing ground cover, crop residues and control sporadic patches of weeds as opposed to blanket spraying with broad-spectrum herbicides (King 1983; Sorrenson et al. 1997) are a few of the notable constraints.

2.3.4 Case Studies

Notable case studies include the following: Basamba et al. (2006) described the results of a long-term field experiment on sustainable crop rotation and ley farming systems on a Colombian acid-savanna oxisol to test the effects of grain legumes, green manures, intercrops and leys as possible components that could increase the stability of systems involving annual crops that were initiated in 1993. In this study, five agropastoral treatments (maize monoculture, maize-soybean rotation, maize-soybean green manure rotation, native savanna, maize-agropastoral rotation) under two tillage systems (no tillage and minimum tillage) were investigated. Lower bulk density and higher total porosity for all treatments and soil layers were found in no-tillage compared to the minimum tillage system. Between the two tillage systems, significantly higher maize grain yields (pB/0.1) were obtained under no-tillage agropastoral treatments compared to the same treatments under minimum tillage. Maize yields on native savanna soils were markedly lower than in the rest of the treatments, indicating the need for improved soil conditions in subsoil layers for root growth of maize.

Abdirashid Elmi et al. (2009) conducted a study on a 2.4 ha field located at Macdonald research farm of McGill university, Montreal, to investigate the relative impacts of long-term Conventional Tillage (CT) and No-Tillage (NT) practices on soil N2O fluxes (FN2O) under grain and silage corn (Zea mays L.) during the 2003 and 2004 growing seasons (May-Sept). Nitrous oxide fluxes were measured using static closed chamber by taking gas samples at 0, 10, 20 and 30 min. An important implication of the findings was that, contrary to many reports in the literature, the adoption of NT may not add to concerns over global atmospheric N2O concentrations. This might be due to a greater rate of N2O reduction to N2 in soils under NT than CT during diffusion up the soil profile because of the higher moisture content under NT system than CT.

Lawrence et al. (1994) carried out a 4 year study (1986–1989) of the effects of tillage practices on profile soil water and crop yield in a Sodosol (Typic Natrustalf) in central Queensland, Australia. The tillage treatments were: zero till fallow (weed control by herbicides), reduced till fallow (chisel plough/scarifier or herbicides) and conventional till fallow (chisel plough/scarifier) At the conclusion of the experiment, soil water accumulation in the 100–180 cm soil layer was 86 mm in zero till, 39 mm in reduced till and 40 mm in conventional till. Results indicated that zero till can be a more productive wheat farming practice than conventional mechanical tillage. The increase in water storage below the root zone of the wheat crop shows that there may be benefit in using a deeper-rooting crop or pasture species in rotation with wheat, particularly after zero till fallows.

2.4 Urban and Periurban Farming

Urban and Periurban agriculture is the practice of cultivating, processing and distributing of producing vegetables and fruits within urban environments for household consumption as well as for sale to the rapidly growing urban population (Bailkey and Nasr 2000). Urban farming is generally practiced for income-earning or food-producing activities though in some communities the main impetus is recreation and relaxation (Fraser 2002). Urban agriculture contributes to food security and food safety in two ways: **first**, it increases the amount of food available to people living in cities, **second**, it allows fresh vegetables, fruits and meat products to be made available to urban consumers.

The recognition of environmental degradation within cities through the relocation of resources to serve urban populations has inspired the implementation of different schemes of urban agriculture across the developed and developing world (Wackernagel and Rees 1994). Great population pressure in and around cities, coupled with the economic crises throughout the region has led to a tremendous increase in the last decade of total city area under food production. This activity is known as urban and peri-urban agriculture. Urban and peri-urban agriculture is practiced for a variety of purposes like commercial reasons, food self-sufficiency and food security (Jianming 2003; Ayenew et al. 2007).

2.4.1 Farming Practice

Land preparation is necessary to provide a good soil for planting, weeding, selection of crop based on the soil type and season, integration of crop and livestock can be done in the free land in and around the city. Rooftop gardens as a specific urban agriculture niche set within a broader system of city gardens enjoy their own set of distinctive benefits. Rooftops are underutilized and rarely-considered urban spaces with great potential for creative development (Atha 2000; Doshi et al. 2003). There are essentially three options for rooftop gardens. The first is container gardening a less formal, cheaper form of roof gardening. In container gardening few to no modifications are made to the existing roof structure, containers i.e. anything from plastic swimming pools to recycled-wood planters -are placed on a rooftop and filled with soil and plants. The second type of roof garden in which the rooftop actually becomes the planting medium which involves more intensive investments but it has its own set of advantages, including greater storm-water retention, building insulation, formation of patchwork offering, temporary habitats to fauna such as birds and butterflies during their long migrations. The third rooftop garden possibility is rooftop hydroponics, in which plants are grown in a soilless medium and fed a special nutrient solution. Rooftop hydroponics can be the lightest of the three options it offers the possibility of faster plant growth and increased productivity (Garnham 2002; Doshi et al. 2003; Price 2000).

2.4.2 Potentials

Urban and Peri-Urban areas expand the economic base of the city by production, processing, packaging and marketing of consumable products. These results in increase of entrepreneurial activities, creation of job opportunities, food costs reduction, products of better quality, food security, increased income, energy efficiency, better health and nutrition. In addition wastewater and organic solid waste can be transformed into resources for growing agriculture products, thus helps to preserve bioregional ecologies from being transformed into cropland and reducing pollution by greening the environment (Wilson 2002; Atha 2000; Wilson and Pelletier 2003; Fleury and Ba 2005).

2.4.3 Constrains

A few important constraints are use of waste water for irrigation without careful treatment result in the spread of diseases, cultivation on contaminated land may result in a health hazard, legal restrictions, lack of security of tenure, uncertainty in land use length, lack of extensive production and economic impediments to accessing land and resources (Smit et al. 1996; CDE 2001; Butler and Moronek 2002).

2.4.4 Case Studies

Blom-Zandstra (2005) study in densely populated areas of the Netherlands indicated that rural and urban areas are strongly interrelated and agriculture has a role in social life, labour and recreation. Fulton's study (2005) in Australia revealed that urban organic community gardens contribute to health promotion and education in sustainability and the weekly direct supply of organic fruits and vegetables to consumers has restored direct farmer-consumer relationships. Wilkins et al. (2005) postulated a closed urban-organic loop in which peri-urban organic agriculture produces the food for the city and in return, recycles organic waste and used water from the city, thus reducing food miles, waste dumps and CO_2 emissions.

2.5 Natural Farming

Do-Nothing Farming also known as Natural Farming (NF) is an alternative farming method to chemical or traditional farming. Natural framing is used to emphasize the importance of "spatially" and "temporarily" overlapping the growing crops, plants and animals so that we can utilize their synergistic effects. Natural Farming with indigenous microorganisms is a distinctive approach to organic farming practiced successfully in more than 30 countries, in home gardens and on a commercial scale (Prell 2010). Natural Farming is unique in that it is not meant to be commercialized but rather practiced by individual farmers with cheap, easily available ingredients and microbes or mycorrhizae indigenous to each locale or farm. These microorganisms are:

- cultured in a simple wooden box of rice
- mixed with brown sugar and stored in a crock
- further propagated on rice bran or wheat mill run
- mixed with soil and cultured again.

The resultant product is then mixed with compost, added to potting soil or spread on beds before planting, the process takes 3–4 weeks (Prell 2010). There are also procedures for water soluble calcium made from eggshells, water-soluble calcium phosphate made from animal bones and vinegar, fish amino acid made from fish waste, lactic acid bacteria, seed soak solution and insect attractants made from rice wine. There are half dozen more inputs that can be made simply and easily at home which are used according to the nutritive and growth cycle of the plants. Many of these inputs are made from waste residues and materials (Prell 2010).

2.5.1 Farming Practice

This method includes the use of crop rotation, minimal irrigation, no or reduced tillage, seed balls, cover crops or mulching and allowing natural regulation of pests (Fukuoka 1978) in order to reproduce natural conditions as closely as possible. Natural Farming pursues agriculture without any specific purpose; it is prohibited to plow, weed, manure or spray chemicals (Katai et al. 2007).

2.5.2 Potentials

Notable advantages are lower costs to the farmer – by 60%, self reliant, farmer friendly, more resilient, diversified products, income and efficient system, stronger, healthier, nutritious plants, higher yield, better quality and zero waste emission.

2.5.3 Constrains

Important drawbacks are lack of wide spread on-farm research and trails, higher demand for manual labour, lack of awareness on the choice and sequencing, availability of green manure crops in rotations, managing ground cover, crop residues and control of sporadic patches of weeds.

2.5.4 Case Studies

Natural Farming is a farming method aiming to produce environmentally safe food and it is characterized by no external input such as chemical fertilizers and biocides. Examination of the long-term change in the productivity under NF will provide us with important information for the low input and sustainable rice production. Yield surveys were conducted in 17 prefectures of three climatically different districts (the Tohoku, Kinki, and Chugoku) in Japan in 1990 and 542 sets of data were collected from NF paddies, differing in years (1–50) after conversion to NF. The average brown rice yields of NF ranged from 352 gm^{-2} in Okayama to 504 gm^{-2} in Akita. The variation in NF yield generally reflected the average regional yield by conventional farming (CF) ($p < 0.05$), but the yield gap between NF and CF (Neera et al. 1999).

2.6 *Eco-Farming*

Eco-farming or site-appropriate agriculture involves treating both regions used for agriculture and individual farms as ecological systems. "Site" restricted to natural conditions like soil, climate and temperature. The demand for stability and sustainability stems from the obligation of each generation to pass on to future generations an environment that remains capable of guaranteeing the fundamentals of human existence. Consideration must also be given to economic development i.e. price–cost ratios, incomes, farm-specific conditions i.e. access to factors of production and the internal forces influencing a farm's operations like self-sufficiency, risk minimization and preservation of soil fertility. Countries must develop forms of agriculture that permit a high degree of self-sufficiency and decentralization at national and regional levels (Egger 1990).

The essential characteristics of these ecofarming systems are:

- maximal but sustainable use of local resources
- minimal use of purchased inputs, only as complementary to local resources
- emphasis on subsistence cropping, combined with complementary production for the market
- ensuring the basic biological functions of soil-water-nutrients-humus
- maintaining a diversity of plant and animal species as a basis for ecological balance and economic stability, with primary emphasis on local species, varieties and races
- conserving life support systems and ecosystem services
- creating an attractive overall land cape which gives satisfaction to the local people

2.6.1 Farming Practice

The farming practices includes restoring appropriate vegetation-trees and shrubs in arable farming, creation of erosion-protection strips parallel to the incline on slopes, planting of hedges to divide a farm into numerous small fields, intercropping, alternating with intensive fallow, organic manuring, integrated livestock husbandry, improved mechanization, supplementary use of mineral fertilizer, integrated plant protection and selective weed control (McNeely et al. 2003).

2.6.2 Potentials

Plant production is geared to natural conditions i.e. site-appropriate preservation of the soil structure, biological processes, soil fertility, prevention of erosion damage, control of contamination of water bodies, preservation of typical landscape features, promotion of nature conservation, protection of species, preservation of ecologically valuable biotopes and making livestock husbandry an integral component of environmentally sound agriculture (Egger 1990).

2.6.3 Constrains

Compulsion of certification, lack of integrated efforts from supporting agencies, centralized input supply, trials on improved methods of composting, public awareness, training capacity, lack of comparable subsidies for organic inputs, and lack of support price to produce, lack of wide spread on-farm research and awareness on high-value crops (Xue 2006) are some of the notable constraints.

2.6.4 Case Studies

Only limited case studies are available on this topic, notable among them are :studies by Xue Dayuan (undated) in China; a review by Kurt Egger (1990) about the ecofarming in areas of Rwanda; eco farming in Africa in www.necofa.org/, the work of Tai-cheol Kim (2008) on eco-farming tourist programs that were developed to utilize the year-round circulation of water through reservoirs, irrigation canals, paddy fields, fish-ways regulating ponds, drainage canals, and pumping stations, that would provide a boom to rural tourism in Korea.

2.7 Permaculture

Permaculture is a design system and philosophy developed by Bill Mollison. Permaculture is the conscious design, maintenance of agriculturally productive ecosystems, stability and resilience of natural ecosystems. It is the harmonious integration of landscape, food, energy and shelter. Without permanent agriculture there is no possibility of a stable social order (Mollison 1988 Design Manual, IX).

2.7.1 Farming Practice

A major feature of permaculture designs is the concept of dividing properties into zones (Mollison 1991). Zone 0: The house would be of passive solar design to provide comfort for the occupants and to minimize power consumption. It could be fitted with its own electricity generating system i.e. solar panels. Zone I: Close to the house this area would contain frequently visited places such as the vegetable garden and workshop. There are only a few particularly useful trees. Zone II: This area is slightly further from the house and contains such things as orchards and poultry. Zone III: This would be most of the land area and would consist of the paddocks used for cropping and grazing. It would contain many trees for windbreaks, fodder production and timber. Zone IV: This area might not be present on every farm but consists of semi-managed areas such as semi-cleared hill country and would be used for some stock and timber production. Zone V: Again this area might not be present on every farm. It would consist of virtually unmanaged native vegetation. Its value would be for aesthetics and wildlife conservation.

The claimed benefits of converting a property to permaculture include a reduction in fuel consumption, increased on-farm fuel production, reduced soil loss, reduced pollution, increased genetic diversity in crops and livestock, improved human, increased employment, and environmental health.

2.7.2 Potentials

It ensures that plant production is geared to natural conditions, i.e. site-appropriate preservation of the soil structure, biological processes, soil fertility, prevention of erosion damage, control of contamination, preservation of typical landscape features, promotion of nature conservation, protection of species, preservation of ecologically valuable biotopes and making livestock husbandry an integral component of environmentally sound agriculture.

2.7.3 Constraints

Compulsion of certification that remains prohibitively costlier for small and marginal framers, lack of integrated efforts from supporting agencies, centralized input supply, trials on improved methods of composting, public awareness and training capacity, lack of comparable subsidies for organic inputs, lack of support price to produce, lack of wide spread on-farm research (Katai et al. 2007; Maposah-Kandemiri et al. 2009) are the notable constraints.

2.7.4 Case Studies

The investigations by Myra Maposah-Kandemiri et al. (2009) on permacultural practices in Muenzaniso Primary School in Zimbabwe promoted environmental citizenship benefits, social and community development, personal development of learners and various school benefits. Similar results were reported by Ben Haggard (1993) in Soly Sombra; Lisa Rayner and Zack Zdinack (2002), in Arizona, New Mexico, southern Colorado and southern Utah.

2.8 Polyculture

Several agronomists recently proposed that traditional multispecies systems models for designing sustainable cropping systems (Gliesmann 2001; Altieri 2002). Polyculture will (a) produce adequate yields of edible grain over several years without tillage or re-sowing; (b) yield as much as or more than an equivalent set of monocultures; (c) manage insects, pathogens and weeds; and (d) compensate for N, phosphorus, and other nutrients removed in harvest (Piper 1999; Jackson and Jackson 1999; Cox et al. 2004). In agroecosystems, biodiversity (1) contribute to constant biomass production and reduce the risk of crop failure in unpredictable environments, (2) restore disturbed ecosystem services, such as water and nutrient cycling, and (3) reduce risks of invasion, pests and diseases through enhanced biological control or direct control of pests (Gurr et al. 2003). Specific features of biodiversity in natural systems may offer a basis for designing multispecies systems (Ewel 1986).

2.8.1 Farming Practice

Polyculture is agriculture using multiple crops in the same space. It improves diversity of natural ecosystems and avoids monoculture. It includes crop rotation, multi-cropping, intercropping, companion planting, beneficial weeds and alley cropping. When compared to monocultures, polycultures have consistently shown lower populations of pest, weeds and it's their indeterminable effects to crop. Increases in the variety of food sources and micro-habitats increased difficulty of pest populations in polycultures to locate dispersed patches of host crop, so thus limits the growth of pest populations and stabilize predator–prey and parasitoids-host population dynamics (Altieri 1999; Chopin and Yarish 1999).

2.8.2 Potentials

Species stability and sustainability (Swift and Anderson 1993; Vandermeer et al. 1998), food, fodder quality and quantity improves (Bulson et al. 1997; Gooding et al. 2007), reducing pests and diseases (Trenbath 1993; Hauggaard-Nielsen et al. 2001). Weed suppression (Welsh et al. 1999; Hauggaard-Nielsen and Jensen 2005), significant reduction in harmful insects in mixed cropping systems compared with monocultures of the same species (Nickel 1973; Perrin 1977; Vandermeer 1989), Environmentally (Donald 2004; Scopel et al. 2005), Economically more stable and sustainable yield (Follis 1993; Igbozurike 1978) are the most significant advantages.

2.8.3 Constrains

The major constraints are competition versus facilitation (Vandermeer 1989) aboveground competition for light (Keating and Carberry 1993), belowground competition for water and nutrients (Casper and Jackson 1997), intercrop and resource allocation (Grime 1977), lack of comparable subsidies for organic inputs (in comparison to artificial agrochemicals), lack of support price to produce, lack of wide spread on-farm research and awareness on high-value crops (Igbozurike 1978).

2.8.4 Case Studies

Andow (1991) reviewed 209 studies on crop mixtures involving 287 different species of parasitic insects. The insects were significantly fewer in 52% of cases (149 species) compared with monocultures, and greater in 15% of cases (44 species).

The field experiments by Gooding et al. (2007) in Europe in the 2002/2003, 2003/2004 and 2004/2005 growing seasons, intercropping wheat with faba bean (Denmark, Germany, Italy and UK) and wheat with pea (France) regularly increased the nitrogen and sulphur concentration in cereal grains, hence increasing the wheat quality for bread making.

Swift and Anderson (1993) proposed a comparable classification of biodiversity in agroecosystems that distinguished productive biota from resource biota i.e. organisms that contribute to pollination, decomposition, recycling and destructive biota i.e. weeds, insect pests and microbial pathogens. It has been reported for numerous taxa in various conditions that associated diversity is positively correlated to planned biodiversity.

2.9 *Integrated Farming Systems*

Combining crops with livestock spatially and temporally has the potential to maintain ecosystem function, prevent agricultural systems from becoming too brittle and increased capability to absorb shocks to the natural resource base promotes multi-functionality i.e. multiple roles assigned to agriculture in Integrated farming systems (Okigbo 1995; Price 2000; Radhammani et al. 2003; Groenfeldt 2005). Multifuctionality agriculture has an activity is entrusted with performing four main functions-economic functions and social functions, environmental functions and food security. The difference between mixed farming and integrated farming is that enterprises in the integrated farming system are mutually supportive and depend on each other (Csavas 1992). Integrated farming systems are effectively systems that have traditionally been undertaken by farmers in countries that include Indonesia, China, Malaysia, Vietnam, Rwanda and Thailand (Gliesman et al. 1981; Csavas 1992; Tokrishna 1992; Choosakul 1999; Praphan 2001). However in many countries these traditional self reliant subsistence systems have been replaced by the establishment of commercial cash crop production systems that have been promoted by governments (Ruaysoongnern and Suphanchaimant 2001).

2.9.1 Farming Practice

The generation of wastes and by-products from the subsystems are transferred between enterprises, thereby reducing the need for external inputs such as feeds and crop nutrients (Csavas 1992; Little and Edwards 2003). Animals on a farm provide inputs to other enterprises and constitute a source of meat and milk, a means of savings and a source of social status (KKU 2001; Schierre et al. 2002; Little and Edwards 2003). However the distinction between the integrated farming system and the commercial farming system is not absolute but is rather a matter of degree of integration of resources in the farm system (Tipraqsa 2006).

2.9.2 Potentials

The significant advantages are – higher species stability and sustainability (Agbonlabor et al. 2003), lesser water usage, energy efficient (Jayanthi et al. 2000), food security, enhancement in -environmental functions, economic functions and social functions (Tipraqsa 2006) and stable and sustainable production (Radhammani et al. 2003).

2.9.3 Constrains

Notable constraints are -animal feed shortages throughout the year, labor bottlenecks, soil degradation (Ngambeki et al. 1992), lack of space, lack of secure land rights, disincentives to adopting integrated farming resulting from government subsidies, credits for fertilizers, herbicides and the long transition period that often occurs when implementing an integrated production system. This lead-in time can vary between 3 and 10 years. Farmers could not afford to forgo declines in food production and income generation over this period (Lightfoot and Minnick 1991).

2.9.4 Case Studies

A farming-system comparison was established on Burgrain Farm to investigate the long term sustainability of farming systems in Switzerland in 1991. The study resulted in less weed dynamics, yield and soil quality are better in IFS than other two farming systems (Hiltbrunner et al. 2008).

Radhammani et al. (2003) describes IFS's as a component of farming systems which takes into account the concepts of minimizing risk, increasing production and profits whilst improving the utilization of organic wastes and crop residues.

Jayanthi et al. (2000) based on experiences from Tamil Nadu, India, described these systems as a mixed animal crop system where the animal component is often raised on agricultural waste products while the animal is used to cultivate the soil and provide manure to be used as fertilizer and fuel.

2.10 *Floating Farming*

Floating farming is an ancient and traditional farming system, practiced in some regions of the world. This form of in-situ hydroponics or soil-less culture is sustainable and comparable with cultivation techniques (Irfanullah et al. 2008). The wetland ecosystems are very important to the economy and lives of the people of the country, as their livelihoods and subsistence are very much linked with the productivity of wetlands. In such places, these floating farming are sustainable in all aspects.

2.10.1 Farming Practice

Chinampa is a small, stationary, artificial island built on a freshwater lake for agricultural purposes. Chinampan was the ancient name for the southwestern region of the Valley of Mexico, the region of Xochimilco, it was there that the technique was – and is still – most widely used. It consists in building up a number of narrow islands each averaging some 6–10 m wide and some 100–200 m long, using layers of vegetation, dirt, and mud. The lake provides the chinampa with moisture laden with decomposing organic wastes that irrigate the crops.

Raised field agriculture is an ancient food production system used extensively by the Aztecs in the Valley of Mexico but also found in China, Thailand and other areas to exploit permanently flooded areas or swamplands bordering lakes called chinampas in the Aztec region, these "islands" or raised platforms (from 2.5 to 10 m wide and up to 100 m long) were usually constructed with mud scraped from the surrounding swamps or shallow lakes. The Aztecs built their platforms up to a height of 0.5–0.7 m above water levels and reinforced the sides with posts interwoven with branches and with trees planted along the edges.

The soil of the platforms is constantly enriched with organic matter produced with the abundant aquatic plants as well as with sediments and muck from the bottom of the reservoirs. A major source of organic matter today is the water hyacinth (Eichornia crassipes) capable of producing up to 900 kg per hectare of dry matter daily. Supplemented with relatively small amounts of animal manure, the chinampas can be made essentially self-sustaining. Animals such as pigs, chickens, ducks are kept in small corrals and fed the excess or waste produce from the chinampas. Their manure is incorporated back into the platforms (Gliesman 1998). On the chinampas, farmers concentrate the production of their basic food crops as well as vegetables. These include the traditional corn-bean-squash or cassava-corn-bean-peppers-amaranth, the fruit trees associated with various cover crops, shrubs or vines. Farmers also encourage the growth of fish in the water courses. High level productivity of chinampas results from several factors.

First, cropping is nearly continuous; only rarely is the chinampa left without a crop. As a result three to four crops are produced each year. One of the primary mechanisms by which this intensity is maintained is the seedbeds in which young plants are germinated before the older crops are harvested. Second, the chinampas maintain a high level of soil fertility despite the continual harvest of crops because they are supplied with high quantities of organic fertilizers. The lakes themselves serve as giant catch basins for nutrients. The aquatic plants function as nutrient concentrators, absorbing nutrients that occur in low concentration in the water and storing them inside their tissue. The use of these plants along with canal mud and muddy water (for irrigation) ensures that an adequate supply of nutrients is always available to the growing crops. Third, there is plenty of water for the growing crop. The narrowness of the chinampas is a design feature that ensures that water from the canal infiltrates the chinampa, giving rise to a zone of moisture within reach of the crop's roots. Even if during the dry season the lake levels fall below the rooting zone, the narrowness of the chinampa allows the chinampero to irrigate from a canoe. Fourth, there is a large amount of individual care given to each plant in the chinampa. Such careful husbandry facilitates high yields (Gliesman 1998).

2.10.2 Potentials

Many other social, economic, agricultural and ecological benefits are associated with the practice of floating gardening (IUCN 2005). Invasive water weeds can be harvested for financial gain, energy production, bio fertilizer or nutrient medium as like a base for mushroom cultivation (Murugesan et al. 1995; Sinkala et al. 2002; Gunnarsson and Petersen 2007).

2.10.3 Constrains

Adequate availability of locally available plant materials (e.g. water hyacinth) for construction of the floating platforms, protecting the floating gardens from wave-action, lack of wide spread on-farm research /trails/awareness on selecting appropriate crops and husbandry for optimizing benefits (Irfanullah et al. 2008) are the important constraints.

2.10.4 Case Studies

A study by Islam (2007) in Bangladesh proved that the soilless cultivation method described and provides an opportunity for income generation during this normally slack season. The waterlogged villages in the southern coastal wetlands of Bangladesh have proved its viability, and the same or similar methods could be used elsewhere. Women seem to find it particularly advantageous, a progressive aspect of development that is generally agreed to be essential for the future of Bangladesh society.

The work of Irfanullah et al. (2008) in Bangladesh has shown that the floating gardening was introduced in 2006 on a pilot-scale in the north-east wetlands of the country, as a contribution to food security and as a supplementary income for the marginalized community. The overall experience of floating cultivation in three selected villages was encouraging. Local people became aware of this new farming system and their level of knowledge improved. Communities were mobilized into groups to make floating platforms, and platform residues were later used to establish winter gardens. Cultivation was successful on both types of plot, and vegetables were both consumed by the producers and sold in the market. The input–output analysis revealed floating gardening to be a feasible alternative livelihood option for the wetland dwellers. The method provided targeted landless people with parcels of land in the monsoon, enabling them to grow vegetables. Floating gardening and associated winter gardening appear to have the potential for introduction to other parts of the world where aquatic weed management is a major problem.

3 Alternative Farming Techniques from India

India has a longer heritage traditional of sustainable natural and organic farming strategies most of which until recently were not properly documented. The country with its rich diversity of ecosystems, soil types and monsoon patterns has evolved a wide variety of alternative farming techniques to suit its local diversity. Among several activists of natural and organic farming, two of them need special attention – Nammazhvar from Tamilnadu and Subhash Palekar from Maharashtra.

3.1 Nammazhvar's Organic Farming Techniques

Nammazhvar born in near Thirukattupalli village, Thanjavur District Tamilnadu, obtained his bachelor's degree in Agriculture, worked for 3 years as a research assistant in a Regional Agricultural Research Station. Disgusted with the working atmosphere that was completely cut off from the ground realities and problems of the local farmers, he was interested in organic farming so he quit his job. He rediscovered and transferred the traditional organic farming techniques by on farm demonstration in innovative farmer's fields. He has written more than 15 books on organic farming in local language Tamil that contain organic farming methods, techniques, strategies, ingredients, procedure and case studies from all over Tamilnadu. He is also giving active training on organic farming to the framers with the help of activists, NGOs and CBOs. He describes the ingredients and preparation of the following mixtures for organic farming.

3.1.1 Preparation of Panchagavya (Organic Liquid Fertilizer)

Cow dung (2 days old)-5 kg; cow urine 3 l; cow milk-2 l; curd-2 l; groundnut cake/Moong dal flour-1.5 kg; well riped banana-10, tender coconut-2 l, Sugarcane/ Jaggery-1 kg. All these ingredients have to be mixed well and kept in a air tight container. The fermented mixture can be used after 14 days. Sprayed weekly once at 300 ml/10 l of water to the crops. Panchagavya is found to enhance the biological efficiency of the crop plants and the quality of fruits and vegetables (Natarajan 2002; Swaminathan et al. 2007; Sangeetha and Thevanathan 2010). Panchagavya is normally advocated as foliar nutrition (Caraka-Samhita 1981; Sustra Samhita 1885; Chauhan 2004; Garg and Chauhan 2003; Saxena et al. 2004).

3.1.2 Preparation of Amirtha Karaisal (Microbial Culture)

Fresh Cow dung, Urine and Jaggery in 1:1:4 ratio fermented for 24 h and then sprayed at a ratio of 1:10.

3.1.3 Preparation of Meein Amilam (Organic Urea)

Waste parts of fish and Jaggery in 1:1 ratio mixed and kept in air tight container for 25 days, then the sprayed at 100 ml/l of water after 3 or 4 days of planting at a weekly interval.

3.1.4 Preparation of Flower Stimulating Solution

Fermented Buttermilk and Fermented coconut milk in 1:1 kept in air tight container for 7 days, sprayed at 1 l/10 l of water.

3.1.5 Preparation of Bio-Insecticide

Caltrops leaf, Adhatoda vasica leaf, Ipomoea carnea leaf, Vitex negundo and Morinda correia. All these leaves are in equal proportion made into fine pieces and soaked in 10 l of Cow urine left in air tight container for 7 days and sprayed to crops at 1 l/10 l of water.

3.1.6 Preparation of Vanamutham (EMO-Effective Micro Organisms)

Papaya-1 kg, Banana-1 kg, Pumpkin-1 kg, Jaggery-1 kg, egg-1. All these ingredients are mixed thoroughly in a pot, kept air tight closed in shade for 30 days, then sprayed for crops/plants at 500 ml/10 l of water.

All these preparations are already field tested and adopted by a larger cross section of highly motivated organic farmers in Tamil Nadu.

3.2 Subhash Palekar's Zero Budget Natural Farming

Subhash Palekar was born in 1949 in Maharashtra, India. He investigated the natural eco system processes of the forest and verified them on his farm for 6 years, since 1989–1995. There were about 154 research projects o on farm research during these period. Based on this, verified research work, he evolved/perfected the technique of Zero Budget Natural Farming. He is now organizing intensive training courses, workshops and seminars for farmers throughout India. His books in English as well as local Indian languages (Marathi, Hindi, Kannad, Telgu, Tamil) on natural farming are widely read. As a consequence, thousands of model natural farms for various crop combinations are established throughout India. His book volumes-The Philosophy of Spiritual Farming (Zero Budget Natural Farming) Part 1, Part II and Part III (2006) in which full description and farming methods and case studies all around the India were given are creating waves among farmers and policy circles.

Zero Budget Natural Farming is a unique method of farming which requires absolutely no monetary investment for purchase of key inputs like seeds, fertilizers and plant protection chemicals from the market. The farmer can grow hardy local varieties of crops without application of fertilizers and pesticides. Since it is a zero budget farming no institutional credit would be required and dependence on hired labor is also reduced to bare minimum.

Among his several varieties of bio-fertilizers/bio-pesticides that can be prepared easily by farmers free of cost using farm animal products and local plant species, the following are widely adopted by farmers throughout India. Mulching, crop rotation, multiple cropping and locally made bio-fertilizers/bio-pesticides are the essential ingredients in his natural farming method.

3.2.1 Preparation of Bijamrita

Ingredients: Water 20 l; Deshi cow dung 5 kg; Deshi cow urine 5 l; One handful soil from the surface of boundary of field; Lime 50 g. 5 kg cow dung in cloth and bound it by small rope as a small bundle. Hang this bundle of cow in the taken 20 l water for a night (12 h).Take a litre of water and add 50 g lime and let stable for a night. Then in morning squeeze this bundle of the cow dung in that water thrice continuously, so that all the essence of cow dung will accumulate in that water. Then add handful of soil in that water solution and stir it well. Then add 5 l of cow urine and lime water and stir it well. This is called Bijamrita. The planting material has to be simply dipped in 'Beejamrutha', taken out and planted. Beejamrutha protects the crop from harmful soil borne and seed borne pathogens during the initial stages of germination and establishment.

3.2.2 Preparation of Jiwamrita (Microbial Culture)

Ingredients: Water 20 l; Deshi cow dung 10 kg; Deshi cow urine 5–10 l; Jaggery (black and red), 2 kg or sugarcane syrup 2 l; Flour of any pulses (beans, blackgram, redgram) 2 kg Handfull of soil from the farm. The above mixture will suffice for one time application on one-acre crop. 'Jeevamrutha' is to be provided once in a fortnight or at least once in a month. It promotes immense biological activity in the soil and makes the nutrients available to the crop. Jeevamrutha is not to be considered as nutrient for the crop but only a catalytic agent to promote biological activity in the soil.

3.2.3 Bio Pesticides for Plant Protection

In the event of outbreak of insects and diseases, the farmer can himself prepare homemade pesticides and use it on the crops.

Fungicide-I- Buttermilk fermentaed for 5 days – 5 l and water 50 l.
Fungicide-II-Deshi cow milk 5 l, Black pepper 200 g and water 200 l.
Insecticide-I-Powder of neem seed or leaves-20 kg and water 200 l,

Insecticide-II-Cow dung-5 kg, cow urine-10 l, neem leaves-10 kg and water 200 l. This mixture is particularly effective against aphids, jassids, mealy bugs and white flies. Insecticide-III-Neem leaves-10 kg, Tobacco powder-3 kg; Garlic paste-3 kg and Green chilies paste-4 kg.

The above ingredients should be soaked in cow urine for 10 days. About 3 l of this mixture can be mixed with 100 l of water and sprayed on crops. The above mentioned fungicides and insecticides can be prepared by the farmer himself and used either as prophylactic or as curative measure for control of crop pests. If the economic injury to crops due to pests is less than 5%, it should be deemed to be 'return to nature' and no plant protection measures should be taken.

3.3 *HOMA Organic Farming*

Agricultural historians re-discovered the most suitable technology to improve modern farming called HOMA Organic Farming, in the Vedas - the most ancient body of knowledge known to humankind. Their origin was in ancient India (http://www.homatherapyindia.com/ accessed on 29th Jan 2010). HOMA means bio-energy denoting the process healing and purifying the atmosphere toxins by fire. In HOMA Organic Farming Technology the basic tool is AGNIHOTRA, the copper pyramid fire tuned to bio-rhythms of sunrise and sunset. The process involves preparing a small fire in a copper pyramid of specified shape and size and putting some whole grain raw brown rice into fire exactly at sunrise and sunset accompanied by chant of two simple mantras for a few minutes. Dried Cow dung and pure cow's ghee (clarified butter) are used to prepare the fire. This completes the process of Agnihotra. The ash that results from the pooja is used to energies composts, plants, animals and humans. It is obviously extremely inexpensive and simple to undertake but requires discipline and regularity (http://www.homatherapyindia.com/ accessed on 29th Jan 2010).

Cow dung contains a substantial amount of Vitamin B12. The cow's dung increases microorganisms and earthworms in the soil improving soil fertility and cow's urine is used as an effective organic pest repellent as it contains sulphur, ammonia vapors, manganese, carbolic acid. Thus application of cow's urine increases immunity of the plants and makes them strong and healthy (http://www.homatherapyindia.com/ accessed on 29th Jan 2010). This has been already demonstrated on large scale in India and abroad (Dhama et al. 2005; Sofia et al. 2006). However more on farm demonstration trials and in depth scientific studies with adequate control are required. The results have to be published in peer reviewed international journals for wider acceptance among the agricultural scientists and extension workers as well as policy makers, planners and the farming communities.

3.4 *Low External Input and Sustainable Agriculture (LEISA)*

In 1983, Netherlands Directorate of International cooperation from the International centre for Low external input and sustainable agriculture (ILEIA) aimed to support farmers in Tropical resource-poor environment, mainly to exchange of information on results on experiences with low external input agriculture. ILEIA introduced the LEISA (Low external input sustainable agriculture) as a concept of sustainable agriculture that brings together environmental awareness and farmers economic and social objectives by developing techniques that enhances optimal use of locally available resources (Kessler and Moolhuijzen 1994; Babou et al. 2009).

This system renounces the use of agrochemicals and seeks to optimize the use of locally available resources by interlinking the components of the farm system.

External inputs shall only be used to provide elements that are deficient in the ecosystem and to enhance available biological, physical and human resources (Babou et al. 2009).

LEISA enables the creation of viable small scale farming, which is a major part of rural livelihoods and thus contributes significantly to developing economies. This system is about finding technical and social options open to farmers who seek to improve productivity and income in an ecologically sound way. The optimal use of local resources and natural processes is preferred and, if necessary, safe and efficient use of external inputs. It is about empowerment of farmers, communities, who endeavor to build their future on their own knowledge, skills, values, culture and institutions. LEISA also practice participatory methodologies to strengthen the capacity of farmers and other actors, to improve their livelihoods and adapt to changing needs and conditions. It seeks to influence policy formulation in a manner conducive to its further development (Kessler and Moolhuijzen 1994; Babou et al. 2009).

Main stones of LEISA (Kessler and Moolhuijzen 1994; Babou et al. 2009) are:

1. Agro-ecology: the knowledge base to apply ecological concepts and principles (for example: synergy, biodiversity, nutrient recycling, natural pest management, complementarily and resilience) to the design and management of sustainable agroecosystems.
2. Optimal and low-cost use of local and external resources: making best use of available local resources and, if necessary, efficient use of modest amounts of modern external inputs.
3. Indigenous knowledge: the knowledge of farmers generated by their own and their ancestors experiences combined with knowledge originating from elsewhere, which has been internalized.
4. Participatory learning, planning and action: participation of development supporters in the farmers development process to strengthen the latter and the own capacity to adapt agriculture to changing needs and conditions and towards sustainability.
5. Social justice and cultural integrity: economic development respecting cultural and social values of those segments of society that have little influence on economic and political decision - making. Special attention is given to specific gender issues, such as access to and control over information, resources and production outputs.

4 Conclusion

10 million ha i.e. more than 1.3% of total cropland are critically damaged each year globally by salinization mostly as a result of irrigation and improper drainage method and use of more chemical fertilizers (Nelson 1991; Melvin Wong 2005; Svotwa et al. 2008). Worlds forest areas are converted every year to additional

cropland needed which were lost. Soils are less fertile, erosion has greatly increased and breakdowns in agro-ecological functions have resulted in poor crop yields, land abandonment and deforestation (IAASTD 2008). The indiscriminate usage of chemical fertilizers has caused severe aquatic and environmental consequences such as - loss of topsoil, decrease in soil fertility, surface and ground water contamination and loss of genetic diversity (Palaniappan and Annadurai 1995; Drinkwater et al. 1998; Vasilikiotis 2000; Svotwa and Jiyane 2006). All these facts and figures emphasize the urgent need for sustaining agricultural production without sacrificing the vital ecosystem services.

There are a few scientists who support the view that biotechnology and Genetic Engineering are crucial for sustaining agricultural production in the future (Harlander 2002; Jauhar 2006; Thomson 2008). However, several other workers argue that genetically modified (GM) crops threaten human health and the environment and will allow large corporations to tighten their grip over agricultural production and thus widen socio-economic disparities. At present most GM crops are grown as high-priced animal feedstock to supply rich nations with meat, rather than to meet the immediate food security needs of local households. The benefits of GM crops are far from certain, risks on health, environmental and socio-economic effects are not yet clearly stated (Altieri 2000; Altieri and Rosset 1999; Newell 2009; Union of Concerned Scientists 2009).

Several researchers concluded that alternative farming techniques can produce enough food to feed the world without increasing the agricultural land base (Uphoff and Altieri 1999; Badgley et al. 2007; LaSalle et al. 2008). There are several papers that prove that biointensively and organically grown food are healthier and more nutritious than that of the food grown by intensive chemical agriculture (Jeavons 2001; Benbrook et al. 2008; Benbrook 2009).

In the future, we have to be concerned more with food and fuel demands of growing global population and expanding economies, rapidly shrinking of natural resources, degrading soil quality and greater climate variability, it becomes evident that the functioning of resilient and sustainable agriculture are dependent on healthy ecosystems (Ericksen et al. 2009). Hence agricultural environmental imperatives must be simultaneously tackled and an ecosystem approach be adopted in agricultural management (FAO 2009).

In continuous coping with extreme weather events and climatic variability, farmers living in harsh environments like regions of Africa, Asia and Latin America have developed and inherited complex farming systems that have the potential to bring solutions to many uncertainties facing humanity in an era of climate change. Recent observations, studies and research suggest that many farmers are prepared for climate change, minimizing crop failure through increased use of drought-tolerant local varieties, water harvesting, extensive planting, mixed cropping, agroforestry, opportunistic weeding, wild plant gathering and a series of other traditional farming system techniques (Osunade 1994; Adesina et al. 1999; Klein and Smith 2003; Nyong et al. 2007).

Hence, there is an urgent need for scientific documentation, verification and validation of the sustainability of traditional agro ecological knowledge base as well as the emerging appropriate agro eco technologies.

Acknowledgements The authors wish to place on record their deep sense of gratitude for the facilities provided by the Pondicherry University administration and the fellowship awarded to one of us (KP).

References

Adesina FO, Siyambola WO, Oketola FO, Pelemo DA, Ojo LO, Adegbugbe AO (1999) Potentials of agroforestry for climate change mitigation in Nigeria: some preliminary estimates. Glob Ecol Biogeogr 8:163–173

Agbonlabor MU, Aromolaran AB, Aiboni VI (2003) Sustainable soil management practices in small farms of Southern Nigeria: a poultry-food crop integrated farming approach. J Sustain Agr 22:51–62

Altieri MA (1999) The ecological role of biodiversity in agroecosystems. Agric Ecosyst Environ 74:19–31. doi:10.1016/S0167-8809(99)00028-6

Altieri MA (2000) Food first special report no. 1 Genetic engineering in agriculture: the myths, environmental risks, and alternatives, Food First/Institute for Food and Development Policy, Oakland

Altieri MA (2002) Agroecology: the science of natural resource management for poor farmers in marginal environments. Agric Ecosyst Environ 93:1–24. doi:10.1016/S0167-8809(02)00085-3

Altieri MA, Rosset P (1999) Ten reasons why biotechnology will not ensure food security, protect the environment and reduce poverty in the developing world. Ag Bio Forum 2:155–162

Amanullah MM, Somasundaram E, Sathyamoorthi K, Thukkaiyannan P (2008) Organic farming in relation to food and environmental security – a review. Green Farming 1(10–11):5–10

Andow DA (1991) Yield loss to arthropods in vegetationally diverse ecosystems. Environ Entomol 20:1228–1235

Andow DA, Hidaka K (1998) Yield loss in conventional and natural rice farming systems. Agric Ecosyst Environ 70(2–3):151–158. doi:10.1016/S0167-8809(98)00122-4

Anonymous (2001) Managing cover crops profitably, sustainable agriculture network. Sustainable Agriculture Publications, Burlington

Araújo SF, Luiz FC, Santos VB, Carneiro RFV (2009) Soil microbial activity in conventional and organic agricultural systems. Sustain 1:268–276. doi: 10.3390/su1020268

Atha A (2000) Container kitchen garden. Sterling Publication, New York

Ayenew YA, Maria W, Azage T, Werner Z (2007) Urban and peri-urban farming systems and utilization of the natural resources in the North Ethiopian Highlands, Conference on International Agricultural Research for Development, Tropentag 2007, University of Kassel-Witzenhausen and University of Göttingen, 9–11 Oct 2007

Babou C, Poyyamoli G, Guna B (2009) Impacts of LEISA based system of rice intensification (SRI) practices on rice grain yield and soil properties in rice – rice – rice cropping system in Puducherry region. Int J agricultural Sci 5(1):43–45

Badgley C, Moghtader J, Quintero E, Zakem E, Chappell M, Aviles-Vazquez K, Samulon A, Perfecto I (2007) Organic agriculture and the global food supply. Renewable Agric Food Syst 22(2):86–108

Badji CA, Guedes RNC, Silva AA, Corre^a AS, Queiroz MELR, Michereff-Filho M (2007) Non-target impact of deltamethrin on soil arthropods of maize fields under conventional and no-tillage cultivation. J Appl Entomol 131(1):50–58. doi:10.1111/j.1439-0418.2006.01118.x

Bahi GS, Pasricha NS (2001) Direct and residual effect of green manuring in relation to fertilizer nitrogen on Toria (Brassica napus L.) and sunflower (Helianthus annuus L.). J Indian Soc Soil Sci 49(1):113–117

Bailkey M, Nasr J (2000) From brownfields to greenfields: producing food in North American Cities, Community Food Security News, Fall 1999/Winter 2000, p 6

Baker BP, Benbrook CM, Groth E, Benbrook KL (2002) Pesticide residues in conventional, integrated pest management (IPM)-grown and organic foods: insights from three US data sets. Food Addit Contam 19:427–446

Balasubramanian AV, Nirmala Devi TD, Merlin FM (2009) (Ed) Use of animal products in traditional agriculture – a pilot project in Southern India. Centre for Indian Knowledge Systems, Chennai

Balfour E (1943) The living soil. Faber & Faber, London

Basamba TA, Amezquita E, Singh BR, Rao IM (2006) Effects of tillage systems on soil physical properties, root distribution and maize yield on a Colombian acid-savanna Oxisol. Acta Agric Scand B 56(4):255–262. doi:10.1080/09064710500297690

Bashan Y, Holguin G, Bashan LE (2004) Azospirillum – plant relationships: physiological, molecular, agricultural and environmental advances. Can J Microbiol 50(8):521–577

Beddington J (2010) (2009) Food security: contributions from science to a new and greener revolution. Phil Trans R Soc B 365:61–71. doi:10.1098/rstb.2009.0201

Ben Haggard (1993) Living community: a permaculture case study at Sol y Sombra, Center for the Study of Community, Santa Fe, p 160

Benbrook C (2009) The impacts of yield on nutritional quality: lessons from organic farming. Hortscience 44(1):12–14

Benbrook C, Xin Z, Yanez J, Davies N, Andrews P (2008) New evidence supports the nutritional superiority of organic plant-based foods. The Organic Center State of Science Review, http://www.organic-center.org/science.nutri.php?action=view&report_id=126

Bene JG, Beall HW, Cote A (1977) Trees, food and people. International Development Research Centre, Ottawa, p 52

Benton TG, Vickery JA, Wilson JD (2003) Farmland biodiversity: is habitat heterogeneity the key? Trends Ecol Evol 18(4):182–188

Bhaskaran, Usha P, Krishna D (2009) effect of organic farming on soil fertility, yield and quality of crops in the tropics. The proceedings of the international plant nutrition colloquium XVI, UC Davis, Davis. Retrieved from: http://www.escholarship.org/uc/item/7k12w04m

BirdLife (2008) Critically endangered birds: a global audit. BirdLife International, Cambridge. Available online at: http://www.birdlife.org/news/news/2008/09/Complete_Critical%20Birds_superlowres.pdf. Accessed 20 Jan 2009

Bisong F, Andrew-Essien E (2010) Indigenous knowledge systems for promoting community conservation education in a Nigerian protected area. Int J Biol 2(2):149–157

Blom-Zandstra M (2005) Trends in organic farming research in the Netherlands. In: Köpke U, Niggli U, Neuhoff D, Cornish P, Lockeretz W, Willer H (eds) Researching sustainable systems. Proceedings of the 1st scientific conference of the international society of organic agriculture research (ISOFAR), held in cooperation with the international federation of organic agriculture movement (IFOAM) and the National Association for Sustainable Agriculture, Australia (NASAA), 21–23 Sept 2005, Adelaide

Blowfield M (2001) Ethical trade and organic agriculture, Tropical Agriculture Association Newsletter, 22–26 March 2001

Bockstaller C, Girardin P, Van der Werf HGM (1997) Use of agroecological indicators for the evaluation of farming systems. Eur J Agron 7:261–270

Bradford J, Wichner C (2009) Sustainable agriculture whitepaper – local organic sustainable profitable, Farmland LP, San Francisco, p 22

Brickle NW, Harper DGC (2002) Agricultural intensification and the timing of breeding of Corn Bunting Miliaria calandra. Bird Study 49(3):219–228

Bugg RL, Trenham PC (2003) Agriculture affects amphibians in: climate change, landscape-scale dynamics, hydrology, mineral enrichment of water. Sustain Agric 15(2):8–11

Bulson HAJ, Snaydon RW, Stopes CE (1997) Effects of plant density on intercropped wheat and field beans in an organic farming system. J Agr Sci 128:59–71

Burgos NR, Talbert RE (1996) Weed control by spring cover crops and imazethapyr in no-till southern pea (Vigna unguiculata). Weed Technol 10:893–899

Butler L, Moronek DM (2002) (eds) Urban and agriculture communities: opportunities for common ground. Council for Agricultural Science and Technology (CAST), Ames, May 2002

Byerlee D, Alex G (2005) Organic farming: a contribution to sustainable poverty alleviation in developing countries? In German NGO forum on environment and development. EED, NABU and WWF, Bonn

Caraka-Samhita (1981) Editor-Translator Sharma P, Chaukhambha Orientalia, Varanasi, pp 1, 213

Carpenter SR, Bennett EM, Peterson GD (2006) Scenarios for ecosystem services: an overview. Ecol Soc 11(1):29, URL: http://www.ecologyandsociety.org/vol11/iss1/art/29

Carpenter SR, Mooney HA, Agard J, Capistrano D, Defries RS, Diaz S, Dietz T, Duraiappah AK, Oteng-Yeboah A, Pereira HM (2009) Science for managing ecosystem services: beyond the millennium ecosystem assessment. Proc Natl Acad Sci USA 106:1305–1312

Carpenter-Boggs L, Kennedy AC, Reganold JP (2000) Organic and biodynamic management: effects on soil biology. Soil Sci Soc Am J 64:1651–1659

Casper BB, Jackson RB (1997) Plant competition underground. Annu Rev Ecol Syst 28:545–570

Cavigelli MA, Hima BL, Hanson JC, Teasdale JR, Conklin AE, Yao-chi Lu (2009) Long-term economic performance of organic and conventional field crops in the mid-Atlantic region. Renewable Agric Food Syst 24(2):102–119. doi:10.1017/S1742170509002555

CBD (2006) Convention of biodiversity. http://www.iisd.ca/process/biodiv_wildlife.htm#cbd. Accessed on 29 Jan 2010

CDE – Chicago Department of Environment (2001) A guide to rooftop gardening. City of Chicago

Chamberlain DE, Fuller RJ, Bunce RGH, Duckworth JC, Shrubb M (2000) Changes in the abundance of farmland birds in relation to the timing of agricultural intensification in England and Wales. J Appl Ecol 37:771–788. doi:10.1046/j.1365- 2664.2000.00548.x

Chan KY (2001) An overview of some tillage impacts on earthworm population abundance and diversity – implications for functioning in soils. Soil Till Res 57:179–191

Chapman GP (2002) The green revolution. The companion to development studies. Arnold, London, pp 155–159

Chauhan RS (2004) Panchagavya therapy (Cowpathy): current status and future directions. Indian Cow 1:3–7

Chigora P, Dzinavatonga N, Mutenheri F (2007) Indigenous knowledge systems and the conservation of small grains seeds: a case of sangwe communal lands of Chiredzi in Zimbabwe. J Sustain Dev Afr 9(3):146–157

Choong MR, Chia-Hui H, Robert L, Joseph K (2005) Study of mechanisms for plant growth promotion elicited by rhizobacteria in Rabidopsis thaliana. Plant Soil 268(1):185–292

Choosakul S (1999) Challenging crisis with sustainable farming. Sustainable resource management project northeast region, Mahasarakham

Chopin T, Yarish C (1999) Seaweeds must be a significant component of aquaculture for an integrated ecosystem approach. Bull Aquacult Assoc Can 99:35–37

Clark MS, Horwath WR, Shennan C, Scow KM (1998) Changes in soil chemical properties resulting from organic and low- input farming practices. Agron J 90:662–671

Cooke RI, Queenborough SA, Elizabeth HA, Mattison, Bailey AP, Sandars DL, Graves AR, Morris J, Atkinson PW, Trawick P, Freckleton RP, Watkinson AR, Sutherland WJ (2009) Integrating socio-economics and ecology: a taxonomy of quantitative methods and a review of their use in agro-ecology. J Appl Ecol 46:269–277. doi:10.1111/j.1365-2664.2009.01615.x1

Costanza R, d'Arge R, de Groot R, Farber S, Grasso M, Hannon B, Limburg K, Naeem S et al (1997) The value of the world's ecosystem services and natural capital. Nature 387: 253–260

Costello M (1999) Native grass species for use as perennial cover crops in San Joaquin Valley Vineyards. Final report. Sustainable agriculture research and education program. http://www.sarep.ucdavis.edu/

Cox TS, Picone C, Jackson W (2004) Research priorities in natural systems agriculture, the land institute, Salina, KS, USA. J Crop Improv 12(1/2):511–531

Craswell ET (1998) Sustainable crop and soil management on sloping lands. Paper presented at the international symposium on Asian agriculture in the 21st century. Food and Fertilizer technology, Center for the Asian and Pacific, Taipei, ROC, 9–12 June

Creamer NG, Bennett MA (1997) Evaluation of cover crop mixtures for use in vegetable production systems. HortScience 32:866–870

Csavas I (1992) Regional review on livestock-fish production systems in Asia. In: Proceedings of the FAO/IPT workshop on integrated livestock-fish production systems, 16–20 Dec 1991, Institute of Advance Studies, University of Malaya, Kuala Lumpur

Daily GC, Matson PA, Vitousek PM (1997) Ecosystem services supplied by soil. In: Daily G (ed) Nature's services: societal dependence on natural ecosystems. Island Press, Washington, DC, pp 113–132

Dalea VH, Polasky S (2007) Measures of the effects of agricultural practices on ecosystem services. Ecol Econ 64:286–296

Defra (2003) Agriculture and biodiversity, USA. Department for Environment, Food and Rural Affairs, London

Delate KM, Duffy C, Chase A, Holste H, Friedrich (2003) An economic comparison of organic and conventional grain crops in a long-term agroecological research (LTAR) site in Iowa, Am J Altern Agric, 18(2):59–69. Journal paper No. 19802 of the Iowa Agriculture and Home Economics Expt. Sta., Ames, 2003

Demeter International e V (2008) A World-wide network. http://www.demeter.net/ (Verified 10 Dec 2009)

Dhama K, Rathore R, Chauhan RS, Tomar S (2005) Panchgavya (Cowpathy): an overview. Int J Cow Sci 1(1):1–15

Ditsch DC, Alley MM, Kelley KR, Lei YZ (1993) Effectiveness of winter rye for accumulating residual fertilizer N following corn. J Soil Water Conserv 48:125–132

Donald P (2004) Biodiversity impacts of some agricultural commodity production systems. Conserv Biol 18:17–38

Dore T, Sadok W, Angevin F, Bergez J-E, Bockstaller C, Colombc B, Guichard L, Reau R (2007) Ex ante assessment of the sustainability of alternative cropping systems: implications for using multi-criteria decision-aid methods, a review. Agron Sustain Dev 27, INRA, EDP Sciences, 2007. Doi:10.1051/agro: 2007043

Doshi RT, Doshi S, Shah V (2003) City farming – the natural alternative, experiences in India. Urban Agric Mag 10:18–19

Dreyfus F, McIntyre BD (2009) Historical analysis of the effectiveness of AKST systems in promoting innovation. In: Dreyfus F, McIntyre BD (eds) International assessment of agricultural knowledge, science and technology for development: global report. Island Press, Washington, DC, pp 57–144

Drinkwater LE, Wagoner P, Sarrantonio M (1998) Legume-based cropping systems have reduced carbon and nitrogen losses. Nature 396:262–265. doi:10.1038/24376

Edge D (2000) Wildlife and agricultural ecosystems. Department of Fisheries and Wildlife, Oregon State University, Corvallis

Egger K (1990) Ecofarming: a synthesis of old and new. Institute of Botany INF 360, University of Heidelberg 0–6900, Heidelberg

Elmi A, Mehdi B, Madramootoo C, Dam R, Smith D (2009) Long-term effect of conventional and no-tillage production systems on nitrous oxide fluxes from corn (Zea mays L.) Field in Southwestern Quebec. Am J Environ Sci 5(3):238–246, ISSN 1553-345X

Ericksen PJ, Ingram JSI, Liverman DM (2009) Food security and global environmental change: emerging challenges. Environ Sci Policy 12:373–377

Evenson RE, Gollin D (2003) Assessing the impact of the Green Revolution, 1960 to 2000. Science 300:758–762

Ewel JJ (1986) Designing agricultural ecosystems for the humid tropics. Ann Rev Ecol Syst 17:245–271

Eyzaguirrea P, Di Gregoriob M, Meinzen-Dick R (2007) Introduction to the special issue on "property rights, collective action, and local conservation of genetic resources", editorial. World Dev 35(9):1481–1488

Falvey L (2005) Reconceiving food security and environmental protection. Asian J Agric Dev 1(2):13–28

FAO (2006) World agriculture: towards 2030/2050 – interim report. Global Perspective Studies Unit, Rome

FAO (2008) World food and agriculture in review part II. http://www.fao.org/docrep/011/i0100e/i0100e00.htm. Retrieved in 11 Nov 2009

FAO (2009) FAO COAG/2009/3 "Agriculture and environmental challenges of the twenty-first century: a strategic approach for FAO", Committee on agriculture twenty-first session, FAO, Rome, 22–25 Apr 2009, p 11

FAO (2009) FAOSTAT. http://faostat.fao.org/. Accessed 09 Apr 2009

FAOSTAT (2008) Statistics Division Resource STAT. FAO, Rome. http://faostat.fao.org/site/405/default.aspx. Accessed 30 Dec 2008

Firbank LG, Petit S, Smart S, Blain A, Robert J, Fuller RJ (2007) Assessing the impacts of agricultural intensification on biodiversity: a British perspective. Phil Trans R Soc B 363:777–787. doi:10.1098/rstb.2007.2183

Fischer G, Shah M, Tubiello FN, van Velhuizen H (2005) Socio-economic impacts on agriculture: an integrated assessment, 1990–2080. Phil Trans R Soc B 360:2067–2083

Fischer G, Tubiello FN, van Velthuizen H, Wiberg DA (2006a) Climate change impacts on irrigation water requirements: effects of mitigation, 1990–2080. Technol Forec Soc Change 74:1083–1107. doi:10.1016/j.techfore.2006.05.021

Fischer J, Lindenmayer DB, Manning AD (2006b) Biodiversity, ecosystem function, and resilience: ten guiding principles for commodity production landscapes. Front Ecol Environ 4:80–86

Fleury A, Ba A (2005) Multifunctionality and sustainability of urban agriculture. Urban Agriculture Magazine, Multiple functions of urban agriculture. Leusden, RUAF, no. 15

Flint, ML (1998) (eds) Pests of the garden and small farm: a growers guide to using less pesticides. University of California Press, Riverside

Follis MB (1993) Economic considerations. In: Nair PKR (ed) Introduction to agroforestry. Kluwer, Dordrecht, pp 385–411

Food and Agriculture Organization (FAO) (1981) Agricultural production statistics. FAO, Rome. Available from http://www.apps.fao.org/

Food and Agriculture Organization of the United Nations (2007) Annual FAO report, The State of Food Insecurity in the World (SOFI). http://www.fao.org/docrep/011/i0100e/i0100e00.htm Retrieved in 11 Nov 2009

Fraser EDG (2002) Urban ecology in Bangkok Thailand: community participation, urban agriculture and forestry. Environments 30(1):2002

Fukuoka M (1985) The natural way of farming. Japan Publications, Tokyo

Fukuoka, Masanobu (1978) The one-straw revolution – Translated from the Japanese by Chris Pearce, Tsune Kurosawa and Larry Korn, Originally published in Japan as Shizen Noho Wara Ippon No Kakumei. Published by Other India Press, Mapusa

Fuller RJ, Norton LR, Feber RE, Johnson PJ, Chamberlain DE, Joys AC, Mathews F, Stuart RC, Townsend MC, Manley WJ, Wolfe MS, Macdonald DW, Firbank LG (2005) Benefits of Organic Farming vary among Taxa. Biol Lett 1:431–434. doi:10.1098/rsbl.2005.0357

Fulton C (2005) Growing sustainable communities: community gardens in the Australian organic movement. In: Köpke U, Niggli U, Neuhoff D, Cornish P, Lockeretz W, Willer H (eds) Researching sustainable systems. Proceedings of the 1st scientific conference of the international society of organic agriculture research (ISOFAR), held in cooperation with the international federation of organic agriculture movement (IFOAM) and the national association for sustainable agriculture, Australia (NASAA), 21–23 Sept 2005, Adelaide

Garg N, Chauhan RS (2003) A cow therapy as an alternative to antibiotics. Indian Science Congress, Bangalore, pp 3–7

Garnham L (2002) Green roofs and the promise of urban agriculture. The Green Roof Infrastructure Monitor 4(2):17–19. www.greenroofs.ca

Geier U, Frieben B, Gutsche V, Köpke U (2001) Ökobilanz des Apfelerzeugung in Hamburg – Vergleich integrierter und ökologischer Bewirtschaftung. Schriftenreihe Institut für OrganischenLandbau. Verlag Dr. Köster, Berlin, p 130

Genghini M, Gellini S, Gustin M (2006) Organic and integrated agriculture: the effects on bird communities in orchard farms in northern Italy. Biodivers Conserv 15:3077–3094

Gianessi L (2009) The potential for organic agriculture to feed the world is being oversold, out looks of pest management. Crop Life Foundation, Washington, DC

Gliessman SR (1998) Agroecology: ecological processes in sustainable agriculture. Ann Arbor Press, Chelsea, MI p 357

Gliesmann SR (2001) Agroecosystem sustainability: developing practical strategies. CRC Press, Boca Raton

Gliesman SR, Garcia R, Amador M (1981) The ecological basis for the application of traditional agricultural technology in the management of tropical agro-ecosystem. Agro Ecosyst 7: 173–185

Godwin RJ, Wood GA, Taylor JC, Knight SM, Welsh JP (2003) Precision farming of cereal crops: a review of a six year experiment to develop management guidelines. Biosyst Eng 84:375–391. doi:10.1016/ S1537-5110(03)00031-X

Goldsmith E, Hildeyard E (1996) (eds) The earth report 2. Mitchell Beazley, London

Gooding MJ, Kasyanova E, Ruske R, Hauggaard-Nielsen H, Jensen ES, Dahlmann C, Von Fragsten P, Dibet A, Corre-Hellou G, Crozat Y, Pristeri A, Romeo M, Monti M, Launay M (2007) Intercropping with pulses to concentrate nitrogen and sulphur in wheat. J Agr Sci 145: 469–479

Greene C, Kremen A (2003) United States organic farming in 2000–2001: adoption of certified systems. Agriculture Information Bulletin No. 780. Economic Research Service, Resource Economics Division, US Department of Agriculture, Washington, DC

Grime JP (1977) Evidence for the existence of three primary strategies in plants and its relevance to ecological and evolutionary strategy. Am Nat 111:1169–1194

Groenfeldt D (2005) Multifunctionality of agricultural water: looking beyond food production and ecosystem service. Paper prepared for the FAO/Netherlands international conference on water for food and ecosystems. The Hague, 31 Jan–5 Feb 2005

Gunnarsson CC, Petersen CM (2007) Water hyacinths as a resource in agriculture and energy production: a literature review. Waste Manage 27:117–129

Gurr GM, Wratten SD, Luna JM (2003) Multi-function agricultural biodiversity: pest management and other benefits. Basic Appl Ecol 4:107–116

Haas G, Köpke U (1994) Vergleich der Klimarelevanz Ökologischer und Konventioneller Landbewirtschaftung. Studie (H) im Auftrag der Enquetekomission des Deutschen Bundestages Schutz der Erdatmosphäre. Economica Verlag, D-Karlsruhe

Haggard B (1993) Living community: a permaculture case study at Sol y Sombra. Center for the Study of Community, Santa Fe, p 160

Halvorson AD, Mosier AR, Reule CA, Bausch WC (2006) Nitrogen and tillage effects on irrigated continuous corn yields. Agron J 98:63–67

Hansen L, Noe E, Højring K (2006) Nature and nature values in organic agriculture. An analysis of contested concepts and value among different actors in organic farming. J Agric Environ Ethics 19:147–168. doi:10.1007/s10806-005-1804-y

Harlander KS (2002) The evolution of modern agriculture and its future with biotechnology. J Am Coll Nutr 21:161S–165S

Hassan RM (2010) Implications of climate change for agricultural sector performance in Africa: policy challenges and research agenda. J Afr Econ 19(Suppl 2):77–105

Hauggaard-Nielsen H, Jensen E (2005) Facilitative root interactions in intercrops. Plant Soil 274:237–250

Hauggaard-Nielsen H, Ambus P, Jensen ES (2001) Interspecific competition, N use and interference with weeds in pea-barley intercropping. Field Crop Res 70:101–109

Hazell P, Wood S (2008) Drivers of change in global agriculture. Phil Trans R Soc B 363:495–515. doi:10.1098/rstb.2007.2166

Hellin J, Higman S (2002) Smallholders and niche markets: lessons from the Andes. Overseas development institute agricultural research and extension network paper no. 118. ITDG Publishing, Oxford, pp 1–6

Hepperly P, Douds D Jr, Seidel R (2006) The Rodale faming systems trial 1981 to 2005: longterm analysis of organic and conventional maize and soybean cropping systems. In: Raupp J, Pekrun C, Oltmanns M, Köpke U (eds) Long-termfield experiments in organic agriculture. International Society of Organic Agriculture Research (ISOFAR), Bonn, pp 15–32

Hiltbrunner J, Scherrer C, Streit B, Jeanneret P, Zihlmann U, Tschachtli R (2008) Long-term weed community dynamics in Swiss organic and integrated farming systems. Weed Res 48:360–369

Hobbs PR, Sayre K, Gupta R (2008) The role of conservation agriculture in sustainable agriculture. Phil Trans R Soc B 363:543–555. doi:10.1098/rstb.2007.2169

Holderbaum JF, Decker AM, Meisinger JJ, Mulford FR, Vough LR (1990) Fall seeded legume cover crops for no-tillage corn in the humid East. Agron Journa 82(1):17–124

Horrigan L, Lawrence RS, Walker P (2002) How sustainable agriculture can address the environmental and human health harms of industrial agriculture. Environ Health Perspect 110(5):445–456

Hussain I, Olso KR, Ebelhar SA (1999) Long term tillage effects on soil chemical properties and organic matter fractions. Soil Sci Soc Am J 63:1335–1341

IAASTD (2008) International assessment of agricultural knowledge, Science and Technology for Development, Global Report

IEA (2008) World energy outlook 2008. Paris, France: International Energy Agency Mitchell, D. 2008 A note on rising food prices. World Bank Policy Research Working Paper Series, No. 4682, World Bank, New York

IFOAM (2006) First IFOAM conference on organic certification. Available www.fao.org/docs/eims/upload/229843/IFOAM-organic-certification.pdf

Igbozurike UM (1978) Polyculture and monoculture: contrast and analysis, although polyculture or interplanting is a dominant production method used by traditional cultivators. GeoJournal 2(5):443–449

Imhoff ML, Bounoua L, Ricketts T, Loucks C, Harriss R, Lawrence WT (2004) Global patterns in human consumption of net primary production. Nature 429:870–873. doi:10.1038/nature02619

IPCC (2007) Climate change: Physical change basis. http://ipcc-wg1.ucar.edu/wg1/wg1-report.html. Retrieved 2007-04-30

Irfanullah HMd, Ahana A, Abdul G, Zakir AK, Abdur Rashid Md (2008) Introduction of floating gardening in the north-eastern wetlands of Bangladesh for nutritional security and sustainable livelihood. Renewable Agric Food Syst 23(2):89–96

Islam T, Atkins P (2007) Indigenous floating cultivation: a sustainable agricultural practice in the wetlands of Bangladesh. Development in Practice 17(1):130–136. doi:10.1080/09614520601092733

IUCN Bangladesh (2005) Baira: the floating gardens for sustainable livelihood. IUCN-the world conservation union. Bangladesh Country Office, Dhaka, p 61

Jackson W, Jackson LL (1999) Developing high seed yielding perennial polycultures as a mimic of mid-grass prairie. In: Lefroy EC, Hobbs RJ, O'Connor MH, Pate JS (eds) Agriculture as a mimic of natural systems. Kluwer, Dordrecht, pp 1–37

Jauhar PP (2006) Modern biotechnology as an integral supplement to conventional plant breeding: the prospects and challenges. Crop Sci 46:1841–1859

Jayanthi C, Rangasamy A, Chinnusamy C (2000) Water budgeting for components in lowland integrated farming systems. Agric J 87:411–414

Jeavons JC (2001) biointensive sustainable mini-farming: I. The challenge. J Sustainable Agr 19(2):49–63

Jianming C (2003) Periurban agriculture development in China, Urban Agriculture Magazine 2009, Retrieved 7 Dec 2007. http://www.ruaf.org/system/files?file=Periurban%20 Agriculture %20Development%20in%20China.pdf

Johnsen K, Jacobsen CS, Torsvik V, Sørensen J (2001) Pesticide effects on bacterial diversity in agricultural soils – a review. Biol Fertil Soils 33:443–453. doi:10.1007/s003740100351

Jose S, Gordon AM (eds) (2008) Toward agroforestry design: an ecological approach, advances in agro forestry 4. Springer, Dordrecht, p 312

Katai O, Minamizono K, Shiose T, Kawakami H (2007) System design of "Ba"-like stages for improvisational acts via Leibnizian space–time and Peirce's existential graph concepts. AI Soc 22:101–112

Keating BA, Carberry PS (1993) Resource capture and use in intercropping – solar radiation. Field Crop Res 34:273–301

Kegley S, Neumeister, Lars, Martin, Timothy (1999) Disrupting the balance. Pesticide action network of North America. http://www.panna.org/resources/documents/disrupting.pdf

Kessler JJ, Moolhuijzen M (1994) Low external input sustainable agriculture: expectations and realities. Netherlands J Agric Sci 42:181–194

Keyzer MA, Merbis MD, Pavel IFPW, van Wesenbeeck CFA (2005) Diet shifts towards meat and the effects on cereal use: can we feed the animals in 2030? Ecol Econ 55:187–202

Khanal RC (2009) Climate change and organic agriculture. The Journal of Agriculture and Environment 10:100–110

Kiers ET, Leakey RRB, Izac AM, Heinemann JA, Rosenthal E, Nathan D, Jiggins J (2008) Agriculture at a crossroads. Science 320:320–321. doi:10.1126/science.1158390

Kim Tai-cheol (2008) Reducing agricultural water use while conserving ecosystems. J Dev Sustain Agric 3:1–8

King AD (1983) Progress in no-till. J Soil Water Conserv. Special issue "Conservation Tilllage", 8:160–161

KKU-Khon Kaen University (2001) New theory: water and land management for agriculture. Study report on agriculture following the Kings concept. Khon Kaen University, Khon Kaen

Klein RJT, Smith JB (2003) Enhancing the capacity of developing countries to adapt to climate change: a policy-relevant research agenda. In: Smith JB, Klein RJT, Huq S (eds) Climate change, adaptive capacity and development. Imperial College Press, London, pp 317–401

Kremen, Ricketts (2000) Global perspectives on pollination disruptions. Conserv Biol 14:1226–1228

Kus J, Stalenga J (2000) Comparison of economic and energy efficiency in ecological and conventional crop production system. In: Alföldi T, Lockeretz W, Niggli N (eds) Proceedings of 13th international IFOAM scientific conference. Vdf Hochschulverlag, Zürich

Lal R (1997) Residue management, conservation tillage and soil restoration for mitigating greenhouse effect by CO_2 enrichment. Soil Tillage Res 43:81–107

Lal R (2008) Soils and sustainable agriculture – a review. Agron Sustain Dev 28:57–64. doi:10.1051/agro:2007025

Landis DA, Wratten SD, Gurr GM (2000) Habitat management to conserve natural enemies of arthropod pests in agriculture. Annu Rev Entomol 45:175–201

LaSalle T, Hepperly P, Diop A (2008) The organic green revolution, Rodale Institute. Available at www.rodaleinstitute.org/files/GreenRevUP.pdf. Accessed 2 Jan2009

Lawrence PA, Radford BJ, Thomas GA, Sinclair DP, Key AJ (1994) Effect of tillage practices on wheat performance in a semi-arid environment. Soil Tillage Res 28:347–364

Lee D (2005) The adoption of low-external input sustainable agriculture in developing Countries. AAEA 87:1325–1334

Letourneau DK, Bothwell SG (2008) Comparison of organic and conventional farms: challenging ecologists to make biodiversity functional. Front Ecol Environ 6:430–438. doi:10.1890/070081

Lichtenberg E (1992) Alternative approaches to pesticide regulation. Northeast J Agric Resour Econ 21:83–92

Liebhardt W (2001) Get the facts straight: organic agriculture yields are good. Organic Farming Research Foundation Information Bulletin Summer 2001 Number 10

Liebman M, Davis AS (2000) Integration of soil, crop and weed management in low-external-input farming systems. Weed Res 40:27–47

Lightfoot C, Minnick DR (1991) Farmer-first qualitative methods: farmers diagrams for improving methods of experimental design in integrated farming systems. J Farming Syst Res Ext 2:11–34

Limpinuntana V, Trelo-ges V, Vityakon P, Patanothai A (2001) Sustainability analysis of existing land-use systems in northeast Thailand. In: Simmons RW, Noble AD, Lefroy, RDB (eds) International workshop on nutrient balance for sustainable agricultural production and natural resources management in Southeast Asia, 20–22 Feb 2001, Bangkok

Lisa R, Zack Z (2002) Growing food in the southwest mountains: a permaculture approach to home gardening above 6,500 feet in Arizona, New Mexico, Southern Colorado and southern Utah, Published Flagstaff Tea Party

Little DC, Edwards P (2003) Integrated livestock-fish systems. FAO, Rome

Lockeretz W (ed) (2007) Organic farming: an international history. CABI, London, p 304

Lotter DW (2003) Organic agriculture. Journal of Sustainable agriculture 21(4):59–128

Luke G (2002) Green roofs and the promise of urban agriculture. The Green Roof Infrastructure Monitor 4(2):17–19, www.greenroofs.ca

Mäder P, Fließbach A, Dubois D, Gunst L, Fried P, Niggli U (2002) Soil fertility and biodiversity in organic farming. Science 296:1694–1697

Macgregao N, Hill SB (2006) A journal whose time has come. Available online at http://www.organicsystems.org/journal/0601/pdf/macgregor-hill.pdf. Verified 14 Dec 2006

Mader P (2002) Soil fertility and biodiversity in organic farming. Science 296:1694–1697

Mader P, Fliessbach A, Dubois D, Gunst L, Fried P, Niggli U (2002) Soil fertility and biodiversity in organic farming. Science 296:1694–1697

Magdoff F, Harold VE (2000) Building soils for better crops, 2nd edn, Sustainable Agriculture Network. Sustainable Agriculture Publications, Burlington

Makatouni A (2001) What motivates consumers to buy organic food in the United kingdom: results from a qualitative study. Department of Agricultural and Food Economics, University of Reading, Reading. http://www.organic-research.com/ Accessed 2 June 2003

Mallik MAB, Williams RD (2008) Plant Growth Promoting Rhizobacteria and Mycorrhizal Fungi in Sustainable Agriculture and Forestry. In: Ren SZ, Mallik AU, Shi ML (eds) Allelopathy in sustainable agriculture and forestry. Springer, New York, pp 321–345

Maposah-Kandemiri M, Peter H, Pat M (2009) A outdoor learning: curriculum imperatives and community relevance in a rural setting. Education 3–13 37(1):15–28

Matriu VN, Dakora FD (2004) Potential use of rhizobial bacteria as promoters of plant growth for increased yield in landraces of African cereal crops. Afr J Biotech 3(1):1–7

Matson PA, Parton WJ, Power AB, Swift MJ (1997) Agricultural intensification and ecosystem properties. Science 277:504–509

McNeely JA, Sara JS (2003) Sourcebook: reconciling agriculture and wild biodiversity: policy and research challenges of Ecoagriculture. Available online www.ecoagriculturepartners.org/home.htm

Mendelsohn R, Williams L (2006) Dynamic forecasts of the sectoral impacts of climate change. In: Schlesinger ME et al (eds) Human-induced climate change: an interdisciplinary assessment. Cambridge University Press, Cambridge

Mertz O, Ravnborg HM, Lövei GL, Nielsen I, Konijnendijk CC (2007) Ecosystem services and biodiversity in developing countries. Biodivers Conserv 16:2729–2737. doi:10.1007/s10531-007-9216-0

Mitchell D (2008) A note on rising food prices. World Bank Policy Research Working Paper Series, No. 4682, World Bank, New York, NY

Milestad R, Hadatsch S (2003) Organic farming and social-ecological resilience: the alpine valleys of Sölktäler, Austria. Conser Ecol 8(1):3. http://www.consecol.org/vol8/iss1/art3/

Mollison B (1988) Permaculture: a designer's manual. Tagari Publications, Tyalgum

Mollison B (1991) Introduction to permaculture. Tagari Publications, Tyalgum, NSW

Muller A (2009) Benefits of organic agriculture as a climate change adaptation and mitigation strategy for developing countries. Environment for development, discussion paper series, EfD DP 09–09, p 15

Murugesan AG, Vijayalakshmi GS, Sukumaran N, Mariappan C (1995) Utilization of water hyacinth for oyster mushroom cultivation. Bioresour Technol 51:97–98

Nair PKR (2008) Agroecosystem management in the 21st century: it is time for a paradigm shift. J Trop Agri 46(1–2):1–12

Natarajan K (2002) Panchagavya – a manual. Other India Press, Mapusa, p 33

Nautiyal S, Bisht V, Rao KS, Maikhuri RK (2008) The role of cultural values in agrobiodiversity conservation: a case study from Uttarakhand, Himalaya. J Hum Ecol 23(1):1–6

Neera P, Katano M, Hasegawa T (1999) Comparison of rice yield after various years of cultivation of natural farming, Plant Prod Sci 2(1):58–64

Nelson PV (1991) In: Nelson PV (ed) Greenhouse operation and management, 4th edn. Prentice Hall, Englewood Cliffs, pp 257–316

Newell P (2009) Bio-hegemony: the political economy of agricultural biotechnology in Argentina. J Lat Amer Stud 41:27–57. doi:10.1017/S0022216X08005105

Ngambeki DS, Deuson RR, Preckel PV (1992) Integrating livestock into farming systems in northern Cameroon. Agr Syst 38:319–338

Nickel JL (1973) Pest situations in changing agricultural systems – a review. Bull Entomol Soc Am 19:136–142

Niemeyer K, Lombard J (2003) Identifying problems and potential of the conversion to organic farming in South Africa. http://www.organicagcentre.ca/OCs/res_policyNiemeyer.pdf. Accessed June 2005

Noble AD, Ruaysoongnern S (2002) The role of indigenous technology and science in rehabilitating degraded light textured soils using high activity clays and bioremediation. International Water Management Institute (IWMI) and Khon Kaen University (KKU), Khon Kaen

Nyong A, Adesina F, Osman EB (2007) The value of indigenous knowledge in climate change mitigation and adaptation strategies in the African Sahel. Mitig Adapt Strat Glob Change 12:787–797. doi:10.1007/s11027-007-9099-0

Okigbo BN (1995) Major farming systems of the lowland savanna of SSA and the potential for improvement. In: Proceedings of the IITA/FAO workshop, Ibadan

Osunade MA (1994) Indigenous climate knowledge and agricultural practices in Southwestern Nigeria. Malays J Trop Geogr 1:21–28

Palaniappan SP, Annadurai K (1995) Nitrate pollution in ground water. Agricultural Inputs and Environment, Scientific Publishers, Jodhpur, pp 1–23

Perrin RM (1977) The role of environmental diversity in crop protection. Prot Ecology 2:77–114

Piesse J, Thirtle C (2009) Three bubbles and a panic: an explanatory review of recent food commodity price events. Food Policy 34:119–129. IEA 2007 World energy outlook 2007. International Energy Agency, Paris

Pimentel D (1995) Amounts of pesticides reaching target pests: environmental impacts and ethics. J Agric Environ Ethics 8(1):17–29

Pimentel D (2006) Impacts of organic farming on efficiency and energy use in agriculture, p 40. www.organicvalley.coop/fileadmin/pdf/ENERGY_SSR.pdf

Pimentel D, Hepperly P, Hanson J, Douds D, Seidel R (2005) Environmental, energetic, and economic comparisons of organic and conventional farming systems. Bioscience 55:573–582

Piper JK (1999) Natural systems agriculture. In: Collins WW, Qualset CO (eds) Biodiversity in agroecosystems. CRC Press, Boca Raton, pp 167–195

Ponting C (2007) A new green history of the world: the environment and the collapse of great civilizations. Penguin Books, New York, p 244. ISBN 978-0-14-303898-6

Poulton C (2004) Economic opportunities for the poor. http://www.dfid-agricultureconsultation.nri.org/economic-opportunity.html. Accessed 16 Apr 2004

Praphan N (2001) Resilient of indigenous knowledge, fight to world crisis. Isan Alternative Farming Network, Ubonratchathani

Prell J (2010) Natural farming- natural farming with indigenous micro organisms. Acres 40:36–37

Pretty JN (1995) Regenerating agriculture. Earthscan Publications Ltd, London

Pretty J (2008) Agricultural sustainability: concepts, principles and evidence. Phil Trans R Soc B 363:447–465. doi:10.1098/rstb.2007.2163, Published online 25 July 2007

Pretty J, Noble A, Bossio D, Dixon J, Hine RE, Penning de Vries P, Morison JIL (2006) Resource conserving agriculture increases yields in developing countries. Environ Sci Technol 40: 1114–1119. doi:10.1021/es051670d

Price T (2000) Cultivation our futures. Final paper. OECD publications no. 2. Organization for economic co-operation and development. Processing ponds and fish product ponds receiving animal wastes, pp 87–97. In: Pullins, RSV; ZH, Shehadeh (eds) Integrated agriculture-aquaculture farming system ICLARM-SEARCH, Manila

Quansah C (2003) The requisite institutional framework to support the development of an organic industry. Department of Crop Science, Faculty of Science and Technology (KNUST), Kumasi, Ghana. Paper presented at workshop on organic industry, 15–17 March, Cape Town

Radhammani S, Balasubramanian A, Ramamoorthy K, Geethalakshmi V (2003) Sustainable integrated farming systems for dry lands: a review. Agric Rev 24:204–210

Raupp J, Ko¨nig UJ (1996) Biodynamic preparations cause opposite yield effects depending upon yield levels. Biol Agric Hortic 13:175–188

Rayner L, Zack Z (2002) Growing food in the southwest mountains: a permaculture approach to home gardening above 6,500 feet in Arizona, New Mexico, southern Colorado and southern Utah, Published Flagstaff Tea Party

Reeve JR, Carpenter-Boggs L, Reganold JP, York AL, McGourty G, McCloskey LP (2005) Soil and winegrape quality in biodynamically and organically managed vineyards. Am J Enology Viticulture 56:367–376

Reganold JP (1995) Soil quality and profitability of biodynamic and conventional farming systems: a review. Am J Altern Agric 10:36–45

Reganold JP, Palmer AS, Lockhart JC, Neil MA (1993) Soil quality and financial performance of biodynamic and conventional farms in New Zealand. Science 260(5106):344–349

Remans R, Spaepen S, Vanderleyden J (2006) Auxin signaling in plant defense. Science 313:171

Rigby D, Young T, Burton M (2001) The development of and prospects for organic farming in the United Kingdom. Food Policy 26(1):599–613

Rigueiro RA, McAdam JH, Mosquera- Losada MR (eds) (2008) Agroforestry in Europe. Advances in agroforestry 6. Springer, Dordrecht, p 452

Roberson EB, Sarig S, Firestone MK (1991) Cover crop management of polysaccharide mediated aggregation in an orchard soil. Soil Sci Soc Am J 55:734–739

Robertson GP, Swinton SM (2005) Reconciling agricultural productivity and environmental integrity: a grand challenge for agriculture. Front Ecol Environ 3:38–46

Robinson T (2008) Output of a food security and engineering conference. See http://www.raeng.org.uk/events/pdf/252/Tom_Robinson.pdf

Robinson RA, Wilson JD, Crick HQP (2001) The importance of arable habitat for farmland birds in grassland landscapes. J Appl Ecol 38:1059–1069

Ruaysoongnern S, Suphanchaimant N (2001) land-use patterns and agricultural production systems with emphasis on changes driven by economic forces and market integration. In: Kam SP, Hoanh CT, Trebuil G, Hardy B (eds) Natural resource management issues in the Korat basin of northeast Thailand: an Overview. Proceedings of the planning workshop on ecoregional approaches to natural resource management in the Korat Basin, Northeast Thailand: Towards further research collaboration, held on 26–29 Oct 1999, Khon Kaen. Los Banos (Philippines): International Rice Research Institute, pp 67–77

Ruttan V (1977) The green revolution: seven generalizations. Int Dev Rev 19:16–23

Ryan M, Ash J (1999) Effects of phosphorus and nitrogen on growth of pasture plants and VAM fungi in SE Australia soils with contrasting fertilizer histories (conventional and biodynamic). Agric Ecosyst Environ 73:51–62

Sachchidananda, Rajiv RJ (1999) Ecological cultivation in the Karanpura Region – a case study. http://ignca.nic.in/cd_07015.htm

Sainju UM, Singh BP (1997) Winter cover crops for sustainable agricultural systems: influence on soil properties, water quality, and crop yields. HortScience 32:21–28

Sanders R (2006) A market road to sustainable agriculture? Ecological agriculture, green food and organic agriculture in China. In: Dev Change 37(1):201–226. Institute of Social Studies. Blackwell Publishing, Malden

Sandhua HS, Wrattena SD, Cullen R, Case B (2008) The future of farming: the value of ecosystem services in conventional and organic arable land. An experimental approach. Ecol Econ 6: 835–848

Sandhua HS, Wratten SD, Cullen R (2010) Organic agriculture and ecosystem services. Environ Sci Policy 13(1):1–7

Sangeetha V, Thevanathan R (2010) Biofertilizer, potential of traditional and panchagavya amended with seaweed extract. Am Sci 6(2):61–67; published online in 8 Aug 2009 at americanscience.org/journals/am-sci/.../07_1124_paper_am0602.pdf

Saubidet MI, Fatta N, Barneix AJ (2000) The effects of inoculation with Azospirillum brasilense on growth and nitrogen utilization by wheat plants. Plant Soil 245(2):215–222

Saxena S, Garg V, Chauhan RS (2004) Cow urine therapy: promising cure for human ailments. Indian Cow 1:25–30

Schierre JB, Ibrahim MNM, van Keulen H (2002) The role of livestock for sustainability in mixed farming: criteria and scenario studies under varying resources allocation. Agr Ecosyst Environ 90:139–153

Scialabba NE (2007) Organic agriculture and food security. Paper presented at International Conference on Organic Farming and Food Security, 3–5 May, Food and Agriculture Organisation, Rome

Scialabba NE-H, Douglas W (2004) The scope of organic agriculture, Sustainable forest management and ecoforestry. In: Protected area management, environment and natural resources working paper no. 18, FAO, Environment and Natural Resources Service, Sustainable Development Department, Rome

Scopel E, Findeling A, Chavez Guerra E, Corbeels M (2005) Impact of direct sowing mulch-based cropping systems on soil carbon, soil erosion and maize yield. Agron Sustain Dev 25:425–432

Segrave R (2005) Community supported agriculture: can it improve farmer to consumer relationships? In: Köpke U, Niggli U, Neuhoff D, Cornish P, Lockeretz W, Willer H (eds) Researching sustainable systems. Proceedings of the 1st scientific conference of the InternationalSociety of Organic Agriculture Research (ISOFAR), held in cooperation with the InternationalFederation of Organic Agriculture Movement (IFOAM) and the National Association for SustainableAgriculture, Australia (NASAA), 21–23 Sept 2005, Adelaide

Setboonsarng S (2008) Can ethical trade certification contribute to the attainment of the millennium development goals? A review of organic and fair-trade certification. ADBI discussion paper 115. Tokyo: Asian Development Bank Institute. Available at http://www.adbi.org/discussion-paper/2008/08/25/2675.organic.fairtrade.certification

Shivay R (2008) Enhancing agricultural productivity through INM approach: to feed the burgeoning population. Green Farming 1(11–17):5–10

Sinkala T, Mwase ET, Mwala M (2002) Control of aquatic weeds through pollutant reduction and weed utilization: a weed management approach in the lower Kafue River of Zambia. Phys Chem Earth 27:983–991

Smit J, Ratta A, Nasr J (1996) Urban agriculture: food, jobs, and sustainable cities. United Nations Development Programme (UNDP), New York

Sofia PK, Prasad R, Vijay VK (2006) Organic farming – tradition – reinvented. Indian J Tradit knowledge 5(1):139–142

Soil Association (2009) Soil carbon and organic farming, p 212

Somers E, Ptacek D, Gysegom P, Srinivasan M, Vanderleyden J (2005) Azospirillum brasilense produces the auxin-like phenylacetic acid by using the key enzyme for indole-3-acetic acid biosynthesis. Appl Environ Microbiol 71(11):1803–1810

Sorrenson WJ, Montoya LJ (1984) Economic implications of soil erosion and soil conservation practices in Paraná, Brazil. Report for the German agency for technical cooperation, GTZ and Fundaçao Instituto Agronómico do Paraná, IAPAR, p 221

Sorrenson WJ, Montoya LJ (1989) Implicaçoes Economicas da Erosao do Solo e do Uso de Algumas Practicas Conservacionistas, Boletin Tecnico No. 21, IAPAR, p 110

Sorrenson W, Lopez PJ, Derpsch R, Nunez M (1997) Economics of no-tillage and crop rotations compared to conventional cultivation cropping systems in Paraguay. 14th ISTRO conference, agroecological and economical aspects of soil tillage, 27 July–1 Aug 1997, Pulawy

Southgate D (2009) Population growth, increases in agricultural production and trends in food prices. Electron J Sustain Dev 1(3):29–35

Southgate D, Graham D, Southgate D (2007) The world food economy. Blackwell Publishing, Malden

Steiner R (1993) Spiritual foundations for the renewal of agriculture: a course of lectures. Bio-Dynamic Farming and Gardening Association, Kimberton, p 310

Steinfeld H, Gerber P, Wassenaar T, Castel V, Rosales M, de Haan C (2006) Livestock'slong shadow: environmental issues and options. Food and Agriculture Organisation of the United Nations, Rome

Sullivan PG, Parrish DJ, Luna JM (1991) Cover crop contributions to N supply and water conservation in corn production. Am J Altern Agric 6:106–113

Sustra Samhita (1885) The medical science of the ancient aryans'. In: Bandopadhyaya AC (Tr and ed), Calcutta

Svotwa E, Jiyane J (2006) Environmental concerns in Zimbabwe's agro-based economy, In: Proceedings of the EEASA international conference 14–18, Aug 2006, BTTC, Harare

Svotwa E, Baipai R, Gwatibaya S (2008) Socieconomic trends and constraints in organic farming in the small holder farming sector of Zimbabwe. J Sustain Dev Afr 10:1

Swaminathan C, Swaminathan V, Vijayalakshmi K (2007) Panchagavya boon to organic farming, International Book Dist vi, p 94, tables, ISBN: 81-8189-164-3

Swift MJ, Anderson JM (1993) Biodiversity and ecosystem function in agroecosystems. In: Shultze E, Mooney HA (eds) Biodiversity and ecosystem function. Springer, New York, pp 57–83

Tebrügge F, Böhrnsen A (1997) Crop yields and economic aspects of no-tillage compared to plough tillage: results of long-term soil tillage field experiments in Germany. In: Tebrügge F, Böhrnsen A (eds) Experiences with the application of no-tillage crop production in the West-European countries. Proceedings of the EC-workshop IV, Boingneville, 12–14 May, pp 25–43

Thamaga-Chitja J, Hendriks SL (2008) Emerging issues in smallholder organic production and marketing in South Africa. Development Southern Africa 25(3):317–326

Thomson JA (2008) The role of biotechnology for agricultural sustainability in Africa. Phil Trans R Soc B 363:905–913. doi:10.1098/rstb.2007.2191

Tilman D, Fargione J, Wolff B, D'Antonio C, Dobson A, Howarth R, Schindler D, Schlesinger WH, Simberloff D, Swackhamer D (2001) Forecasting agriculturally driven global environmental change. Science 292:281–284. doi:10.1126/science.1057544

Tilman D, Cassman KG, Matson PA, Naylor R, Polasky S (2002) Agricultural sustainability and intensive production practices. Nature 418:671–677

Tipraqsa P (2006) Opportunities and constraints of integrated farming system in Northeast Thailand. A case study of the Huai Nong Ian catchment, Khon Kaen Province. Ecology Development Series No. 35. University of Bonn. Cuvillier Verlag, Göttingen

Tokrishna R (1992) Integrated livestock-fish farming systems I Thailand. In: Mukherjee TK, Moi PS, Panadam JM, Yang YS (eds) Proceedings of the FAO/IPT Workshop on Integrated Livestock-Fish Production Systems. Environ 93:1–24

Trenbath BR (1993) Intercropping for the management of pests and diseases. Field Crop Res 34:381–405

Tugel A, Lewandowski A, Happe-vonArb D (eds) (2000) Soil biology primer, Revised edn. Soil and Water Conservation Society, Ankeny

Turinek M, Grobelnik-Mlakar S, Bavec M, Bavec F (2009) Biodynamic agriculture research progress and priorities. Renewable Agric Food Syst 24(2):146–154

Tyler DD, Wagger MG, McCracken DV, Hargrove WL, Carter MR (1994) Role of conservation tillage in sustainable agriculture in the southern United States, conservation tillage in temperate agroecosystems. Lewis Publishers Inc, Boca Raton, pp 209–229

UNEP/GRID-Arendal (2009) Impacts on biodiversity and ecosystems from conventional expansion of food production. Accessed at http://www.grida.no/publications/rr/food-crisis/page/3569. Aspx on 23 Aug 2010

Union of Concerned Scientists (2009) Failure to yield – evaluating the performance of genetically engineered crops, p 43

Uphoff N, Altieri MA (1999) Alternatives to conventional modern agriculture for meeting world food needs in the next century (Ithaca: Cornell international institute for food, agriculture and development, 1999); Altieri MA, Applying agroecology to enhance productivity of peasant farming systems in Latin America. Environ Dev Sustain 1:197–217

Uri ND (1999) Factors affecting the use of conservation tillage in the United States. Water Air Soil Pollut 116:621–638. doi:10.1023/A:1005168928627

U.S. Census Bureau (2008) Total midyear population for the world: 1950–2050. http://www.census.gov/ipc/www/idb/worldpop.html. Accessed July 2009

USDA (2003) Linking Land Quality, Agricultural Productivity, and Food Security, by Keith Wiebe, Agricultural Economic Report No. (AER823), p 63

Vakali C (1999) Possibilities and limits of reduced primary Tillage in organic farming. http://www.iol.unibonn/

Vandermeer JH (1989) The ecology of intercropping. Cambridge University Press, Cambridge

Vandermeer J, Van Noordwijk M, Anderson J, Ong C, Perfecto I (1998) Global change and multispecies ecosystems: concepts and issues. Agr Ecosyst Environ 67:1–22

Vasilikiotis C (2000) Can organic farming "Feed the World"? University of California, Berkeley ESPM-Division of Insect Biology, 201 Wellman-3112 Berkeley

Von Koerber K, Kretschmer J (2006) Ernährung nach den vier Dimensionen. Wechselwirkungen zwischen Ernährung und Umwelt, Wirtschaft, Gesellschaft und Gesundheit, Ernährung and Medizin2006; 21:178–185. Medizinverlage Stuttgart GmbH and Co.

Wackernagel M, Rees W (1994) Ecological footprint and appropriated carrying capacity: a tool for planning toward sustainability. University of British Columbia, Vancouver

Walaga C (2006) The development of organic agriculture sector in Africa: potential and challenges. Report on organic agriculture worldwide. http://www.rafiusa.org/pubs/OrganicReport.pdf Accessed 13 Oct 2006

Watson CA, Kristensen ES, Alroe HF (2006) Research to support the development of organic food and farming. In: Kristiansen P, Taji A, Reganold J (eds) Organic agriculture: a global perspective. CSIRO Publishing, Collingwood, pp 361–383

WBCSD (2008) Agricultural ecosystems: facts and trends, WBCSD/IUCN, Conches, p 19

Welsh JP, Philipps L, Bulson HAJ, Wolfe M (1999) Weed control for organic cereal crops. In: Proceedings of the Brighton crop protection conference – weeds, Brighton, pp 945–950

Wen D, Liang W (2004) Soil fertility quality and agricultural sustainable development in the black soil region of Northeast China. Environ Dev Sustain 3(1):31–43

Wetterich F, Haas G (1999) Ökobilanz Allgäuer Grünlandbetriebe - Intensiv, Extensiviert, Ökologisch. Schriftenreihe Institut für Organischen Landbau, ISBN 3-89574-365-8, Verlag Dr. Köster, Berlin, 94S

Wilkins C, Mead B, Dwyer A (2005) Adelaide green city – creating an urban organic closed loop. In: Köpke U, Niggli U, Neuhoff D, Cornish P, Lockeretz W, Willer H (eds) Researching sustainable systems. Proceedings of the 1st scientific conference of the international society of organic agriculture research (ISOFAR), held in cooperation with the international federation of organic agriculture movement (IFOAM) and the national association for sustainable agriculture, Australia (NASAA), 21–23 Sept 2005, Adelaide

Willer H, Youssefi M (2007) The world of organic agriculture – statistics and emerging trends 2007. International Federation of Organic Agriculture Movements IFOAM/Research Institute of Organic Agriculture FiBL, Bonn/Frick

Wilson C (2000) Environmental and human costs of commercial agricultural production in South Asia. Int J Soc Econ 27(7/8/9/10):816–846

Wilson G (2002) Can urban rooftop microfarms be profitable? Urban Agric Mag 7:22–24

Wilson A, Pelletier MR (2003) A garden overhead: the benefits and challenges of green roofs. Environmental Building News, Special reprint 10(11) Nov 2001, revised: May 2003, pp 1–10

Winterborne J (2005) Hydroponics: indoor horticulture, Published by Pukka Press, 2005, United Kingdom, p 113

Wong M (2005) Salinity effects in nursery and landscape plants, Soil Crop Manag, Jan 2005, SCM-12

World Food Summit (1996) Rome declaration on World Food Security and World Food Summit plan of action. FAO, Rome

Xu H, Parr JF, Umemura H (eds) (2000) Nature farming and microbial applications. Food Products Press, Binghamton, p 402. ISBN 1-56022-083-X

Xue D (2006) China's Eco-farming – an effective approach for conservation and sustainable use of agricultural bio diversity. https://www.cbd.int/doc/case-studies/agr/cs-agr-cn.pdf

Index

E. Lichtfouse (ed.), *Genetics, Biofuels and Local Farming System*, Sustainable Agriculture Reviews 7, DOI 10.1007/978-94-007-1521-9,